Energy Systems and Sustainability

Edited by Godfrey Boyle, Bob Everett and Janet Ramage

OXFORD
UNIVERSITY PRESS

Oxford University Press in association with The Open University

OXFORD
UNIVERSITY PRESS

Great Clarendon Street, Oxford OX2 6DP

Published by Oxford University Press, Oxford in association with
The Open University, Milton Keynes

The Open
University

Oxford University Press is a department of the University of Oxford.

It furthers the University's objective of excellence in research,
scholarship, and education by publishing worldwide in

Oxford New York Auckland Bangkok Buenos Aires Cape Town
Chennai Dar es Salaam Delhi Hong Kong Istanbul Karachi
Kolkata Kuala Lumpur Madrid Melbourne Mexico City Nairobi
São Paulo Shanghai Taipei Tokyo Toronto

Oxford is a registered trade mark of Oxford University Press in the UK
and in certain other countries

Published in the United States by Oxford University Press Inc., New York

The Open University, Walton Hall, Milton Keynes MK7 6AA

First Published in the United Kingdom 2003

Copyright © The Open University 2003, reprinted 2004

British Library Cataloguing in Publication Data

Data available

Library of Congress Cataloguing in Publication Data

Data available

ISBN 0-19-926179-2

1 3 5 7 9 10 8 6 4 2

Edited, designed an typeset by The Open University

Printed in the United Kingdom by Bath Press, Glasgow.

T206 Energy for a Sustainable Future

This book was produced as a major component of the Open University's second level undergraduate course *T206 Energy for a Sustainable Future* and, through co-publication with Oxford University Press, is also available to students and staff at other Universities worldwide.

A companion text, *Renewable Energy: Power for a Sustainable Future*, is also co-published by Oxford University Press and The Open University.

To order copies of either book, contact:

Oxford University Press, Saxon Way West, Corby, Northamptonshire, NN18 9ES or visit their web site at www.oup.com

T206 Energy for a Sustainable Future also includes two other textbooks, a set of Study Guides and a variety of other supplementary printed materials. This additional course material is accompanied by several hours of material on video, produced by the BBC in partnership with the Open University, and a number of software-based exercises and simulations on CD-ROM. A selection is available from Open University Worldwide Ltd. at http://www.ouworldwide.com or by writing to Open University Worldwide, Walton Hall, Milton Keynes, MK7 6AA.

T206 Energy for a Sustainable Future and other Open University courses are available to residents of the European Union and other countries where The Open University has agreed to allow registration. Further details of *T206 Energy for a Sustainable Future* and how to enrol on the course are available via our T206 Taster Web Site: http://www.open.ac.uk/T206/index.htm

Alternatively, write to: The Course Manager, T206 Energy for a Sustainable Future, Faculty of Technology, The Open University, Milton Keynes MK7 6AA, United Kingdom.

Preface

If current trends continue, by the end of the twenty-first century the world's population is likely to have almost doubled and its wealth increased by a factor of perhaps ten times. World energy demand will probably have doubled and possibly quadrupled, despite major improvements in energy efficiency. How can this enormous demand be supplied, cleanly, safely and sustainably? This is the key question underlying the content of this book and its companion text *Renewable Energy: Power for a Sustainable Future*.

Even at current consumption levels, existing energy systems have many deleterious effects on human health and the natural environment. In particular, carbon dioxide and other greenhouse gases released by fossil fuel burning threaten to cause unprecedented changes in the earth's climate, with mainly adverse consequences. There is now a broad consensus that in the long term the world will need to shift to low- or zero-carbon energy sources if the impacts of climate change are to be mitigated.

For fossil fuels, the most promising approach may be to convert them to hydrogen, capturing and sequestering their carbon content so that it cannot enter the atmosphere, and combine the hydrogen with atmospheric oxygen in fuel cells to produce electricity and heat, cleanly and efficiently.

Although nuclear fuels are carbon-free, concerns persist about the cost, safety, proliferation and waste disposal problems associated with nuclear power. Nevertheless, new fission reactor designs could emerge that might help allay such concerns and, in the longer term, nuclear fusion could prove a technologically, commercially and environmentally viable option.

The renewable energy sources, in the main, appear more sustainable than fossil or nuclear fuels, though many technologies are still under development and the costs of some are currently high. With increased investment in research, development and deployment, the renewables could supply a very much larger share of the world's energy. Nevertheless, as our companion text *Renewable Energy* makes clear, they also have impacts on humanity and the environment.

This book and its companion have been written initially for undergraduates studying the course T206 *Energy for a Sustainable Future* at The Open University. They are also aimed at students and staff in other universities and at professionals, policy-makers and members of the public interested in sustainable energy futures. We hope that both books will contribute to a better understanding of the sustainability problems of our present energy systems and of potential solutions. We also hope they convey something of the enthusiasm we feel for this complex, fascinating and increasingly important subject.

Godfrey Boyle
Course Team Chair, T206 *Energy for a Sustainable Future*

Contents

CHAPTER I   INTRODUCTORY OVERVIEW

1.1 **Introduction**   **3**
         Why sustainable energy matters   3

1.2 **Definitions: energy, sustainability and the future**   **6**

1.3 **Present energy sources and sustainability**   **7**
         Fossil fuels   7
         Nuclear energy   16
         Bioenergy   19
         Hydroelectricity   21
         Summary   23

1.4 **Renewable energy sources**   **23**
         Solar energy   23
         Indirect use of solar energy   28
         Non-solar renewables   31
         Sustainability of renewable energy sources   34

1.5 **Energy services and efficiency improvement**   **34**
         Energy services   34
         Energy efficiency improvements   36
         The rebound effect   46

1.6 **Energy in a sustainable future**   **47**
         Changing patterns of energy use   47
         Long-term energy scenarios   49

         References   **54**

CHAPTER 2   PRIMARY ENERGY

2.1 **World primary energy consumption**   **57**
         What is primary energy?   57
         What is energy consumption?   57
         'Energy arithmetic'   58
         Watts   59
         Kilowatt-hours   59

2.2 **Quantities of energy**   **61**
         Units based on oil   61
         Units based on coal   63
         The BTU and related units   63
         The calorie and related units   63

2.3 **Interpreting the data**   **64**
         Definitions   64
         Conversions   64
         Conventions   65

2.4 **World energy sources**   **66**
         International comparisons   69

2.5	**Primary energy in the UK**	**71**
	Britain's changing energy scene	72
	Renewables in the UK	74
2.6	**Primary energy in Denmark**	**75**
	Renewables in Denmark	78
2.7	**Primary energy in the USA**	**79**
	Renewables in the USA	82
	Review	83
2.8	**Primary energy in France**	**83**
2.9	**Primary energy in India**	**85**
	Renewables in India	86
2.10	**Summary**	**88**
	References	**89**
CHAPTER 3	WHAT DO WE USE ENERGY FOR?	
3.1	**Primary, delivered and useful energy**	**93**
3.2	**The expanding uses of energy**	**95**
	Food	96
	Fertilizers	97
	Domestic energy	100
	Industry	104
	Transport	107
	Services	108
3.3	**Energy uses today**	**109**
	The energy balance for the UK	109
	International comparisons	118
	Comparisons of delivered energy	121
3.4	**Conclusions**	**125**
	References and data sources	**127**
	Further reading	**128**
CHAPTER 4	FORMS OF ENERGY	
4.1	**Introduction**	**131**
4.2	**Kinetic and potential energy**	**131**
4.3	**Heat**	**136**
4.4	**Electrical energy**	**136**
	Electrons	138
4.5	**Electromagnetic radiation**	**140**
4.6	**Chemical energy**	**145**
	The nuclear atom	147
4.7	**Nuclear energy**	**148**
	Protons and neutrons	148
	Isotopes	149
	The nuclear force	149
	Radioactivity and fission	150
4.8	**Energy and mass**	**151**
4.9	**Summary**	**152**
	References	**153**
	Further reading	**153**

CHAPTER 5   COAL

5.1   The fossil fuels   **157**

5.2   From wood to coal   **158**
The early years   158
The Industrial Revolution   160
The nineteenth century   162

5.3   The resource and its use   **163**
Types of coal   163
Reserves and production   166
The uses of coal   170

5.4   Coal combustion   **171**
The composition of coal   171
The combustion process   172
Proximate analysis   173
Combustion products   174

5.5   Fires, furnaces and boilers   **176**
Power station boilers   177
Flue gases   181

5.6   Summary   **183**

References   **184**

CHAPTER 6   HEAT TO MOTIVE POWER

6.1   Introduction   **187**

6.2   Steam engines   **188**
The Early Years   188
Savery and Newcomen   190
James Watt   194

6.3   The principles of heat engines   **197**
Carnot's law   198
The Carnot engine   198
Atoms in motion   200
Heat flow   202

6.4   The age of steam   **202**
Improving the efficiency   203
Mobile power   205

6.5   Steam turbines   **209**
Steam, speed and rpm   209
Parsons' turbo-generator   211

6.6   Power station turbines   **215**
The turbines   216
A 660-MW turbine   218

6.7   Futures   **222**

References   **223**

CHAPTER 7   OIL AND GAS

Part 1   Oil and gas as primary fuels   227

7.1   **Introduction**   **227**

7.2   **The origins and geology of petroleum**   **228**

7.3   **The origins of the oil and gas industry**   **230**
       Petroleum for illumination   231
       Petroleum for transport   234
       The natural gas industry   235

7.4   **Finding petroleum**   **236**
       How do we get it?   236
       Where do we get it?   240

7.5   **Refining and products**   **245**
       Introduction   245
       The fractions   246
       Getting more of what you want   247
       The many products – a summary   248

7.6   **UK demand for oil products: past, present and future**   **250**
       Introduction   250
       Time comparison, 1967 and 2000   250

7.7   **UK demand for gas: past, present and future**   **254**
       Measurement   254
       Demand by sector   254

7.8   **Why so special?**   **258**
       Cheap and readily available   258
       Indigenous production/security of supply   258
       Convenience and ease of use   258
       Clean to burn   259
       Ease of distribution/storage/portability   260
       Energy density   260

7.9   **Substitutes for oil and gas?**   **261**

Part 2   Oil and gas as secondary fuels   262

7.10   **Introduction**   **262**

7.11   **Obstacles to coal conversion**   **262**
       Technology   262
       Environment   262
       Conversion efficiencies   263
       Cost and price   263
       Capital cost   264
       Summary   265

7.12   **Gas from oil**   **265**

7.13   **Oil from gas**   **266**

7.14   **Gas from coal**   **267**
       A little history   267
       Gas from coal – the future   270

7.15   **Oil from coal**   **273**
       Oil from coal – the future   277

7.16   **Non-conventional sources of petroleum**   **278**
       Introduction   278

	Oil shale	278
	Tar sands	280
	Heavy oil deposits	281
	Summary	282
7.17	**The wider future**	**282**
	Introduction	282
	Lesson from America	284
	First America… now the world	286
	Beyond Hubbert's peak	290
	References	**291**

CHAPTER 8	OIL AND GAS ENGINES	
8.1	**Introduction**	**295**
8.2	**The petrol or spark ignition engine**	**295**
	The birth of the car engine	298
	The motorization of the US	302
	Aircraft petrol engines	303
	Compression ratio and octane number	304
	Lead additives	304
8.3	**The diesel engine**	**305**
	Diesel power for ships	306
	Diesel engines for road, rail and air	307
8.4	**Petrol and diesel engines – reducing pollution**	**309**
	Emissions from petrol engines	311
	Obtaining best efficiency	314
8.5	**The gas turbine**	**314**
	The German jet engine	315
	The British jet engine	316
	Post-war developments	318
	Modern jet engines	319
	Industrial gas turbines	321
	Improving power and efficiency	321
	Gas turbines for cars	321
8.6	**The Stirling engine**	**322**
	Principles	323
	The Philips engine	324
	The quest for the Stirling car engine	325
8.7	**Conclusion**	**327**
	References and sources	**329**
	Further Reading	**329**

CHAPTER 9	ELECTRICITY	
9.1	**Introduction**	**333**
9.2	**Making electricity in the nineteenth century**	**334**
	Batteries and chemical electricity	334
	Magnetism and generators	335
	The rise of electric lighting	337
	AC or DC?	339
	High voltage or low voltage?	343
	Metering and tariffs	347

9.3   **The continuing development of electric lighting**   **348**
Gas fights back   348
Improving the incandescent light bulb   348
The fluorescent lamp   349
The light emitting diode (LED)   352

9.4   **Electric traction**   **353**
Electric trams and trains   353
Battery electric vehicles   355
Electric transmissions and hybrid electric drives   357

9.5   **Expanding uses**   **358**
Telecommunications and information technology   358
Cooking and heating   361
Refrigeration   362
Electric motors everywhere   364
Where electricity is used today   365

9.6   **Large scale generation**   **365**
Thermal power stations   365
Hydroelectricity   369
Combined heat and power generation   369

9.7   **Transmission and distribution**   **371**
The National Grid   371
Coal by wire   373
The Channel link   374
The grid today   374

9.8   **Running the system**   **375**
What exactly is being optimized?   376
Ownership of the system   377
Balancing supply and demand   377
Peak demands and pumped storage   379
The privatized UK system   380

9.9   **The dash for gas**   **381**

9.10   **Electricity around the world**   **384**
United Kingdom: a summary   385
United States of America   385
France   386
Denmark   387
India   388

9.11   **Conclusion**   **390**

References and sources   **390**

Further reading   **392**

CHAPTER 10   NUCLEAR POWER

10.1   **Introduction**   **395**

10.2   **Radioactivity**   **396**
Alpha particles   397
Beta particles   397
Gamma particles   399
Radioactive decay and half-life   399
An effect without a cause   400

10.3	**Nuclear fission**	**402**
	Experiments with neutrons	402
	Fission	402
10.4	**1939–1945: reactors and bombs**	**405**
	The first reactor	405
	New elements	408
	Atomic bombs	410
	'Swords into ploughshares'	410
10.5	**Thermal fission reactors**	**411**
	The reactor core	414
	Structures	414
	Safety	414
	Types of thermal fission reactor	416
10.6	**Nuclear fuel cycles**	**422**
	Mining and extraction	422
	Enrichment and fuel fabrication	423
	Spent fuel	423
10.7	**Fast reactors**	**427**
	The liquid-metal fast breeder reactor	428
10.8	**Power from fusion**	**429**
	Approaches to a fusion reactor	431
	References	**433**
	Further reading	**434**
	Useful web sites	**434**

CHAPTER 11   THE FUTURE OF NUCLEAR POWER

11.1	**Introduction**	**437**
11.2	**Reasons for decline**	**438**
	Nuclear accidents	439
	Nuclear economics	442
	Nuclear decline worldwide	446
11.3	**Nuclear power: a long-term answer to climate change?**	**447**
	Nuclear wastes	449
	Uranium reserves	450
	Stretching reserves: the fast breeder reactor	454
	Nuclear fusion: the ultimate answer?	455
11.4	**New nuclear developments**	**456**
	Safer nuclear power	457
	Waste management and reprocessing	460
	Cheaper nuclear power	464
	Development issues and proliferation	466
1.5	**The future: conflicting views**	**468**
11.6	**Conclusion**	**470**
	References	**471**

CHAPTER 12   COSTING ENERGY

12.1 Introduction   **477**

12.2   **Energy prices today**   **478**
Petrol and diesel fuel   478
Domestic energy prices   479
Industrial energy prices   480

12.3   **Inflation, real prices and affordability**   **482**
The value of money   482
Affordability and fuel poverty   487

12.4   **Investing in energy**   **489**
Price and cost   489
Balancing investment against cash flow   490
Discounted cash flow analysis   493

2.5   **Real world complications**   **509**
Security and diversity of supply   509
Externality costs of pollution and disaster   511
New technologies, economies of scale and 'market washing'   511

12.6   **Conclusions**   **512**

References   **514**

Statistical data sources   **515**

Further reading   **516**

CHAPTER 13   PENALTIES: ASSESSING THE ENVIRONMENTAL AND HEALTH
IMPACTS OF ENERGY USE

13.1   **Introduction**   **519**

13.2   **Classifying the impacts of energy use**   **519**
Classification by source   519
Classification by pollutant   520
Classification by scale   522

13.3   **Household-scale impacts**   **523**
Wood burning in developing countries   523

13.4   **Workplace-scale impacts**   **525**
Biomass harvesting and forestry   525
Hydro and wind power   525
Coal, oil and gas   525
Nuclear power   527

13.5   **Community-scale impacts**   **531**
Urban air pollution in developed countries   532
Urban air pollution in developing countries   536
Community impacts of hydroelectricity   536

13.6   **Regional-scale impacts**   **537**
Acid deposition   537
Public impacts of nuclear power   537

13.7   **Global-scale impacts**   **539**
Global climate change   539

13.8   **Accidents and risk**   **545**

13.9   **Comparing the impacts of electricity generating systems**   **553**

13.10   **Comparing the external costs of electricity generating systems**   **561**

13.11   **Summary and conclusions**   **565**

References and further reading   **566**

CHAPTER 14   REMEDIES: MAKING FOSSIL FUEL USE MORE SUSTAINABLE

14.1   Introduction   **573**

14.2   Reducing pollutant emissions from fossil fuel combustion   **574**
Fuel switching   574

14.3   Capturing and sequestering carbon emissions from fossil fuel combustion   **577**
Carbon sequestration in forests   577
Carbon capture and sequestration beneath the earth's surface   578
Ocean sequestration of carbon   580

14.4   The fuel cell: energy conversion without combustion   **584**
Fuel cell types   586

14.5   A fossil-fuel based hydrogen economy   **587**
Hydrogen storage and use in transport   590
Hydrogen safety   592

14.6   Can fossil fuel use be made more sustainable?   **594**

References   **596**

Appendix A   **597**
A1 Orders of magnitude   597
A2 Units and conversions   598

Reference   **600**

ACKNOWLEDGEMENTS   601

INDEX   607

Chapter 1

Introductory Overview

by Godfrey Boyle

Figure 1.1 Oil wells on fire in Kuwait during the Gulf War in 1990–91

Figure 1.2 Tanker drivers block the entrance to a UK refinery in September 2000, to protest against high fuel prices

Figure 1.3 A fire on the Piper Alpha gas rig in the North Sea in 1988 killed 167 people

Figure 1.4 An oil spillage from the Exxon Valdez tanker in 1989 contaminated 2100 km of beaches in Alaska and caused extensive harm to wildlife. The cost of cleanup was estimated at some $3 billion

1.1 Introduction

One of the greatest challenges facing humanity during the twenty-first century must surely be that of giving everyone on the planet access to safe, clean and sustainable energy supplies.

Throughout history, the use of energy has been central to the functioning and development of human societies. But during the nineteenth and twentieth centuries, humanity learned how to harness the highly-concentrated forms of energy contained within fossil fuels. These provided the power that drove the industrial revolution, bringing unparalleled increases in affluence and productivity to millions of people throughout the world. As we enter the third millennium, however, there is a growing realization that the world's energy systems will need to be changed radically if they are to supply our energy needs sustainably on a long-term basis.

This introductory overview aims to survey, in very general terms, the world's present energy systems and their sustainability problems, together with some of the possible solutions to those problems and how these might emerge in practice during the twenty-first century.

Why sustainable energy matters

The world's current energy systems have been built around the many advantages of fossil fuels, and we now depend overwhelmingly upon them. Concerns that supplies will 'run out' in the short-to-medium term have probably been exaggerated, thanks to the continued discovery of new reserves and the application of increasingly advanced exploration technologies. Nevertheless it remains the case that fossil fuel reserves are ultimately finite. In the long term they will eventually become depleted and substitutes will have to be found.

Moreover, fossil fuels have been concentrated by natural processes in relatively few countries. Two-thirds of the world's proven oil reserves, for example, are located in the Middle East and North Africa. This concentration of scarce resources has already led to major world crises and conflicts, such as the 1970s 'oil crisis' and the Gulf War in the 1990s. It has the potential to create similar, or even more severe, problems in the future.

Substantial rises in the price of oil also can cause world-wide economic disruption and lead to widespread protests, as seen in the USA and Europe in 2000.

The exploitation of fossil fuel resources entails significant health hazards. These can occur in the course of their extraction from the earth, for example in coal mining accidents or fires on oil or gas drilling rigs.

They can also occur during distribution, for example in oil spillages from tankers that pollute beaches and kill wildlife; or on combustion, which generates atmospheric pollutants such as sulphur dioxide and oxides of nitrogen that are detrimental to the environment and to health.

Fossil fuel combustion also generates very large quantities of carbon dioxide (CO_2), the most important anthropogenic (human-induced) greenhouse gas.

The majority of the world's scientists now believe that anthropogenic greenhouse gas emissions are causing the earth's temperature to increase at a rate unprecedented since the ending of the last ice age. This is very likely to cause significant changes in the world's climate system, leading to disruption of agriculture and ecosystems, to sea level rises that could overwhelm some low-lying countries, and to accelerated melting of glaciers and polar ice.

Figure 1.5 Rising sea levels due to global warming could overwhelm some low-lying nations, such as Tuvalu, a group of nine coral atolls in the Pacific

Figure 1.6 Rising global temperatures have already caused significant melting of ice around the North Pole, which is now accessible to ships at certain times of the year. This does not affect global sea levels, since most Arctic ice is floating. But if ice at the South pole, much of which is based on land, were to melt, this would cause very substantial rises in sea levels

Nuclear power has grown in importance since its inception just after World War II and now supplies some 7% of world primary energy. A major advantage of nuclear power plants, in contrast with fossil fuelled plants, is that they do not emit greenhouse gases. Also, supplies of uranium, the principal nuclear fuel, are sufficient for many decades – and possibly centuries – of supply at current use rates. However the use of nuclear energy, as we shall see, gives rise to problems arising from the routine emissions of radioactive substances, difficulties of radioactive waste disposal, and dangers from the proliferation of nuclear weapons material. To these must be added the possibility of major nuclear accidents which, though highly unlikely, could be catastrophic in their effects. Although some of these problems may be amenable to solution in the longer-term, such solutions have not yet been fully developed.

Extracting energy from fossil or nuclear fuels, in the course of providing energy-related services to society, generates significant environmental and social impacts. These impacts are greater than they need be because of the low efficiency of our current systems for delivering energy, converting it into forms appropriate for specific tasks, and utilizing it in our homes, machinery, appliances and vehicles. An important way of mitigating the environmental impacts of current fuel use is therefore to improve the efficiency of these systems. Over the past few decades, significant efficiency improvements have indeed been made, but further major improvements are feasible technologically – and are, in many cases, attractive economically.

Of course, not all energy sources are of fossil or nuclear origin. The renewable energy sources, principally solar energy and its derivatives in the form of bioenergy, hydroelectricity, wind and wave power, are increasingly considered likely to play an important role in the sustainable energy systems of the future. The 'renewables' are based on energy flows that are replenished by natural processes, and so do not become depleted with use as do fossil or nuclear fuels – although there may be other constraints on their use. The environmental impacts of renewable energy sources vary, but they are generally much lower than those of conventional fuels. However, the current costs of renewable energy sources are in many cases higher than those of conventional sources, and this has until recently retarded their deployment.

All these considerations suggest that in creating a sustainable energy future for humanity during the coming decades, it will be necessary:

1   to implement greatly improved technologies for harnessing the fossil and nuclear fuels, in order to ensure that their use, if continued, creates much lower environmental and social impact;

2   to develop and deploy the renewable energy sources on a much wider scale; and

3   to make major improvements in the efficiency of energy conversion, distribution and use.

These three general approaches will be explored further below, and in greater detail in the remaining chapters of this volume. The subject of Renewable Energy will be dealt with in the companion volume, *Renewable Energy* (Boyle, 1996, 2003).

1.2 **Definitions: energy, sustainability and the future**

Energy is Eternal Delight

William Blake, 1757–1827, (1994)

Figure 1.7 William Blake

What do we mean by 'energy'? What does the concept of 'sustainability' entail? And what, for that matter, do we mean by the 'future' in this context?

The eighteenth century poet and artist William Blake, quoted above, probably expressed our personal experience of energy as we feel it in our day-to-day lives more accurately than any scientific definition. Indeed, the word energy, when it first appeared in English in the sixteenth century, had no scientific meaning at all. Based on a Greek word coined by Aristotle, it meant forceful or vigorous *language*.

It was not until the early 1800s that the concept of energy in the modern sense was developed by scientists to describe and compare their observations about the behaviour of such diverse phenomena as the transfer of heat, the motion of planets, the operation of machinery and the flow of electricity. Today, the standard scientific definition is that **energy is the capacity to do work**: that is, to move an object against a resisting force.

In everyday language, the word 'power' is often used as a synonym for energy – and indeed in this book and its companion volume the two words may occasionally be used in this rather loose way merely as substitutes for each other. But when speaking scientifically, **power** is defined as **the rate of doing work**, that is, the *rate* at which energy is converted from one form to another, or transmitted from one place to another. The main unit of measurement of energy is the joule (J) and the main unit of measurement of power is the watt (W), which is defined as a rate of one joule per second. The twin concepts of energy and power will be discussed in more detail in Chapter 2.

The term 'sustainability' entered into common currency relatively recently, following the publication of the report *Our Common Future* by the United Nations' Brundtland Commission in 1987. The commission defined **sustainability**, and in particular **sustainable development**, as '**development that meets the needs of the present without compromising the ability of future generations to meet their own needs.**' (United Nations, 1987).

In the context of energy, sustainability has come to mean the harnessing of those energy sources:

- that are not substantially depleted by continued use;
- the use of which does not entail the emission of pollutants or other hazards to the environment on a substantial scale; and
- the use of which does not involve the perpetuation of substantial health hazards or social injustices.

This is, of course, a very broad ideal. Although a few energy sources can come close to fulfilling these conditions, most fall considerably short of the optimum. This means that, in practice, sustainability is a relative rather than an absolute concept. It is not so much that some energy sources are sustainable and others not; it is more that some energy sources, in certain contexts, are more sustainable than others. Determining the relative sustainability of one energy system *vis-à-vis* another is usually a complex process, involving detailed consideration of the specific processes and technologies proposed, the context in which they are being used and the differing values and interests of the various parties involved.

For example, suppose the Government of a country is proposing to construct a large hydro-electric power plant like the one shown in Figure 1.25 in Section 1.3 below. The villagers whose homes would be flooded by the associated reservoir would probably take a different view of the plant's sustainability to that taken by the city-based planners in the electricity utility proposing its construction, whose homes would be unaffected and whose careers would probably stand to benefit from such a major capital project.

When we speak of 'the future' in the context of a 'sustainable future', what do we mean? Next year? One or two decades hence? The end of the twenty-first century? The end of the third millennium? Forever?

Ideally, in view of the Brundtland Report's injunction that humanity should not compromise the needs of future generations, we should judge the sustainability of all energy systems on an indefinite time scale – far into the very distant future. In practice, however, this might be realistically interpreted as endeavouring to ensure that energy systems become sustainable (or at the very least, much less un-sustainable) over the next century or so – with the additional proviso that, even beyond that time horizon, few substantial difficulties can presently be envisaged. Future generations will be justified in blaming us for creating problems that were foreseeable; but they can hardly hold us responsible for eventualities that none of us could have anticipated.

1.3 **Present energy sources and sustainability**

So what are the principal energy systems used by humanity at present, and how sustainable are they?

Until quite recently, human energy requirements were modest and our supplies came either from harnessing natural processes such as the growth of plants, which provided wood for heating and food to energize human or animal muscles, or from the power of water and wind, used to drive simple machinery.

Fossil fuels

But the nineteenth and twentieth centuries saw a massive increase in global energy use, based mainly on burning cheap and plentiful **fossil fuels**: first coal, then oil and natural gas. These fossil fuels now supply nearly 80% of the world's current energy consumption (see Box 1.1).

Fossil fuels are extremely attractive as energy sources. They are highly concentrated, enabling large amounts of energy to be stored in relatively small volumes. They are relatively easy to distribute, especially oil and gas which are fluids.

BOX 1.1 **World primary energy consumption**

The population of the world rose nearly four-fold during the twentieth century, from 1.6 billion in 1900 to approximately 6.1 billion in 2000. However, world primary energy use increased at a much faster rate. Between 1900 and 2000, it rose more than 10-fold (Figure 1.8). *(Primary Energy will be defined in Chapter 2)*

For most of the nineteenth century the world's principal fuel was firewood (or other forms of traditional 'bioenergy'), but coal use was rising fast and by the beginning of the twentieth century it had replaced wood as the dominant energy source. During the 1920s, oil in turn began to challenge coal and by the 1970s had overtaken it as the leading contributor to world supplies. By then, natural gas was also making a very substantial contribution, with nuclear energy and hydro power also supplying smaller but significant amounts.

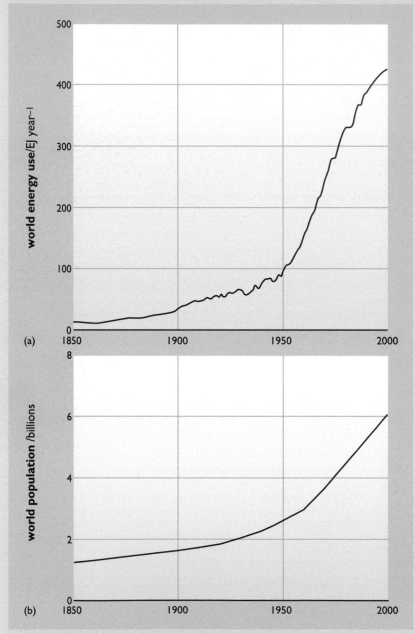

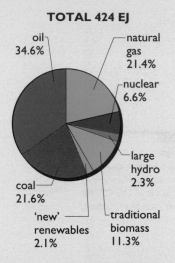

Figure 1.9 Percentage contributions of various energy sources to world primary energy consumption, 2000. Total consumption in 2000 was 424 exajoules, equivalent to just over 10 000 million tonnes of oil. The average *rate* of consumption was some 13.4 million million watts (13.4 terawatts). Note that the actual amounts of electricity produced by nuclear and hydro power were almost the same, but due to a statistical convention in the definition of primary energy, the nuclear contribution is multiplied by a factor of 3 (see Chapter 2)

Figure 1.8 (a) Growth in world primary energy use, 1850–2000; (b) Growth in world population, 1850–2000 (source: (a) United Nations Development Programme, 2000; (b) US Census Bureau, 2000)

As Figure 1.9 shows, total world primary energy use in 2000 was an estimated 424 million million million joules, i.e., 424 exajoules, equivalent to some 10 100 million tons of oil. All of these quantities and units are explained in more detail in Section 2.2 of the next chapter (see also Appendix A).

Figure 1.10 A typical supertanker used to ship oil around the globe. To transport the c.3500 million tonnes of oil used by the world in 2000 would require some 14 000 tanker journeys, assuming a typical tanker capacity of 250 000 tonnes. Total world primary energy use in 2000 was equivalent to the capacity of some 40 000 supertankers of this size

By the year 2000, oil was still the largest single contributor to world supplies, providing about 35% of primary energy, with gas and coal supplying roughly equal shares at around 21–22%, nuclear providing nearly 7% and hydropower 2%. In 2000, traditional biofuels still supplied an estimated 11%, while more modern forms of 'bioenergy' provided around 2%, with other 'new renewables' like wind power contributing a very small (though rapidly growing) fraction of world demand.

On average, world primary energy use *per person* in 2000 was about 70 thousand million joules (70 gigajoules), including non-commercial bioenergy. This is equivalent to about one and two-thirds tonnes of oil per person per year, or about 5 litres (just over one Imperial gallon) of oil per day.

But this average conceals major differences between the inhabitants of different regions. As Figure 1.11 illustrates, North Americans annually consume the equivalent of about 8 tonnes of oil per head (about 20 litres per day), whereas residents of Europe and the former Soviet Union consume about half that amount, and the inhabitants of the rest of the world use only about one-tenth.

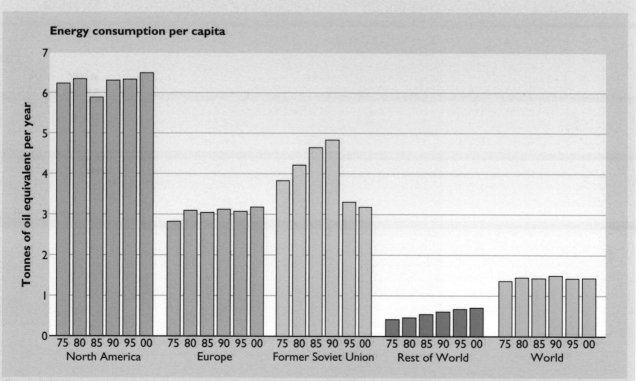

Figure 1.11 *Per capita* primary energy consumption, in tonnes of oil equivalent per year, for different regions of the world and for the world as a whole, 1975–2000. World consumption *per person* has shown almost no growth over the past 20 years. North American consumption *per capita* is more than twice that of Europe and the former Soviet Union, and almost 10 times the level in the Rest of the World. Note that these figures include only commercially-traded fuels (i.e. they exclude traditional biofuels) (source: BP, 2001)

During the twentieth century, these unique advantages enabled the development of increasingly-sophisticated and effective technologies for transforming fossil fuel energy into useful heat, light and motion; these ranged from the oil lamp to the steam engine and the internal combustion engine. Today, at the beginning of the twenty-first century, fossil fuel-based systems reign supreme, supplying the great majority of the world's energy.

The fossil fuels we use today originated in the growth and decay of plants and marine organisms that existed on the earth millions of years ago. Coal was formed when dead trees and other vegetation became submerged under water and were subsequently compressed, in geological processes lasting millions of years, into concentrated solid layers below the earth's surface. Oil and associated natural gas originally consisted of the remains of countless billions of marine organisms that slowly accreted into layers beneath the earth's oceans and were gradually transformed, through geological forces acting over aeons of time, into the liquid and gaseous reserves we access today by drilling into the earth's crust. The fossil fuels are composed mainly of carbon and hydrogen, which is why they are called hydrocarbons.

Figure 1.12 Parc Colliery in Cwm Park, Rhonda Valley, Glamorgan, South Wales, (photo, 1960). Coal was the fuel that powered the industrial revolution. Its combustion produces relatively large amounts of carbon dioxide (CO_2) compared with other fuels. It also results in particulates (soot), and sulphur dioxide emissions, though these can be reduced by various techniques. The use of coal in UK homes and industry has now been largely superseded by natural gas, but it is still used for electricity generation. Huge world-wide coal reserves remain, enough for several hundred years' use at current rates

Figure 1.13 A North Sea oil drilling platform. Oil is the world's leading energy source. Its high energy density and convenience of use are particularly advantageous in the transport sector, where it is the dominant fuel. Oil combustion produces less CO_2 per unit of energy released than burning coal, but more CO_2 than burning natural gas. Proven world oil reserves are sufficient for about 40 years of use at current rates

Figure 1.14 The offshore rig Semac 1, a natural gas drilling platform in the North Sea. Natural gas combustion produces significantly lower CO_2 emissions per unit of energy than the combustion of other fossil fuels. Emissions are also free from sulphur dioxide or particulates. The relative cleanliness and convenience of natural gas have made it the preferred fuel for heating and, increasingly, for electricity generation in Western Europe. Proven world gas reserves are sufficient for about 60 years of use at current rates

Since the fossil fuels were created in specific circumstances where the geological conditions were favourable, the largest deposits of oil, gas and coal tend to be concentrated in particular regions of the globe (see Figure 1.15) – although less appreciable deposits are remarkably widespread. The majority of the world's oil reserves are located in the Middle East and North Africa, while the majority of our natural gas reserves are split roughly equally between the Middle East/North Africa and the former Soviet Union. (BP, 2002) Although coal deposits are rather more evenly spread throughout the world, three-quarters of world coal reserves are concentrated in just four countries: Australia, China, South Africa and the United States of America. (United Nations, 2000; BP, 2002) (Figure 1.15).

Although human society now consumes fossil fuels at a prodigious rate, the amounts of coal, oil and gas that remain are still very large. One simple way of assessing the size of reserves is called the reserves/production (R/P) ratio – the number of years the reserves would last if use continued at the current rate.

Coal has the largest R/P ratio. Present estimates suggest the world has more than 200 years' worth of coal left at current use rates. For oil, current R/P estimates suggest a lifetime of about 40 years at current rates. For gas, the R/P ratio is somewhat higher, at around 60 years. (BP, 2002) (Figure 1.16)

Fossil fuel reserves/production ratios need to be interpreted with great caution, however. They do not take into account the discovery of new

(a)

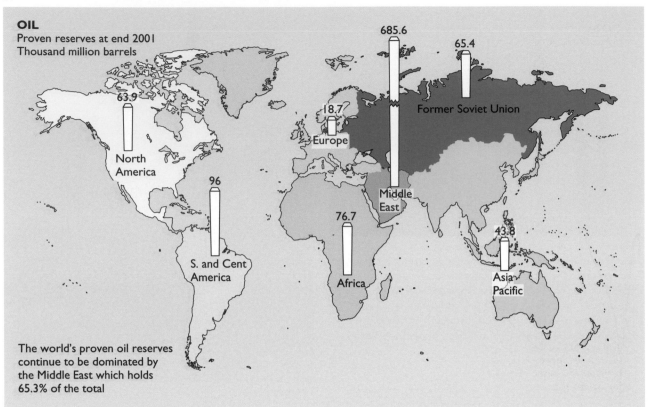

OIL
Proven reserves at end 2001
Thousand million barrels

The world's proven oil reserves
continue to be dominated by
the Middle East which holds
65.3% of the total

(b)

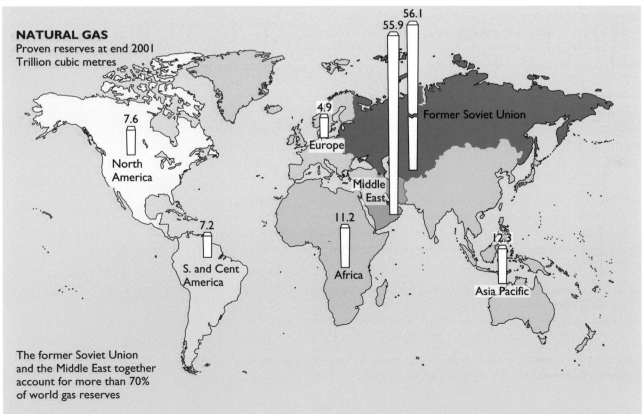

NATURAL GAS
Proven reserves at end 2001
Trillion cubic metres

The former Soviet Union
and the Middle East together
account for more than 70%
of world gas reserves

(c)

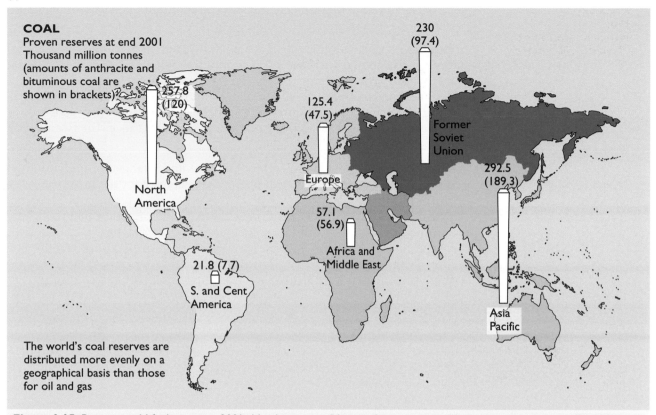

COAL
Proven reserves at end 2001
Thousand million tonnes
(amounts of anthracite and
bituminous coal are
shown in brackets)

230
(97.4)

257.8
(120)

125.4
(47.5)

Former
Soviet
Union

North
America

Europe

292.5
(189.3)

57.1
(56.9)

Africa and
Middle East

21.8 (7.7)

S. and Cent
America

Asia
Pacific

The world's coal reserves are
distributed more evenly on a
geographical basis than those
for oil and gas

Figure 1.15 Proven world fuel reserves, 2001: (a) oil reserves; (b) natural gas reserves; (c) coal reserves (source: BP, 2002)

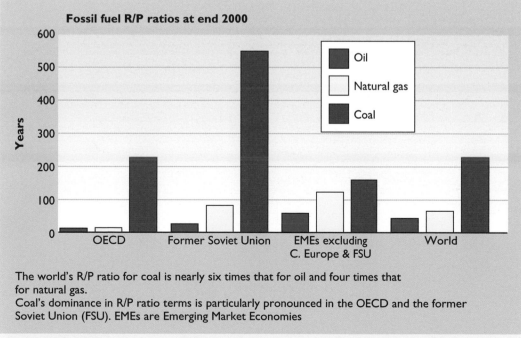

Fossil fuel R/P ratios at end 2000

Oil

Natural gas

Coal

Years

OECD

Former Soviet Union

EMEs excluding
C. Europe & FSU

World

The world's R/P ratio for coal is nearly six times that for oil and four times that
for natural gas.
Coal's dominance in R/P ratio terms is particularly pronounced in the OECD and the former
Soviet Union (FSU). EMEs are Emerging Market Economies

Figure 1.16 Reserves/production (R/P) ratios (in years) for oil, natural gas and coal, 2000, for various
regions of the world and the world as a whole (source: BP, 2002)

proven reserves, or technological developments that enable more fuel to be extracted from deposits or improve the economic viability of 'difficult' deposits.

Despite these developments, it seems likely that, at least in the case of oil from conventional sources, world production will reach a peak in the first decade of this century. From then on, although vast quantities of conventional oil will still remain, the resource will be on a declining curve. (United Nations, 2000; Campbell and Laherrere, 1998). This seems likely to lead to increased instability and potential for conflict as the twenty-first century proceeds.

The massive use by our society of coal, oil and gas has, literally, fuelled enormous increases in material prosperity – at least for the majority in the industrialized countries. But it has also had numerous adverse consequences. As already mentioned, these include air and water pollution, mining accidents, fires and explosions on oil or gas rigs, conflicts over access to fuel resources and, perhaps most profoundly, the global climate change that is likely to be the result of increasing atmospheric carbon dioxide concentrations caused by fossil fuel combustion (see Box 1.2).

BOX 1.2 The greenhouse effect and global climate change

The greenhouse effect in its natural form has existed on the planet for hundreds of millions of years and is essential in maintaining the Earth's surface at a temperature suitable for life. Without it, we would all freeze.

The sun's radiant energy, as it falls on the earth, warms its surface. The earth in turn re-radiates heat energy back into space in the form of infra-red radiation. The temperature of the earth establishes itself at an equilibrium level at which the incoming energy from the sun exactly balances the outgoing infra-red radiation.

If the earth had no atmosphere, its surface temperature would be approximately minus 18 °C, well below the freezing point of water. However our atmosphere, whilst largely transparent to incoming solar radiation

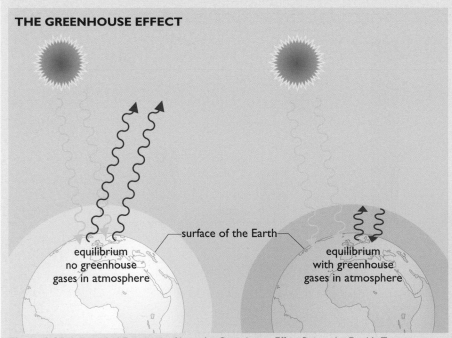

THE GREENHOUSE EFFECT

surface of the Earth

equilibrium no greenhouse gases in atmosphere

equilibrium with greenhouse gases in atmosphere

Figure 1.17 A Simplified Depiction of how the Greenhouse Effect Raises the Earth's Temperature

in the visible part of the spectrum, is partially opaque to outgoing infra-red radiation. It behaves in this way because, in addition to its main constituents, nitrogen and oxygen, it also contains very small quantities of 'greenhouse gases'. Put simply, these enable the atmosphere to act like the panes of glass in a greenhouse, allowing the sun to enter but inhibiting the outflow of heat, so keeping the earth's surface considerably warmer than it would otherwise be. The average surface temperature of the earth is in fact around 15 °C, some 33 °C warmer than it would be without the greenhouse effect.

The most important greenhouse gases are water vapour, carbon dioxide and methane, though other gases such as the synthetic Chlorofluorocarbons (CFCs) also play significant but lesser roles.

Water vapour evaporating from the oceans plays a major part in maintaining the natural greenhouse effect, but human activities have very little influence on the vast processes through which water cycles between the oceans and the atmosphere.

Carbon dioxide (CO_2) is also primarily generated by natural processes. These include the process of

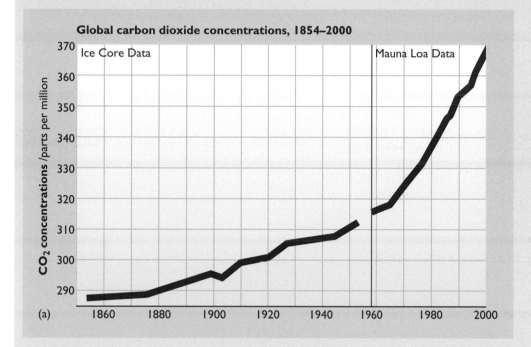

(a)

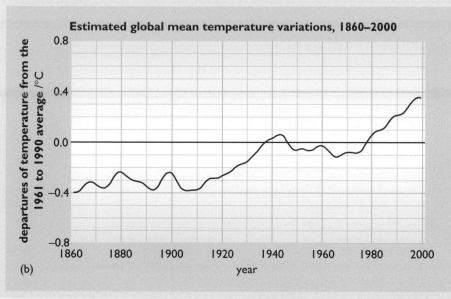

(b)

Figure 1.18 (a) Atmospheric concentrations of carbon dioxide (CO_2), 1854–2000; (b) estimated global mean temperature variations, 1860–2000. Carbon dioxide data from 1958 were measured at Mauna Loa, Hawaii; pre-1958 data are estimated from ice cores (source: Intergovernmental Panel on Climate Change, 2001)

respiration, in which organisms 'breathe out' carbon dioxide; and the emissions of CO_2 that occur when organisms die and the carbon compounds of which they are composed decay. But since the industrial revolution, the burning of fossil fuels by humanity has been adding substantial quantities of CO_2 to our atmosphere. The fossil fuels are essentially compounds of carbon and hydrogen. Coal consists mostly of carbon, the chemical symbol for which is C. Natural gas, the chemical name for which is *methane*, consists of carbon and hydrogen (symbol:H). Each carbon atom is surrounded by four hydrogen atoms, so in chemical shorthand its symbol is CH_4. Oil is a more complex mixture of many different hydrocarbon molecules. When any of these fuels is burned, carbon dioxide is produced, along with water.

The concentration of CO_2 in the atmosphere in pre-industrial times was around 280 parts per million by volume (ppmv) but levels have been steadily rising since then, reaching nearly 370 ppmv in 2000.

The other main greenhouse gas, methane, is given off naturally when vegetation decays in the absence of oxygen – for example, under water. However various human activities, including increasing rice cultivation, which causes methane emissions from paddy fields, and leaks of fossil methane from natural gas distribution systems, have caused the levels of methane in the atmosphere to increase sharply. Concentrations have risen from about 750 parts per billion by volume (ppbv) in pre-industrial times to around 1750 ppbv in 2000.

These additional emissions of carbon dioxide and methane are the main causes of the so-called *anthropogenic* – that is, human-induced – greenhouse effect. Unlike the operation of the natural greenhouse effect, which is benign, the anthropogenic greenhouse effect is almost certainly the cause of a global warming trend that could have very serious consequences for humanity. Though a small minority dissents, the majority of scientists now believe that the anthropogenic effect, acting to enhance the natural processes, has already caused the mean surface temperature of the earth to rise by about 0.6 °C during the twentieth century (Intergovernmental Panel on Climate Change, 2001). Moreover, if steps are not taken to limit greenhouse gas emissions, atmospheric CO_2 levels will probably rise by 2100 to between 540 and 970 ppmv (depending on the assumptions made). These levels would be between two and three times the pre-industrial CO_2 concentration, and would be likely to lead to rises in the earth's mean surface temperature of between 1.4 and 5.8 °C by the end of the century. Such temperature rises would be unprecedented since the ending of the last major Ice Age, more than 10 000 years ago.

These temperature rises would be very likely to result in significant changes to the earth's climate system. Such changes would probably include more intense rainfall, more tropical cyclones, or long periods of drought, all of which would disrupt agriculture. Moreover, ecosystems might be damaged with some species unable to adapt quickly enough to such rapid changes in climate.

In addition, due to thermal expansion of the oceans, sea levels would be expected to rise by around 0.5 metres during the twenty-first century, sufficient to submerge some low-lying areas and islands. In the longer term, further sea level rises would result if the Antarctic ice sheets were to melt significantly.

Nuclear energy

Nuclear energy is based on harnessing the very large quantities of energy that are released when the nuclei of certain atoms, notably uranium-235 and plutonium-239, are induced to split or 'fission'. The complete fission of a kilogram of uranium-235 should produce, in principle, as much energy as the combustion of over 3000 tonnes of coal. In practice, the fission is incomplete and there are other losses, but nevertheless nuclear fuels are more highly-concentrated sources of energy than fossil fuels.

The heat generated by nuclear fission in a nuclear power station is used to raise high-pressure steam which then drives steam turbines coupled to electrical generators, as in a conventional power station.

The development of 'peaceful' nuclear electricity generation after its use for military purposes in World War 2 was initially heralded as ushering-in a new era of virtually-limitless, clean power that some predicted would be 'too cheap to meter'. In practice, however, nuclear electricity has proved to

Figure 1.19 Queen Elizabeth II opening Calder Hall nuclear power station in Cumbria, UK, in 1956

be more expensive than that from fossil fuels. Since the UK opened the world's first grid-connected nuclear power station at Calder Hall in Cumbria in 1956, nuclear electricity generation has expanded to a point where it now accounts for nearly 7% of world primary energy, and for over 17% of the world's electricity. In some countries, it is the principal source of electricity generation. France, for example, derives three-quarters of its electricity from nuclear power.

Figure 1.20 The Civaux Pressurized Water (PWR) nuclear Reactor near Poitiers, Vienne, France. In 2000, France produced 77% of its electricity from nuclear power

A major advantage of nuclear energy is that the operation of nuclear power plants results in no emissions of CO_2 or of other 'conventional' pollutants like sulphur dioxide. However, there are some emissions from the fossil fuel used in uranium mining, nuclear fuel manufacture, and the construction of nuclear power plants.

There seems little danger of the world 'running out' of nuclear fuel in the near future. Uranium reserves have been identified in many countries and are sufficient for many decades of use at current rates, and there are probably enough additional deposits to extend this to several centuries. Furthermore, advanced nuclear technologies such as the 'fast breeder reactor' (FBR) could enable uranium deposits to be used even more effectively, thus extending the lifetime of reserves. In an FBR, the plentiful but non-fissile isotope uranium-238 is transformed into fissile plutonium-239, which can then be used as reactor fuel. But the development of FBRs has been inhibited by substantial technical and safety problems, and by the low price of uranium which currently makes the technology un-competitive economically.

Although the majority of nuclear reactors in most countries have operated without serious safely problems, a number of major accidents, like those at Windscale in the UK in 1957, Three Mile Island in the USA in 1979 and Chernobyl in the Ukraine in 1986, have created widespread public unease about nuclear technology in general – despite the opinion of nuclear-industry experts who argue that such anxieties are irrational.

Less spectacular are the continued releases of harmful radioactive by-products, in small but insidious and cumulative quantities, to the atmosphere and oceans during the routine operation of nuclear power plants and fuel manufacturing or reprocessing facilities.

Figure 1.21 The Chernobyl nuclear power plant following the accident in 1986

There is also the problem of how, ultimately, to dispose of nuclear waste products, some of which remain hazardous for many thousands of years; and the problem of proliferation of nuclear materials such as plutonium-239 and uranium-235, which could fall into the hands of 'rogue states' or 'terrorists' capable of creating crude but devastating atomic weapons from them. Nuclear power stations and reprocessing facilities may themselves be vulnerable to terrorist attacks, which could result in the release of very large quantities of radioactive substances into the environment.

Despite these difficulties, the nuclear industry is attempting to develop more advanced types of nuclear reactor which, it claims, will be cheaper to build and operate, and inherently safer, than existing designs. These are being promoted as an improved technological option for generating the carbon-free electricity that will be required later in the twenty-first century if global climate change is to be mitigated.

Another potentially important nuclear technology is that of nuclear *fusion*. As its name implies, this involves the fusing together of atomic nuclei, in this case those of deuterium (so-called 'heavy' hydrogen). This process, similar to that underlying the generation of energy within the sun, also results in the release of very large amounts of energy. However, in order to create fusion on earth

Figure 1.22 The Joint European Torus (JET) experimental nuclear fusion reactor, located at Culham, Oxfordshire, UK

it is necessary to create conditions in which the nuclei of special forms (called isotopes) of hydrogen interact in an extremely confined space at extremely high temperatures, and so far scientists have only been able to make this happen for a few seconds.

Moreover, the energy required to power the process currently greatly exceeds the energy generated. Research into fusion power continues, with substantial funding, but most experts consider that the technology, even if eventually it can be demonstrated successfully, is very unlikely to become commercially available for many decades.

Bioenergy

Not all fuel sources are, of course, of fossil or nuclear origin. From prehistoric times, human beings have harnessed the power of fire by burning wood to create warmth and light and to cook food.

Wood is created by photosynthesis in the leaves of plants. Photosynthesis is a process powered by solar energy in which atmospheric carbon dioxide and water are converted into carbohydrates (compounds of carbon, oxygen and hydrogen) in the plant's leaves and stems. These, in the form of wood or other 'biomass', can be used as fuels – called **biofuels**, which are sources of **bioenergy**.

Wood is still very widely used as a fuel in many parts of the 'developing' world. In some countries, other biofuels such as animal dung (ultimately also derived from the growth of plants) are also used. As described in Box 1.1, such traditional biofuels are estimated to supply some 11% of world primary energy, though the data are somewhat uncertain.

If the forests that provide wood fuel are re-planted at the same rate as they are cut down, then such fuel use should in principle be sustainable. When forests are managed sustainably in this way, the CO_2 absorbed in growing replacement trees should equal the CO_2 given off when the original trees are burned. However, this is only true when complete combustion of the wood occurs and all the carbon in the wood is released as carbon dioxide. Although near-complete combustion can be achieved in the best available wood stoves and furnaces, most open fires and stoves are not so efficient. This means that not only is carbon dioxide released (albeit in somewhat smaller quantities if the combustion is incomplete) but other combustion products are also emitted, some of which are more powerful greenhouse gases than CO_2. In particular, these can include methane, which on a molecule-for-molecule basis has 20 times the global warming potential of CO_2 over a 20 year period. The *incomplete* combustion of wood can therefore release a mixture of greenhouse gases with a greater overall global warming effect than can be offset by the CO_2 absorbed in growing replacement trees. This suggests an urgent need to improve the efficiency of traditional wood burning processes (Smith *et al.*, 2000). However it should be stressed that the overall global effect of greenhouse gas emissions arising from incomplete biomass combustion in developing countries is probably much less than that of emissions from burning fossil fuels, which occurs mostly in the 'developed' countries.

A further problem is that in many 'developing' countries wood fuel is being used at a rate that exceeds its re-growth, which is not only unsustainable

Figure 1.23 The 10 MW 'Arbre' plant at Eggborough, Yorkshire, Britain's first wood-fired power station. Commissioning has been delayed by financial and technical problems

but also results in villagers having to travel ever-increasing distances, often involving great hardship, to gather sufficient firewood for their daily needs. Also, when it has been gathered, firewood is often burned very inefficiently in open fires – as was the case in Britain and many other 'developed' countries until quite recently. This not only results in excess greenhouse gas emissions, as we have seen, but also gives much less effective warmth than if an efficient stove were used. Moreover, it usually results in high levels of smoke pollution, with very detrimental health effects.

Not all bioenergy use is in the form of traditional biofuels. As noted in Box 1.1, a significant contribution to world supplies now comes from so-called 'modern' bioenergy power plants. These feature the clean, high-efficiency combustion of straw, forestry wastes or wood chips from trees grown in special plantations. The heat produced is either used directly or for electricity generation, or sometimes for both purposes.

Municipal wastes, a large proportion of which are biological in origin, are also widely used for heat or electricity production. However, there is considerable controversy over whether or not energy from waste should be regarded as 'sustainable'. Waste-to-energy plants have been opposed by some environmental groups on the grounds that, in order to be economically viable, they need to be fed with a steady stream of waste over many years, which discourages better solutions to the waste problem, such as material re-use or recycling. There are further concerns over possible emissions of dioxins, which are carcinogenic, from the combustion of chlorine compounds present in municipal waste.

Figure 1.24 A substantial proportion of vehicles in Brazil are fuelled by alcohol derived from sugar cane

Another modern source of bioenergy is alcohol (ethanol) produced by fermenting sugar cane or maize, which is quite widely used in vehicles in Brazil and some states of the USA. The alcohol is often blended with conventional petroleum to form a mixture known as 'Gasohol'.

Hydroelectricity

Another energy source that has been harnessed by humanity for many centuries is the power of flowing water, which has been used for milling corn, pumping and driving machinery. During the twentieth century, its main use has been in the generation of hydro-electricity, and hydropower has grown to become one of the world's principal electricity sources. It currently provides some 2.3% of world primary energy. However, the relative contribution of hydroelectric power (and of other electricity-producing renewables) is under-stated by a factor of about three in most national and international compilations of energy statistics. This is due to a convention whereby the heat produced by thermal power stations (both fossil and nuclear-fuelled) is included as part of their primary energy

Figure 1.25 The large hydro-electric power plant at Glen Canyon Dam, Lake Powell, Arizona, which has an electrical generating capacity of 1300 MW. Lake Powell was formed with the construction of the Glen Canyon Dam in 1963, on the Colorado River in Page, Arizona. It is the second largest man-made lake in the United States and is 187 miles long

contribution, even though that heat is normally wasted. (A more detailed explanation of such conventions is given in Chapter 2.) The annual *electricity* outputs of the world's hydro and nuclear power plants are actually about the same, but due to this statistical convention the primary energy contribution of nuclear is calculated to be about 7% whereas that of hydro is only about one-third of this. Hydro power in 2000 contributed over 17% of world electricity.

The original source of hydroelectric power is solar energy, which warms the world's oceans, causing water to evaporate from them.

In the atmosphere, this forms clouds of moisture which eventually falls back to earth in the form of rain (or snow). The rain flows down through mountains into streams and rivers, where its flow can be harnessed using water wheels or turbines to generate power.

When harnessed on a small scale, hydropower creates few, if any, adverse environmental impacts.

However, many modern hydro installations have been built on a very large scale, involving the creation of massive dams and the flooding of extremely large areas. This often entails the re-location of many thousands of indigenous residents who are usually, to say the least, reluctant to move from their homes. Other impacts include adverse effects on fish and other wildlife, reductions in water-borne nutrients used in agriculture downstream, increases in water-borne diseases – and not least, the rare but catastrophic effects of dam failures. A further problem with large dams is that in certain circumstances trees and other vegetation trapped below water when a reservoir is flooded can decay 'anaerobically' (i.e. in the absence of oxygen). This produces methane which, as we have seen, is a more powerful greenhouse gas than the CO_2 that would have been produced if the tree had decayed normally in the presence of oxygen from the atmosphere.

However, the current consensus is that greenhouse gas emissions from hydropower generation are likely to be at least an order of magnitude lower than those from fossil fuel generated electricity (United Nations, 2000).

Summary

This section has described how fossil fuels provide the majority of the world's energy requirements, with bioenergy, nuclear energy and hydropower also making major contributions. The other 'renewable' energy sources currently supply only a small fraction of world demand, although the contribution of these 'renewables' seems likely to grow rapidly in coming decades, as we shall see in the following section.

1.4 Renewable energy sources

Fossil and nuclear fuels are often termed **non-renewable** energy sources. This is because, although the quantities in which they are available may be extremely large, they are nevertheless finite and so will in principle 'run out' at some time in the future.

By contrast, hydropower and bioenergy (from biofuels grown sustainably) are two examples of **renewable** energy sources – that is, sources that are continuously replenished by natural processes. Renewable energy sources are essentially *flows* of energy, whereas the fossil and nuclear fuels are, in essence, *stocks* of energy.

World-wide, there has been a rapid rise in the development and deployment of renewable energy sources during the past few decades, not only because, unlike fossil or nuclear fuels, there is no danger of their 'running out', but also because their use normally entails no (or few) greenhouse gas emissions and therefore does not contribute to global climate change.

The companion volume, *Renewable Energy*, describes in more detail the renewable energy sources, which range from solar power in its various forms, through bioenergy and hydropower to wind, wave, tidal and geothermal energy. (Figure 1.26)

The general nature and scope of the various 'renewables' can be briefly summarized as follows, beginning with the most important renewable source, solar energy.

Solar energy

Solar energy, it should firstly be stressed, makes an enormous but largely unrecorded contribution to our energy needs. It is the sun's radiant energy, as noted in Box 1.2, that maintains the Earth's surface at a temperature warm enough to support human life. But despite this enormous input of energy to our civilization, the sun is virtually ignored in national and international energy statistics, which are almost entirely concerned with consumption of commercial fuels.

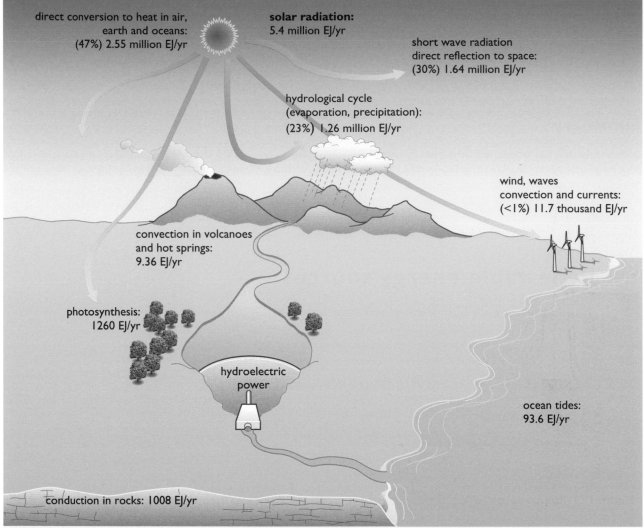

direct conversion to heat in air,
earth and oceans:
(47%) 2.55 million EJ/yr

solar radiation:
5.4 million EJ/yr

short wave radiation
direct reflection to space:
(30%) 1.64 million EJ/yr

hydrological cycle
(evaporation, precipitation):
(23%) 1.26 million EJ/yr

wind, waves
convection and currents:
(<1%) 11.7 thousand EJ/yr

convection in volcanoes
and hot springs:
9.36 EJ/yr

photosynthesis:
1260 EJ/yr

hydroelectric
power

ocean tides:
93.6 EJ/yr

conduction in rocks: 1008 EJ/yr

Figure 1.26 The various forms of renewable energy depend primarily on incoming solar radiation, which totals some 5.4 million Exajoules (EJ) per year. Of this, approximately 30% is reflected back into space. The remaining 70% is in principle available for use on Earth, as shown, and amounts to approximately 3.8 million EJ. This is some 10 000 times the current rate of consumption of fossil and nuclear fuels, which in 2000 amounted to some 360 EJ. Two other, non-solar, renewable energy sources are shown in the figure. These are the motion of the ocean tides, caused principally by the moon's gravitational pull (with a small contribution from the sun's gravity); and geothermal heat from the earth's interior, which manifests itself in convection in volcanoes and hot springs, and in conduction in rocks

The sun has a surface temperature of 6000 °C, maintained by continuous nuclear fusion reactions between hydrogen atoms within its interior. These nuclear reactions will gradually convert all of the hydrogen into heavier elements, but this is a relatively slow process and the sun should continue to supply power for another 5 billion years.

The sun radiates huge quantities of energy into the surrounding space, and the tiny fraction intercepted by the Earth's atmosphere, 150 million km away, is nonetheless equivalent to about 15 000 times humanity's present rate of use of fossil and nuclear fuels. Even though approximately one-

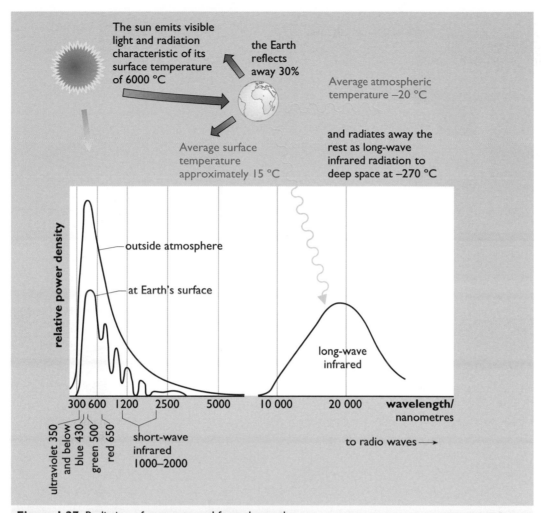

The sun emits visible light and radiation characteristic of its surface temperature of 6000 °C

the Earth reflects away 30%

Average atmospheric temperature −20 °C

Average surface temperature approximately 15 °C

and radiates away the rest as long-wave infrared radiation to deep space at −270 °C

relative power density

outside atmosphere

at Earth's surface

long-wave infrared

300 600   1200   2500   5000   10 000   20 000   **wavelength/** nanometres

ultraviolet 350 and below

blue 430

green 500

red 650

short-wave infrared 1000–2000

to radio waves ⟶

Figure 1.27 Radiation of energy to and from the earth

third of the intercepted energy is reflected away by the atmosphere before reaching the earth's surface, this still means that a continuous and virtually-inexhaustible flow of power amounting to 10 000 times our current rate of consumption of conventional fuels is available in principle to human civilization.

Solar energy, when it enters our buildings, warms and illuminates them to a significant extent. When buildings are specifically designed to take full advantage of the sun's radiation, their needs for additional heating and for artificial lighting can be further reduced.

Solar power can also be harnessed by using solar collectors to produce hot water for washing or space heating in buildings.

Such collectors are in widespread use in sunny countries such as Israel and Greece, but are also quite widely used in less sunny places such as Austria. Even in cloudy Britain there are more than 40 000 roof-top solar water heating systems.

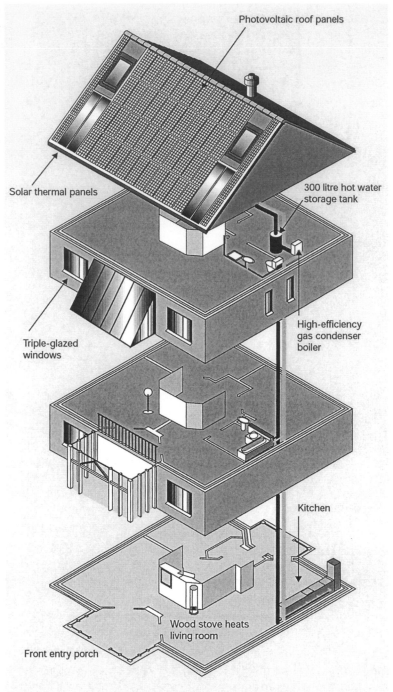

Photovoltaic roof panels

Solar thermal panels

300 litre hot water
storage tank

Triple-glazed
windows

High-efficiency
gas condenser
boiler

Kitchen

Wood stove heats
living room

Front entry porch

Figure 1.28 The roof of this solar house in Oxford (view from garden, left; cutaway view, above) has a grid-linked 4 kW array of photovoltaic panels. These generate enough electricity to supply its annual requirements, plus a surplus which is used to provide part of the power to run a small electric car. The roof also incorporates a 5 m^2 array of solar water heating panels which provide three-quarters of the house's hot water requirements. The house is well-insulated and includes a conservatory that contributes 'passive' solar energy to space heating in Spring and Autumn, supplemented by a natural gas boiler and a small wood-burning stove used on very cold days

Figure 1.29 The largest 'solar thermal-electric' installation of its kind in the world, the Luz project in California's Mojave Desert, has a peak output of some 350 megawatts and occupies several square kilometres of land

In regions such as Southern California, where solar radiation levels are more than twice those of the UK and skies are clearer, the sun's rays are strong enough to make it practicable to generate high-temperature steam using arrays of concentrating mirrors. The steam can then be used to power a turbine that drives a generator to produce electricity (Figure 1.29).

Harnessing solar energy to provide electricity directly involves the use of a different and more sophisticated technology called solar photovoltaics (PV). Photovoltaic 'modules' are made of specially-prepared layers of semi-conducting materials (usually silicon) that generate electricity when photons of sunlight fall upon them. Arrays of PV modules are normally mounted on the roofs or facades of buildings, providing some or all of their electricity needs. (Figures 1.28 and 1.30)

Photovoltaic technology is growing very rapidly and several countries have initiated major development and demonstration programmes. Germany, for example, plans to install 100 000 PV roofs and building facades by the end of 2003.

Photovoltaics may well make a significant contribution to world needs in coming decades, but at present its share of world consumption is extremely small. This is mainly due to the very high cost of PV modules, which are currently produced in relatively small quantities. Studies have shown that if the annual output of the manufacturing plants that produce PV modules were increased by a factor of about 20, the cost of PV-generated electricity could be reduced to a point at which it would be competitive with electricity from conventional sources in many industrialized countries.

Figure 1.30 This 3500 m² solar office building at the Doxford International Business Park near Sunderland in the UK incorporates 646 m² of photovoltaic modules. These have a peak power output of 73 kW and generate some 55 000 kWh of electricity per year. The building is well insulated and uses passive solar design to maximise the use of natural daylight and to minimise space heating and air conditioning needs. It is also designed for natural ventilation and night-time cooling

Indirect use of solar energy

The above examples illustrate the *direct* harnessing of the sun's radiant energy to produce heat and electricity. But the sun's energy can also be harnessed via other forms of energy that are *indirect* manifestations of its power. Principally, these are bioenergy and hydropower, already discussed in Section 1.3 above, together with wind energy and wave power.

Wind energy

When solar radiation enters the earth's atmosphere, because of the curvature of the earth it warms different regions of the atmosphere to differing extents – most at the equator and least at the poles. Since air tends to flow from warmer to cooler regions, this causes what we call winds, and it is these air flows that are harnessed in windmills and wind turbines to produce power.

Wind power, in the form of traditional windmills used for grinding corn or pumping water, has been in use for centuries. But in the second half of the twentieth century, and particularly in the past few decades, the use of modern wind turbines for electricity generation has been growing very rapidly. Installed wind generating capacity has doubled every two and a half years since 1991, and at the end of 2001 the world total was over 23 000 MW. (*Windpower Monthly*, 2002) Denmark derives more than 15% of its electricity from wind, and in other countries such as Germany, Spain and the United States of America turbines have in recent years been installed at a rate of rate of more than a thousand megawatts per year.

Figure 1.31 This wind farm, at Carno in mid-Wales, is one of the largest in Europe. It incorporates 56 wind turbines, each with a rotor diameter of 44 metres and a tower height of 31.5 metres. The total installed capacity is 33.6 MW, sufficient to provide power for some 25 000 homes

At present, most of these turbines have been installed on land. But several countries have ambitious plans to install thousands of wind turbines offshore. Denmark, for example, has three offshore wind farms and plans many more, as part of its aim of deriving 30% of its electricity from wind by 2020– though these plans are subject to future political approval.

Figure 1.32 The Middelgrunden wind farm, completed in 2001, is located in the sea just off Copenhagen harbour in Denmark. It includes 20 two megawatt wind turbines, which provide 3% of the electricity consumption of the Copenhagen municipality

Figure 1.33 Britain's first offshore wind turbines, located 1 km away from the coast at Blyth harbour, Northumberland. The twin 2 MW turbines were installed in 2000 by a consortium including AMEC, Border Wind, Shell Renewables and the Dutch electricity utility Nuon

The UK, too, has ambitious offshore wind power proposals. Britain's first two offshore wind turbines were installed off Blyth harbour in Northumberland in 2000, and sites have been identified for 13 offshore wind farms that could be built in the coming decade. These would have a total installed capacity of 1600 MW.

Wave power

When winds blow over the world's oceans, they cause waves. The power in such waves, as they gradually build up over very long distances, can be very great — as anyone watching or feeling that power eventually being dissipated on a beach will know.

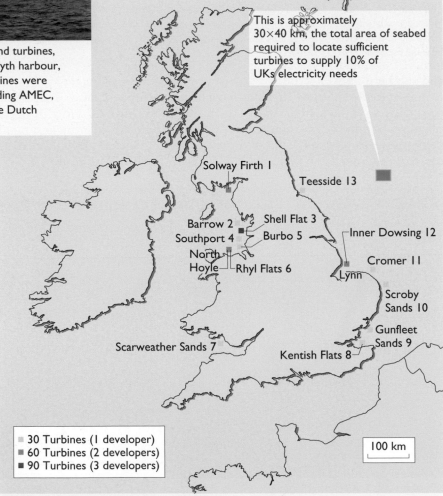

Figure 1.34 Proposed locations of Britain's first 13 offshore wind farms. Also shown is the area of the North Sea that would be needed for offshore wind farms to produce 10% of the UK's current annual electricity demand

Various technologies for harnessing the power of waves have been developed over the past few decades, of which the 'oscillating water column' (OWC) is perhaps the most widely used. In an OWC, the rise and fall of the waves inside an enclosed chamber alternately blows and sucks air through a special kind of air turbine, which is coupled to a generator to produce electricity.

Wave energy technology is not as fully developed as wind power or photovoltaics, but its potential has recently been re-emphasized by several governments, including that of the UK. Rapid advances in developing and demonstrating the technology can be expected over the coming decade.

Figure 1.35 The 500 kW 'Limpet' wave energy plant installed in 2001 on the Scottish island of Islay

All of the renewable energy sources described above — solar, bioenergy, hydropower, wind and wave — are, as we have seen, either direct or indirect forms of solar energy. However there are two other renewable sources, tidal and geothermal energy, that do not depend on solar radiation.

Non-solar renewables

Tidal energy

The energy that causes the slow but regular rise and fall of the tides around our coastlines is not the same as that which creates waves. It is caused principally by the gravitational pull of the moon on the world's oceans. The sun also plays a minor role, not through its radiant energy but in the form of its gravitational pull, which exerts small additional effect on tidal rhythms.

The principal technology for harnessing tidal energy essentially involves building a low dam, or barrage, across the estuary of a suitable river. The barrage has inlets that allow the rising sea levels to build up behind it. When the tide has reached maximum height, the inlets are closed and the impounded water is allowed to flow back to the sea in a controlled manner, via a turbine-generator system similar to that used in hydroelectric schemes.

The world's largest tidal energy scheme is at La Rance in France, which has a capacity of 240 MW (Figure 1.36).

There are a few other, smaller, tidal plants in various countries, including Canada, Russia and China. The United Kingdom has one of the world's best potential sites for a tidal energy scheme, in the Severn Estuary. If built, its capacity would be around 8600 MW, much larger than any other single power plant, and it could provide about 6% of current UK electricity demand. But the scheme has not yet been implemented, mainly due to its very high capital cost and concerns about the effects on wildlife in the Severn estuary.

Figure 1.36 The 240 MW tidal barrage installed at the Rance Estuary in France

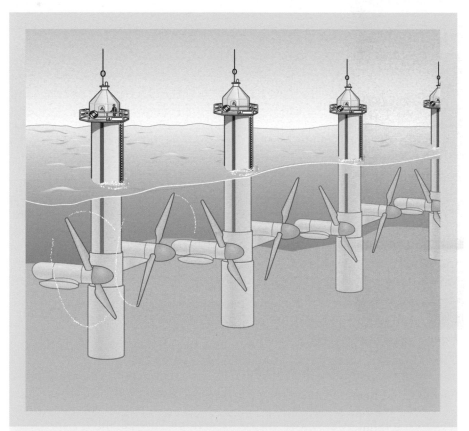

Figure 1.37 Artist's impression of an array of undersea tidal current turbines. The twin-rotor turbines can be raised to the surface to avoid the need for underwater maintenance

Another, newer tidal energy technology involves the use of underwater turbines (rather like submerged wind turbines) to harness the strong tidal and oceanic currents that flow in certain coastal regions. A 10 kW prototype tidal current turbine was tested at Loch Linne, in Scotland, in 1994, and a larger, 300 kW prototype was tested off the Devon coast in 2002.

The technology is still under development, but its prospects are promising.

Geothermal energy

Geothermal energy is another renewable source that is not derived from solar radiation. As the name implies, its source is the earth's internal heat, which originates mainly from the decay of long-lived radioactive elements. The most useful geothermal resources occur where underground bodies of water called aquifers can collect this heat, especially in those areas where volcanic or tectonic activity brings the heat close to the surface. The resulting hot water, or in some cases steam, is used for electricity generation where possible, for example in Italy, New Zealand and the Philippines, and for direct heating use in more than 60 other countries. Geothermal energy is already making a minor but locally useful contribution to world energy supplies.

Figure 1.38 One of the geothermal power plants at Larderello, Italy, used to provide electricity and hot water

If geothermal heat is extracted in a particular location at a rate that does not exceed the rate at which it is being replenished from deep within the earth, it is a renewable energy source. But in many cases this is not so: the geothermal heat is in effect being 'mined' and will 'run out' locally in perhaps a few years or decades.

Sustainability of renewable energy sources

Renewable energy sources are generally sustainable in the sense that they cannot 'run out' – although, as noted above, both biomass and geothermal energy need wise management if they are to be used sustainably. For all of the other renewables, almost any realistic rate of exploitation by humans would be unlikely to approach their rate of replenishment by nature, though of course the use of all renewables is subject to various practical constraints.

Renewable energies are also relatively 'sustainable' in the additional sense that their environmental and social impacts are generally more benign than those of fossil or nuclear fuels. However, the deployment of renewables in some cases entails significant environmental and social impacts. Renewable energy sources are generally much less concentrated than fossil or nuclear fuels, so large areas of land (or building surfaces) are often required if substantial quantities of energy are to be collected. This can lead to a significant visual impact, as in the case of wind turbines.

Also, the monetary costs of many renewable sources are at present considerably higher than those of conventional fuels. Until this imbalance is reduced, either by reducing the costs of renewables or through increases in the costs of conventional sources, renewables may be unable to succeed in capturing a substantial fraction of the world market.

Renewables may seem attractive in many ways, but how large a contribution might they make to world energy needs in the future? This is an important question to which we shall return, initially in the final section of this introductory overview, and in more detail in the companion volume, *Renewable Energy*.

1.5 Energy services and efficiency improvement

Energy services

Except in the form of food, no one needs or wants energy *as such*. That is to say, no one wants to eat coal or uranium, drink oil, breathe natural gas or be directly connected to an electricity supply. What people want is *energy services* – those services that energy uniquely can provide. Principally, these are: heat, for warming rooms, for washing and for processing materials; lighting, both interior and exterior; motive power, for a myriad of uses from pumping fluids to lifting elevators to driving vehicles; and power for electronic communications and computing.

When Thomas Edison set up the world's first electric power station in New York in 1882, it was not electricity he sold, but light. He provided the electricity and light bulbs, and charged his customers for the *service* of illumination. This meant he had a strong incentive to generate and distribute electricity as efficiently as possible, and to install light bulbs that were as efficient and long-lasting as possible.

Unfortunately, the early Edison approach did not survive, and the regulatory regime under which most utilities operate today simply rewards them for selling as much energy as possible, irrespective of the efficiency with which

Figure 1.39 Dynamo Room of Edison's Electric Lighting Station, Pearl Street, New York, 1882

it is used or the longevity of the appliances using it. In a few countries, however, governments have changed the way energy utilities are regulated by setting up mechanisms to reward them for providing *energy services* rather than mere energy. In this case, customers benefit by having lower overall costs, the utility makes as much profit as before, and the environment benefits through reduced energy wastage and the emission of fewer pollutants.

Linking supply and demand

But apart from these relatively few enlightened examples, the efficiency with which humanity currently uses its energy sources is generally extremely low. At present, only about one-third of the energy content of the fuel the world uses emerges as 'useful' energy, at the end of the long supply chains we have established to connect our coal and uranium mines, our oil and gas wells, with our energy-related needs for warmth, light, motion, communication, etc.

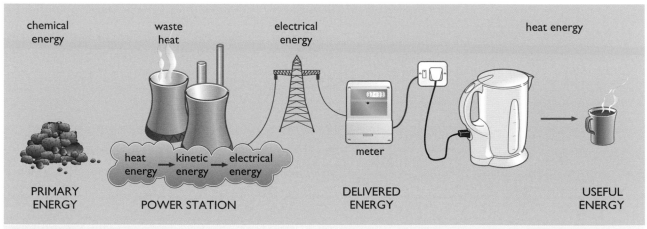

Figure 1.40 An example of one of the energy 'chains' linking primary energy with delivered energy and useful energy, via various energy transformations

The remaining two-thirds usually disappears into the environment in the form of 'waste' heat. One of the reasons for our continuing inefficiency in energy use is that energy has been steadily reducing in price, in real terms, over the past 100 years (see Figure 1.41).

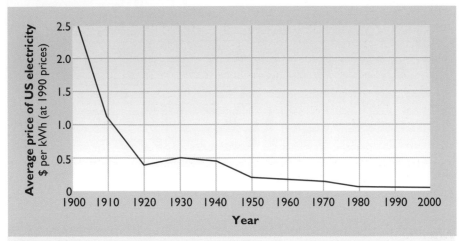

Figure 1.41 Average household rates for US electricity, 1900–2000, expressed in real terms, i.e. taking into account the effects of inflation (source: Smil, 2000)

Energy's decreasing cost means that our society has only a relatively-weak financial incentive to use it more wisely.

The chains that link energy supplies with users' demands are lengthy and complex, as Figure 1.40 illustrates. Each link in the chain involves *converting* energy from one form or another, for example in the burning of coal to generate electricity; or *distributing* energy via some kind of transmission link or network, such as a national electricity grid or gas pipeline infrastructure.

Energy efficiency improvements

Supply-side measures

On the *supply* side of our energy systems, there is a very large potential for improving the efficiency of electricity generation by introducing new technologies that are more efficient than older power plant. The efficiency of a power plant is the percentage of the energy content of the fuel input that is converted into electricity output over a given time period. Since the early days of electricity production, power plant efficiency has been improving steadily.

The most advanced form of fossil-fuelled power plant now available is the Combined Cycle Gas Turbine (CCGT). CCGTs are more than 50% efficient, compared with the older steam turbine power plant that is still in widespread use, where the efficiency is only about 30%, and thus two-thirds of the energy content of the input fuel is wasted in the form of heat, usually dumped to the atmosphere via cooling towers.

CCGTs are more 'climate friendly' than older, coal-fired steam turbine plant, not only because they are more efficient but also because they burn natural gas, which on combustion emits about 40% less CO_2 than coal per unit of energy generated. Overall, taking into account both the higher efficiency and natural gas's lower CO_2 emissions, when compared with traditional coal-fired plant CCGT-based power plants release about half as much CO_2 per unit of electricity produced. Most of the reductions that occurred in Britain's CO_2 emissions during the 1990s were due to the so-called 'dash for gas' as a substitute for coal in power generation.

In some countries, the 'waste' heat from power stations is widely used in district heating schemes to heat buildings. In 2000, some 72% of Denmark's electricity was produced in such 'Combined Heat and Power' systems.

After fuels have been converted to electricity, whether in CCGTs or steam turbine only plant, further losses occur in the wires of the transmission and distribution systems that convey the electricity to customers. In the UK, these amount to around 8%. Overall, this means that even when a modern, high-efficiency CCGT is the electricity generator, less than half the energy in its input fuel emerges as electricity at the customers' sockets. In the case of older power stations the figure is around one-quarter.

Clearly, there is room for further improvements in the supply-side efficiency of our electricity systems, by further increasing the efficiency of generating plant and by ensuring the whatever 'waste' heat remains is piped to where it can be used.

Coal, oil and gas, when they are used directly rather than for electricity generation, are also subjected to processing, refining and cleaning before being distributed to customers. Some energy is also lost in their distribution, for example in the fuel used by road tankers or the electricity used to pump gas or oil through pipelines. However, these losses are much lower, typically less than 10% overall. This means that over 90% of the energy content of coal, oil and gas, if used directly, is available to customers at the end of the processing and distribution chain. The scope for further *supply*-side efficiency improvements is obviously much more limited here than in the case of electricity.

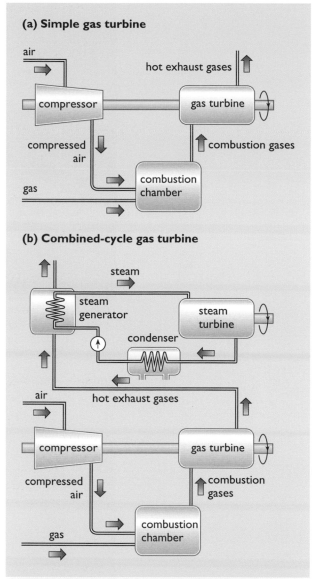

Figure I.42 Diagram comparing the operation of a simple gas turbine power plant (a) with that of a combined cycle gas turbine plant (b). In the latter, the hot exhaust gases from the gas turbine are used to raise steam to power a steam turbine. The steam turbine and gas turbine are coupled to a generator to produce electricity. In a conventional, steam turbine-only power plant, the heat required to produce the steam comes from a boiler. Steam turbines and gas turbines are described in more detail in Chapters 6 and 8. CCGTs are described in more detail in Chapter 9

Figure 1.43 This combined-cycle gas turbine power station at Deeside in the UK was commissioned in 1994 and has an output of 500 MW

Figure 1.44 This coal fired power station at Didcot, Oxfordshire, UK, was commissioned in 1972 and has a capacity of 2000 MW. Two-thirds of the energy content of the fuel burned in such power stations is dissipated by the cooling towers to the atmosphere in the form of steam

Demand-side efficiency improvements

Let us now look briefly at what can done to improve the efficiency of energy use at the *demand* side – that is, in our buildings, industries and vehicles.

Improving the sustainability of energy use by applying demand-side measures involves two distinct approaches, one **technological**, the other **social**.

The technological approach involves installing improved energy conversion (or distribution) technologies that require less input energy to achieve a given level of useful energy output or energy *service*.

The social approach involves re-arranging our lifestyles, individually and collectively, in minor or perhaps major ways, in order to ensure that the energy required to perform a given service is reduced in comparison with other ways of supplying that service.

For example, you may live in a densely populated town with shops, offices, schools and other amenities scattered evenly around. You may be able to do your shopping, go to work, and take the children to school without using a car, simply by walking relatively short distances. Or you may find it convenient to catch a bus, as bus services are usually more frequent and efficient in higher-density settlements.

On the other hand, you may live in a town with a similar population, but one that has been designed (as have many new towns) to have a low population density (i.e. fewer residents per hectare of land), with shops and offices concentrated in the town centre. In this case, you may well use a car for many of your local journeys, consuming fossil fuels and generating emissions of greenhouse gases and other pollutants. In both towns, the residents receive exactly the same levels of service: shopping, working, schooling etc. But in the high-density town the residents can use energy services more sustainably than in the low-density town – all other things being equal.

In Government energy statistics, energy demand is usually broken down into four main sectors:

The domestic sector

This obviously consists of individual households, within which the main categories of energy use are for space heating, water heating, cooking, lighting and other electrical appliances.

The commercial and institutional sector

(often termed the **Services Sector**). This sector consists of offices, shops, schools, hospitals, banks etc. The energy requirements of this sector are very similar to those of the domestic sector: space heating, water heating, cooking, lights and appliances. Air conditioning, however, is more prevalent in this sector than in the domestic sector – at least in countries with temperate climates, like the UK. In this sector, as in the Domestic sector, most of consumption is within **buildings**.

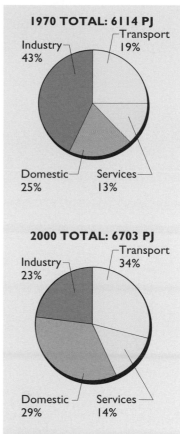

Figure 1.45 UK delivered energy, 1970 and 2000, by Sector. Between 1970 and 2000, overall delivered energy use rose by just under 10%. Transport energy use rose by 96%, while energy use in the domestic and services sectors rose by 27% and 18% respectively. Industrial consumption, by contrast, fell by 41% over the 30 year period. (One petajoule, PJ, is one thousand million million joules, see Appendix.) (source: Department of Trade and Industry, 2001)

The main technological measures that can be taken to conserve energy and use it more efficiently within buildings include:

- improved levels of insulation in walls, roofs and floors, to reduce heat losses through these elements;

- energy-efficient windows, designed to allow less heat to escape whilst still admitting large amounts of sunlight;

- draught-proofing and heat recovery systems to reduce heat lost through ventilation whilst retaining sufficient fresh air within the building;

- more efficient boilers that require a smaller fuel input to achieve a given level of space or water heating, together with improved insulation of pipes to reduce heat distribution losses;

- energy-efficient lights that require much smaller amounts of power to provide a given level of illumination;

- energy-efficient appliances, such as refrigerators, cookers, washing machines, dishwashers, TV sets and hi-fi equipment in the domestic sector; or more efficient computers, copiers and other business equipment in the commercial and institutional sector. These consume less energy while delivering the same level of service as their inefficient predecessors;

- improved control systems, to ensure that energy-consuming equipment is not switched on when not needed, and that power output levels match the requirements of users.

Figures 1.46 Some energy-efficient appliances. Compact fluorescent light bulbs use one-quarter of the electricity consumed by their incandescent counterparts. New energy-efficient refrigerators can use less than half the electricity of their predecessors

Figures 1.47 This supermarket in London is designed to use half the electricity of a conventional new food store of similar size

Figures 1.48 This building at the University of East Anglia consumes less than half the energy of a conventional air-conditioned building of comparable size and function

The industrial sector

This sector mainly covers manufacturing industry, service industries being categorized under 'Services'. Much of industrial energy use also occurs within buildings, and consists of requirements for space heating, water heating, cooking, lights and appliances, as in the Domestic and Commercial & Institutional Sectors. But in addition, many industries, such as the steel industry, use substantial quantities of high temperature heat and large amounts of electricity to power various specialized processes. These demands in many cases exceed those of the buildings where the activities are housed and of the people within them.

So apart from improving the energy efficiency of the buildings and appliances in the industrial sector, where the approaches are similar to those in the domestic and services sectors, there are other measures that apply specifically to industry. In particular, these include 'cascading' of energy uses, where 'waste' heat from a high-temperature process is used to provide energy for lower temperature processes; and the use of high-efficiency electric motors, pumps, fans and drive systems, with accurate matching of motors to the tasks they are required to perform, and accurate sizing of pipes and their associated pumps.

Dematerialization

The measures that can be adopted by industry also include reductions in the material content of products, for example in car bodies or drinks cans, where thinner metals can be used without any reduction in the required strength; or the substitution of less energy-intensive materials, as in the use of plastics instead of steel for car bumpers.

These measures are one form of what has been termed 'dematerialization' – a reduction in the material-intensity (and hence the energy-intensity) of production.

Another form of dematerialization involves changes that are more social than technological. It occurs when the structure of a country's entire economy shifts towards less energy- and materials-intensive activities. For example, in the UK the steel industry today accounts for a much smaller share of the country's gross domestic product (GDP) than it did 20 years ago. By contrast, the UK services sector now constitutes a much bigger fraction of GDP than two decades ago. Since the service sector usually requires less energy than the steel industry for every pound's worth of production, Britain's overall energy demands have been less than they would otherwise have been. However, if the steel that was formerly manufactured in Britain is now manufactured abroad but still imported to the UK in similar quantities, all that has happened is that the energy input, with its associated CO_2 emissions and their implications for global warming, has been transferred to another country.

The transport sector

Motor vehicles (cars, vans, buses, trucks, motor cycles) dominate the transport sector in developed countries. But this sector also encompasses many other modes of transport, including rail, air and shipping, and non-motorized transport forms such as cycling and walking.

As can be seen from Figure 1.49, the various forms of transport vary enormously in their energy requirements per passenger-kilometre travelled. Cycling and walking, of course, require no fuel input apart from food.

In most developed countries there has been an enormous increase in transportation, measured in passenger-kilometres travelled annually, over the past few decades (Figure 1.50). Most of this has involved motorized transport, mainly fuelled by oil, and so energy use has also increased greatly, as have the associated CO_2 emissions.

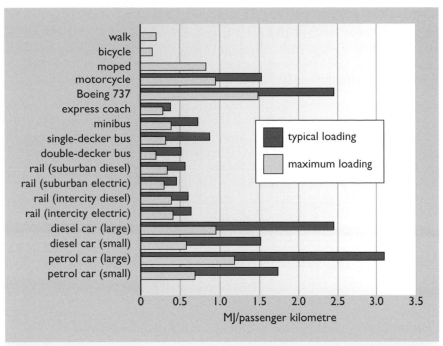

Figure 1.49 Energy efficiency of different modes of transport in the UK (source: Hughes, 1993)

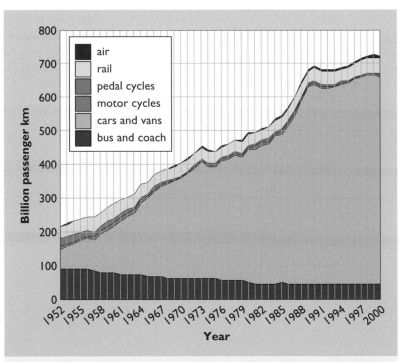

Figure 1.50 Annual passenger-kilometres travelled in the UK, 1952–2000, by transport mode. Note: air travel data refers to internal flights only (source: DTLR, 2001)

Transport energy demand reduction: social measures

Clearly, one social way of reducing the energy required by the transport sector is to shift a proportion of people's journeys away from the energy-intensive modes and towards the more energy-frugal modes. This process is sometimes termed 'modal shift'.

This could be achieved without reducing the total number of journeys, or the overall distance travelled, so that the amenity or service enjoyed by the traveller would remain the same. If, for example, a greater proportion of long-distance journeys within Europe were made by inter-city train rather than by air, the overall energy demand involved could be reduced substantially. Or if urban commuters made more journeys to work by rail or bus instead of using their cars, the effects would be similar. And if householders walked to their local shops instead of taking their cars, no fossil fuels at all would be used for those journeys. Of course, if people are to undertake transport modal shifts of these kinds, they will need to be encouraged by fast, comfortable, efficient services – or penalized into switching by such measures as congestion charging, which is being implemented in central London and other major cities.

Transport energy demand reduction: technological measures

In addition to such social measures, there are numerous technological options for improving the energy-efficiency of transport energy use. Improving vehicle fuel economy is one obvious measure, and the average fuel economy (in miles per gallon, or litres per 100 km) of vehicles has indeed improved very substantially in most developed countries over the past few decades. However, this improvement has been largely offset (in the UK at least) by an increase in the total number of vehicle-miles travelled, and by increases in the average speeds of vehicles, both of which result in increased fuel consumption.

Figure 1.51 The Toyota Prius, a 'hybrid' petrol-electric car

Nevertheless, manufacturers continue to introduce new models with steadily improving fuel economy, partially spurred by legislation requiring them to do so. New approaches include 'hybrid' petrol-electric cars such as the Toyota Prius (Figure 1.51).

In addition to such incremental improvements, there are also more radical possibilities, such as the 'hypercar', proposed by engineers at the Rocky Mountain Institute in the USA (Figure 1.52).

Figure 1.52 The 'Hypercar', designed by engineers at the Rocky Mountain Institute, Colorado, USA, would be streamlined and made of strong but ultra-light, composite materials

This approach involves the use of strong but ultra-lightweight composite materials such as carbon fibre or Kevlar, combined with a highly streamlined body shell. The drive system is would either be of the 'hybrid' type, consisting of a small gasoline-fuelled engine augmented by electric motors and a small battery store; or a more advanced system employing a fuel cell powered by hydrogen. Fuel cells are rather like conventional batteries, except that they are continuously re-charged by supplying fuel – usually hydrogen gas – that reacts electrolytically with oxygen from the atmosphere to produce an electric current. In the hypercar, the fuel cell would generate electricity for electric motors that provide power to the wheels. The hydrogen fuel would either be stored in tanks in its pure form, or generated on-board by 're-forming' fossil fuels. The oxygen would come from the surrounding atmosphere. Hypercars, their proponents claim, could achieve between three and five time the fuel economy of current models, with emissions levels approaching zero in the case of the hydrogen-fuel cell version.

Hypercars may still be some way off, but major manufacturers such as Daimler-Chrysler and Ford have recognized the need to make dramatic reductions in vehicle CO_2 emissions in the long term, and are investing many hundreds of millions of dollars in the production of fuel-celled vehicles (Figure 1.53).

Figure 1.53 This Mercedes A-Class car is powered by a fuel cell running on hydrogen. The manufacturers, Daimler-Chrysler, and other car-makers such as Ford and General Motors, have announced plans to introduce similar cars on the market around 2004

The rebound effect

When individuals or organizations implement energy efficiency improvements, they usually save money as well as saving energy. However, if the money saved is then spent on higher standards of service, or additional energy-consuming activities that would not have otherwise been undertaken, then some or all of the energy savings may be eliminated. This tendency is sometimes known as the 'rebound effect'. For example, if householders install improved insulation or a more efficient heating boiler, they should in principle reduce their heating bills. However, if they instead maintain their homes at a higher temperature than before, or heat them for longer periods, the savings may be wholly or partly negated. Alternatively, they may decide to spend the money saved through lower heating bills by taking a holiday involving air travel. Since air travel is quite energy-intensive (see Figure 1.49) once again the energy savings will be offset by increased consumption, albeit of a different kind.

In devising national policies to encourage energy efficiency improvement, Governments need to take the rebound effect into consideration. In some cases, it may mean that the energy savings actually achieved when energy efficiency measures are implemented are less than expected. Another policy implication is that citizens should be given incentives to spend any savings they make when they implement energy efficiency measures in ways that are energy-frugal rather than energy-intensive.

1.6 **Energy in a sustainable future**

In this chapter we have briefly introduced three key approaches to improving the sustainability of human energy use in the future:

■ 'Cleaning-up' fossil and nuclear technologies

■ Switching to renewable energy sources

■ Using energy more efficiently

(a) 'Cleaning-up' fossil and nuclear technologies

This means mitigating some of the adverse 'environmental' consequences of fossil and nuclear fuel use through the introduction of new, 'clean' technologies that should substantially reduce pollution emissions and health hazards. These include 'supply-side' measures to improve the efficiency with which fossil fuels are converted into electricity in power stations; cleaner and more efficient combustion methods; the increasing use of 'waste' heat in combined heat-and-power schemes; and 'end of pipe' technologies to intercept and store pollutants before they enter the environment. This approach also includes 'carbon sequestration' [Box 1.3] and 'fuel switching' – shifting our energy use towards less-polluting fuels, for example from coal to natural gas. It may also be possible to 'clean up' nuclear power by adopting more advanced technologies that are safer and entail the emission of fewer radioactive substances over the entire nuclear fuel cycle.

(b) Switching to renewable energy sources

The use of renewable energy usually involves environmental impacts of some kind, but these are normally lower than those of fossil or nuclear sources.

Approaches (a) and (b) are essentially 'supply-side' measures – applied at the supply end of the long chain that leads from primary energy production to useful energy consumption.

(c) Using energy more efficiently.

This, as we have seen, involves a mixture of social and technological options, applied at the demand-side of the energy chain.

How might these three approaches to improving the future sustainability of our energy systems be combined in future? What are the various possibilities, and what are the main factors that will determine the ultimate outcomes?

Changing patterns of energy use

Before considering the feasibility, and the plausibility, of radical changes in patterns of energy production and consumption, of the kind that will be needed during first half of the twenty-first century if we are to progress towards sustainability, it is useful to recall the profound changes that have already occurred in our energy systems during the latter half of the twentieth century.

BOX 1.3 **Carbon sequestration**

One way of mitigating climate change that could be important is called 'carbon sequestration'. To sequester means to 'put away', and sequestration of carbon essentially involves finding ways of removing the carbon generated by fossil fuel burning and storing it so that it cannot find its way back into the atmosphere.

One way of sequestering carbon is to plant additional trees which 'soak up' CO_2 from the atmosphere while they are growing. However, whilst this could provide a partial response to the problem of rising CO_2 levels, the sheer magnitude of world emissions is now so great that sequestration in forests alone is probably impractical. It has been estimated that to sequester in trees the carbon produced by world fossil fuel combustion over the next 50 years would require the afforestation of an area the size of Europe from the Atlantic to the Urals. (RCEP 2000). Also, when these trees eventually decayed and died, they would emit a similar quantity of CO_2 to that which they absorbed during growth, so it would be necessary to replace the old trees with new ones on a indefinite basis.

However wood fuel from fast-growing plantations, managed sustainably, could be harvested and used as a substitute for fossil fuels, instead of simply being allowed to grow to maturity and then decay. This would offset the carbon emissions that would otherwise have been generated by burning the fossil fuels.

Another approach to sequestering CO_2 is to extract it after combustion in, for example, a power station and store it in some suitable location. It appears to be technically possible to transport by pipeline large quantities of post-combustion CO_2 and store it indefinitely in disused oil or gas wells or in saline aquifers beneath the sea bed (Figure 1.54). Further research is required to confirm the feasibility, security, safety and economic viability of such techniques. They would only be a realistic option in the case of power stations or similar large installations: it would hardly be practicable to apply this approach to emissions from vehicles or homes.

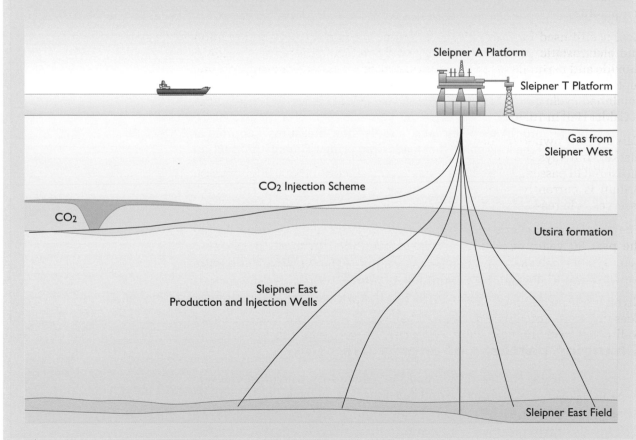

Figure 1.54 Norwegian Statoil's Sleipner field project. Gas from this field has a very high CO_2 content. Excess CO_2 is pumped into a saline aquifer, the Utsira formation, about 800 m below the sea bed. A million tonnes per year of CO_2 are 'sequestered' in this way

In Britain just after World War II most homes and other buildings were heated by coal. Most electricity generation was coal-fired, and most rail transport was propelled by coal-burning steam engines. Coal combustion caused major pollution problems, including the notorious London 'smogs' which in most winters caused the premature deaths of hundreds (and occasionally thousands) of people until the introduction of the Clean Air Act in 1956.

Coal miners perished in their dozens, and sometimes hundreds, in mining accidents every year, and many others died slowly of lung diseases caused by inhaling coal dust. Open coal fires in most houses were so inefficient that, despite consuming large quantities of energy, they only heated a few rooms effectively whilst the rest remained cold.

Motor cars were still owned only by a minority and air travel was confined to a small elite. Most people travelled by bus, train, cycle or on foot. Journeys were relatively few, compared with today, and usually over quite short distances.

Since the late 1940s, the UK's energy systems have been transformed. Natural gas, which burns much more cleanly and efficiently, was introduced very rapidly to British homes and buildings from the 1970s, after its discovery beneath the North Sea, and has now replaced coal as the main heating fuel for buildings. Most homes now have gas-fired central heating systems which ensure that the whole house is maintained at a comfortable temperature.

Coal is still used for electricity generation, but flue gas desulphurization and electrostatic precipitators now greatly reduce emissions of sulphur dioxide and particulates. In new power stations, coal is increasingly being replaced by gas, which can be burned very cleanly and efficiently using combined cycle gas turbines. Nuclear power, since its modest beginnings at Calder Hall in 1956, now contributes around one-quarter of UK electricity.

Cars are now owned by the majority, air travel overseas has become a mass market, railways are powered mainly by electricity, and travel overall, measured in passenger-kilometres, has tripled since the 1950s (Figure 1.50). Britain is currently a net exporter of oil, thanks to its large North Sea reserves, whereas before the 1970s all its oil was imported.

The dramatic changes that have occurred in Britain's energy systems during the past 50 years have, broadly, been paralleled in most 'developed' countries over the same period. Examples of changing patterns of energy use in other EU countries are given in Chapters 2 and 3.

Given the scale and profundity of the changes over the past half-century, it does not seem unrealistic to suggest that equally-profound changes could well occur over the next 50 to 100 years, as we attempt to improve the sustainability of our energy systems, nationally and globally.

Long-term energy scenarios

To begin to understand the range of long-term future possibilities, let us look briefly at two major studies of future sustainable energy options, the first addressing the UK situation, the second taking a world perspective.

The Royal Commission on Environmental Pollution scenarios

The UK's Royal Commission on Environmental Pollution produced its 22nd report *Energy: the Changing Climate* in June 2000. The commission examined what changes would be needed in Britain's energy systems if, as suggested by the various reports of the Intergovernmental Panel on Climate Change (IPCC, 2001), it should prove necessary to reduce the country's emissions of greenhouse gases by about 60% by 2050.

The Commission investigated the various possibilities very thoroughly and summarized them in four 'scenarios' for 2050. Scenarios are not predictions of what *will* happen, but plausible outlines of what *could* happen, under given conditions. In all four scenarios, the overall contribution from fossil fuels is reduced to approximately 40% of current consumption, consistent with the 60% reduction in fossil fuel use required to achieve a 60% cut in CO_2 emissions.

The RCEP scenarios are summarized in Box 1.4. They demonstrate that it would be feasible for the UK to progress towards much greater sustainability (in terms of reducing CO_2 emissions) in its energy systems over the next 50 years. They also show that there are a number of ways in which this could be achieved.

The actual outcome over coming decades will depend on the extent to which we change our lifestyles and our technologies in order to conserve energy; how effective we are in generating and using it more efficiently; how rapidly we choose to develop and deploy renewable energy sources; how large a role we choose to give to nuclear power; and whether or not we decide to implement carbon sequestration and other technologies for 'cleaning-up' fossil fuels.

The World Energy Council scenarios

What are the possibilities for radical changes in our energy systems when viewed from a world perspective? There have been numerous studies of the various future options for the world's energy systems. One of the most recent and most comprehensive was produced in 1998 by the International Institute for Applied Systems Analysis (IIASA) and the World Energy Council (WEC), a version of which was published in 2000 as part of the United Nations' *World Energy Assessment* (United Nations Development Programme, 2000). IIASA is a leading 'think tank' based in Austria, whilst the WEC is a body that represents the world's main energy producers and utilities. For simplicity, we shall refer to their scenarios here as the World Energy Council (WEC) Scenarios.

There are six WEC scenarios in all, and these have been grouped into three 'cases', A, B and C. Case B includes only one scenario, termed 'Middle Course'. Case A consists of three 'High Growth' scenarios, and case C includes two 'Ecologically-Driven' scenarios.

Each scenario incorporates different assumptions about rates of economic growth and the distribution of that growth between rich and poor countries; about the choices that are made between different energy technologies and the rapidity with which they are developed; and regarding the extent to which ecological imperatives are given priority in coming decades. They

BOX 1.4 **Four energy scenarios for the UK in 2050**

Four scenarios were constructed to illustrate the options available for balancing demand and supply for energy in the middle of the twenty-first century if the UK has to reduce carbon dioxide emissions from the burning of fossil fuels by 60%.

Scenario 1: no increase on 1998 demand, combination of renewables and either nuclear power stations or large fossil fuel power stations at which carbon dioxide is recovered and disposed of.

Scenario 2: demand reductions, renewables (no nuclear power stations or routine use of large fossil fuel power stations).

Scenario 3: demand reductions, combination of renewables and either nuclear power stations or large fossil fuel power stations at which carbon dioxide is recovered and disposed of.

Scenario 4: very large demand reductions, renewables (no nuclear power stations or routine use of large fossil fuel power stations).

The key parameters for these four scenarios are as follows:

	Scenario 1	Scenario 2	Scenario 3	Scenario 4
Percentage reduction in 1997 carbon dioxide emissions	57	60	60	60
DEMAND (percent *reduction* from 1998 final consumption)				
low-grade heat	0	50	50	66
high-grade heat	0	25	25	33
electricity	0	25	25	33
transport	0	25	25	33
Total	0	36	36	47
SUPPLY (GW) (annual average rate)				
fossil fuels	106	106	106	106
intermittent renewables	34	26	16	16
other renewables	19	19	9	4
baseload stations (either nuclear or fossil fuel with carbon dioxide recovery)	52	0	19	0

Source: Royal Commission on Environmental Pollution, 2000

all assume that world population will increase from its current (2000) level of around 6.1 billion to 10.1 billion by 2050 and 11.7 billion by 2100. (More recent UN projections, however, suggest that these figures may be over-estimates, with 9 billion as the new median population estimate for 2050 (United Nations, 2001). Other recent research also suggests that world population is likely to peak before the end of the twenty-first century and then begin to decline. (Lutz *et al.*, 2001)).

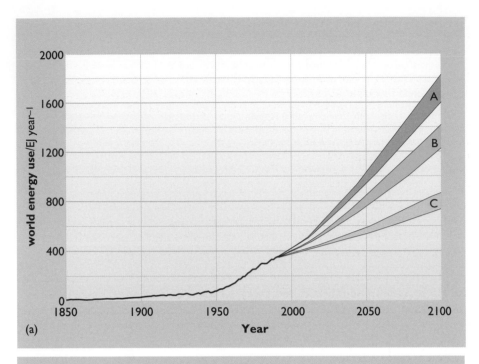

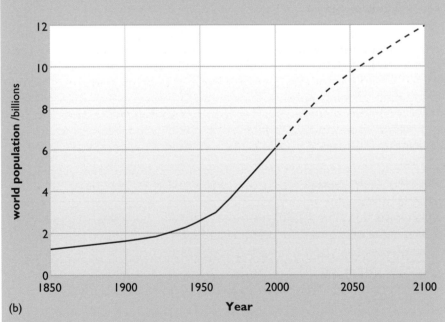

Figure 1.55 (a) Global primary energy requirements, 1850–1990, and projected requirements 1990–2100 in the three World Energy Council scenario 'cases', A, B and C. World energy use here includes commercially-traded energy only;
(b) World population, 1850–2000 and projected population, 2000–2100 (see text)
(source: United Nations Development Programme, 2000)

The results of these assumptions are shown in Figure 1.55 which also shows world population growth from 1850 to 2000 alongside the various scenario projections to 2100.

In all three High Growth scenarios, the world's economy expands very rapidly, at an annual average rate of 2.5% per annum – significantly faster than the historic growth rate of about 2% per year. In all of them, primary energy intensity (the amount of primary energy required to produce a dollar's worth of output in the economy) reduces quite rapidly, reflecting a fairly strong commitment to energy efficiency measures and/or dematerialization. The three scenarios differ mainly in their choices of energy supply technologies. One is based on ample supplies of oil and gas; another envisages a return to coal; and the third has an emphasis on non-fossil sources, mainly renewables with some nuclear. By 2100, the High Growth scenarios all envisage world primary energy consumption rising to over 1800 exajoules, more than four times the 2000 level.

In the single Middle Course scenario, economic growth is lower than in the High Growth scenarios, averaging around 2.1% per annum, close to the historic average rate. Primary energy intensity improves rather more slowly, reflecting a slightly lower world-wide emphasis on energy efficiency improvement. Energy supplies come from a wide variety of fossil, nuclear and renewable sources, and by 2100 total primary energy consumption has reached more than 1400 EJ, over three times the 2000 level.

In the two Ecologically-Driven scenarios, world economic growth is 2.2% per annum, slightly higher than in Middle Course, but there is a very high emphasis on improving energy efficiency, reflected in substantially lower primary energy intensity figures. Both scenarios feature a strong development of renewables, alongside a continued use of oil, coal and natural gas. In one scenario, nuclear energy is phased out by 2100 whereas in the other some nuclear power is retained. Overall primary energy consumption increases to some 880 EJ by 2100, just over twice the 2000 level.

The WEC authors conclude that, judged in terms of their sustainability, one of the High Growth scenarios (the third) includes many elements favouring sustainable development, though the other two High Growth scenarios do not. The Middle Course scenario, however, falls short of fulfilling most of the conditions for sustainable development.

The Ecologically-driven scenarios, unsurprisingly, are much more compatible with sustainable development criteria, although one of them requires a more radical departure from current policies since it envisages a phasing-out of nuclear energy.

The overall message of the WEC scenarios, examining possible solutions at a world scale, is similar to that of the RCEP scenarios for Britain: that progress to much greater sustainability in our energy systems is feasible over the next 50–100 years; that there are a number of different paths to sustainability; and that some paths are probably better than others.

The WEC scenarios, and a number of other similar studies, will be examined in more detail in the companion volume, *Renewable Energy*.

Meanwhile, in this volume we now turn away from this general overview to examine in more detail our current energy systems and their sustainability, starting with a look at our primary energy sources.

References

Anon (2002) 'The Windicator', *Windpower Monthly,* January, p. 50.

Blake, William (1994) *The Marriage of Heaven and Hell*, Dover Publications.

Boyle, G. (1996, 2003) *Renewable Energy*, Oxford, Oxford University Press in association with the Open University.

BP (2002) *BP Statistical Review of World Energy* [online], BP. Available at http://www.bp.com/centres/energy2002/index.asp [accessed 12 July 2002].

Campbell, C. J., and Laherrere, J. H (1998) 'The End of Cheap Oil', *Scientific American*, March, pp. 60–65.

Department of Trade and Industry (2001) *Digest of UK Energy Statistics*, 2000.

Department of Transport, Local Government and Regions (DTLR) (2001) *Transport Statistics, Great Britain* (27th edn), The Stationery Office, Table 9.1, p. 161.

Hughes, P. (1993) *Personal Transport and the Greenhouse Effect: a Strategy for Sustainability,* London, Earthscan.

Intergovernmental Panel on Climate Change (2001) *Climate Change 2001: the Scientific Assessment*, Cambridge University Press.

International Energy Agency (2000) *Key World Energy Statistics from the IEA*, Paris, International Energy Agency.

Lutz, W., Sanderson, W., and Scherbov, S. (2001) 'The End of World Population Growth' *Nature* 412, 543–545, August.

Royal Commission on Environmental Pollution (2000) *Energy: the Changing Climate*, London, Stationery Office.

Smil, V. (2000) 'Energy in the Twentieth Century: Resources, conversions, costs, uses, and consequences', *Annual Review of Energy and the Environment,* vol. 25, pp.21–51.

Smith, K. R., Uma, R., Kishore, V. V. N., Zhang, J., Joshi, V., and Khalil, M. A. K. (2000) 'Greenhouse Implications of Household Stoves: An Analysis for India' *Annual Review of Energy and the Environment*, vol. 25, pp. 741–763.

United Nations (2001) *World Population Prospects; the 2000 Revision,* UN Population Division, Department of Economic and Social Affairs, 18 pp.

United Nations Development Programme (2000) *World Energy Assessment*, New York, United Nations.

United Nations: World Commission on Environment and Development (1987) *Our Common Future* (the Brundtland Report), Oxford, Oxford University Press.

Chapter 2

Primary Energy

by Janet Ramage and Bob Everett

2.1 **World primary energy consumption**

> 'The world is consuming primary energy at a rate of about 13.4 terawatts.'

This statement raises a number of questions.

- What is primary energy?
- How does the world *consume* energy?
- What are 13.4 terawatts, and how are they related to the world annual consumption of 10 100 Mtoe quoted in Chapter 1?
- How do we *know* the world's energy consumption? What are the sources of energy data, and how reliable are they?

The first part of this chapter addresses these essential questions about the nature and basis of our knowledge of the world's use of energy. We will then be in a position to look at the situation in more detail. The chapter continues with accounts of the changing contributions to world primary energy over the past century or so, and the varying patterns of energy consumption and energy production across the different regions of the world. To illustrate these contrasts, it concludes with some details of the recent energy history of a few selected countries.

What is primary energy?

Essentially, **primary energy** is the total energy 'content' of the original resource. Our main present resources are the fossil fuels (coal, oil and natural gas) and the biofuels such as wood, straw, dried dung, etc. (The energy content of the food we eat is not customarily included in the count.) To these we can add the energy provided by nuclear power stations and by hydroelectric or geothermal plant and other 'renewables' such as solar or wind power.

The rather arbitrary nature of the definition becomes evident if we consider solar energy. When special systems such as solar panels or photovoltaic cells are used, their output may be included in the primary energy total, but the daily contribution of solar energy in warming and illuminating our buildings does not normally appear in national or international statistics. Section 2.3 below discusses in more detail the methods used to assess and compare the various contributions to primary energy production or consumption.

What is energy consumption?

One of the most fundamental scientific laws states that energy is conserved. The total quantity stays constant. You cannot create energy or destroy it. If you have ten units of energy at the start, you have ten units of energy – somewhere – at the finish. In this sense, we never consume energy.

It is however a matter of great practical importance that energy can take many different forms, and what we *can* do – and have done at least since our ancestors first used fire – is to devise means of converting from one form to another. When we talk of consuming energy this is what we mean: converting from the chemical energy stored in fuels such as wood, coal, oil or gas, from the energy stored in atomic nuclei, from the gravitational energy of water in a high reservoir, the kinetic energy of moving water or the wind, and the radiated energy of sunlight, into heat or electrical energy or light or the kinetic energy of a moving vehicle. We'll discuss all these forms of energy later, but the first important point is this: *Consumption is conversion.*

BOX 2.1 'Conservation of energy'

Why should I switch off lights to conserve energy, when there is a law which states that energy is <u>always</u> conserved?

The question is of course mischievous, deliberately confusing two different meanings of 'conserve'. It is however important to appreciate that phrases such as 'conservation of energy' do have these two meanings, both of which are common in discussions of our uses of energy. The scientific law is fundamental. It underlies all our reasoning, even when it is not specifically stated. In the other sense of the term, when we are told to 'conserve energy' by switching off unnecessary lights, we are really being asked to conserve *energy resources* – in this case, the fuel consumed in power stations. Fortunately the context usually makes it clear which meaning is intended, and in practice the two senses of the word rarely lead to problems.

'Energy arithmetic'

Any serious discussion of energy must be quantitative:

> 'My car uses very little petrol.'

In driving a thousand miles, or standing in the garage? Compared with Saudi Arabian oil exports or with a bicycle?

The trivial example illustrates two requirements. In order to compare quantities we must be able to measure them, i.e. we need *units* (litres, or gallons, or tonnes); and then we must know which type of quantity we are discussing (litres per kilometre, litres per year or just litres).

In 1960 the scientific world reached agreement on a single consistent set of units: the Système Internationale d'Unités. The SI system uses three main base units: the metre, the kilogram and the second, and the units for many other quantities are derived from these. Some of the derived units, such as *metres per second* for speed, reveal their base units immediately, whilst for others the combination of base units has been replaced by a specific name. The name of the SI unit for energy is the **joule**, abbreviated **J**, and you'll find more details of this and other SI units in Appendix A. In everyday terms, one joule is a rather small quantity of energy – roughly the amount needed to toss a medium-sized apple just one metre vertically upwards.

One of the happier consequences of the energy debates of the past few decades has been a growing appreciation of the advantages of using this universal unit for all amounts of energy. Nevertheless, if you open a book or technical paper, you can still find yourself in a rather less tidy world. Quantities of energy are quoted in tonnes (or tons) of oil or coal, cubic metres of gas, kilowatt-hours, terawatt-years, therms, calories and Calories, and if we are to follow the real world debate, we must come to terms with these. Accordingly, one aim in this chapter is to introduce the art of 'energy arithmetic' – of converting between different ways of specifying quantities of energy.

The need to use extremely large numbers can also lead to problems. Most of us can visualize a dozen objects, perhaps even a hundred, but who can picture 13 400 000 000 000? We cannot avoid such very large numbers, but they can be made more manageable by using special names, or more compact ways of writing them. Appendix A explains these methods in detail, and Table 2.1 is a short summary of the prefix names used in this chapter.

Table 2.1 Prefixes

symbol	prefix	multiply by[1,2]
k	kilo-	one thousand
M	mega-	one million
G	giga-	one billion (one thousand million)
T	tera-	one trillion (one million million)
P	peta-	one quadrillion (one billion million)
E	exa-	one quintillion (one billion billion, or one million million million)

1 Note that each multiplier is one thousand times the previous one.

2 The multipliers beyond one million have the now usual USA meanings: one billion is one thousand million and not one million million as in the older British usage.

Watts

A terawatt is one million million watts – but what is a watt? The important point is that a watt is not a unit of energy, but a *rate* at which energy is being transformed or converted from one form to another. Technically a watt is a unit of **power**, of energy per second:

- One **watt** is by definition one joule per second.

Thus a 600 W heater is converting electrical energy into heat at a rate of 600 joules in each second. And we, the population of the world, with our 13.4 TW rate of consumption, are converting 13.4 million million joules of primary energy every second into the forms of energy we want (and a great deal of waste heat).

Kilowatt-hours

The kilowatt-hour (kWh) is a unit of *energy*.

- One **kilowatt-hour** is the amount of energy converted in one hour at a rate of one kilowatt.

The heater in a 3 kW clothes dryer, for instance, used for 40 minutes (two-thirds of an hour), converts 2 kWh of electrical energy into heat energy.

Like any quantity of energy, a kilowatt-hour must of course be equal to a certain number of joules. The reasoning in Box 2.2 shows that one kilowatt-hour is 3.6 megajoules.

It is important to appreciate that the kilowatt-hour and the watt are *general* units for energy and power respectively. Although many of us meet them first in the context of electricity, they are equally applicable to the energy you use and the power you develop in running up a flight of stairs.

BOX 2.2   kW and kWh

We note that 1 kW is 1000 watts (Table 2.1), and that there are 3600 seconds in an hour.

Power

1 watt = 1 joule per second

1 kilowatt = 1000 joules per second

1 kilowatt = 3 600 000 joules per hour

Energy

1 kilowatt-hour = 3 600 000 joules

1 kWh = 3.6 MJ

BOX 2.3   Per capita consumption

It can be useful to convert very large numbers into more manageable quantities. Instead of total world energy consumption, we might consider the average consumption per person.

The world rate of primary energy consumption is 13.4 TW, which is 13.4 million million watts, and the world population is about 6100 million people. The average *per capita* rate of consumption is therefore

$$13\,400\,000 \div 6100 = 2197 \text{ watts.}$$

On average, therefore, we are each consuming primary energy at a steady rate of about 2.2 kW.

There are 24 hours in a day, so the daily consumption per person is

$$2.2 \text{ kW} \times 24 \text{ hours} = 53 \text{ kWh.}$$

On average, therefore, we each consume just over 50 kWh of primary energy every day.

Remembering that 1 kWh is 3.6 MJ, this becomes about 190 MJ, which is the energy content of a little over five litres of oil – about one and a quarter UK gallons.

So the average person – man, woman and child – uses the energy equivalent of just over a gallon of oil a day. This must of course supply *all* our energy needs: food production and a water supply, the provision of housing, heat for cooking and to keep us warm, clothing and manufactured goods, transport of people and freight, communications and entertainment, and the medical, educational and other services that we expect.

2.2 **Quantities of energy**

The publication of national or international energy data was largely a development of the second half of the twentieth century, but records of dealings in *commodities* are as old as trade itself. During the eighteenth and nineteenth centuries, coal became an extremely important commodity for developing countries such as Britain. As it was also the dominant energy source, the data on coal production and consumption came to serve as national energy data for much of the period. When new energy sources such as oil began to appear, it was natural to assess their contributions in terms of the quantity of coal they could replace, and Britain continued to do this into the 1980s, expressing all national energy data in *tonnes of coal equivalent*.

Meanwhile, some of the most accessible international energy data were being assembled and published by the major oil companies, and not surprisingly their favoured unit for energy was the *tonne of oil equivalent*. In the UK, where oil has been the major fuel since 1970, the national statistics now tend also to use tonnes of oil equivalent.

Units based on oil

When oil is burned, whether in a furnace or an internal combustion engine, its chemical energy is converted into heat energy. One **tonne of oil equivalent (toe)** is simply the heat energy released in the complete combustion of 1000 kg of oil. This varies between crude oils from different sources, but a commonly used world average is 41.88 GJ (41 880 MJ). When the data do not justify this precision, 42 GJ is a useful approximation. World annual primary energy consumption then becomes 10 100 Mtoe (Box 2.4).

Figure 2.1 Filling barrels at a Pennsylvania oil well in 1870

This of course includes all forms of energy, so the actual world oil consumption of 3500 million tonnes a year accounts for just over a third of the total.

Another measure of quantity of oil and correspondingly of energy is the *barrel*. This odd unit, alien in a world of pipelines and super tankers, comes from the size of the barrels used to carry oil from the world's first drilled well in Pennsylvania in the 1860s. One barrel is 42 US gallons or 35 Imperial (British) gallons – about 160 litres.

How is a barrel of oil related to a tonne of oil? A barrel is a certain *volume*, whereas a tonne is of course a *mass*, and crude oils from different sources have different densities, so more barrels would be needed to hold one tonne of a 'light' crude than for a 'heavier' one. The solution has again been to adopt a world average: 7.33 barrels to the tonne. Using this we find that the energy content of one barrel is approximately 5.71 GJ, and this is one **barrel of oil equivalent** (**boe**). The oil industry commonly expresses data in **million barrels daily** (**Mbd**). In 2000, for instance, world oil consumption was 73.9 Mbd, and world total primary energy was equivalent to 203 Mbd (Box 2.4).

Finally, we have the everyday units for the fuel used in our vehicles: the litre and the gallon. Petrol (gasoline) has a slightly higher energy content *per tonne* than crude oil – about 44 GJ. But it has an appreciably lower density, and in terms of volume, the energy content is about 150 MJ per Imperial gallon, or 33 MJ per litre, compared with nearly 36 MJ per litre for crude oil.

BOX 2.4   World energy in Mtoe, Mbd and TW

In terms of the accepted SI unit for energy, world annual primary energy consumption for the year 2000 was **424 EJ** (exajoules).

We have also expressed this rate of consumption as 10 100 Mtoe per year, 13.4 TW and 203 Mbd. This box shows the energy arithmetic that relates all these figures.

Millions of tonnes of oil equivalent

The value given in the text for a tonne of oil equivalent is 42 GJ.

> 1 Mtoe is therefore 42 million GJ.

> 424 EJ is the same as 424 000 million GJ

World consumption in Mtoe is therefore

> 424 000 ÷ 42 = **10 100 Mtoe**

Millions of barrels of oil daily

There are 365 days in a year, so the daily primary energy is

> 424 ÷ 365 = 1.16 EJ.

The value given in the text for a barrel of oil equivalent is 5.71 GJ.

1 Mboe is therefore 5.71 million GJ.

1.16 EJ is the same as 1160 million GJ.

World consumption in Mbd is therefore

> 1160 ÷ 5.71 = **203 Mbd**.

Terawatts

The conversion from exajoules a year to terawatts starts with the definition of the watt:

> 1 watt is 1 joule per second …

> which is 3600 joules per hour …

> or 24 × 3600 = 86 400 joules per day …

> or 365 × 86 400 = 31 536 000 joules per year …

> which is 31.5 MJ per year.

> 1 terawatt (TW) is one million million watts …

> which is 31.5 million million MJ per year …

> or 31.5 EJ per year.

World consumption in TW is therefore

> 424 ÷ 31.5 = **13.4 TW**.

Units based on coal

One **tonne of coal equivalent** (**tce**) is the heat energy released in burning one metric tonne of coal (Box 2.5). Coal is a much more variable material than crude oil, and world-wide its energy per tonne ranges from less than 20 GJ to over 30 GJ. The figure of 28 GJ per tonne is often adopted in energy statistics, and is the one used in this book unless otherwise specified.

BOX 2.5 **Tonnes, tons and short tons**

As mentioned above, national or even international energy data do not yet appear in one agreed set of units, and whilst the approved SI units for mass are the kilogram and its multiples such as the metric tonne (1000 kg), you will still find other 'tons' in use. This box describes the relationships.

1   The **tonne**, or metric tonne, is 1000 kg and is equal to about 2205 lb (pounds).

2   The **ton**, a unit in the pre-metric system of weights and measures of the UK and many other countries, is still widely used. The hundredweight (cwt) is an intermediate unit, equal to one-twentieth of a ton. One ton is 2240 lb – about 1.6% more than a tonne.

3   The **short ton** is still found occasionally as the unit for quantity of coal or wood in some countries. One short ton is 2000 lb – about 10% less than a tonne.

Figure 2.2 Filling a London coal cellar. Coal was delivered in hundredweight sacks, and the 'coal holes', often with attractive iron covers, can still be identified in many eighteenth- or nineteenth-century streets

The BTU and related units

Before the general adoption of the joule, the *British thermal unit* (BTU) and its multiples were in common use, in the English-speaking world in particular.

■ One **BTU** is the heat energy needed to warm one pound of water by one degree Fahrenheit and is equal to 1055 joules. Multiples of the BTU include the therm (100 000 BTU) and the quad.

■ One **quad** is a quadrillion British thermal units (see Table 2.1) and is equal to 1.055 EJ.

These units are still used, notably in the USA where the common unit for energy quantities on the national scale is the quad. As can be seen, the BTU and the quad are slightly larger than the kilojoule and exajoule respectively.

The calorie and related units

In most of Europe, and many other countries, the common unit for heat in the past was the calorie.

■ One **calorie** is the heat energy needed to warm one gram of water by one degree Celsius and is equal to 4.19 joules.

  For many purposes the **kilocalorie**, written kcal or Calorie (with capital C), has proved more convenient, and it remains familiar as the unit for the energy content of food.

This gives us yet another way of looking at our energy consumption. Nutritionists tell us that the *food energy* needed to support an adult is about 2000 Calories a day, which is a little over 8 MJ. In Box 2.3 we saw that world average daily primary energy consumption per person is about 190 MJ. It appears therefore that the energy we each use in non-food forms is, on average, more than twenty times the amount we each need to feed ourselves. As we shall see, this is by no means universally the case.

2.3 Interpreting the data

There remains a final question about world primary energy, or indeed any energy data. How do we know? Before venturing further into the sources of energy, we should perhaps discuss the sources of *data*. Where do the figures come from? The first answer is that we find them in official statistics, technical journals and similar publications. However, one shouldn't believe everything one reads in books (or anything in newspapers) and care is always needed in interpreting published figures, for reasons which we can characterize under three headings: *definitions*, *conversions* and *conventions*.

Definitions

World data usually start as national statistics, and with some 200 countries it is hardly surprising that the terminology doesn't always match at the seams. Does 'production' include energy used by the producer? Does 'consumption' include energy used for transmission of energy? Unless we know the answers to such questions, how are we to interpret the statement that 75.102% of Britain's coal in 2000 was used for electricity generation? In the absence of pages of explanation, it is surely better to say, 'About 75% …', or even, 'Roughly three-quarters …'.

A further mismatch appears in comparing figures for *production* and *consumption*. One would hope that any difference would be accounted for by changes in stocks, but when production data necessarily come from producers and consumption data from consumers this is by no means always the case. Recent world data, for instance, include 15.8 million tonnes of 'unidentified' crude oil exports. Some of it may be on the high seas – in ships, one hopes – but the figures again illustrate the problem.

Conversions

We have seen several examples of conversion between different energy units, but have not bothered too much about the nature of these relationships. On inspection we find that the term *equivalent* has been used in a number of different ways.

First there are cases where the conversion between units is *exact*. One watt is exactly one joule per second because that is how it is defined; and 1 kWh is therefore exactly 3.6 MJ. Then there are relationships which although not exact are known very precisely and may be regarded as *universal*. The conversions between joules, British thermal units and Calories are examples.

When we come to quantities such as the heat content of a fuel, matters are not quite so simple. The heat content of a particular specimen of oil can be measured to great accuracy under laboratory conditions, and with similar care we might measure the solar energy reaching a particular roof in the course of a particular day. But it is hardly practicable to use these methods for the total output during the lifetime of an entire oil well or solar panel. In the real world it becomes essential to use *average* values. The problem is that not everyone uses the same average. If your tonne of oil equivalent and daily solar energy are not the same as mine, our discussion is likely to end in confusion. Once more, the rule is to make sure we know what the figures mean before using them.

Conventions

Finally, there is the rather different question of the output from power stations. The difficulty is not in measuring it, as most national data include the annual kilowatt-hours produced, and conversion of these to joules is no problem. The question is whether this output should count as 'primary energy'? Shouldn't that be the *input*? Unfortunately there are difficulties with this. Recording the input of coal, oil or gas is relatively straightforward, but measuring the total 'water energy' entering a country's hydroelectric plants in a year, or the total wind energy sweeping across its wind turbines, is not practicable. And nuclear plants, whose input is the result of a complex series of processes, pose a similar problem.

Figure 2.3 shows the essential facts for the world's main types of power station. In most **thermal power stations**, where heat from the fuel produces steam or hot gases to drive the turbine, about two-thirds of the energy input

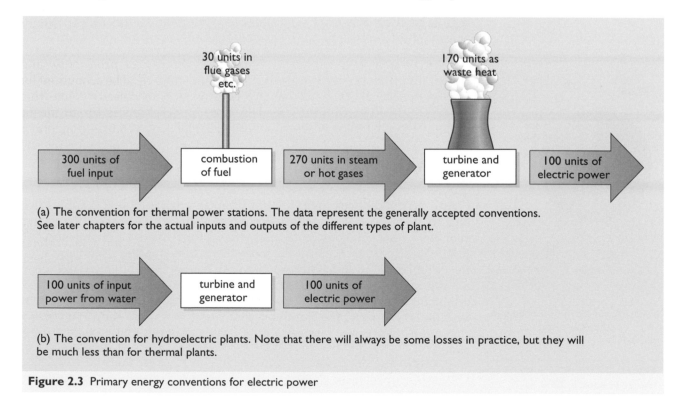

(a) The convention for thermal power stations. The data represent the generally accepted conventions. See later chapters for the actual inputs and outputs of the different types of plant.

(b) The convention for hydroelectric plants. Note that there will always be some losses in practice, but they will be much less than for thermal plants.

Figure 2.3 Primary energy conventions for electric power

becomes waste heat. This is not the case for hydroelectric plants, where the output is only a little less than the input. These facts have led to a fairly straightforward convention for the main types of power station.

- The **notional primary energy** input for all main types except hydroelectric is taken to be the electrical output divided by 0.33 (roughly the same as multiplying it by 3).

- The primary energy input for hydroelectric plants is taken to be equal to the electrical output.

This dual convention has become dominant and is now used by bodies such as the United Nations Department of Economic and Social Affairs and the International Energy Agency (IEA) of the OECD, in the annual BP Statistical Review of World Energy and in UK energy statistics. It is therefore the method adopted in this book, and all data have been converted in accordance with the above rules unless we specify otherwise. A consequence of this is that when using the data or looking at diagrams such as Figure 2.4 below, it is important to bear in mind, for instance, that although the world primary contribution from nuclear power is shown as three times that of hydroelectricity, the annual *electrical outputs* from the two are in fact nearly the same.

Other conventions than this one are unfortunately still in use, or have been until recently. In the past, for instance, the UK used to multiply *all* power station outputs by about 3, including hydroelectricity. Some countries still do this, including France (but with the output divided by 0.386 rather than by 0.33). In contrast, some countries and international organisations treat nuclear power in the same way as hydroelectricity – using the electrical output as the primary energy contribution, with no multiplier. And further questions arise with the growing contributions from renewable sources such as geothermal energy, wind and solar power – and before long, perhaps wave and tidal power as well. (The conventions used for such sources in this book are given in the relevant tables or graphs.)

It is always wise to read the small print when using statistical data, but it should be clear from the above account that it is essential with energy data. The principal sources of the data in Sections 2.4–2.9 below are given in the list of references at the end of this chapter.

2.4 World energy sources

Figure 2.4 reproduces Figure 1.9 of Chapter 1, and Table 2.2 shows the data in more detail. The picture is clear enough. We see a situation in the year 2000 where oil contributes more than a third of all our energy and the fossil fuels together account for more than three-quarters. The largest of the other contributions is almost certainly energy from biomass, although for the reasons outlined in Box 2.6, its exact magnitude is difficult to establish. If we consider only the 'commercially traded' sources, excluding traditional biomass, the dominance of fossil fuels becomes even more striking. They account for nearly ninety percent of the world's total traded energy.

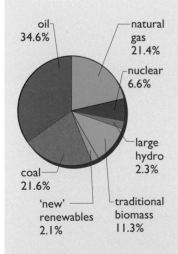

oil 34.6%
natural gas 21.4%
nuclear 6.6%
large hydro 2.3%
coal 21.6%
'new' renewables 2.1%
traditional biomass 11.3%

Total: 424 EJ or 10 100 Mtoe per year, or an average rate of 13.4 TW

Figure 2.4 World primary energy by source, 2000

Table 2.2 World primary energy, 2000

Energy source	Quantity in customary units	Energy in EJ	Percentage contribution	
			to total energy	to commercial energy only
Fossil fuels				
oil	3504 million tonnes	146.7	34.6%	39.1%
natural gas	2407 billion cubic metres	90.7	21.4%	24.1%
coal	3125 million tonnes	91.6	21.6%	24.4%
Fossil fuels total		**329.0**	**77.6%**	**87.6%**
Electric power				
nuclear electricity	2590 billion kWh[1]	28.0[2]	6.6%	7.5%
large hydro	2670 billion kWh[1]	9.6[2]	2.3%	2.6%
Other renewables				
traditional biomass[3]		48.0	11.3%	
'new' renewables[4]		9.0	2.1%	2.4%

1 Actual power station output.

2 See Conventions, above.

3 See Box 2.6.

4 About 7 EJ from 'new' bioenergy, 1.6 EJ of geothermal heat and the remainder from wind, small-scale hydro and solar power.

(Principal sources: BP, 2001; UN, 2000)

BOX 2.6 Bioenergy

Bioenergy is the general term for energy derived from biomass: materials such as wood, plant and animal wastes, etc., which – unlike the fossil fuels – were living matter relatively recently. Such materials may be burned directly to produce heat or power, but can also be converted into solid, liquid or gaseous **biofuels**.

Estimates of the contribution of biomass to world primary energy are subject to considerable uncertainty. The fuels are often 'non-commercial' – they may be gathered in surrounding forests or fields, or arise as waste by-products of other activities, and are often used on site, or bartered for other goods or services. In other words, they are not formally traded, so the economists' methods of keeping track of quantities are not available. It has become customary to refer to these resources as traditional biomass, distinguishing them from 'new' bioenergy sources such as purpose-grown wood or other 'energy crops', forestry wastes and municipal solid waste. These are often commercially traded and can be treated in the same way as other 'new' renewables such as wind power or solar PV.

In recent years, detailed studies have suggested total bioenergy contributions ranging from 10% to 15% of world primary energy (Hall *et al.*, 1993; WEC, 1998; IEA, 1998). 'New' bioenergy – in the world context mainly urban waste used as fuel for power stations – currently contributes about 7 EJ, and the 48 EJ shown in Table 2.2 for traditional biomass brings the total bioenergy contribution to 55 EJ, or a little less than 13% of total primary energy.

Figure 2.5 reveals how consumption of these non-renewable and carbon dioxide-producing energy sources increased during the second half of the twentieth century. The dramatic growth in the use of oil is even more obvious over the longer period shown in Figure 2.6, although its most rapid rate of rise occurs between 1965 and 1973, with an average increase of nearly 8% a year. Had this rate been maintained, as seemed likely in the early 1970s, the annual output required by the end of the century would have exceeded *20 billion* tonnes, or about six times the actual output in 2000. It is no surprise that sudden doubts about future supplies led to panic and disarray. The crises of the 1970s, with oil prices doubling in 1973 and rising steeply again at the end of the decade, followed by economic recession in the early 1980s, did eventually bring the growth in oil consumption to a halt – but only after some delay, and only temporarily. The mid-eighties saw a return to annual increases, and despite all the crises, the world has consumed more oil since 1985 than in the whole of history before that date.

Natural gas production, with its steady, almost linear growth, stands in marked contrast, and its history is very different. In 1950, use was almost

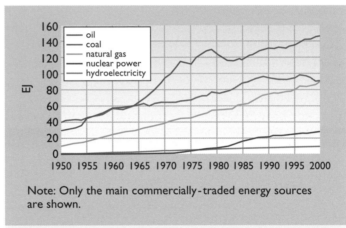

Note: Only the main commercially-traded energy sources are shown.

Figure 2.5 World annual primary energy consumption by source, 1950–2000. (source: Times, 1980; BP, various issues)

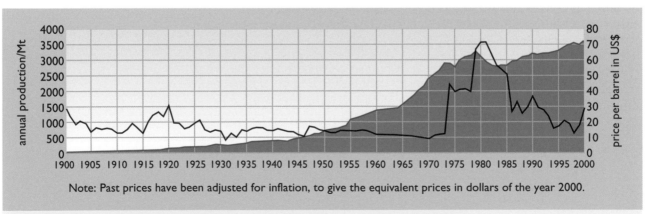

Note: Past prices have been adjusted for inflation, to give the equivalent prices in dollars of the year 2000.

Figure 2.6 World oil production and the price of oil, 1900–2000. (source: Ramage, 1997)

confined to the USA, and as late as 1970, North America was still responsible for two-thirds of world production (and consumption). But new sources elsewhere, already being developed in the 1960s, account for much of the rise of the next thirty years. By the year 2000, the output from Russia and other countries of the former Soviet Union was only a little less than that of North America, each providing about a third of the total. (The gas fields of the North Sea, the Middle East and the Far East supplied almost all the remainder.) Natural gas, as the 'cleanest' of the fossil fuels, has become the energy source of choice for heating and power generation; but the increasing reliance of a number of Western European countries on gas delivered by pipeline from Russia and her neighbours is starting to raise strategic concerns.

Coal, the dominant fuel for two centuries, fell to second place in the 1950s, and despite a slight rise at the very end of the twentieth century, seems destined to fall to third place during the early years of the twenty-first. Nevertheless, it is worth noting that world coal consumption more than doubled in the fifty-year period of Figure 2.5. Moreover, as we shall see later, coal can be a source of synthetic liquid fuels as substitutes for petroleum, and future oil crises might yet bring a reversal in the fortunes of this least desirable fossil fuel, as happened briefly in the 1980s.

As Table 2.2 shows, nuclear and large-scale hydroelectric plants each contribute some 2600 billion kilowatt-hours of electrical output a year. Both have experienced reductions in their rates of growth in recent decades, and in both cases environmental concerns have been in part responsible.

More detailed discussions of the primary energy sources, the histories of their use and accounts of the associated technologies appear in later chapters, or for the renewable resources, in the companion volume to this book, *Renewable Energy* (Boyle, 1996).

International comparisons

13.4 TW divided equally between all the inhabitants of the world gives each of us, as we have seen (Box 2.3), a little over 2 kW of continuous primary power.

It will have been no surprise to learn in Chapter 1 that world energy is not distributed in uniform shares, and the further comparisons in Table 2.3 surely provide food for thought. A seventh of the world's population is currently consuming nearly half the world's primary energy. The average daily energy used by an individual in the world's wealthiest two dozen countries is over six times that in the rest of the world. It is a salutary thought that to bring the remaining hundred or more countries even to the present European level of per capita energy provision would require world primary energy production to rise to almost *twice* the current level. And it is worth noting in this context that the energy data in the table include the contribution from traditional biomass, usually an important component of total energy in the less developed countries. It follows that for these countries to reach European levels in the use of 'modern' energy sources would require an even greater increase – and of course a corresponding increase in the environmental consequences of the use of these resources.

The average citizen of the USA earns nearly thirty times as much as the average Bangladeshi and consumes over sixty times as much energy each day. These are extremes, but comparison of *per capita energy consumption* and *per capita income* does show a not unexpected correlation. However, we should be wary of the easy conclusion that rising living standards necessarily mean the consumption of ever more energy each year. As we'll see in later chapters, many of our energy systems have improved in

Table 2.3 International comparisons, 2000

	Percentage of world total ...			Comparison with world average per capita ...	
	population	energy produced[1]	energy consumed[1]	primary energy consumption as a multiple of world average	Gross Domestic Product[2] as a multiple of world average
Wealthiest countries[3]	14%	35%	48%	3.3	3.7
Rest of world	86%	65%	52%	0.6	0.5
Selected regions					
USA + Canada	5%	21%	26%	5.0	4.6
Western Europe	15%	11%	16%	2.3	3.1
Middle East	2%	12%	3%	1.3	0.9
Africa	13%	8%	5%	0.4	0.3
Selected countries[4]					
USA	4.6%	17.2%	23.5%	5.0	4.7
China	21.4%	10.5%	10.8%	0.5	0.5
Russian Federation	2.3%	9.4%	6.0%	2.5	1.1
Japan	2.1%	1.0%	5.1%	2.4	3.6
India	16.5%	4.3%	5.0%	0.3	0.3
France	0.9%	1.3%	2.7%	2.8	3.5
Canada	0.5%	3.6%	2.4%	4.7	3.9
United Kingdom	1.0%	2.8%	2.3%	2.3	3.2
Brazil	2.8%	1.3%	1.8%	0.6	1.0
Australia	0.3%	2.1%	1.1%	3.4	3.8
Poland	0.6%	0.8%	0.9%	1.5	1.4
Greece	0.2%	0.1%	0.3%	1.5	2.1
Switzerland	0.1%	0.1%	0.3%	2.2	4.0
Bangladesh	2.2%	0.1%	0.2%	0.1	0.2
Denmark	0.1%	0.3%	0.2%	2.3	3.8
Kenya	0.5%	0.1%	0.1%	0.3	0.1

1 Annual primary energy, including bioenergy contributions.

2 The **Gross Domestic Product** (**GDP**) of a country is the total annual production of the nation's economic system, i.e. the value of everything the country produces in a year. The data used here are adjusted to take into account the local purchasing power of the GDP per person in each country. Many goods are normally cheaper in the poorest countries, so the contrasts would be even greater if normal exchange rates were used to convert to US dollars.

3 USA and Canada, Western Europe, Australia and New Zealand, Japan

4 In descending order of total primary energy consumption.

(Sources: BP, 2001; Gazetteer, 2001; IEA, 2001; UN, 2000)

efficiency by large factors over the years – supplying the same quantity of heat, light or driving power for a much smaller energy input. Unfortunately, the response has at times been an increased demand for the output, leading in some cases to an *increase* in the total demand for energy.

In the remainder of this chapter we compare the primary energy production and consumption data for a few of the countries in Table 2.3, to see the detail behind some of the differences. For the world as a whole, total annual primary energy production and consumption differ so slightly that we have implicitly taken them to be equal, but this is by no means true for individual countries. The difference between energy consumption and indigenous production determines of course whether a country is a net energy importer or exporter, a matter of considerable economic and strategic importance. The recent histories of our chosen countries provide some interesting contrasts, and we'll look briefly at each in turn.

2.5 Primary energy in the UK

The main contributions to primary energy production and consumption for the UK in the year 2000 are shown in Figure 2.7. The 'renewables' category includes biofuels, hydroelectricity, wind and solar energy (see also Table 2.4 below). The captions show the energy totals, expressed in some of the units introduced above, and Figure 2.7(b) includes per capita consumption data.

When we compare Figure 2.7 with the corresponding world data (Figure 2.4) the most striking difference is in the role of natural gas, which accounts for two-fifths of UK consumption – nearly twice its proportion for the world as a whole. The histories of primary energy production and consumption

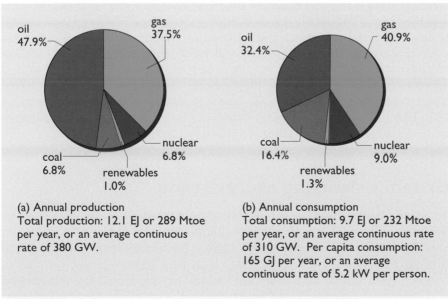

(a) Annual production
Total production: 12.1 EJ or 289 Mtoe per year, or an average continuous rate of 380 GW.

(b) Annual consumption
Total consumption: 9.7 EJ or 232 Mtoe per year, or an average continuous rate of 310 GW. Per capita consumption: 165 GJ per year, or an average continuous rate of 5.2 kW per person.

Figure 2.7 UK: Primary energy production and consumption, 2000. (source: BP, 2001; DTI, 2001.)

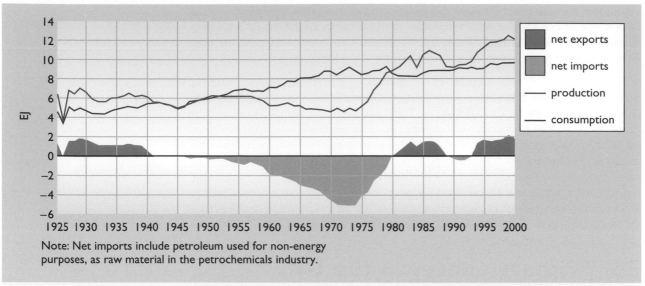

Note: Net imports include petroleum used for non-energy purposes, as raw material in the petrochemicals industry.

Figure 2.8 UK annual primary energy production and consumption and net exports and imports, 1925–2000

reveal further significant differences, particularly over the past fifty years. As Figure 1.8(a) showed, world total primary energy consumption experienced an almost continuous rise throughout the twentieth century, pausing only briefly during oil crises or periods of recession, and leading to an annual consumption in the year 2000 that was nearly five times that of 1950.

The UK pattern of primary energy consumption was rather different (Figure 2.8), particularly over the second half of the century. There was a fairly steady increase from the end of World War II until the first oil crisis in 1973, although at a much slower rate than for the world as a whole. This rise then virtually ceased, and consumption remained fairly constant almost to the end of the century, resulting in an annual consumption in the year 2000 that was not even twice the 1950 value. Britain's energy *production* exhibits a very different pattern again, falling by some 25% over a period when consumption continued to grow, and bringing a country which had been an energy exporter for well over a century to a position where more than half her energy needs were being met by imports. The situation then reversed again, with the rather steady consumption of the final quarter of the century accompanied by rapidly growing production.

Britain's changing energy scene

To see the background to these fluctuating fortunes we need to look at the data for individual energy sources (Figure 2.9). Until the 1950s, Britain was not merely a coal-producing country but a coal-*based* country, with almost all primary energy production and over ninety percent of consumption coming from coal. Coal fuelled the railways, the power stations and industrial machinery, and together with the 'town gas' that was derived from coal, met almost all the country's heating needs. But this was about to change. Coal production, which had started to rise again after the war years, began its long decline. Meanwhile, as in other industrialized countries, oil consumption in the UK was growing rapidly, with an average annual increase of over 10% a year from 1950 to 1970 (Figure 2.9(b)). These two factors are sufficient to account for the transformation during the 1950s from net energy exporter to net importer, a state that was to continue for more than thirty years.

However, the late 1960s already saw the start of yet another change, with the first contributions from Britain's North Sea gas fields. In the early years, from 1967 to 1972, output more than doubled each year – a remarkable annual average growth rate of over 100%. Consumption rose in step as the change from town gas to natural gas spread across the nation. Coal was of course the main energy source being displaced, but coal production was dropping in parallel with the falling demand. In consequence, the overall effect of Britain's natural gas resource on energy imports was very slight in these early years, and it was only with the discovery of yet another new resource that the trend eventually reversed.

The story of the exploration and subsequent development of North Sea oil will be told later, but its immediate consequences for the UK can be seen in Figure 2.9. The first significant deliveries came in 1976, and within three years oil had outstripped coal in its contribution to primary energy production. (As a proportion of primary *consumption,* oil – imported – had overtaken coal in the late 1960s.) By 1980 Britain was self-sufficient in oil, and despite the continuing fall in coal output, was about to become

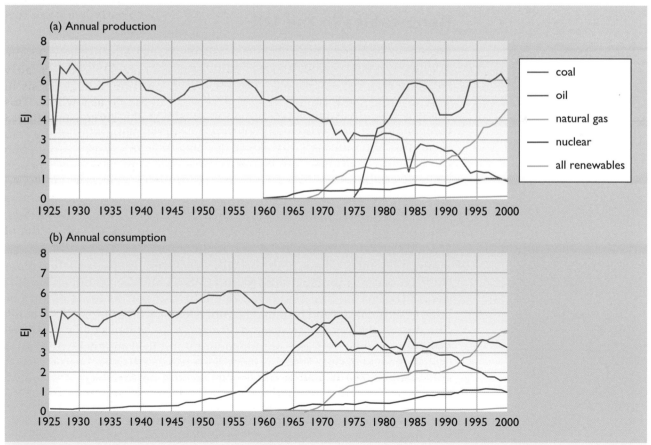

Figure 2.9 UK: Primary energy by source, 1925–2000

again a net energy exporter. A brief reversal occurred when oil production fell by a third following the steep drop in world oil price in the early 1980s (see Figure 2.6 above), but by the turn of the century output reached its highest level ever, and as we have seen, oil accounted for almost half the nation's primary energy production.

The period of rapidly increasing oil output saw a levelling-off in gas production, and with completion of the national change from town gas, a much slower annual rate of rise in consumption. But then came the 'dash for gas' of the 1990s, as the electricity generating industry started to take advantage of the technical and financial merits of gas turbine plants. This development and its environmental and other consequences are treated in more detail in later chapters.

Nuclear power is the only other energy source to make a significant contribution to UK primary energy. Growth at a rather modest 3% average annual rate over the last quarter of the twentieth century brought its contribution, on the 'notional primary input' basis, to almost exactly that of coal in the year 2000. Whether or not the slight fall at the end of the century will continue in the coming decades remains a current issue at the time of writing.

Later chapters treat the history of Britain's energy industries in more detail and discuss the social, economic and technical factors that have determined the changes described here.

Renewables in the UK

In the year 2000, renewable resources contributed just 1% of primary energy produced in the UK. Hydropower, for instance, with an output sufficient to meet the electricity demand of about a million households, contributed only 1.5% of Britain's electricity and less than 0.2% of the country's primary energy – too little to appear on the above diagrams. World-wide, the hydro contribution is proportionately ten times as great, but there are obvious geographical reasons why the UK is unlikely ever to approach that level. Nevertheless, technical improvements to existing systems and possibly an increased output from very small-scale hydro plants may lead to modest UK growth in the future.

Biofuels, which virtually disappeared from Britain's energy supply two centuries ago with the decline of wood as a fuel, have reappeared in recent years in a very different form. As Table 2.4 shows, some 80% of the renewables total comes from organic waste materials, and the main contributors are **municipal solid wastes (MSW)**, and their product, **landfill gas (LFG)**. For the past century or more, Britain has disposed of 90% or more of its domestic and commercial rubbish in landfills, where the organic component, decaying over a period of years in the absence of air, produces a gas that is relatively rich in methane (the main component of natural gas). In the past decade or so this landfill gas has increasingly been collected and used, mainly as fuel for small-scale electric power plants.

The UK does obtain energy from other renewable sources, but their annual contributions remain too small to appear separately on the diagrams above. (For more detailed accounts of renewables in the UK, see Boyle, 1996.)

Table 2.4 UK: renewable energy contributions to primary energy, 2000

Energy source	Used to generate electricity[1]	Used to generate heat	Totals by source	Percentage of all renewables
HYDROELECTRICITY				
large scale	17.5			
small scale	0.86			
Hydro total			**18.4**	14.7%
BIOFUELS				
MSW combustion	23.4	3.20		
landfill gas	30.0	0.57		
sewage sludge	5.03	1.72		
wood wastes[3]		20.7		
straw[3]		3.00		
other wastes	13.1	2.01		
Biofuels total			**102.8**	82.2%
WIND	3.4		**3.4**	2.72%
GEOTHERMAL[2,3]		0.03	**0.03**	0.03%
SOLAR ENERGY[2]				
solar heating[3]		0.44		
photovoltaics	0.004			
Solar total			**0.44**	0.35%
TOTALS	93.4	31.6	125.1	

Data: All energies are in petajoules (PJ). The population of the UK is about 59.2 million and the land area is 245 thousand square kilometres.

1 For hydro, wind and photovoltaics the figures represent the electrical output. All others represent the energy content of the fuels.

2 Brief accounts of these sources appear in the following sections.

3 Approximate data based on surveys carried out at various times during the 1990s.

(Source: DTI, 2001)

2.6 Primary energy in Denmark

Denmark's primary energy production and consumption in the year 2000 are shown in Figure 2.10, and comparison with the UK data in Figure 2.7 reveals some interesting similarities. Denmark (population 5.3 million) consumes slightly less than one-tenth of the primary energy of the UK (population 59.2 million), so the *per capita* consumption is very similar for the two countries. In both countries, fossil fuels account for just over 92% of the primary energy produced and 89% of the energy consumed. And as the diagrams show, both countries produce more energy than they consume (see also Box 2.7 overleaf).

BOX 2.7 **Self-sufficiency**

A useful measure of the extent to which a country depends on energy imports is its degree of **self-sufficiency**. This is defined as the total primary energy production divided by the total primary energy consumption, expressed as a percentage. A self-sufficiency greater than 100% obviously implies that the country has an energy surplus and is therefore likely to be a net exporter. This is the case for both Denmark, with a self-sufficiency of 138%, and the UK with 123%. It is important, however, to note that self-sufficiency in an individual energy resource such as oil can be even more important than overall self-sufficiency.

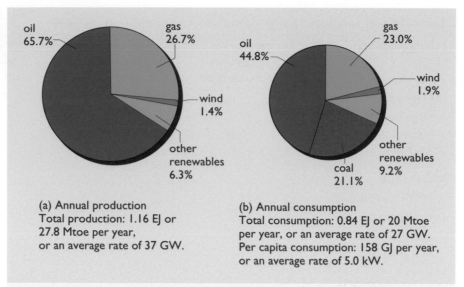

(a) Annual production
Total production: 1.16 EJ or 27.8 Mtoe per year, or an average rate of 37 GW.

(b) Annual consumption
Total consumption: 0.84 EJ or 20 Mtoe per year, or an average rate of 27 GW. Per capita consumption: 158 GJ per year, or an average rate of 5.0 kW.

Figure 2.10 Denmark: primary energy production and consumption, 2000

A significant difference between the two countries in the year 2000 is the nature of the non-fossil fuel contribution to energy production (and consumption). In the UK, as we have seen, most of this is from nuclear power, with renewables contributing only 1% of total primary energy. Denmark, in contrast, has no nuclear contribution and renewables account for nearly 8% of total production, with wind alone providing a greater percentage than all renewables in the UK.

When we compare the recent histories of the two countries (Figures 2.8 and 2.11), other major differences appear, and Figures 2.9(a) and 2.12(a) reveal the details behind the contrast. In 1960, Britain was reaching the end of a long period of energy self-sufficiency sustained by her coal exports. Denmark in 1960 had virtually no primary energy resources. A limited amount of lignite (a low quality form of coal; see Chapter 5) was the only fossil fuel, and the country had been almost totally dependent on imported fuel throughout modern times. By 1970, the continuing need for imported coal together with a steep rise in oil consumption (Figure 2.12(b)) was bringing her to a serious position strategically and economically. The

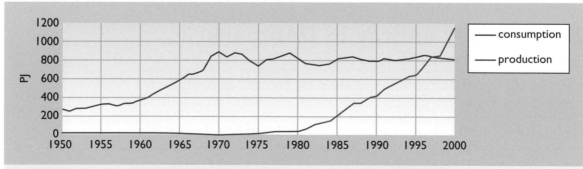

Figure 2.11 Denmark: Annual primary energy production and consumption, 1960–2000

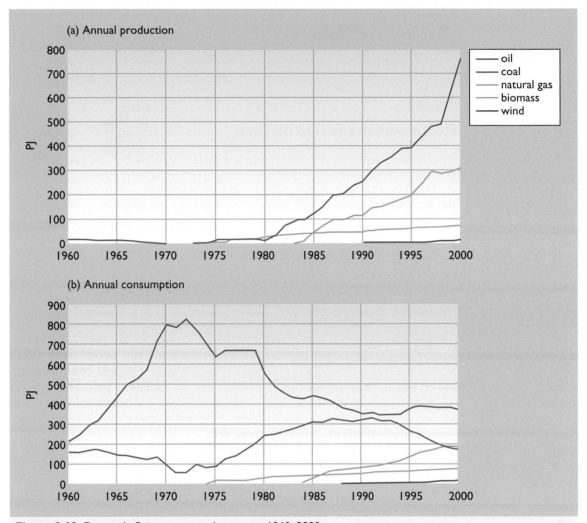

Figure 2.12 Denmark: Primary energy by source, 1960–2000

measures taken by successive Danish governments to ameliorate this situation are discussed in the next chapter, but their main result can be seen in the graphs: a total primary energy consumption in 2000 that is slightly *less* than in 1970, and a remarkable reduction in oil consumption to less than half its peak value. As Figure 2.12 shows, initially this was achieved in part by increased consumption of imported coal, but the final decade of the century saw domestic production of oil and gas making an appreciable contribution, and in 1997 Denmark became self-sufficient in energy for the first time in modern history.

The details appear in Figure 2.12(a). In the late 1970s – at a time when the largest contribution to domestic energy production was 20–30 PJ from renewables – the development of oil in Denmark's sector of the North Sea began. By 1985, output was growing at an average annual rate of over 12%, and with consumption still falling, Denmark became self-sufficient in oil in 1993. In the year 2000, as much oil was exported as was consumed internally.

Denmark's development of her natural gas resource came about five years after the oil – a reversal of the UK sequence. Closer study of Figures 2.9 and 2.12 reveals another difference. Until 1995 the UK remained a net importer of gas, piping it directly from the Norwegian fields, whereas Denmark has since the start of production always *exported* about half her gas by pipeline to Germany and Sweden.

Renewables in Denmark

As we have seen, the renewable energy sources play a much greater role in Denmark than in the UK, and in comparing the data in Tables 2.4 and 2.5 we must bear in mind the difference in population of the two countries. We then see that Denmark derives nearly fifty times as much energy *per capita* from the wind, and more than seven times as much from biofuels, as the UK. Municipal solid wastes account for similar proportions of the biofuels in both countries, but agricultural, or rural, wastes play a larger role in Denmark.

Table 2.5 Denmark: renewable energy contributions to primary energy, 2000

Energy source	Contribution	Totals by source[1]	Percentage of all renewables
HYDROELECTRICITY		**0.103**	0.12%
BIOFUELS			
MSW combustion	30.34		
wood[2]	22.55		
straw	13.05		
biogas	2.91		
other wastes	0.05		
Biofuels total		**68.91**	77.4%
WIND		**15.99**	17.96%
GEOTHERMAL[3]		**0.06**	0.07%
SOLAR		**0.33**	0.37%
HEAT PUMPS[3]		**3.66**	4.11%
TOTAL		**89.00**	

Data: All energies are in petajoules (PJ). The population of Denmark is about 5.3 million and the land area is 43 thousand square kilometres.

1 For hydro and wind the figures represent the electrical output. All others represent the energy content of the fuels.

2 Comprises wood chips and pellets, wood sold as 'firewood' and wood wastes. (Imports in the form of chips and pellets increase the consumption of wood to about 5 PJ more than the production shown here.)

3 Additional heat energy extracted from the ground or water (see the main text).

(Source: DEA, 2001)

Geothermal energy, heat drawn from regions below the Earth's surface, contributes a tiny fraction of the renewables total in both Britain and Denmark. The availability of this resource obviously depends on local geology, and neither country expects a major increase in its input. However, the final item in Table 2.5, the energy to be gained from surroundings that are at normal ambient temperature through the use of *heat pumps*, should be of greater interest. A heat pump, as the name suggests, 'pumps' heat from a cooler region into a warmer one, against the natural direction of heat flow. The principle is discussed later, in Chapters 6 and 9, but the result is obviously useful, warming buildings in cold weather or, in reverse, cooling them on hot days. Denmark, quite justifiably, includes such gains in the renewables total, and the contribution shown in the table, although a small fraction of the whole, represents an appreciable heat gain. With a proportionate annual contribution, the UK gain would be enough to heat half a million typical houses.

2.7 Primary energy in the USA

With the USA we come, not surprisingly, to a very different situation: a country that consumes a quarter of the world's primary energy, has twice the per capita energy consumption of Denmark or the UK and whose self-sufficiency is well *under* 100%. Primary energy production supplies less than three-quarters of annual consumption, a shortfall that leaves the US as the world's major energy importer as well as its major consumer. The pattern of consumption is not however very different from the other two countries, showing the dominance of fossil fuels that is common to all the world's industrialized countries and a *percentage* contribution from nuclear power similar to that of the UK. The actual nuclear output is of course

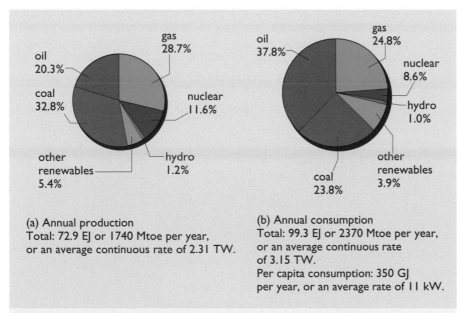

(a) Annual production
Total: 72.9 EJ or 1740 Mtoe per year,
or an average continuous rate of 2.31 TW.

(b) Annual consumption
Total: 99.3 EJ or 2370 Mtoe per year,
or an average continuous rate
of 3.15 TW.
Per capita consumption: 350 GJ
per year, or an average rate of 11 kW.

Figure 2.13 USA: Primary energy production and consumption, 2000

much greater than for the UK, and at some 750 TWh (trillion kWh) a year it puts the USA easily in first place world-wide.

Fifty years ago the USA was self-sufficient in energy, as it had been throughout most of its recorded energy history. This was also the case for the UK, as we've seen, and comparison of Figures 2.14 and 2.8 reveals two countries whose thirst for oil in the period from about 1950 to 1970 was leading to similarly growing gaps between energy production and consumption. However, the similarity is only superficial and hides one crucial difference. In the early 1970s the UK was just starting to develop her indigenous oil resources. The US, in contrast, was beginning to exhaust hers, as Figure 2.15 shows. These two facts account in large part for the difference in self-sufficiency of the two countries at the end of the century.

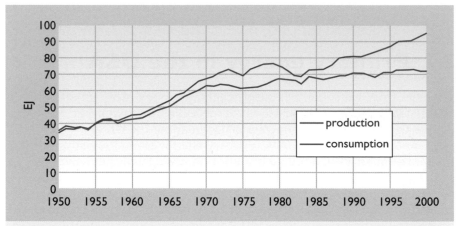

Figure 2.14 USA: Annual primary energy production and consumption, 1950–2000

Figure 2.15 bears closer study. At the start of the 1970s, production from the existing oil fields in the US reached its peak, and output was already falling when the first dramatic increase in world oil prices occurred in 1973. The extent to which the country had become dependent on the oil producers of the Middle East was made obvious to everyone, and for a couple of years consumption fell. It soon resumed its rise, however, encouraged in part by the development of the Alaskan oil fields, whose

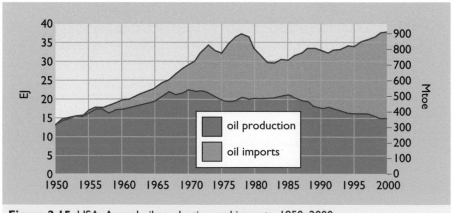

Figure 2.15 USA: Annual oil production and imports, 1950–2000

output delayed the fall in national production for about a decade. The further oil price increase in 1978 and the recession of the early 1980s brought about a more serious fall in consumption. However, as the graphs show, this was only temporary, and the steady rise in consumption throughout the 1990s resulted in the greatest shortfall ever by the end of the century.

Comparison of Figure 2.16(a) with Figures 2.9(a) and 2.12(a) shows at least one other striking contrast. We have looked at Denmark, with essentially no coal industry at any time, and the UK, once living on coal but with the industry in decline for the past fifty years. Now we see the US, with energy contributions from oil and from gas overtaking coal as early as the 1950s but coal *reversing* this situation in the 1990s. In this case, however, the USA is more characteristic of the world as a whole. As we'll see in later chapters, the demand for electric power rose rapidly throughout the second half of the twentieth century in almost every country, and coal, still the principal fuel for most of the world's power stations, experienced a corresponding growth in output. Britain, with its 'dash for gas' is therefore an exception, and Denmark, for reasons discussed in the next chapter, is another. (Our next country, France, is a third, for yet other reasons.)

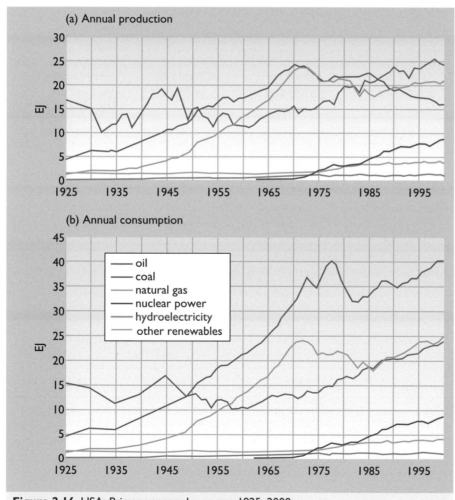

Figure 2.16 USA: Primary energy by source 1925–2000

Renewables in the USA

On a *per capita* basis, the USA is remarkably similar to Denmark in the broad features of its use of renewables. The average American and the average Dane both use the equivalent of about 4700 kWh a year of primary energy from renewable sources, and in both cases roughly three-quarters of this comes from the combustion of waste products – either for electricity production or for direct use of the heat. In both cases again, most of the remainder comes from one source, with the difference that in the US this source is water and in Denmark it is wind. The actual magnitudes of these contributions are of course very different – as are the differences in scale of the two types of plant, from the megawatt or so of the latest 'large' wind turbine, to a thousand times this for a large hydroelectric plant such as the Grand Coolee on the Columbia River. The US renewables total does include a contribution from wind power (mainly from wind turbines imported from Denmark), and comparison of Tables 2.5 and 2.6 shows that the annual output from this source is in fact very similar for the two countries (and about five times the corresponding figure for the UK.)

For a country with very large sunny areas, we might be surprised that the solar contribution is a mere 1.5% of renewable energy – about a thousandth

Table 2.6 USA: renewable energy contributions to primary energy, 2000

Energy source	Used to generate electricity[1]	Used to generate heat	Totals by source	Percentage of all renewables
HYDROELECTRICITY	899		899	18.8%
BIOFUELS				
wastes[2]	268	302		
wood[3]	431	2308		
alcohol fuels[4]		147		
Biofuels total			3455	72.3%
GEOTHERMAL	315	22	337	7.0%
SOLAR ENERGY				
solar heating		42		
photovoltaics	31			
Solar total			74	1.5%
Wind	16		16	0.3%
TOTALS	**1960**	**2821**	**4781**	**100%**

Data: All energies are in petajoules (PJ). The population of the USA is about 280 million and the land area is 9.36 million square kilometres.

1 For hydro, wind and photovoltaics the figures represent the electrical output. All others represent the energy content of the fuels.

2 Includes MSW, landfill gas, sewage, agricultural wastes (except woody wastes), tyres, etc.

3 All wood, including wastes.

4 Ethanol from vegetable matter, to be blended into motor gasoline. (See Chapter 7 and Boyle, 1996.)

(Source: EIA, 2001)

of the country's total primary energy. In terms of energy produced per square kilometre of land area it is however some five times greater than the Danish figure (and thirty-five times Britain's). As the data in Table 2.6 show, the US, unlike the other two countries, does have an appreciable geothermal resource. Geothermal power plants in the western states extract some ninety billion kilowatt-hours of heat from the ground annually in the form of super-heated steam, providing sufficient electricity for a few million households.

Review

We have tended in this discussion to emphasize the contrasts between countries, but a review of Figures 2.7, 2.10 and 2.13 shows that, whilst the total quantities of energy and the degrees of self-sufficiency may be very different, the general patterns of *consumption* are not dissimilar. Each country meets roughly two-thirds of its demand for primary energy from the two premium fossil fuels, oil and gas, and about another fifth from coal. And as we saw above, the pattern for the world as a whole is again similar – not surprisingly, as world consumption is dominated by industrialized countries such as these.

However, not every country has chosen – or has been able – to adopt this pattern, and we end the present survey with very brief accounts of two whose primary energy consumption includes major contributions from sources that play only a small role in those considered above.

2.8 Primary energy in France

Even the most casual comparison of Figure 2.17 with the corresponding pie charts for the previous three countries reveals striking contrasts.

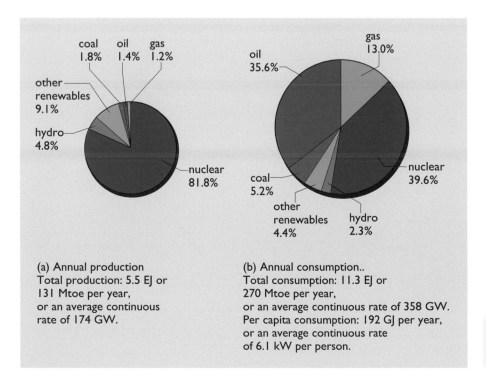

(a) Annual production
Total production: 5.5 EJ or
131 Mtoe per year,
or an average continuous
rate of 174 GW.

(b) Annual consumption..
Total consumption: 11.3 EJ or
270 Mtoe per year,
or an average continuous rate of 358 GW.
Per capita consumption: 192 GJ per year,
or an average continuous rate
of 6.1 kW per person.

Figure 2.17 France: Primary energy production and consumption, 2000

This time they are not in the quantities, as primary energy consumption *per capita* in France is only a little greater than in Denmark and the UK. The differences are in the extremely low self-sufficiency – less than 50% –

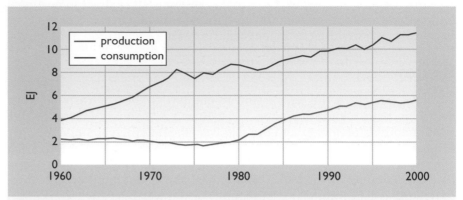

Figure 2.18 France: Annual primary energy production and consumption, 1960–2000

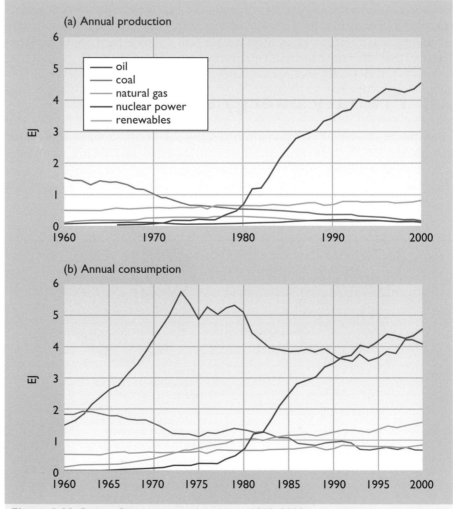

Figure 2.19 France: Primary energy by source, 1960–2000

and in the major role played by nuclear power. The reasons are easy to see in the energy production data. France has few indigenous fossil fuels but has chosen to develop her nuclear power industry. This can of course only partially compensate for the missing oil or gas resources, and the country is left with imported oil and gas accounting for half her primary energy consumption.

Figures 2.18 and 2.19 show the development of this situation over recent decades. We see the now familiar growth in oil consumption during the 1960s, accompanying a slight fall in indigenous coal production; but France, unlike the UK or Denmark, was not about to be saved by the discovery of oil or gas. Another significant factor was that at the start of this period large areas of rural France still lacked a mains electricity supply. The consequences of these factors, and the development of the French electricity industry, are discussed in more detail in the next few chapters.

2.9 **Primary energy in India**

India is the world's second most populous country and its sixth largest consumer of energy. Over the past twenty years, the population has been growing at just under 1% a year and annual primary energy consumption at about 2.5% for the 'conventional' resources only, a discrepancy between the two growth rates that is not unusual for a developing country. In the year 2000, India's per capita consumption, for the conventional resources only, was slightly over 12 GJ a year – effectively an average continuous supply of just under 400 watts per person, or about a thirtieth of the US average.

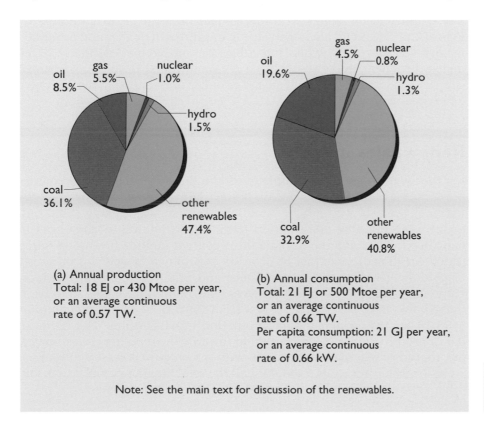

(a) Annual production
Total: 18 EJ or 430 Mtoe per year, or an average continuous rate of 0.57 TW.

(b) Annual consumption
Total: 21 EJ or 500 Mtoe per year, or an average continuous rate of 0.66 TW.
Per capita consumption: 21 GJ per year, or an average continuous rate of 0.66 kW.

Note: See the main text for discussion of the renewables.

Figure 2.20 India: Primary energy production and consumption, 2000

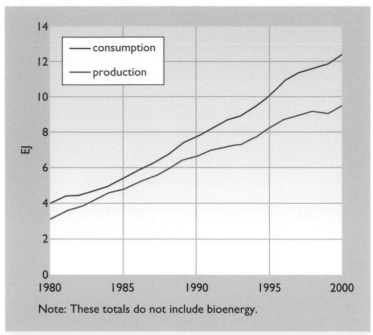

Figure 2.21 India: Annual primary energy production and consumption, 1980–2000

India's overall energy self-sufficiency is 86%, but if the renewables are excluded, the figure falls to 80%, and as Figure 2.21 shows, the shortfall in primary energy has been increasing steadily over the past few decades. The country has a large coal resource, and the coal imports implied by Figure 2.22 (a) and (b) are largely due to questions of quality and economic factors. Not so with oil, where the indigenous output met less than half the demand in 2000, and had not risen for some years. Despite government support, exploration – for both oil and gas – has not resulted in the discovery of any major new fields.

Renewables in India

For the average citizen – certainly in rural India – the renewable resources are at least as significant as the fossil fuels. Exactly how significant on the national scale is difficult to establish. As discussed above (Box 2.6), quantifying the total contribution from 'local' renewables is subject to considerable uncertainty. In India these sources include fuel wood, dung (dried for direct use as fuel or digested to produce 'biogas' for burning), bagasse (sugar cane fibre), rice husks and other agricultural residues. Estimates of the total annual bioenergy contribution have tended to lie in the range 5–15 GJ per person per year (Hall 1991, EIA 2002), which implies 5–15 EJ for the current Indian population of one billion. The figure of 8.5 EJ used in Figure 2.20 is weighted towards the more recent estimates, but should be regarded as uncertain within a range of at least ± 20%. Fuel wood probably accounts for about half the total bioenergy, with dung and other agricultural residues contributing most of the remainder.

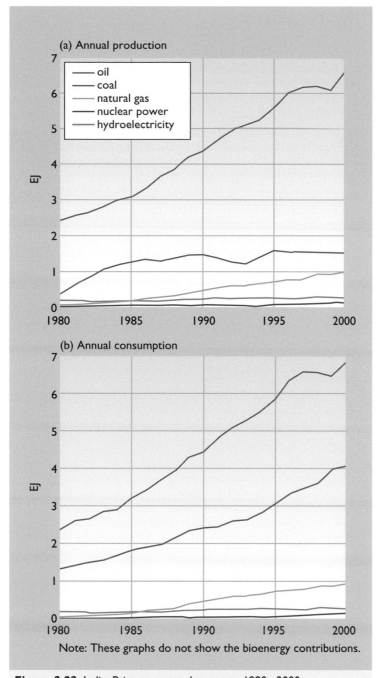

Figure 2.22 India: Primary energy by source, 1980– 2000

Some renewables are of course better documented, including large-scale hydroelectricity (shown separately in the graphs here), and smaller contributions from small-scale hydro, solar energy and wind power. India's installed capacity places her amongst the world's top half a dozen countries in terms of wind power, but the output accounts for no more than a percent or so of the renewables total.

2.10 **Summary**

The main purpose of this chapter has been to establish a basis for later discussions throughout the book. Not the scientific basis: that comes later, starting in Chapter 4. Here, after introducing the essential terminology and discussing the nature and reliability of the data, we have mainly been concerned to establish a foundation of facts about the quantities of energy used today and the sources of that energy. We have also looked at the past, at how we reached the present situation. And, probably of greatest importance for the future, we have seen the disparities in energy availability between different regions of the world. How the world – and the individual countries discussed above – use the energy is the subject of the next chapter, and again it starts with a question: Why do we need energy?

References

Boyle, G. (ed.) (1996) *Renewable Energy*, Oxford, Oxford University Press/ The Open University.

BP (various) *BP Statistical Review of World Energy*, London, The British Petroleum Company plc.

BP (2001) *BP Statistical Review of World Energy* [online], BP. Available at http://www.bp.com (and select 'energy reviews') [accessed 24 May 2002]

DEA (2001) *Danish Energy Agency, Energy Statistics* [online], Danish Energy Agency. Available at http://www.ens.dk (and click the tiny Union Jack for the English version) [accessed 24 May 2002]

DTI (various) *Digest of UK Energy Statistics*, London, Department of Trade and Industry.

DTI (2001) *Digest of UK Energy Statistics, Department of Trade and Industry* [online], DTI. Available at http://www.dti.gov.uk/epa/dukes [accessed 24 May 2002]

EIA (2001) *US Energy Information Agency, Annual Energy Review 2000* [online], US Energy Information Agency. Available at http://www.eia.doe.gov/fuelrenewable.html [accessed 24 May 2002]

EIA (2002) *US Energy Information Agency, Country Briefings, India* [online], US Energy Information Agency. Available at http://www.eia.doe.gov/emeu/cabs/india.html [accessed 24 May 2002].

Eurostat (1990) *Energy 1960–1988*, Luxembourg, Office des publications officielles des Communautés Européeannes, Luxembourg.

Gazetteer (2001) T*he World Gazetteer* [online], The World Gazetteer, Available at http://www.gazetteer.de/st/stata.htm [accessed 24 May 2002].

Hall, D.O. (1991) 'Biomass Energy', *Energy Policy*, vol. 19, no. 8, pp.711–737.

Hall, D.O. *et al.* (1993) 'Biomass for Energy: Supply Prospects', in Johansson, T.B. *et al.*, *Renewable Energy: Sources for Fuels and Electricity*, Washington D.C, Island Press.

IEA (1998) *International Energy Agency Proceedings of the Conference on Biomass Energy: Data, Analysis and Trends*, Paris, International Energy Agency.

IEA (1998) *Key Energy Indicators for India*, Paris, International Energy Agency.

IEA (2001) *Key World Energy Statistics*, International Energy Agency, Paris, 2001 Edition.

MINEFI (2001) *L'Energie en France – Repères*, Paris, Ministère de l'Economie, des Finances et de l'Industrie.

Ministry of Power (1962) *Digest of Statistics*, London, Ministry of Power.

Ramage, J. (1997) *Energy: A Guidebook*, 2ed., Oxford, Oxford University Press.

Times (1980) *The Times World Atlas*, London, Times Books Limited.

UN (2000) *World Energy Assessment: Energy and the Challenge of Sustainability*, New York, United Nations Development Programme and World Energy Council.

WEC (1998) *Survey of Energy Resources,* London, World Energy Council.

Chapter 3

What do we use Energy for?

by Bob Everett and Janet Ramage

3.1 Primary, delivered and useful energy

The previous chapter has described the flows of large amounts of primary energy, yet as pointed out in Chapter 1, what we as consumers really require are **energy services**, such as warm homes, cooked food, illumination, transport, and manufactured articles. A great deal of industrial and commercial activity is involved in supplying these 'end products' from the primary energy sources discussed in Chapter 2. The questions we need to ask are: what are the essential 'useful' forms of energy that we need; how much energy is lost in supplying this; and what exactly are we finally paying for?

Lighting is a very good example. If you want to read at night you need reasonable illumination. You could light a candle and produce light directly from fuel (candle-wax). Candles are not very bright, of course, and since a candle converts only about three ten-thousandths (0.03%) of its fuel into light, it is not a very efficient way of doing things. You are more likely to turn on an electric light.

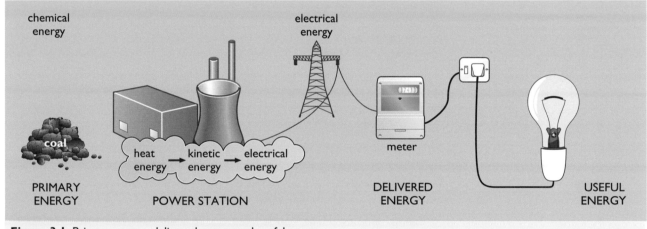

Figure 3.1 Primary energy, delivered energy and useful energy

As we have seen, what happens when you do this is a succession of energy processes, with energy being lost as waste heat at every stage. At the input stage is the primary energy source, which might be coal, oil or gas in the naturally occurring state in the ground or under the sea. Alternatively, the primary energy source might be the heat produced in a nuclear reactor. Taking coal as our example, we find that by the time the coal reaches the power station typically about 2.5% of this energy has already been used in mining and transporting it. Similar losses apply to oil, which may have been transported thousands of miles by sea and will also have been refined, and to natural gas, which may have been pumped to western Europe through pipelines stretching from distant Siberia.

In the power station the chemical energy of the coal is converted to heat, then into the kinetic energy of the steam turbines and finally into electrical energy. As we have seen, only about 30–40% of the energy of the fuel emerges as electricity. Some of this is likely to be lost as heat in the transformers and wires of the electricity distribution system, before it reaches your electricity meter. At this point it becomes **delivered energy**. This is what you actually receive and what you pay for – about a third of the primary energy extracted at the coal mine.

The wastage does not stop there. The electrical energy has to be converted into light of a suitable combination of wavelengths for you to read your book. Ultimately this light is the **useful energy** that you need.

If you were determined to be energy efficient, you could read by the orange light of a street lamp. These can convert about 30% of the electrical energy into visible light. But you are more likely to choose the more acceptable light of an incandescent lamp (a normal filament bulb). This however will produce only about 5 watts of visible light for every hundred watts of electric power, the rest being lost as heat. So, overall, for every gigajoule of primary energy that left the coal mine, only about 16 MJ have been converted into light (see Box 3.1); the other 984 MJ have become waste heat. This may sound appallingly wasteful, but is still fifty times more efficient than using candles!

BOX 3.1 Conversion efficiency

The conversion **efficiency**, often simply called the **efficiency**, of any energy conversion system is defined as **the useful energy output divided by the total energy input**. In practice it is very common to express this as a percentage of the input.

$$\text{Efficiency} = \frac{\text{Energy output}}{\text{Energy input}} \times 100\%$$

What is the efficiency of the complete coal-to-light conversion process described in the text?

We'll consider the fate of 1 tonne of coal: 28 GJ (gigajoules) of primary energy in the ground.

If 2.5% of this energy is used in mining and transporting the coal, the energy entering the power station is only 97.5% of this:

Energy entering power station = $0.975 \times 28 = 27.3$ GJ

If we take the fuel-to-electricity efficiency of a modern coal-fired power station (commonly referred to as the **thermal efficiency**) to be 35%:

Electrical energy leaving the power station = $0.35 \times 27.3 = 9.56$ GJ

On average, about 7.5% of this will be lost as heat in transmission in the wires and transformers or the way to the user, who receives only 92.5%, so

Delivered electrical energy = $0.925 \times 9.56 = 8.84$ GJ

But an incandescent light bulb turns only 5% of this into light (see the main text), so

Useful light output = $0.05 \times 8.84 = 0.44$ GJ

Thus an input of 28 GJ of primary energy produces an output of 0.44 GJ of useful light energy

$$\text{Overall energy efficiency} = \frac{0.44}{28} \times 100\% = 1.6\%$$

This theme of increasing efficiency of energy use is a recurring one. Although it may seem that we are currently living in a 'gas-guzzling' society, the truth is that we would probably be guzzling even more if it hadn't been for many significant improvements in energy efficiency, particularly over the last 150 years.

3.2 The expanding uses of energy

As we've seen in the previous chapter, the world use of energy has risen enormously over the past two centuries. This has been the result of a growing world population multiplied by an increasing energy use per person. For those who live in an industrialized world of cheap, readily available energy, it can be difficult to imagine what life would be like without it. It is worth, therefore, reflecting on exactly what society uses energy for and what changes have taken place over the broad time scale.

There are many ways of classifying these uses. Modern statisticians like to categorize them by the sectors of the economy; industrial, commercial, domestic and transport. We will look at these figures and the detailed changes over the last 40 years later in this chapter. Looking back further in time, Vaclav Smil (Smil, 1994) has made estimates of energy use per head of population (or *per capita*) for different societies in history. He has categorized the uses according to food, household, production, transport and services, although in practice, as in any other classification system the boundaries may blur into each other. Figure 3.2 shows figures for the energy consumption of the UK in 2000, together with Smil's estimates for five past societies:

■ Europe in 10 000 BC - a stone-age society of hunter-gatherers living in wooden huts.

■ Egypt in 1500 BC - a bronze-age culture with settled agriculture, organized irrigation and enough spare time to build the Pyramids.

■ Han Dynasty China in 100 BC - another agricultural society, also with organized irrigation, but which had mastered the art of making cast iron.

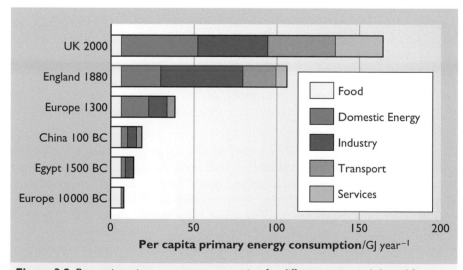

Figure 3.2 Per capita primary energy consumption for different societies (adapted from Smil, 1994). In the UK 2000 bar, energy loses such as those in electricity generation are included under each sector

■ Europe in 1300 AD - an advanced agricultural society, using wrought iron smelted with charcoal, trading coal by sea and capable of building large cathedrals. This is the world of Chaucer's Canterbury Tales. The UK population was then about 6 million.

■ England in 1880 - Victorian industrial society, fuelled by coal and driven by steam, criss-crossed with new railways, exporting enormous amounts of coal to the world, busy inventing a host of new chemical and industrial processes, and just about to embark on exploiting mains electricity. The UK population was then about 30 million, about half the current figure.

Each society consumes more energy than the previous one, yet it is not a matter of using more energy for every category of use, rather that new uses arise and expand. Moreover, they are likely to arise in one country, such as agriculture in Egypt or industrial society in Victorian England, and then spread across the world. Let us look at each category of energy use in turn.

Food

This is the first and ultimately most important category in Figure 3.2. The actual edible food energy or dietary energy needed by a human being is about 10 MJ per day, about 3.6 GJ per year. Obviously if you are involved in hard physical labour, then you will need more than this. This figure does not normally appear in energy statistics, though the energy to grow, harvest, package, transport and cook food does.

For prehistoric European humans in 10 000 BC, life was a matter of hunting wild animals and gathering naturally growing food plants. The energy used to do this was your own personal hard effort. If you could gather enough food energy to supply this, you prospered, if you couldn't you would starve. In addition to this the energy of wood was used to cook the food and to heat the relatively primitive homes, wood huts or caves.

The first agricultural societies grew up around 3000 BC in Mesopotamia, planting crops and using domesticated animals such as oxen to carry heavy loads and augment the hard human physical labour of ploughing and raising water for irrigation. These skills slowly spread to surrounding countries. By 1500 BC, the landscape of the Nile Valley of ancient Egypt was intensively cultivated, involving ploughing and irrigation. The farming base of society was sufficiently productive to generate a surplus of food for the ruling elite and (amongst others) the army of Pyramid builders.

The Han Dynasty Chinese of 100 BC had a similar intensive agriculture, but had improved the productivity of their ploughing with mass-produced cast-iron ploughshares. The country's extensive canal system allowed the trade of agricultural produce.

Stepping forward to Europe in 1300 AD, we come to another agricultural society, again slightly more advanced, with oxen being replaced by stronger horses and more efficient designs of ploughs. The vast bulk of the population lived on, and worked, the land. They existed in a **subsistence economy**, eating what they grew and trading what little surplus was left over after the feudal landowners had taken their cut. The basic methods that they used had changed little for over a thousand years, though the quality of tools and ability to harness and use draught animals had improved. By this time, water power had also been harnessed for irrigation and grinding corn and the earliest windmills were starting to become common.

Figure 3.3 A traditional irrigation windmill at work in Crete (photo Bob Everett)

By the time we reach England in the 1880s society was radically different. Agriculture was embarking on changes every bit as drastic as those taking place at the same time in industry. The new job opportunities in urban factories, coupled with competition from imported food brought by steam ship and railway, led to a continuing decline in rural population. By this time the bulk of the English population were living in rapidly growing manufacturing cities, swelled by a continuing drift of manpower from the countryside. Yet all these people had to be fed. Although farming efficiency had been improved in the eighteenth century by introducing larger and more powerful breeds of horse, the nineteenth century saw the introduction of artificial fertilizers and farm mechanization. The idea of modern farms as 'agribusiness' presiding over 'corn deserts' provokes angry debate today, but it is the end product of a set of trends set in motion over a hundred years ago.

Fertilizers

Eighteenth and nineteenth century chemists had discovered that sun and rain were not the only ingredients required to grow crops. Plants also needed phosphorus, potassium and nitrogen in suitable forms.

From the 1840s onwards various deposits of rocks were quarried which could supply the modest amounts of phosphorus and potassium required. Canals and then railways made the transport of these bulk fertilizers practicable, even into rural areas. But supplying large quantities of nitrogen remained a problem. The main supplies in Europe during the late nineteenth century were sodium nitrate (saltpetre) imported by sea from Chile, and ammoniacal liquors – a by-product of town gas production from coal (see Chapter 7 – Oil and Gas).

In 1914 the First World War created a nitrate supply crisis for Germany. Its chemical industry came to the rescue with the Haber-Bosch process, manufacturing ammonia (NH_3) using nitrogen in the air and hydrogen from town gas. Initially this was needed for the manufacture of explosives for the war, but afterwards it was seen that the process had a genuine 'swords into ploughshares' application by providing nitrogenous fertilizer. Initially the process was extremely energy intensive. In 1930 an ammonia plant would have consumed the energy equivalent of 9 tonnes of oil to make one tonne of ammonia. Since then there have been extensive improvements. A modern (1994) plant only requires 23 GJ (about 0.7 tonnes of oil equivalent) per tonne of ammonia. Nevertheless the energy involved in fertilizer manufacture remains a major input in modern intensive agriculture.

The immediate rewards of using nitrate fertilizer were increased yields with existing strains of crops. In 1300 typical English yields of wheat were less than 1 tonne per hectare of land. By 1880 they had been more than doubled. Yet even these yields have been increased by the breeding of special strains of crops capable of making use of very high levels of nitrogen in the soil. British wheat production reached yields of 4 tonnes per hectare in the 1970s and at the time of writing the figure stands at around 8 tonnes per hectare.

This massive use of synthetic nitrogen fertilizer is now world-wide, and it has been estimated (Smil, 1994) that at least one-third of the protein in the current global food supply is derived from the Haber-Bosch process. The disadvantages of this high nitrogen input have been the **eutrophication** or poisoning of watercourses with high nitrate levels, and emissions of nitrous oxide, a greenhouse gas. The latter is discussed in Chapter 13, Penalties.

Looked at one way, this extra fossil fuel energy input is simply an aid to improving the efficiency of take-up of solar radiation into useful farm produce. A more pessimistic viewpoint would see this as '*a sad hoax, for industrial man no longer eats potatoes made from solar energy, now he eats potatoes partly made of oil*' (Odum, 1971). The question of whether or not an agricultural process can produce more energy in the crop than the fossil fuel inputs taken to grow and process it is an important one, particularly for the development of biofuels such as ethanol and bio-diesel.

Farm mechanization

While Victorian industry ran on coal, agriculture ran on horses and human muscle-power and this continued into the 20th Century. By 1901 Great Britain had 3.5 million horses of which of which 1.1 million were employed on farms. About 30% of lowland farm area was devoted to their keep, 10% just for farm horses. This horse population was matched almost one to one with about a million full-time farm workers plus an army of casual labour, mainly from the cities, at harvest time. Mechanical assistance was limited to a few low-powered steam engines used for threshing and grinding. Although horse-drawn reaping machines had been introduced in the UK in the early years of the nineteenth century their use was limited by the small size of fields and narrow lanes. In the US, space was not a problem. By the end of the nineteenth century fully automatic horse drawn combine harvesters were in use. The largest needed 40 horses to operate but could harvest a hectare of wheat in 40 minutes.

Farm mechanization in the UK was very slow to develop. By 1920 there were only 10 000 farm tractors in the country. They were outnumbered nearly 100:1 by horses. There was little incentive for mechanization. During the 1920s and '30s there was high unemployment and labour was cheap. British agriculture was in a very depressed state, competing with cheap imported food.

However, in 1939 all that changed. World War II brought a serious food blockade and food rationing. Meat became a tightly rationed luxury, and land that had been pasture for livestock was ploughed up to grow more vegetables for direct human consumption. The war drained manpower from the land, and productivity could only increase by more mechanization.

After the War, British society had changed. Post-war reconstruction required man-power in factories. Labour prices rose compared to the price of farmland and mass-production techniques had reduced the price of farm machinery. It no longer made economic sense to employ whole armies of agricultural labourers when a few men and machines could do the same job. A modern tractor can plough a field ten times faster than its horsedrawn equivalent. Even so, it is perhaps surprising that it was not until the 1950s that the number of tractors in Britain exceeded the number of farm horses.

Figure 3.4 A modern combine harvester rapidly chews its way through an Essex wheat field

Again, looked at one way, the mechanization of farms has meant an injection of fossil fuel energy. On the other hand, removing the horse has freed all their grazing land for further agricultural production. Smil has estimated that feeding America's farm horses required 25% of their cultivated land. One key advantage of the tractor is that you can switch it off! Once the ploughing or harvesting is done it can be put away in a barn for next year. A horse keeps eating, whether or not you use it! Another 'benefit' is that the large workforce is no longer needed, theoretically freeing it for other uses. This is fine if there is alternative employment, but in many cases it has just continued the trend of the depopulation of the countryside and growth of cities.

The food shortages of World War 2 have left a deep mark on UK (and EU) agricultural policies. There are extensive subsidies to promote farm production which have implied high levels of energy and fertilizer use. At the time of writing, the UK is self-sufficient in a wide range of basic foodstuffs. The energy price paid for this amounts to a modest 1 GJ per year per capita for the fuel energy to run UK farms and about the same amount again in manufactured fertilizers.

Domestic energy

Compared to the modern centrally-heated, electrically-lit society, the past was cold in winter, draughty, smoky, dirty and very dark after sunset. For prehistoric man, home was likely to be a cave or a wooden hut. Later societies developed their building skills with bricks and mortar and metal woodworking tools to produce the house as we know it today. Yet central to the home was the *fire* for cooking, heating and lighting.

Heating, washing and cooking

The Roman elite may have had central heating, but most of their expertise disappeared in the dark ages. In mediaeval England, the use of fire was a rather unruly and dangerous affair (especially so in a wooden house) and must have been extremely energy inefficient. According to Bowyer (1973), there appears to be no evidence of chimneys in England before the 12th Century. Neither the London Building Byelaws of 1189 nor of 1212 mention fireplaces or chimneys, even though the main reason for their introduction was to reduce the risk of fire. Buildings simply had permanently open smoke-holes in the roof. The indoor air quality must have left a lot to be desired and cooking could be a serious fire hazard. The Great Fire of London in 1666, which destroyed over 13 000 houses, started in a pie shop in Pudding Lane. As with other, later, energy-related catastrophes, it led to a series of Acts of Parliament, in this case, specifying how chimneys were to be built, at least to be safe, if not necessarily efficient.

Figure 3.5 The coal fire was the heart of the home for heating and cooking – a reconstruction in the Beamish Museum, Co. Durham

The modern fireplace is a product of careful design, originating in the sixteenth century. The nineteenth century fireplace shown in Figure 3.4 is also a masterpiece of cast iron construction, providing **space heating** (that is heating of the living spaces) as well as an oven for baking and a place to boil kettles and saucepans. **Domestic hot water** for washing and cleaning also had to be produced in this manner – today we would use a dedicated gas or electric water heater.

In rural Britain, wood would have been the normal fuel used in such a fireplace, but by the seventeenth century, coal was widely used in cities (see Chapter 5). The hearth was the centre of the home, and especially after dark, not least because of the expense of lighting any other room (see below). This fire effectively needed to be kept burning all year round for cooking and water heating. In summer it would make the living room too hot, and in winter it would be too cold. The joys of warm rooms and running hot water, although technically feasible, passed most of Britain by until the latter part of the twentieth century. This was stoically accepted as part of life. The Reverend Francis Kilvert, a Herefordshire clergyman, describes getting up on Christmas Day, 1871:

> …It was an intense frost. I sat down in my bath upon a thick sheet of ice which broke in the middle into large pieces whilst sharp points and jagged edges stuck all round the sides of the tub …, not particularly comforting to the naked thighs and loins, for the keen ice cut like broken glass. The ice water stung and scorched like fire. I had to collect the floating pieces of ice and pile them on a chair before I could use the sponge and then I had to thaw the sponge in my hands for it was a mass of ice…

The open coal fire, with a thermal efficiency of about 25%, remained the normal mode of heating in UK homes until the 1960s, much to the derision of visitors from the Continent and US where central heating was considered normal. Central heating only reached the bulk of UK homes after the introduction of North Sea natural gas in the late 1970s.

The introduction of town gas made from coal in the early nineteenth century was initially for lighting. It soon became apparent that it was an ideal controllable fuel for cooking. Unlike a coal stove, a gas cooker could be turned on and off quickly. Cooking, with all its attendant smells and mess, could be carried out wholly in the kitchen, instead of spilling over into the living room with its coal fire devoted to space heating. This in turn influenced house design, making the kitchen and living/dining rooms into completely separate entities.

The Victorian housewife's life was not easy. Even middle-class families would aspire to have a female domestic servant (or 'scivvy') whose life was even harder. Before the advent of modern detergents and washing machines, cleaning clothes was mostly a matter of hard physical labour, scrubbing them in hot water with soap. Before the electric vacuum cleaner, house cleaning was an endless inefficient chore of chasing after dust and grit with a dustpan and brush.

The First World War in 1914 was a turning point. Millions of men were drafted into the Armed Forces creating a severe labour shortage. Women left domestic service to work in the munitions factories where they enjoyed considerable equality with men. After the war they did not want to return to the old ways as domestic drudges. There was a 'servant shortage'. The market was ripe for selling 'labour saving appliances', vacuum cleaners, washing machines, electric irons, etc., a process which has continued to the present day.

Preserving and processing food

Today, 'food' is something that we buy from a supermarket, usually wrapped in plastic. We are likely to deposit it into our domestic refrigerator and, at mealtimes, transfer it rapidly to the plate via a microwave oven. At the extremes it has become an industrial product, processed and packaged by machines and transported vast distances.

In the past the preserving and processing of food was something that largely took place in the home. Before about 1000 AD, even basic tasks such as the grinding of corn would have been done by hand in the home with a small hand grindstone. The introduction of

Figure 3.6 Hard female labour – before washing machines and modern detergents washing clothes meant hard scrubbing with hot water and soap

water, wind and horse-powered mills turned this from a home activity to a (local) industrial one. Also, given the fire risks in a mediaeval home, baking was a task that was safest to leave to a professionally-run bake-house.

Dealing with meat was a particular problem. Animals could either be slaughtered and preserved with salt or kept alive for as long as possible before being served up. In the countryside this didn't pose too much problem. In the expanding cities, it created chaos. Even up to the mid-nineteenth century herds of cattle were driven on the hoof through the streets to city-centre markets to be sold and then slaughtered. This was a road traffic nightmare, but also meant that a whole host of food-rendering industries naturally grew up, also in the cities. All these required energy for process heat. Animal fats were rendered down to make, amongst other things, tallow for lighting fuel and bones boiled to make glue. One of the smelliest processes was considered to be boiling down blood to make fertilizers.

The arrival of railways and cold storage by the end of the nineteenth century was a godsend. Cattle markets, the new cold stores and all the smelly processing industries were removed to their current locations in industrial estates well away from the noses of the more well-to-do householders.

This freed up city centres as the home for the services sector, in particular banking and shopping.

Although 'machine' refrigeration as we know it today is a late nineteenth century invention, it did not really take off until the early twentieth century because bulk ice was a globally traded commodity. In 1900 the UK imported half a million of tonnes of ice from Norway and even some from the US (see Weightman, 2001). The food distribution chain relied on blocks of ice bought along with the food. As refrigeration technology improved, and especially as electricity became cheaper through the twentieth century, commercial ice-making plants replaced imported ice. Finally, towards the end of the 20th century, the mass-produced commercial and domestic refrigerator arrived, killing off the ice trade, but contributing to the large increase in electricity consumption (see below).

Lighting

The past was very dark because artificial lighting was extremely expensive. This in turn was primarily due to the inefficiency of technology which depended on the naked flame. The basic process was one of combustion of a fuel, be it candle wax, lamp oil, or coal gas, so that the particles of soot in the flame glowed and gave off a tolerable light. Perceived light intensity is measured in *lumens*. A modern hundred-watt filament lamp emits about 1400 lumens, whereas a wax candle only emits about 13 lumens. A lumen is actually a unit of light energy, but one which uses the human eye as the meter. 'What is the brightness of a light?' may sound as vague a question as 'how long is a piece of string?' but by careful research it has been pinned down to an 'average' response of a large number of individuals. From this it has been possible to lay down standards of acceptable levels of illumination.

A reasonable modern standard for desk lighting is about 300 lumens per square metre, which would translate into 23 candles per square metre of desk surface! You may like to try reading this book by the light of a single candle to appreciate the problem. But in the past you might not have been able to afford more than this. For example a budget study of a 1760s Berkshire family estimated that they would have spent nearly 1% of their annual income to get a mere 28 000 lumen-hours of light (see Nordhaus, 1997). Put another way that is equivalent to two candles (and probably rather smelly tallow ones) for three hours per day!

Eighteenth and nineteenth century developments of oil and kerosene lamps managed to boost the output of a single lamp to 10 or 20 candlepower, but this was largely done by increasing the fuel throughput rather than the efficiency. It did not significantly improve the affordability of artificial light; nor did the introduction of lighting by town gas made from coal, which was priced to be competitive with oil and kerosene.

In the 1870s, electric arc lighting first appeared, but it was very harsh and bright and only really suitable for lighthouses and street lighting. Then in the 1880s the first commercial

Figure 3.7 A naked gas flame is an extremely inefficient light source (an example in Judge's Lodgings Museum, Presteigne)

incandescent electric lamps became available. These had the right lighting qualities to compete with oil and gas lamps. However, almost immediately the invention of the fabric gas mantle by Carl Auer von Welsbach improved the efficiency of gas lighting by a factor of four. This allowed it to compete with electric lighting well into the twentieth century (see Chapter 9). Since then there has been a continuous development both in new light sources and in the efficiency of electricity generation used to power electric lamps. The effectiveness of lights or lighting systems is expressed in terms of their **efficacy** (or 'usefulness') in lumens per watt. Table 3.1 below charts the progress over the years:

Table 3.1 Efficacy of different lighting sources

Lighting source	Approximate efficacy in lumens per watt
Candle	0.07
Gas mantle	0.9
Modern incandescent electric lamp	5*
Compact fluorescent lamp	25*
Low pressure sodium street lamp	60*

* Per watt delivered by the primary source, taking electricity generation efficiency as 33%

It is quite remarkable that the complex process of choosing to burn a litre of kerosene in an engine, to drive a generator, to power a fluorescent lamp, can produce 250–450 times more useful light than burning the same amount in an oil lamp. The overall result is that the real cost of useful light has fallen dramatically. It is thus not surprising that we now light our homes in a way that would have been impossible to afford in the past. The 1760s Berkshire family mentioned above spent 1% of their income to obtain 28 000 lumen-hours of light a year. Compare that with a sample 1960s US family who used over 13 000 000 lumen-hours per year for a paltry 0.3% of household income (Darmstadter, 1972).

Even so, it is worth reflecting that two billion people in developing countries do not have electric light and are effectively still technologically in the nineteenth century.

Industry

Physical labour

The word 'industry' normally conjures up pictures of smoking chimneys. However, it should not be forgotten that industry also includes a vast range of *physical* activities, digging, sawing, polishing, grinding, which are now carried out by machines.

The scene from the French workshop in Figure 3.8 shows labourers who are providing the physical work. They are likely to have kept this up for a whole working day, six days a week! Today their function would have been replaced by an 800 watt electric motor (about the rating of a heavy-duty DIY electric drill). Some tasks, such as endlessly sawing tree-trunks into planks or polishing flat sheets of glass by hand must have been mind-numbingly dull as well as physically exhausting.

The development of textile mills in Lancashire in the eighteenth century required mechanical energy to drive the spinning machines. The practical choices were: power from water mills, horses running in treadmills or the developing steam engines (see Chapter 6). By the 1840s it was clear that steam power was the way ahead and by the 1880s it was the normal source of motive power in UK factories.

But steam engines were heavy and difficult to move. They were ideal for ships, where weight was not a problem, but for motion on land they needed carefully laid rails with no steep gradients. By 1880, Victorian England was criss-crossed with railways on viaducts and cuttings, but all of this civil engineering construction work was done by the hand labour of a large army of 'navvies' (or navigators) and thousands of horses. The steam shovel, the bulldozer and the JCB digger are all twentieth century developments.

Neither was the steam engine very suitable for small mechanical tasks. The domestic sewing machine, for example, was traditionally a hand or treadle-powered device. It had to wait until the twentieth century for the development of small, cheap, electric motors and the availability of cheap electricity to drive it. Now, it seems that almost everything has an electric motor in it. Even the average DIY tool box is now likely to contain an electric drill, an electric saw and probably even an electric screwdriver. All of this has contributed to the continuing rise in electricity demand (see below).

Figure 3.8 Eight men rotating a vertical capstan in a mid-eighteenth century French Workshop. The rope is pulling gold wire through a die (Diderot and D'Alembert (1769–1772))

Iron and steel manufacture

The Europeans of 10 000 BC lived in the Stone Age. The Ancient Egyptians lived in the Bronze Age and we (from about 1000 BC onwards) live in the Iron Age. Extracting metals such as iron from their ores is not easy. It needs a fuel which is both capable of burning to produce a high temperature and chemically reacting with iron oxide to produce metallic iron. Traditionally this high quality fuel was charcoal – almost pure carbon made from wood. The charcoal-making process was not very efficient. It took about ten tonnes of wood to make one tonne of charcoal. In the UK the wood supplies held out until the seventeenth century when the switch to coal took place. Even then, the coal usually needed to be refined to coke to obtain the high temperatures (see Chapter 5).

Most early furnaces could not make pure iron. They produced a lumpy mixture of iron and slag which then had to be physically beaten out with hammers to produce 'wrought iron'. Thus eighteenth century iron foundries grew up where there were three resources – iron ore, coal, and water power to drive the bellows for the furnaces and the hammers. At that time it took ten tonnes of coke to produce one tonne of iron. Since then there have been continuous improvements in the processes beginning with the introduction of the blast furnace in the 1830s with steam engines to drive the bellows.

Wrought iron is not normally suitable for making tough machine components. For that you need steel which is made from iron by carefully burning off the carbon content. In Victorian Britain iron and steel production increased massively, swelling the energy needs of the 'production' sector. By 1903 it was using nearly 30 million tonnes of coal a year, over one-sixth of the country's total consumption. By this time it was possible to produce a tonne of iron with less than two tonnes of coke. Today, a century on, this figure has dropped to less than one tonne of coke per tonne of iron. However, in terms of total output, the UK iron and steel industry has now faded to a shadow of its former glory.

Aluminium smelting

Victorian steel was fine for steam engines, but not much use for 20th century aeroplanes, which required something much lighter. Aluminium as an element was isolated in 1824 but proved extremely difficult to purify chemically. Smelting the metal requires six times more energy than smelting iron. It was found that it could be separated from aluminium oxide using electrolysis, but mass production had to wait for the availability of cheap electricity. In the 1880s smelting 1 tonne of aluminium metal required more than 50 000 kilowatt-hours of electricity. Even though by the end of the 20th Century this figure had been reduced by more than two-thirds, the price of aluminium is still largely determined by the cost of the electricity used to make it. The 'embodied energy' of aluminium is amongst the highest of common household materials available today, which is why recycling of aluminium cans is an important energy-saving activity.

BOX 3.2 What is embodied energy?

Many materials require a large amount of energy to produce them. This is called their 'embodied energy'. It can be an important factor in choosing materials or even whole processes. Below are some example values.

Material	Energy Cost (MJ kg^{-1})	Production process
Aluminium	227-342	Metal from aluminium ore
Cement	5-9	From raw materials
Copper	60-125	Metal from copper ore
Plastics	60-120	From crude oil
Glass	18-35	From sand and other materials
Iron	20-25	From iron ore
Bricks	2-5	Baked from clay
Paper	25-50	From standing timber

Source: Smil, 1994

Processes that need high temperature heat

There are many other industries with a long history that need large amounts of high temperature heat. Brick making, for example, is simple enough, but requires enough energy to drive all the water out of what are essentially just lumps of mud. The other ingredients of traditional building practice are lime mortar and cement. The essential ingredient of lime mortar is quicklime or calcium oxide, made by heating chalk to drive off the carbon dioxide content. Cement is slightly more complex and is made by heating silicate clays to drive off the water content. The Romans used cement to build the 43 metre domed concrete roof of the Pantheon in Rome in 120 AD, but the secret of its manufacture disappeared and was only rediscovered in the nineteenth century. Its first major use in the UK was in the construction of the London sewers in the 1860s.

Glass making is another energy-intensive activity that has been practised for over 4000 years. It requires sufficient heat to melt sand. Traditionally this required a high quality fuel such as charcoal, or later coke, and a lot of hard work with the bellows to fan the flames. Since the whole point about glass is that it should be transparent, it is vital that the soot and grit from the fuel is kept out of the mixture. The ideal modern solution is to use electricity to provide the heat. As with aluminium the high embodied energy makes glass recycling important for energy saving.

In addition there are a whole host of other chemical processes that need process heat. One of the most important introduced in the early years of the nineteenth century was the manufacture of soaps and detergents, essential for the booming wool and cotton industries. Soap manufacture needed sodium hydroxide. This was produced by the Leblanc process from common salt (sodium chloride), limestone, coke and sulphuric acid using coal for process heat. Ironically, although the end product was supposed to be clean white textiles, this process was notorious for its air pollution consequences, emitting large quantities of hydrochloric acid vapour. It was this problem that led to the formation of the Alkali Inspectorate and some of the first UK legislation on industrial air pollution. Today, sodium hydroxide is made using electricity to separate the sodium and chlorine of common salt. Like aluminium smelting this is an industry that depends for its profits on cheap electricity.

If you look back at Figure 3.2 you will see that the total UK per capita energy use for industry in 2000 is little changed from 1880. This reflects the increased energy efficiency of production, but also a shift in the UK economy away from heavy energy-intensive industry to other high-value products such as electronics, and to earning more from the services sector (see below).

Transport

The modern growth in transport energy use is commented on at the end of this chapter, but it is worth pointing out that this can only occur because we are 'free' to travel. In many past societies (and some modern ones!) this was not an option. In the UK in 1300 AD feudal lords 'owned' the peasants as well as the farmland. They were every bit as tied to the land as the animals. Most people never travelled more than a day's walking distance from their homes in their whole lives. They were also limited by the atrocious state of the roads. The characters in Chaucer's Canterbury Tales were the lucky few - a pilgrimage to see Canterbury Cathedral was the closest they would get to a 'holiday'. The energy inputs to the transport sector were simple enough, wind power for sailing ships and copious quantities of hay and oats for horses.

By the 1880s society had changed entirely. People were, theoretically at least, free to do what they wanted and travel as they wished. Although road travel had improved during the eighteenth century, railways swept aside the expensive stage coach competition for long distance land travel. Steam ships were transporting freight not just on short coastal routes, but also on regular long-distance ocean-going routes. However, in the cities the horse bus and horse tram ruled the road. It was these and competing cheap railway fares that ushered in modern concepts of 'commuting' and 'suburbia'; it was no longer necessary to live next to your workplace. It was also permissible for the urban work force to have 'summer holidays' and they had the spare cash to afford them. The railway and steamship companies were only too happy to provide the transport arrangements to new seaside resorts. The transport energy sources were now still hay and oats for horses, but also enormous amounts of coal; by 1903 UK railways were using 13 million tonnes of coal a year and coastal shipping another 2 million tonnes. Photographs of the main roads at this time show them to be strangely empty by modern standards. Everything went by train and continued to do so well into the 20th century.

However, the urban horse did not last into the new century. In cities the electric tram and the petrol-engined bus had almost completely substituted for their horse equivalents by the First World War in 1914. This freed up large areas of hayfields around cities, which had provided the transport energy supply, for yet more suburbia and yet more commuting.

In the US, the motor car took off in 1907 with the famous Model 'T' Ford. Although Adolph Hitler introduced the Autobahn and his 'people's car', the Volkswagen, to Germany in the 1930s, the real explosion in car ownership in Europe did not come until after the Second World War. In the UK this was marked by the massive programme of motorway building in the 1960s and 70s. Although the growth in energy use for UK land-based transport flattened out in the late 1990s, that for international air transport, has continued to rise, encouraged by cheap air fares.

Services

If you live in a subsistence farming economy, as most people did until the 14th Century, your life style is very limited and most of the food and goods that you need are supplied locally. There is not much need for trade, distribution, or for that matter, money. By the time we reach the nineteenth century, England had become 'a nation of shopkeepers' as Napoleon put it. Farm produce had to be sold in cities and manufactured goods made in cities were traded world-wide. Distribution companies, banks and insurance companies became every bit as important as manufacturing industry, employing more and more clerks and increasing volumes of paper. In 1880 England, the 'Service sector' was limited but growing and it has continued to grow until the present day. Food and goods are things increasingly made by machines, while the sales, distribution and surrounding financial investment are activities done by people in offices and shops.

The Service sector includes almost all activities that aren't in the others. It includes all office activities in commercial offices and in public administration. It also includes education (and the writing of books such as this one). Educational expectations have increased enormously since the 1880s. The actual energy inputs to education are probably quite small – a recent study suggested that travel was the major energy cost of Open University courses, outweighing the energy costs of paper and printing (Roy *et al*, 2002).

'Services' also include newspapers and mass entertainment. In 1300 AD, there were no newspapers (Thomas Caxton only started printing books in the 1470s) and mass entertainment was limited to travelling story tellers. By 1880, Britain had high levels of literacy, a thriving newspaper and book publishing industry, all of which in turn depended on high volume paper manufacture, itself quite an energy-intensive industry. The development of radio and television in the 20th century has been a major spur to the spread of mains electricity from the 1930s onwards. It is not that the receivers actually require large amounts of energy; it is rather that the alternatives of battery power have always proved expensive and inconvenient.

Telecommunications is another important area in the services sector. By the 1880s the electric telegraph had been developed and criss-crossed Europe. There were even transatlantic cables. The telephone as we know it had just been invented. It is perhaps worth reflecting on how much useful communication the telephone and its modern descendent, the Internet, give for very little actual energy input.

Increasingly society is dependent on various electric and electronic devices for control and regulation. If the traffic lights fail in a modern city, then the result is grid-lock and chaos. But these are really only 'automatic traffic cops' substituting energy for a human presence. The early gas-lit version from the 1860s shown in Figure 3.9 still needed to be rotated manually. Whilst Victorian railways could physically propel 500 tonne trains at 100 kilometres per hour, they could only do so *safely* by the use of signalling, and this only became effective after the introduction of the electric telegraph in the 1850s. Similarly, modern airports can only function through the extensive use of radar and air traffic control. Given that in these examples so much can be achieved with such relatively small amounts of energy, it is perhaps amazing why we waste so much very crudely in other applications.

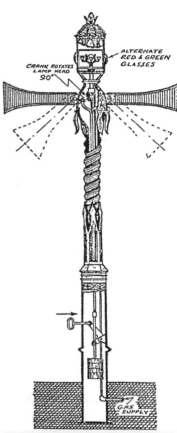

Figure 3.9 This early gas-lit traffic light, installed outside the Houses of Parliament, was described in the Times in 1868. The red and green lantern was rotated through 90° by a handle at the base, allowing it to control both traffic and pedestrians.

3.3 Energy uses today

The energy balance for the UK

By the end of the twentieth century, in the year 2000, total UK primary energy consumption had risen to almost 10 EJ, equivalent to almost a quarter of a billion tonnes of oil per year. The population had doubled from its 1880 figure to almost 60 million. Per capita annual primary energy use had risen to just over 165 GJ or almost 4 tonnes of oil equivalent. Whilst in the past the UK depended almost entirely on coal, 'diversity of supply' including oil, gas and nuclear power is now a key aim of energy policy.

The previous chapter has described the range of different primary energy sources available, but where is all this energy used? The picture is potentially quite confusing, since there are many transformations that take place within the energy system, notably in the refining of oil and the generation of electricity. Since physics tells us that energy is conserved, the total annual energy consumption of a nation, a factory or a household must be equal to its total annual energy supply. If we draw up a balance-sheet with energy income on one side and expenditure on the other, then the books must balance.

Table 3.2 shows a much simplified energy balance sheet for the UK for the year 2000. On the supply side, we have home production. We can then add in imports and subtract the exports. Next we also subtract the non-energy uses of oil and gas (for example as lubricants, or raw material for plastics, etc.) and fuel used by ships travelling overseas, which is traditionally not included in national consumption.

Since the aim is to find the amounts of energy used in this particular year, we must allow for any changes in stocks: subtracting fuel stored and adding fuel used from the stocks. The bottom line is then the actual consumption of each fuel. (Note that 'primary electricity production' refers to the output of UK nuclear, hydro and wind generators, and 'imported electricity' is that which comes from France through the link under the English Channel). Finally, adding all the contributions, we have a total national primary energy consumption of about 9700 PJ.

Table 3.3, now describes where all this energy was used. This, of course, is now **delivered energy** as received by the consumer. The categories of consumer are fairly straightforward. **Transport** covers both public and private, carrying both people and goods, in road vehicles, trains, and planes (internal flights only). **Industry** is obvious; but note that this item doesn't include energy used by the energy industries themselves. **Domestic** includes all the buildings that people live in; and **Services** is everything else: shops and offices, schools and colleges, museums, etc. Public lighting and energy used in agriculture have been included in this table as well, though fertilizer production comes under industry.

In Table 3.2 we talked about 'coal' and 'oil'. In the final-use stage of Table 3.3, the categories of energy need to be changed slightly: 'liquid' now refers to all the oil products (diesel, petrol, heating oil, aviation fuel, etc.); 'solid' includes coke, smokeless fuels and all other solid fuels as well as coal; and 'electricity' now means all electricity, from power stations of every type.

Having carefully tracked all the delivered energy purchased by everyone in each sector, we obtain a figure of just over 6700 PJ. This is a great deal less than the 9700 PJ on the supply side, so it appears that our balance-sheet doesn't balance. However, the two tables are not talking about the same thing. The difference between the *primary energy* in Table 3.2 and the *delivered energy* in Table 3.3, well over a quarter of the original energy, is the energy 'lost' by the energy industries in converting primary energy into the convenience forms of energy that we, the consumers, want to use.

The consumption table allows us to look at the final energy use in two ways. The 'total' column on the right of Table 3.3 shows the amount going to each of the four sectors. It reveals, for instance, that we use more energy in transporting ourselves and our possessions from place to place than we do in manufacturing, or just living at home.

BOX 3.3 **UK energy balance for 2000**

Table 3.2 UK primary energy production and consumption

Supply	Coal[1] /PJ	Oil /PJ	Natural gas/PJ	Primary electricity[2]/PJ	Total[3] /PJ
Production	920	5767	4486	844	12 017
Imports (+)	673	3135	94	47	3949
Exports (−)	−34	−5231	−527	−1	−5792
Non-Energy	0	−466	−47	0	−513
Marine Bunkers	0	−92	0	0	−92
To Stock (−)	138	34	−34	0	138
Consumption	1697	3147	3972	891	9707

[1] 'Coal' here includes other solid fuels such as Municipal Solid Waste

[2] Nuclear electricity is treated as the equivalent primary energy input to produce it at an efficiency of 30%. Actual electricity production is effectively about 30% of primary energy, but conventions elsewhere in this book use 33%.

[3] Some totals may differ slightly from the sums of the items due to rounding.

Source: DTI, 2001a

Table 3.3 UK delivered energy consumption

	Solid /PJ	Liquid /PJ	Natural gas/PJ	Delivered electricity[1]/PJ	Total[2] /PJ
Transport	0	2280	0	32	2311
Industry	117	267	722	409	1515
Domestic	91	136	1332	403	1961
Services, including public lighting and agriculture	16	101	456	341	915
All Sectors	223	2783	2511	1184	6702

[1] Here treated as the energy content of the delivered electricity (i.e. 3.6 PJ per TWh)

[2] Some totals may differ slightly from the sums of the items due to rounding.

Source: DTI, 2001a

Another way to divide up the delivered energy use is shown in the four entries along the bottom line, 'all sectors'. A striking feature is how little anybody wants in the form of solid fuels. This category would have looked completely different back in 1880, but now even industry only uses a tiny 8% of its energy in this form. The demand for liquid fuel for internal combustion engines and the preference for gas central heating in the domestic sector are obvious, as is the ever-growing demand for electricity.

Figure 3.10 shows the data of Tables 3.2 and 3.3 in a graphical form. In the top bar is the primary energy supply and below it the delivered energy consumption divided by fuel, by sector and by also end use.

The second bar of this chart clearly shows the magnitude of the energy losses in conversion and delivery. The minute amount of solid fuel shown in this bar indicates that the bulk of UK coal consumption goes for electricity generation.

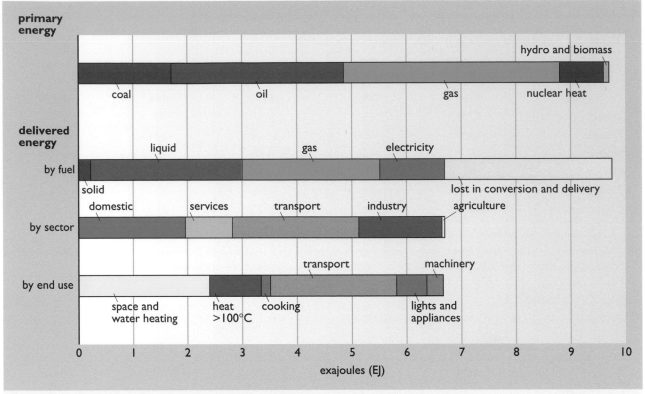

Figure 3.10 UK primary and delivered energy use for 2000 (sources DTI, 2001a; DTI, 2001b)

The third bar shows the breakdown between the different sectors. The energy use in the commercial sector is roughly equally divided between commercial offices in the private sector and public administration, government, schools, hospitals, etc. We will look at the transport energy use breakdown a little later on. Modern industry covers a whole range of activities. Their relative energy use is shown in Figure 3.11. Iron, steel and the manufacturing of metal products consumes over a quarter of the industrial energy use. The chemical industry, which includes the production of plastics, uses a further 21%.

The processing of food and drink, which might have been seen as a 'domestic' activity back in the 1300s is now a significant part of 'industry'.

Right at the end of the bar is a tiny section of 50 PJ labelled 'Agriculture'. This is only the energy required to run the UK's farms. The energy content of the fertilizers they use is probably about the same amount again, but its production is classified under 'Industry'.

In the final bar of Figure 3.10 we can see a breakdown in terms of end use. This kind of data can only be determined from surveys of sample groups of consumers. As such, the results can be subject to errors. Nevertheless it is useful information. It is, perhaps, striking that over a half of delivered energy is used as heat. For example, in the domestic sector, the bulk of the delivered energy is used for space heating (see Figure 3.12). The same is true in the Services sector. It is also remarkable that the 170 PJ we use for cooking food is far larger than the 100 PJ or so used on farms to grow it.

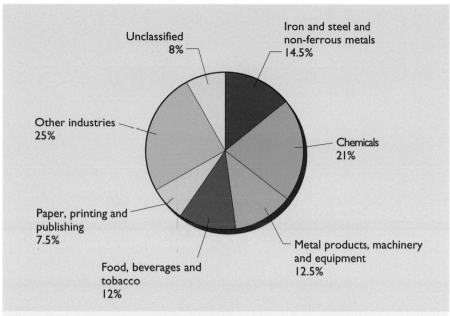

Figure 3.11 Breakdown of UK industrial energy use (source: DTI, 2001a)

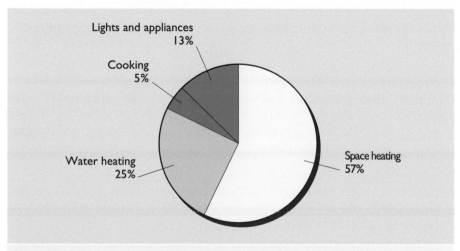

Figure 3.12 Breakdown of UK domestic delivered energy use – 1995 estimate (source: DTI, 1997)

Electricity conversion losses

Tables 3.2 and 3.3 tell us nothing about the reason for the difference between the total primary energy consumption and the total delivered energy consumption. For that we need more detailed statistics, showing all the intermediate conversion processes. There are many of these, each involving some 'lost' energy. For example, about 10% of the energy content of the petroleum in the 'primary energy' bar of Figure 3.10 is lost in the oil refinery by the time it has been converted to the 'delivered' liquid fuels in the second bar of the chart. Other losses are incurred in the conversion of coal to other solid fuels and pumping gas through pipelines. But the chief culprit is the generation of electricity. The 1.2 EJ of electricity we consume each year -

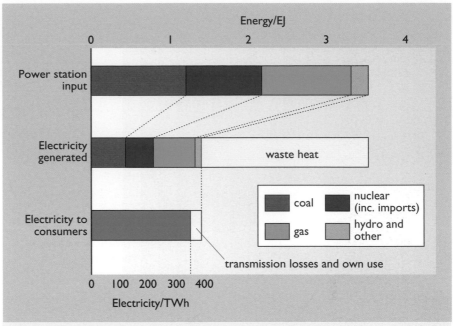

Figure 3.13 UK generation and distribution of electricity – 2000

5500 kWh per head of the population - accounts for about 2000 of the lost petajoules.

Figure 3.13 gives the details. Notice that it has two horizontal scales, one in terawatt-hours (10^9 kWh) and the other in exajoules for comparison with the tables above. The first bar shows the main fuels used to generate electricity:

- The total coal input to power-stations amounts to a little under 1200 PJ, or about 75% of the total annual coal consumption shown in Table 3.2
- Then there is the 850 PJ of nuclear heat input.
- Natural gas contributes 1160 PJ. This figure represents a 50-fold increase since 1990 - the so called 'dash for gas'. This is a subject that we will return to in later chapters.
- Other sources contribute about 180 PJ. These include oil, waste materials and also hydro and wind power.

We have then a total input to all the public supply power stations of about 3400 PJ. The second bar shows the actual electrical output, approximately 1350 PJ or 40% of the input. Nearly all the rest has become waste heat either dumped into the sea or into the sky via large cooling towers. A further 4% of the generated power is used by the power-stations themselves, and therefore doesn't enter the distribution network. Finally, around 8% of the energy leaving the power stations is lost on the way to the consumers in the electrical resistance of the cables and losses in transformers. Ultimately, we find the consumers receiving the 1184 PJ shown in Table 3.3. Just over a third of the input energy has become useful output, or to put it another way, the overall 'system efficiency' is 35%.

Couldn't the waste heat be put to use? Indeed it could. In countries such as Denmark, power stations don't just make electricity; their waste heat output

Figure 3.14 The UK electricity industry annually disposes of over 2 EJ of waste heat into the sky or the sea

is recycled into a massive network of insulated pipes carrying it into about half the buildings in the country. This large-scale distribution of heat is known as **district heating**. When the heat comes from power stations, rather than just large boilers, the process is known as **Combined Heat and Power Generation**, **CHP**, **or co-generation**. In 2000, the bulk of Danish electricity came from CHP units. The beneficial effect of this on the Danish national energy balance is described at the end of this chapter.

Trends in UK energy consumption

As we've seen above, in 2000 the UK used 6700 PJ of delivered energy. Distributed between the 60 million inhabitants, this works out at about 110 GJ per capita per year. At the beginning of the 20th century, in the year 1903, the UK used 167 million tonnes of coal and about 2 million tonnes of oil (Jevons, 1915). The vast bulk of this energy, probably over 85%, was delivered energy. The fledgling electricity industry was tiny and the bulk of 'conversion losses' were in the mining industry's own use of coal and in the provision of town gas. Divide this amongst the 37 million inhabitants (in 1903) and the figure for per capita annual delivered energy consumption is 110 GJ, identical to that nearly 100 years later.

There are key differences between 1903 and 2000. In 1903 the fuel (almost all coal) was delivered to the factory and home and used with the poor efficiencies of the time. The heating efficiency of a domestic coal fire is typically about 25% and we already have seen the poor efficiency figures for gas lighting. The fuel had to be burnt on site and the losses took place right there. In 2000, we were using the convenience fuel of electricity with a high efficiency at the point of use. The useful heat output of an electric fire is almost 100% of the electrical input. Modern incandescent and fluorescent lamps give far more useful lumens per watt than their gas counterparts. We get far more useful *energy service* out of the delivered energy. The price we pay is the enormous energy losses at power stations.

Although looked at on the long term time perspective, UK delivered energy on a per capita basis hasn't changed much in a century; there have been ups and downs and shifts between the different sectors. Figure 3.13 and Table 3.4 below chart the changes in the sectors since 1970.

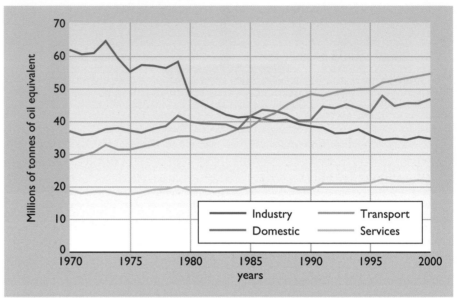

Figure 3.15 Sectoral changes in UK energy consumption 1970–2000 (source: DTI, 2001c). Remember 1 Mtoe = 42 PJ

Table 3.4 Trends in UK delivered energy

Sector	1970 delivered energy (PJ)	2000 delivered energy (PJ)	% change 1970–2000
Transport	1179	2311	+96%
Industry	2612	1515	−42%
Domestic	1544	1961	+27%
Service Sector	779	915	+18%
TOTAL	**6114**	**6702**	**+10%**

Sources: DTI, 1971; DTI, 2001a

Since 1970, total delivered energy consumption has been rising. Consumption in both the domestic and services sector has gone up. The increase in domestic heating energy consumption reflects the increasing standards of comfort in UK homes, spurred on by the availability of cheap North Sea Gas. In 1970 only 31% of homes had central heating. By 2000 this figure had risen to 89%. There has also been a continuing rise in domestic electricity consumption, feeding a whole host of new electrical appliances.

The rise in consumption in the services sector also reflects increased standards of heating. Lighting makes up a significant proportion of the electricity demand in this sector. There has also a continued increase in electricity consumption as offices are filled with computers and photocopiers. Indeed in many office buildings, especially those with large areas of glazing, the problem is that of dealing with the *surplus* of heat. Even in the UK, air conditioning is seen as essential by new office developers, much to the dismay of those interested in energy conservation.

The decline in industrial energy consumption reflects the 'dematerialization' of the economy, the shift away from heavy industry, such as steel making and car manufacturing, to lighter 'high value' industries such as electronics. The monetary value of industrial output has continued to rise, but the energy used in the process has been falling. This is part of a much longer trend (see below).

The most important change has been the large rise in transport energy use. Figure 1.50 in Chapter 1 shows the rise in UK travel since 1952 in terms of passenger-kilometres. What is most noticeable is the massive increase in travel by car. Even this chart understates the overall increase in transport use since it only includes *internal* UK air travel. Figure 3.16 shows UK transport energy consumption from 1970 onwards, including road freight and the energy used in refuelling international aircraft. It can be seen that the bulk of the energy consumption is now in road and air transport. Railway energy use is just a thin line on the bottom of the chart. However, it is worth reflecting that in 1903 the UK railways used 13 million tonnes of coal (about 9 Mtoe). The large decrease in railway energy use reflects both the contraction of the railway network since then, and the switch from steam to diesel and electric traction that took place in the 1960s (see Chapter 9).

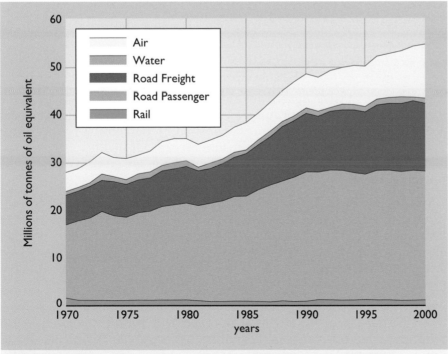

Figure 3.16 UK transport energy consumption 1970-2000. Source: DTI, 2001b

The growth in road transport energy consumption was greatest during the 1980s but has now slowed almost to a halt. However, air transport consumption continues to rise encouraged by air fares that are cheap in relation to earnings. In 1999 London's three airports alone dealt with 986 000 take-offs and landings.

International comparisons

Energy and GDP

The Gross Domestic Product (GDP) of a country is the monetary value of everything the country produces in a year. We would expect that energy consumption and GDP might be related and indeed it is generally true that the higher the GDP of a country, the higher its energy consumption (see Table 2.3 in Chapter 2). This has led energy forecasters in the past to assume that growth in GDP must automatically require more energy.

Table 3.5 shows the relation between energy consumption and GDP on a per capita basis for our five sample countries.

Table 3.5 Energy and GDP for five countries (1999 figures)

	UK	Denmark	US	France	India
Population (millions)	59.5	5.3	273	60.3	998
Annual per capita GDP:					
at current exchange rate[1]	\$21 100	\$37 700	\$31 500	\$28 200	\$450
at local purchasing power[2]	\$20 500	\$25 100	\$31 500	\$21 800	\$2173
Annual per capita energy consumption	162 GJ	159 GJ	348 GJ	177 GJ	21 GJ
Energy/GDP ratio[2]	8.0 MJ/\$	6.3 MJ/\$	11 MJ/\$	8.0 MJ/\$	9.0 MJ/\$

Source: IEA (2001)

[1] GDP converted to US dollars at the 1999 monetary exchange rate

[2] GDP converted at a rate which takes into account its local purchasing power.

The first row of GDP figures are obtained by converting the GDP in the local currency into dollars at the 1999 monetary exchange rate. The next row of the table shows the figures expressed in 'purchasing power' terms taking into account the differences in the cost of goods in different countries. To make the comparisons easier, all the quantities are given per capita, i.e. the totals are divided by the populations.

Let us look at Denmark and the US. We see that the goods, services, etc., produced per person in the US in 1999 were worth \$31 500 and those in Denmark \$37 700 at the current exchange rate. Yet the per capita primary energy consumption in Denmark was less than a half of that in the US. So it appears that Denmark managed to produce goods, etc., worth more per person but uses far less energy per person to do it.

A useful way to compare the 'energy efficiencies' of different countries is to divide the energy consumed by the GDP. The result is the **energy/GDP ratio**, the amount of energy used in producing each dollar's worth of national product expressed in MJ per dollar (MJ/\$). The final line in the table shows this for our five countries. There are clearly considerable differences in this figure from country to country. Why? The following are just a few of many possible answers:

■ They may obtain their primary energy in less wasteful forms than others (hydroelectric rather than fossil fuelled power-stations, for instance).

■ They may earn their GDP by less energy-intensive types of activity than others: agriculture rather than heavy industry, or commercial rather than industrial activity, and so on.

■ They may use less energy for the same purposes (greater industrial energy efficiency, better insulation of buildings, etc.)

■ They may use more energy than others to support economically non-productive activities (watching TV, going for a drive, reading a book).

The historical picture is also interesting. Figure 3.17 shows the energy intensity of different countries from 1880 onwards, expressed in constant 1972 US$, i.e. corrected for inflation (see Chapter 12, Costing Energy for details of this).

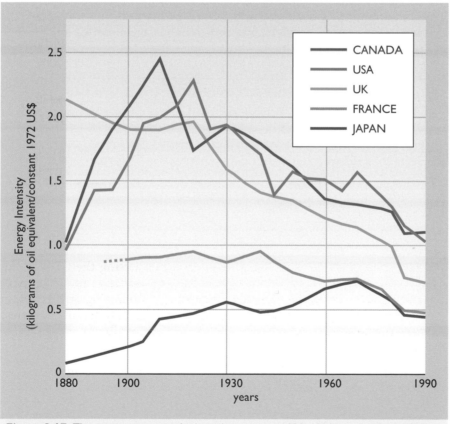

Figure 3.17 The energy intensity of selected economies 1880–1990 (source: Smil, 1994)

This diagram shows that when countries industrialize their Energy/GDP ratio rises to a peak and then declines. The UK was the country that industrialized first and its peak occurred around 1850. Canada's peak occurred in 1910 and that for the US occurred in 1920. As other countries such as Japan industrialized, they were able to do so in a more energy efficient manner by drawing on the past experience of the UK, US and Canada. Between 1880 and 1990 the UK's Energy/GDP ratio fell by a factor of three and is still continuing to fall. Put another way, modern UK citizens manage to produce goods and services of equivalent value with only one-third of the energy of their counterparts 120 years ago.

The clearest example that increased GDP does not necessarily require increased energy is in the case of Denmark. In the 22 years from 1977 to 1999, GDP increased by nearly 50%, yet primary energy consumption did not increase at all (Dal and Jensen, 2000).

More and more electricity

It is perhaps extraordinary that although mains electricity was first deployed in the 1880s we still don't seem to be able to get enough of it. Figure 3.16 below shows how per capita electricity consumption has risen dramatically over the last 40 years, spurred on by falling prices and a host of new electric gadgets and applications. As might be expected, US per capita electricity consumption is the highest in our sample of five countries. Although in the UK and Denmark the rise in consumption has flattened off in recent years, the trend in both the US and France is onwards and upwards. The US rise is particularly disturbing because the main generation fuel is coal, which has serious implications for CO_2 emissions.

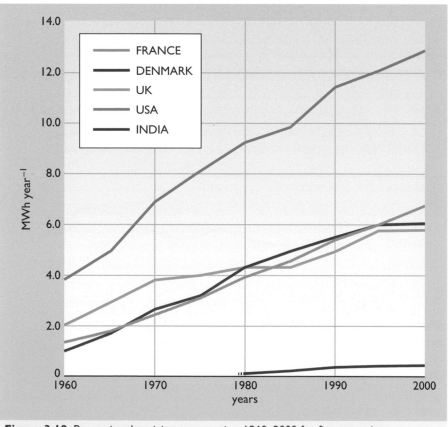

Figure 3.18 Per capita electricity consumption 1960–2000 for five countries

In France, since 1960, per capita consumption has risen by a factor of 5. The rise for the domestic and commercial sector of the economy is impressive. In 1960 it used 17.5 GWh; in 2000 it used 246 GWh – fourteen times as much! This, in part, has reflected the general improvement in housing conditions. A census of 1954 described the primitive state of French housing. Less than 60 percent of households had running water, only a quarter had an indoor toilet, and only one in ten had a bathroom and central heating. By 1990 all of these had become almost universal (Prost, 1991).

The Danish rise in electricity use is also remarkable, because since the mid 1970s it has been achieved without any increase in national primary energy consumption. A key ingredient of this success has been the increasing use of CHP, making sure that the waste heat from the power stations is put to good use.

Although electrification in India has also increased, it has barely kept pace with the booming population, so that the per capita consumption remains very low and has hardly increased over the past decade.

More and more travel

Transport energy use and car ownership are major growth areas world-wide. Car ownership in Los Angeles reached a level of one car to every three people in 1923 and has kept on rising. The rest of the world has been trying to catch up ever since.

Some sample statistics for recent years are given in Table 3.6.

Table 3.6 Recent transport statistics for five countries

	Transport energy growth, 1980–2000	Per capita transport energy use, 2000 (GJ y^{-1})	Cars per 1000 inhabitants, 1998
UK	+55%	39	404
France	+46%	38	456
Denmark	+35%	37	343
US	+35%	100	>500
India	Not quantified but probably very large	About 4	approx 4

In practice even the large US figure is likely to be an underestimate because of a rather tight definition of a 'car'. A large number of US road vehicles are not 'cars' but 'Sports Utility Vehicles' (SUVs for short). The number of 2-axle 4 wheeled vehicles per 1000 inhabitants is closer to 750, i.e. three to every four people. The problems of reducing pollution from car engines are discussed in Chapter 8.

Comparisons of delivered energy

Broadly speaking, the pattern of energy use in the UK is not significantly different from that in Denmark, France, or the US. However, there are some differences that are worth pointing out.

UK and Denmark

While the UK has had a history of relative fuel security, with plenty of coal, and, since the 1970s, an abundance of oil and gas, Denmark has been in a very different position. As pointed out in the previous chapter, in 1972, Denmark was almost 95% reliant on imported oil (see Figure 2.11). The world price of this rose by a factor of 7 between 1973 and 1980 (see Figure 2.5), making a large dent in the Danish national budget. Drastic measures had to be taken. Initially there was a lot of belt-tightening – at one point the driving of cars on Sundays was banned. Subsequently, a number of more medium- to long-term policy decisions were implemented:

- Switching power stations from oil to coal
- Expanding the use of combined heat and power generation (CHP)
- Phasing out individual oil-fired central heating units, many of which had poor thermal efficiencies by expanding the district heating networks.
- Implementing new insulation standards in Building Regulations.
- Expanding renewable energy supplies, especially biomass and wind power.

The results were quite impressive. The rapid growth in national primary energy consumption halted and consumption has remained almost stable at about 800 PJ per year ever since (see Figure 2.10). Between 1975 and 1985 domestic energy consumption for space heating was reduced by 50%. Since 1985 it has been held constant by continuing the expansion of the district heating networks and replacing even more oil-fired boilers with new gas-fired ones, using the country's new-found gas supplies (Dal and Jensen, 2000). By 2000 well over a half of Danish homes used district heating. The results of this attention to energy efficiency and renewable energy can be seen in a comparison of the breakdowns in delivered energy for the year 2000 between the UK and Denmark (see Figure 3.19).

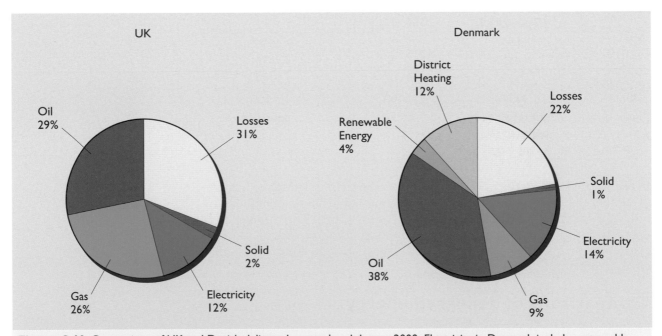

Figures 3.19 Comparison of UK and Danish delivered energy breakdowns, 2000. Electricity in Denmark includes renewable electricity

In Denmark district heating made up 12% of the delivered energy supply and the extensive use of CHP kept the national figure for losses down to 22% compared to the UK figure of 31%. Although Denmark is still heavily reliant on imported oil, its energy mix is very varied with a 4% contribution from renewable energy.

One other point is that Danish agriculture uses 7% of the country's delivered energy as opposed to 0.8% for the UK.

United States

US energy use is high – and is still increasing. Primary energy consumption increased by over 40% between 1970 and 2000. This growth was shared between all sectors of the economy. Energy use in the commercial sector doubled over this period. As in the UK, natural gas is widely used as a heating fuel, but as pointed out above, electricity use is a major growth area. About a half of US electricity comes from coal-fired power stations; coal consumption for electricity generation increased by almost a factor of three between 1970 and 2000.

Since, naturally, everything is bigger in the US, it is no surprise that the conversion losses in their electricity industry are 22 EJ, i.e. more than twice the total UK primary energy demand!

Transport energy use in the US is another major growth area. In 2000 it made up 36% of US delivered energy use and of this 97% was oil. The 35% increase in transport energy use between 1980 and 2000 amounted to 7.3 EJ per year, i.e. more than the whole of UK delivered energy use. This growth has contributed to the widening import gap between US oil consumption and production (shown in the last chapter in Figure 2.14) which stood at 23 EJ in 2000. At $25 a barrel, 23 EJ represents an import bill of approximately $100 billion per year.

France

French energy use has also been steadily increasing. Their primary energy consumption increased by 70% between 1970 and 2000. As in the US this growth has been shared across all sectors of the economy. The key factor has, of course, been their increasing use of nuclear power. In 2000 almost 80% of their electricity came from this source. As for delivered fuels, the mix is similar to the UK, though there is less use of natural gas and more use of oil, coal, wood and electricity.

India

India is a large country with a population of a billion people, of whom only about 25% live in cities or large towns. As pointed out in Chapter 2, about a half of the country's energy supply consists of biofuel in the form of wood, crop wastes, dung, etc. Since these materials are never 'traded', the precise amounts used don't enter the normal energy statistics. Although average Indian per capita energy and electricity use is very low, it is not uniformly so. The major cities are highly industrialized, with an increasing demand for electricity, cars and transport fuel.

The figures for the use of commercially traded fuels are straightforward enough and are shown in Table 3.7.

Table 3.7 India: percentages of commercially traded energy used by different sectors

Sector	Percentage	Breakdown
Industry	49%	73% coal, 14% petroleum, 2% gas, 11% electricity
Transport	22%	98.5% petroleum, 1.5% electricity for railways
Agriculture	5%	10% petroleum, 1% gas, 89% electricity for irrigation
Residential	10%	71% petroleum, 28% electricity, 1% gas
Others	14%	61% petroleum, 34% gas, 5% electricity

Source: TERI, 2002

Indian industry has much the same mix of iron and steel, chemical and manufacturing components as other countries. Coal being an indigenous fuel is widely used. The transport sector uses the same mix of petrol and diesel as other countries, and the railways consume a modest amount of electricity.

It is when we come to agriculture that the curious nature of the Indian energy economy shows itself. The bulk of agricultural electricity, indeed a third of all Indian electricity, is sold at subsidized rates for irrigation and water pumping. Indian agriculture uses nine times more energy as electricity than it does in petroleum for farm machinery. This can perhaps be explained by a 1991/2 survey which showed that the 1.3 million tractors on Indian farms were far outnumbered by the 84 million draught animals.

Although the 'residential' sector is shown as consuming 10% of the commercially traded energy, this is only a part of the story. There are really three groups of people. The richer city dwellers are likely to cook with LPG or kerosene stoves and have access to electric light. Poorer city dwellers are likely to use LPG or kerosene lamps for lighting and firewood for cooking. At the bottom of the league are the rural poor, who are likely to use firewood or cow dung for cooking and, if they are lucky, have LPG or kerosene lighting. According to the Tata Energy Research Institute (TERI, 2002), overall 40% of India's household energy needs are met by traditional biofuels and 90% of rural households depend primarily on them. The poor efficiencies of fuel-based lighting have been already discussed. The efficiencies of cooking stoves using firewood and dung are also very low, less than 10%, compared to figures of 50% or more using kerosene or LPG stoves.

Although there is a policy of national electrification, by 1991 only 76% of urban households and 31% of rural households had an electricity connection. This is about the level of connection in the UK in 1930.

3.4 Conclusions

In the few pages of this chapter we have covered over 10 000 years and spanned the economies of a number of different countries.

There are a number of basic trends. As we have risen from the culture of hunter-gatherer through to subsistence farmer and then to modern industrial society, energy sources have been tapped:

(a) to allow the basic tedious tasks of life to be carried out with less human labour,

(b) to allow human activities to carry on into the hours of darkness,

(c) to enable new products to be manufactured and distributed, and

(d) to allow new activities to take place, such as mass travel and communication.

The process of growing food, which even in the fourteenth century occupied the full-time labour of most of the population, is now carried out in Europe and the US by a relatively few people and machines, aided by artificial fertilizers. This has allowed the bulk of the population to concentrate on manufacturing and service activities, and to have time for 'leisure'.

The efficiency of basic activities, such as the heating of houses and cooking of food, has been improved by the development of better chimneys and stoves. New fuels such as town gas, oil, electricity and natural gas have been developed and deployed so that life at the point of use is much cleaner and more pleasant. Artificial lighting, which 150 years ago would have been considered very expensive, is now so cheap that it hardly merits thinking about.

The process of industrialization, especially in the UK, has involved the use of a massive amount of fossil fuels. As other countries industrialized they did so in a more energy efficient manner, leading to falling ratios of energy use per unit of GDP.

Over the last 40 years there has been a very large growth in electricity consumption. Electricity appears to be the clean, controllable, end-use fuel of choice, even though its generation may mean the production of considerable pollution at the power station (and out of sight of the consumer) and the disposal of large amounts of waste heat.

The current main growth area in energy use is in transport. Following in the footsteps of the US, there has been a large growth in car ownership and use in Europe, leading to pollution problems and congested cities. In the UK the rise in road transport energy consumption has levelled off since the 1990s, but the energy consumption of air transport continues to rise.

However, the majority of the world's population are still living in societies without the benefits of all these energy services. In India a large proportion of the urban and rural poor still live in conditions equivalent in energy terms to those in China in 100 BC.

Overall, looking back in time we would probably not wish to go back to an earlier era. Indeed there may be an underlying fear that if energy supplies were to run short we would automatically be propelled back to a 14th century lifestyle. However, when we look at Victorian society in the 1880s, we can see a culture furiously manufacturing products and inventing exciting new ones, but doing so by using large amounts of energy in what we would see now as an inefficient manner, and living in a pall of urban air pollution that we are no longer prepared to tolerate.

It is worth asking whether the future inhabitants of the year 2120 might look back to us here in the first decade of the 21st century with similar feelings.

References and data sources

Bowyer, J. (1973) *History of Building*, Crosby, Lockwood & Staples.

Burnett, John (1969) *A history of the cost of living*, Harmondsworth, Penguin.

Dal, P. and Jensen, H. S. (2000) *Energy Efficiency in Denmark*, Danish Energy Ministry.

Danish Energy Authority (2001) *Energy Statistics 1972–2000*. Available at http://www.ens.dk [accessed 04 June 2003].

Darmstadter, J. (1972) 'Energy Consumption: Trends and patterns' in Schurr, S. H. (ed.) *Energy, economic growth, and the environment*, Baltimore Md, John Hopkins University Press.

Department of Trade and Industry (1997) *Energy Consumption in the United Kingdom (Energy Paper 66)*, HMSO.

Department of Trade and Industry (2001a) *Digest of UK Energy Statistics (DUKES), 2000*, HMSO.

Department of Trade and Industry (2001b) *UK Energy Sector Indicators 2001*, HMSO.

Department of Trade and Industry (2001c) *UK Energy in Brief: July 2001*, HMSO

Diderot, D, and D'Alembert, J. L. (1769–72) *L'Encyclopedie ou Dictionnaire Raissonne des Sciences des Arts et des Metiers*. Avec approbation et privilege du Roy, Paris.

Energy Information Administration (2000) *Annual Energy Review 2000*, US Energy Information Administration. Available at http://www.eia.doe.gov/emeu/aer/overview.html [accessed 04 June 2003].

Eurostat (1990) *Energy 1960–1988*, Office des publications officielles des Communautés européeannes, Luxembourg, ISBN 92-826-1696-7.

International Energy Agency (2001) *Key World Energy Statistics from the IEA*, Paris, International Energy Agency.

Jevons, H. S. (1915) *The British Coal Trade*, David and Charles reprints (reprinted 1969).

Kandlikar, M. and Ramachandran, G. (2000) 'The Causes and Consequences of Particulate Air Pollution in Urban India', *Annual Review of Energy and Environment*, 25, 629–84.

McGowan, T. (1989) 'Energy-efficient lighting' in Johansson T. B. *et al* (eds) *Electricity: efficient end-use and new generation technologies, and their planning implications*, Lund, Lund University Press.

MINEFI (2001) *L'Energie en France – Repères*, Ministère de l'Economie, des Finances et de l'Industrie.

Odum, H. T. (1971) *Environment, Power and Society*, New York, Wiley-Interscience

Nordhaus, W. D. (1997) 'Do Real-Output and Real-Wage Measures Capture Reality? The History of Lighting Suggests Not' in Bresnahan, T. F. and Gordon, R. J. (eds) *The Economics of New Goods,* London, University of Chicago Press.

Plomer, W. (ed) (1977) *Kilvert's Diary*, Penguin Books.

Prost, A. (1991) 'Public and Private Spheres in France' in Prost, A. and Vincent, G. (eds) *A History of Private Life, Volume 5: Riddles of Identity in Modern Times,* Cambridge, Mass., Harvard University Press, pp. 1–143.

Roy, R., Potter, S., Smith, M. and Yarrow, K. (2002) 'Towards Sustainable Higher Education: environmental impacts of conventional campus, print-based and electronic distance/open learning systems', *Report DIG-07*, Design Innovation Group, Milton Keynes, The Open University.

Tata Energy Research Institute (TERI) (2002), *Teri Energy Data Directory and Yearbook, 2001/2002,* New Delhi, TERI.

Weightman, G. (2001) *The Frozen Water Trade*, HarperCollins.

Further reading

Smil, V. (1994) *Energy in World History*, Westview Press, ISBN 0-8133-1901

Department of Trade and Industry (1997) *Energy Consumption in the United Kingdom (Energy Paper 66),* Government Statistical Service.

Chapter 4

Forms of Energy

by Janet Ramage

4.1 Introduction

Look up *energy* in a dictionary and you will usually find at least two definitions. There are the common everyday uses of the word: vigorous activity, strenuous exertion, etc. Then there is the scientific definition, typically telling us that energy is *the capacity to do work*. This may be technically correct (under certain conditions) but probably means little to most readers. It certainly fails to convey two properties that make energy such an important feature of the real world beyond the physics textbook.

■ Energy can take many different forms.

■ Energy can be quantified.

A third feature, without which the concept of energy would be of little interest, is expressed in the **Law of Conservation of Energy**, which may be stated as follows.

■ In any change from one form of energy to other forms, the total quantity of energy remains constant.

The energy input to a power station, for example, may come from a fuel (fossil, bio- or nuclear), from moving water or air, from the Sun or even from the centre of the Earth. Whatever the source, we can in principle measure the input during a period of time – an hour, say, or a year – and follow the fate of exactly this quantity of energy as it is transformed through all the processes of the power plant.

The development of this concept of energy as one entity – one actor who can put on many faces – coincides almost exactly with the nineteenth century, from the first use of the scientific term *energy* in 1807 to Einstein's identification of energy with mass in 1905. Its history therefore provides a suitable framework for this account. Short case studies illustrate present-day uses of the different forms of energy and also serve as brief introductions to some of the systems studied in more detail in later chapters and in the companion volume, *Renewable Energy* (Boyle, 1996).

4.2 Kinetic and potential energy

> The grand modern ideas of Potential and Kinetic Energy cannot be too soon presented to the student.
>
> Peter Guthrie Tait (1831–1901), mathematician and physicist, professor of natural philosophy at Edinburgh (Tait, 1870)

The scientific concept of energy arose from questions about the nature of motion. What is special about a moving object? What keeps it moving? In the seventeenth century, the great period that saw the birth of modern science, people such as Galileo, Huygens and Newton developed the idea that moving objects must possess something that stationary objects lack. This 'something' appeared under various different names, such as *vis viva* ('living force'), until in 1807 Thomas Young proposed the term **energy**.

The name may have been changing but the concept itself was gradually becoming clearer. The first step was Isaac Newton's proposal that the natural behaviour of a moving object is to continue in its current state of motion (**Newton's First Law of Motion**). We shouldn't, it seems, be asking what

Figure 4.1 Thomas Young (1773–1829) spent most of his life in London although born in Somerset. A man of wide-ranging interests, he spoke most European and many oriental languages and made the first steps towards deciphering the hieroglyphics on the Rosetta Stone. He trained as a doctor, and an interest in eyes led him to explanations of accommodation and astigmatism and a theory of colour vision. In 1801 he became professor of physics at the Royal Institution. Most famously, he demonstrated the effects produced when light passes through narrow apertures, proving that it must be a type of wave.

Figure 4.2 Isaac Newton (1642–1727) was born in Lincolnshire and grew up on a farm, a bright child who delighted in making models. He is generally held to be the greatest scientific genius ever. His achievements make impressive reading: he developed the three laws of motion and discovered the law of gravitation, showed that white light consists of many colours, invented a new telescope – which he made, grinding the mirror from an alloy he also invented, discovered the binomial theorem and invented the calculus, reformed the national coinage and was a Member of Parliament (twice).

keeps something moving, but rather what stops it – or more generally, what changes its motion. Newton again offered an answer: the motions of objects are changed by *forces* (**Newton's Second Law of Motion**). It is the downward force of gravity that causes objects to accelerate towards the Earth and, as Galileo had shown, everything dropped from a given height will reach the ground in the same time and with the same final speed (in the absence of air resistance).

These falling objects must be gaining the 'something' that came to be called energy. Where does it come from? Notice the significance of this question. Once people ask where the energy comes from, they are beginning to treat it as something real, not just some ghostly entity that might appear or disappear into nothingness. They are starting to think of energy as *a quantity that is conserved*.

Early explanations of the energy gain suggested that the downward force, the gravitational pull of the Earth, performs *work* on the object, and this work in some way provides the energy associated with the motion. This was a useful concept, but the real revolution in ideas came in the 1850s. It started with the opposite question. An object thrown upwards slows down and eventually stops, losing its 'motion' energy. Where does this energy go? In 1853, William Rankine, observing that the object is gaining the potential to move, suggested that, '*by the occurrence of such changes, actual energy disappears and is replaced by Potential or Latent Energy.*' (Rankine, 1881) Rankine's term was soon adopted, and within a few years **potential energy** came to mean not just a potentiality, but a new *form* of energy. The energy of a moving body, now called **kinetic energy**, was another form, and there might be still more.

Figure 4.3 William Rankine (1820–1872) was a Scottish engineer. Born and educated in Edinburgh, he became Professor of Civil Engineering at Glasgow at the age of 35. He was one of the founders, with Carnot and Joule, of the subject of thermodynamics – the study of energy in all its forms. His textbooks on thermodynamics and applied mechanics and his *Manual of the Steam Engine*, widely read and reprinted many times, were major influences in the engineering developments of the nineteenth century.

To complete the picture it was necessary to know how to quantify these two forms of energy. How do they depend on the mass, speed or position of an object? The reasoning starts with a return to the idea of the *work done* by a force. By equating the change in potential energy of an object to the work done in lifting it, Rankine related potential energy to mass and height. The requirement that total energy must be conserved as the object falls then leads to a formula for kinetic energy in terms of mass and speed (Box 4.1).

BOX 4.1 **Work, potential energy and kinetic energy**

Newton's explanation of the behaviour of falling apples, the moon and the planets rests on his brilliant separation and clarification of the laws of mechanics and gravitation. Expressed in modern terms, two essential features are that:

- the force (F) needed to produce a particular acceleration of any object is proportional to the acceleration (a) and to the mass of the object (m): $F = m \times a$,
- the gravitational force pulling an object towards the Earth (technically its **weight**, W) is also proportional to its mass.

Note that the *unit* for force must be the unit for mass times acceleration, which is **kg m s^{-2}**. However, as with the joule for energy, the unit for force has been given a specific name: the **newton (N)**.

From Galileo's observations it follows that all objects in free fall on Earth must have the same downward acceleration. This is called the **acceleration due to gravity** (g). The value of g is determined by the gravitational pull of the Earth, and varies slightly from place to place, the average at sea level being 9.81 m s^{-2} (metres per second per second). The approximation $g = 10$ m s^{-2} is often used when greater precision is not required.

The downward force causing this acceleration is the weight of the object, so it follows that:

$$W = m \times g$$

The weight of an object, in newtons, is thus about 10 times its mass in kilograms. (Or we can reverse the statement and say that an object whose weight is 1 N has a mass of about 0.1 kg – a nice large apple.)

Work

Rankine proposed that the **work done** by a force as it moves something is equal to *the force multiplied by the distance moved in the direction of the force*, i.e.:

work = force × distance.

Potential energy

Following Rankine, the change in potential energy on raising an object is to be equated to the work done by the force raising it.

To raise an object at a steady speed, you need an upward force that is just equal to its weight, so the work done in raising the object against gravity through a height H is:

work = force × distance = $W \times H = m \times g \times H$

This final quantity, usually written mgH, is therefore the increase in potential energy.

It should be noted that whilst the term 'potential energy' is still used informally for this case, the

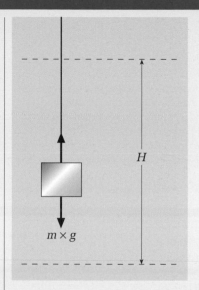

Figure 4.4
Raising a mass

quantity associated with height differences should strictly be called the **gravitational potential energy**, to distinguish it from other types of potential energy that are now recognised. (See for instance Section 4.3.)

Kinetic energy

If energy is conserved, the potential energy lost by a *freely falling* object must appear as a gain in its kinetic energy, so an object that falls from rest (zero kinetic energy) through a height H will acquire kinetic energy equal to $m \times g \times H$.

How is this final kinetic energy related to the final speed of the object, v?

Consider three facts:

- If the time taken to fall is t, and the acceleration is g, the final speed is $g \times t$.
- The *average* speed during the fall is equal to the total distance travelled divided by the time taken: $H \div t$.
- With steady acceleration, the average speed is just half the final speed: $\frac{1}{2}v$.

It follows from these that $g \times H = \frac{1}{2}v^2$. (Note that this relationship is a logical consequence of the definitions of speed and acceleration: a piece of mathematics, not a law of physics.)

Finally, therefore, the kinetic energy of the moving object becomes $\frac{1}{2}mv^2$.

Summary

The change in potential energy of a mass m displaced through a vertical distance H is mgH.

The kinetic energy of a mass m moving with speed v is $\frac{1}{2}mv^2$.

BOX 4.2 Using kinetic energy: a wind turbine

How much power does the wind deliver to a turbine with a diameter of 20 metres when the wind speed is 12 metres per second (27 miles per hour)?

Air has mass, so the wind – air in motion – carries kinetic energy, and a modern wind turbine is designed to transform this into electrical energy. In siting a turbine or a wind farm it is obviously important to know how much wind energy is available, and the aim of this case study is to show in principle how this may be calculated. The variability of the wind must of course be taken into account in practice, but the calculation is simplified here by considering just one wind speed.

First we must clarify the question. The wind delivers energy continuously, so we need to ask how much energy arrives in a given time: not the number of joules, but the number of joules per second. This is technically the **power**. As we saw in Chapter 2 (Box 2.1), power is the *rate* at which energy is delivered or transformed, and its unit is the **watt** (W), one watt being a rate of one joule per second.

To find the rate at which energy is delivered by the wind, we need to know the mass of air reaching the turbine per second. The reasoning is as follows.

■ If the wind speed (v) is 12 metres per second, all the air in a cylinder with a length of 12 metres will pass through the swept area (A) in each second (Figure 4.5).

■ The diameter is 20 metres, so A is the area of a circle with radius 10 metres. This is $\pi \times 10^2$ square metres.

■ The volume (V) of the cylinder is therefore

$V = v \times A = 12 \times \pi \times 10^2 = 3770$ cubic metres.

■ To find the *mass* of air, we need to know the **density** of air: the mass per cubic metre. At normal air pressure this is 1.29 kg m^{-3}, so the mass of air arriving per second is:

$m = 1.29 \times 3770 = 4863$ kg.

Figure 4.5
A wind turbine

■ Knowing that the kinetic energy of a mass m moving with speed v is $\frac{1}{2}mv^2$ (Box 4.1), we find that the *energy* arriving per second is

$$\frac{1}{2} \times m \times v^2 = \frac{1}{2} \times 4863 \times 12^2$$
$$= 350\ 000 \text{ joules per second}$$
$$= 350\ 000 \text{ watts}$$

and this is the **input power: 350 kW**.

Using the same reasoning, it is easy to show that halving the wind speed would reduce the input power to a mere 44 kW, an eighth of the above, whilst doubling it would lead to eight times the input. The power, in other words, is proportional to the *cube* of the wind speed. This strong dependence on wind speed is obviously very important, both in assessing a site and in considering the effect of variability of winds.

Of course, not all the power delivered by the wind can be extracted. The moving air behind the turbine will carry away some kinetic energy, and in any machine there are energy 'losses' in the form of heat produced by friction. Even under ideal wind conditions the electrical power output of a wind turbine is likely to be less than half the wind power input. For the 20-metre turbine in a 12 metres per second wind, this would mean an output power of 100–150 kW.

A more detailed account of wind power appears in the companion volume, *Renewable Energy*.

BOX 4.3 **Using potential energy: pumped storage**

The upper reservoir of a pumped storage system has a surface area of 8 square kilometres, and lies 370 metres (about 1200 ft) above the lower reservoir. How much energy is stored if overnight pumping raises the water level in the top reservoir by one metre?

A pumped storage system stores the surplus overnight output of a power station by pumping water into a high reservoir, holding it there until demand rises again, and then releasing it to generate electric power. The purpose of this case study is to show how the energy stored by such a system is related to the area and location of the upper reservoir.

We need to find how much water is pumped up.

The surface area of the reservoir, 8 square kilometres, is 8 million square metres, and the volume of water needed to raise this by 1 metre is 8 million cubic metres: 8×10^6 m^3.

■ The density of water is 1000 kilograms per cubic metre: 1×10^3 kg m^{-3}.

■ So the stored mass is:

volume × density = $8 \times 10^6 \times 1 \times 10^3 = 8 \times 10^9$ kg, which is 8 billion kilograms.

The *potential energy* gained when a mass m is raised through a height H is mgH (Box 4.1).

Using the approximation $g = 10$ m s^{-2}, the gain in potential energy is

$8 \times 10^9 \times 10 \times 370 = 30 \times 10^{12}$ J = 30 TJ (30 million million joules).

This is enough to provide just over half a million households with electricity for a day.

More details of pumped storage appear in *Renewable Energy*.

Figure 4.6 The upper reservoir of the Cruachan pumped storage plant, 1200 ft above Loch Awe, in Scotland

Figure 4.7 Francis Bacon (1561–1626), born in the Strand, London, educated at Cambridge and as a barrister, led an adventurous life seeking advancement in the courts of Elizabeth I and James I. He became Lord Chancellor but was found guilty of taking bribes. He published a confession (claiming that accepting bribes hadn't changed his decisions), was fined and sent to the Tower. King James paid the fine and Bacon was released after a few days. He wrote widely on many topics, but is best known for the *Advancement of Learning* and the *Novum Organum*, in which he argued that science must proceed not by contemplating the works of the Greek philosophers but by observation of the natural world.

4.3 Heat

Writing nearly 400 years ago, Francis Bacon thought heat to be a 'brisk agitation' of the particles of matter:

> It must not be thought that heat generates motion or motion heat (although in some respects this is true), but the very essence of heat, the substantial self of heat, is motion, and nothing else.
>
> (Bacon, 1620)

Beautifully expressed, and very much like our present view. But unfortunately Bacon was far ahead of his time. Quite different theories, regarding heat as some kind of fluid, arose during the next 200 years, and dominated well into the nineteenth century. The final acceptance of Bacon's view came only at the end of the century, with the success of the **atomic theory of matter**. The atomic theory held that everything consisted of arrangements of indivisible fundamental particles, called atoms – and that there was only a limited number of types of atom. (The name came from the Greek $\alpha\tau\omu o\sigma$, meaning uncuttable.)

As this picture of matter became established, the **kinetic theory of heat** developed rapidly. A simple gas such as helium was seen to consist of a very large number of identical atoms moving constantly in random directions. When the average kinetic energy of the atoms was found to depend only on the temperature, it was a relatively short step to identifying heat itself as kinetic energy. For more complex materials, such as molecules or metals, the energy associated with the forces between atoms must also be taken into account, but it remains the case that heat energy is not some new fundamental form. The 'heat content' of a simple gas is no more than the kinetic energy of its atoms, and an increase in temperature is an increase in the average speed of these atoms: *hotter means faster*.

4.4 Electrical energy

James Joule was a man with a passion for science. Working alone in his private laboratory, he carried out a series of beautifully precise experiments in the 1840s. In the best known of these he used a falling weight to drive a paddle wheel whose rotation heated a fluid, establishing that the heat produced was proportional to the work done by gravity in pulling down the falling weight. Another version was even more ingenious: the falling weight was used to drive a small generator that effectively ran a simple immersion heater. The relationship between work and heat was the same in both experiments – and the same again in others. By the end of the 1840s Joule was convinced that the heat output was not merely related to the work input, but that work was converted into heat. Unfortunately, his results and views were largely ignored for about ten years.

The second of Joule's experiments is extremely interesting because, rather than direct conversion from work to heat, there is an intermediate stage where *electrical energy* must be playing a role.

Electrical effects were already familiar, of course, in the nineteenth century. It had been long known that there were two types of electricity: rubbing

BOX 4.4 **Using heat: storage heaters**

How large a volume of water, heated from 20 °C to 80 °C, would be needed to store the surplus power station output of Box 4.3 in the form of heat?

The idea of storing surplus energy in the form of heat is probably very old. (Did hunter-gatherers use rocks heated by the sun or on a fire to keep themselves warm in their caves?) The domestic night-store heater was introduced in the 1950s as a means of 'storing' cheaper overnight electricity to provide heating during the day. And more recently, growing interest in solar heating has generated many ideas for storing heat in the masonry of buildings.

This case study, on a rather larger scale, should perhaps be regarded more as a comparison of the energy storage capacities associated with different forms of energy than as a practicable scheme.

We start with a definition and some data on water.

- The heat energy needed to raise the temperature of 1 kg of any substance by 1 °C is called its **specific heat capacity**.

- The specific heat capacity of water is 4200 J kg^{-1} K^{-1} (joules per kilogram per kelvin[*]).

- The density of water is 1000 kg m^{-3} (kilograms per cubic metre).

Raising the temperature of one cubic metre of water from 20 °C to 80 °C therefore requires

$$1000 \times 4200 \times 60 = 252 \text{ MJ (million joules).}$$

The total heat energy stored by the reservoir of Box 4.3 is 30 million million joules, so the volume of water to be heated is 30 million million divided by 252 million, which is about 120 000 cubic metres.

The pumped storage system of Box 4.3 needed 8 million cubic metres. Comparing the methods, we see that the energy stored by heating each cubic metre of water through 60 °C is about *seventy* times the energy stored by raising the same cubic metre through 1200 ft.

Electrical heating is simple, and if the energy stored is so much greater, why is this method not used instead of pumped storage? The answer comes when we ask how the energy is to be retrieved, and lies with the Second Law of Thermodynamics, to be discussed in Chapter 6.

[*] The accepted scientific unit for temperature is the kelvin (K), but an increase of 1 K is exactly the same as an increase of 1 °C, so we can use this more familiar unit in the calculations. The kelvin scale of temperature is introduced in Chapter 6.

---

cat's fur with an amber rod leaves the rod with one type and the cat with the other. Both rod and cat have acquired **electric charge** but of opposite types, which came to be called positive and negative. Delicate experiments showed that objects with the same type of charge repelled each other whilst those with opposite charges attracted. The parallel with the gravitational force between two masses was recognised, and led to the idea of **electrical potential energy** arising from forces between two or more electric charges.

Electrical circuits, involving moving rather than static electric charges, were also investigated. Alessandro Volta invented the battery – the 'voltaic cell' – in 1799, and about twenty years later, Georg Ohm began to create order from rather diffuse ideas of what we now call **current** and **potential difference**. An electric current, he said, is a flow like the flow of heat, and potential difference is like the temperature difference that causes the flow. The analogy is good, but further clarification had to wait for the next advance in theories of matter: the demise of the indivisible atom.

Figure 4.8 James Joule (1818–1889) was born in Salford, where his father owned a brewery. Except for a few lessons from the elderly John Dalton, he was almost entirely self-educated in science and remained an amateur, outside the scientific establishment. At an Oxford meeting in 1847, the chairman, regarding his work as unimportant, asked him to cut short his paper and invited no discussion. However, by Joule's account, a young man rose and 'created a lively interest' by his comments. The young man was William Thomson (Lord Kelvin), whose own account tells of 'a very unassuming young man, who betrayed no consciousness that he had a great idea to unfold. I was tremendously struck with the paper …'

Figure 4.9 Alessandro Volta (1745–1827) was born in Como where he taught physics before moving to Pavia as Professor of Natural Philosophy. He became interested in the experiments of his compatriot Galvani, who had shown that the muscles of a frog leg contracted when a brass hook attached to them touched the iron plate on which the leg lay. Galvani attributed this to 'animal electricity', but Volta realised that the frog was unnecessary: two different metals in fluid contact were sufficient. Using strips of zinc and copper dipping into brine, he constructed the first continuous source of electric current.

Electrons

The death knell of the 'uncuttable' atom came with the discovery of the **electron**. Between about 1870 and 1890, experiments with a forerunner of the TV tube revealed that a red-hot wire emitted particles with very interesting properties.

- They were very light, with about one two-thousandth of the mass of the lightest atom.
- They were all completely identical: the same mass and the same electric charge.
- Their electric charge was of the type called negative.

The discovery of these very light, electrically charged particles which must come from *inside* the atoms dramatically changed the picture of matter. It also offered a new model for the electric current in a metal. It could be seen that metals are good conductors of electricity because they have **free electrons**. Each atom of the metal releases one or two electrons, becoming an **ion**, and the picture of a piece of metal is now of a lattice of these ions through which the free electrons are constantly moving, in random directions at extremely high speeds.

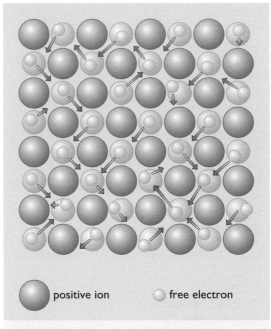

positive ion   free electron

Figure 4.10 Ions and electrons in a metal

When a length of wire is connected to an electrical supply, the drift of this entire swarm along the wire is the electric current whose effects we observe. (See also Box 4.5.)

Ohm famously determined that the current in a wire is proportional to the potential difference across it (**Ohm's Law**), and Joule almost equally famously discovered that the rate at which heat is developed by a flowing current is proportional to the square of the current (the phenomenon known as **Joule heating**). Together, these results lead to useful expressions for electrical energy and power (Box 4.6).

Figure 4.12 Georg Simon Ohm (1787–1854) was the son of a Bavarian locksmith, who struggled hard to pay for his education. He eventually became a teacher in Cologne, where he carried out his electrical experiments, publishing his conclusions as a book in 1827. This was very badly received ('a web of naked fantasy'), and when a minor official persuaded the Minister that the author was 'unworthy to teach science', Ohm resigned. Fortunately, scientists elsewhere recognised the value of his work, and his reputation was soon re-established. In 1852 he became professor of experimental physics at the University of Munich.

BOX 4.5 **A power supply**

It is important to appreciate that an 'electrical supply' does not *supply* electrons; it recycles them.

Consider a simple circuit such as a battery and torch bulb (Figure 4.11). The free electrons, as they move along the wire filament of the bulb, are constantly colliding with the ions of the metal. In these collisions the ions gain energy, which we observe as heating of the metal. The electrons, gradually losing energy as they pass along the wire, eventually return to the battery, which replaces the lost energy. As long as the circuit is complete, the battery supplies energy continuously, and this is the real function of any electrical supply. It is, correctly, a **power supply**.

Figure 4.11 A simple circuit with battery and torch bulb

BOX 4.6 **Current, voltage, resistance, power and energy**

Current

Electric current could in principle be measured by counting the number of electrons per second passing any point, but even a modest current would involve numbers in the trillions. The normal unit for electric current is the more practical ampere, or **amp (A)**, with its sub-units milliamp (**mA**) and microamp (μA).

Voltage

The quantity commonly referred to as the **voltage** of a supply is the difference in *electrical potential* that it maintains between its terminals: a measure of the energy supplied to electric charges as they pass through. The unit, the **volt**, is defined so that if a current of one amp flows for one second from a one-volt supply, the energy supplied is exactly one joule.

The potential difference between the two ends of a wire such as the bulb filament in Figure 4.11 is also commonly referred to as *the voltage across the wire*. In the circuit shown this is equal to the supply voltage. If two or more such components are connected in a simple loop with a supply, the *sum* of the individual component voltages must be equal to the supply voltage.

Resistance

If two quantities are proportional to each other, the result of dividing one by the other is always the same. Ohm's Law therefore allows us to define the resistance (*R*) of a component as the voltage (*V*) across it divided by the current (*I*) flowing through it:

$$R = V \div I$$

The resistance of a particular component, defined in this way, may not in practice have a constant value. Increasing the current in a wire, for instance, may cause its temperature to rise, and this usually increases its resistance. But even if this occurs, it remains useful to define resistance as voltage divided by current, and indeed, many scientists, if asked to 'state Ohm's Law', will offer either the above definition or a well-known rearrangement of it:

$$V = I \times R$$

The unit for resistance is the **ohm**, with the symbol Ω (Greek letter omega).

Electric power

Power is the rate at which energy is transformed. So Joule's rule can be reformulated: the *power dissipation* in a component is proportional to the square of the current (I^2). This rule only holds, however, if Ohm's Law also holds. The more fundamental, general rule is that the power is proportional to *the product of the voltage and the current*.

It follows from the above definition of a volt that the power in watts is simply *equal* to the voltage in volts times the current in amps:

$$P = V \times I$$

This relationship holds equally for the electric power provided by a supply or the power being dissipated as heat in a component such as the lamp filament.

BOX 4.7   Storing electrical energy: batteries

A 12-volt car battery has a capacity of 100 ampere-hours. How much energy does it store when fully charged?

The most commonly used systems for storing electrical energy are rechargeable batteries. Strictly, the electrical energy is converted into chemical energy as a battery is charged, with the reverse process taking place when it is used to deliver electric power. Nevertheless, it seems appropriate to consider this method of storage in the context of electrical rather than chemical energy.

Batteries, or sets of batteries, offer a wide range of energy storage capacities and have the advantage of being portable. This case study considers the potential of the common lead-acid car battery in two very different contexts.

The storage capacity of a car battery is usually expressed in ampere-hours. A battery with a capacity of 100 ampere-hours could supply a current of 2 amps for 50 hours, half an amp for 200 hours, and so on.

We take the first case. If the current is 2 A and the voltage is 12 V, the power output (see Box 4.6) is

$$P = V \times I = 12 \times 2 = 24 \text{ W}.$$

One watt is one joule per second, so the energy supplied in 50 *hours* is

$$24 \times 50 \times 60 \times 60 = 4\,320\,000 \text{ J} = 4.32 \text{ MJ}$$

and this is the energy stored when the battery is fully charged.

Comparison with the 1350 MJ stored by the contents of a full 40-litre (9-gallon) petrol tank highlights the major problem for electric cars. The achievement of a

reasonable range of travel requires the development of rechargeable batteries with much greater storage capacity per kilogram or per litre of occupied space.

It is also worth noting the impracticability of battery storage for really large-scale systems. To store the overnight power-station output of Boxes 4.3 and 4.4 would require the equivalent of nearly seven million car batteries.

Figure 4.13 Recharging the battery of an electric car

4.5   Electromagnetic radiation

In the early nineteenth century, long before the discovery of the electron, experiments began to show that there was an extremely close relationship between *electricity* and *magnetism* – two effects that had previously been regarded as quite independent. In 1820 Hans Oersted showed that a compass needle was deflected when brought near an electric current. In the same year André Ampère demonstrated that there is a *magnetic* force between two parallel wires carrying electric currents, and this is how the unit of current, the ampere, is defined. In 1821 Michael Faraday showed that a wire carrying a current would rotate around a magnet – the first electric motor; and in 1832 he found that moving a bar magnet rapidly through a coil of wire caused a current to flow briefly in the coil – the first generator (see Chapter 9). In 1864 James Clerk Maxwell published his theory of electromagnetism, drawing together all these results as a set of mathematical equations. Its consequences were revolutionary.

Figure 4.14 Michael Faraday (1791–1867) was born in London. His father, a blacksmith, was often ill, and the family lived in poverty. At thirteen he became a bookbinder's errand-boy, and later his apprentice. Reading the books that he bound, he became fascinated by science, and saved a shilling (5p) a week to attend an evening course. When a customer gave him tickets for four lectures by Sir Humphrey Davy, he was delighted, and made careful notes afterwards. Determined to become a scientist, he wrote to the President of the Royal Society, but received no reply. Increasingly desperate, he eventually appealed to Davy, sending his written-up notes as support. Davy was impressed, and Faraday became laboratory assistant at the Royal Institution. Here he stayed, eventually succeeding Davy as Director. In addition to his wide-ranging research activities, he was an accomplished lecturer, founding the popular Friday evening discourses which still continue.

Figure 4.15 James Clerk Maxwell (1831–79), born into a prominent Scottish family, was initially educated at home, but was sent to school after his mother died. It was not a success. He was shy, and his clothes, designed by his father, were odd. The boys called him 'dafty'. Then, to general amazement, he starting winning prizes for mathematics and poetry, and at fourteen his first mathematical paper was published. At Edinburgh and Cambridge he continued to win prizes and write poetry. After a period in Aberdeen, in 1860 he became professor at Kings College London, where his most important ideas were developed. He resigned in 1865, partly through ill-health, and spent some years in Scotland writing his great book on electromagnetism. In 1871 he was persuaded to accept the first chair in experimental physics at Cambridge, and devoted the remainder of his life, until his early death from cancer, to establishing the Cavendish Laboratory.

Maxwell's laws of electromagnetism are arguably as great an achievement as Newton's laws of motion. (Ludwig Boltzmann, another great scientist, was prompted to quote Goethe, '*Was it God who wrote these lines …*') Maxwell adopted the idea of **fields** of force, spreading throughout space. An electric field surrounds any electric charge, and another charge experiences a force when it is placed in this field. A magnetic field surrounds a magnet or an electric current, and another magnet or current experiences a force when placed in this field. Maxwell's laws incorporate Faraday's discovery that a voltage (which implies an electric field) can be produced by a changing magnetic field, and they introduce the converse effect, that a magnetic field is produced by a changing electric field. Their great triumph was the remarkable prediction that there can be **electromagnetic waves** that consist of nothing but oscillating electric and magnetic fields, and that these can travel through totally empty space. Crucially, the theory also predicted the speed of the waves – and it was equal to the speed of light.

The nature of light had been debated for centuries, most famously between Newton, who thought it to be a stream of particles, and others who supported a wave theory. By the nineteenth century the evidence in favour of waves was incontrovertible, but the nature of these waves remained unclear. Maxwell's theory now identified them as electromagnetic waves. The question then arose of the possibility of other types of electromagnetic wave. The radiation lying just outside the visible spectrum (ultra-violet and infra-red at shorter and longer wavelengths respectively) was presumably **electromagnetic radiation**, and heat radiation (now known to lie mainly at still longer wavelengths) could be included. Then in 1887 Heinrich Hertz demonstrated the first man-made 'signal' carried by electromagnetic waves – the precursor of radio, TV and microwave transmissions. The full range of today's known electromagnetic spectrum, from the gamma rays of radioactivity to the longest radio waves, would appear only gradually during the next century.

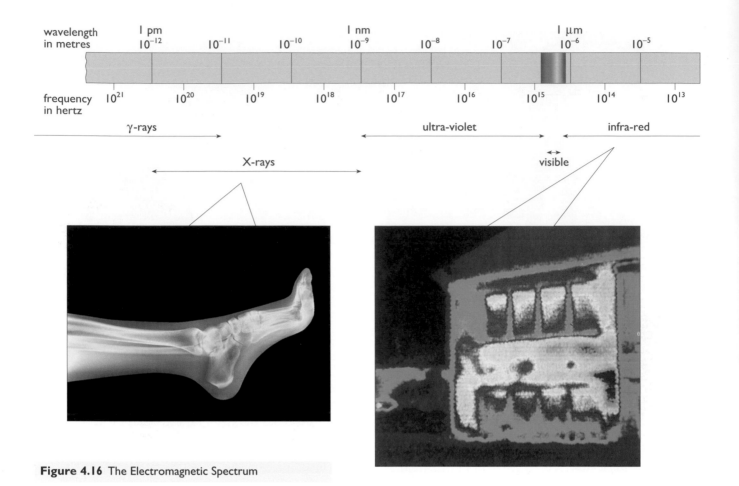

Figure 4.16 The Electromagnetic Spectrum

The sources of *all* electromagnetic waves are oscillating electric charges. In the case of light these are the electrons in atoms; but any oscillating charge or alternating current necessarily generates electromagnetic radiation. As Maxwell's theory predicted, the waves carry *energy*. Indeed, one might regard electromagnetic radiation, travelling through space in the absence of any material substance, as 'pure energy'.

There was to be still another twist. Max Planck in 1900 and Albert Einstein in 1905 showed that the energy carried by electromagnetic radiation is *quantized*. It is transmitted or received not as a continuous flow, but in discrete well-defined units or *quanta*, called **photons**. The energy of each photon depends only on the rate at which the wave oscillates. A photon of red light carries less energy than one of blue light, which has a shorter wavelength and oscillates at a higher frequency, and a photon of ultra-violet is more energetic still – matters of some importance for photovoltaic systems. (Perhaps the great Newton was right after all, and light really is a stream of particles? Unfortunately not, and we must live with the fact that light behaves sometimes as a wave and sometimes as discrete packets of energy. It *literally* depends how you look at it.)

One final, rather theoretical point may be worth noting. Electromagnetic radiation and the 'electrical energy' discussed above are both included in

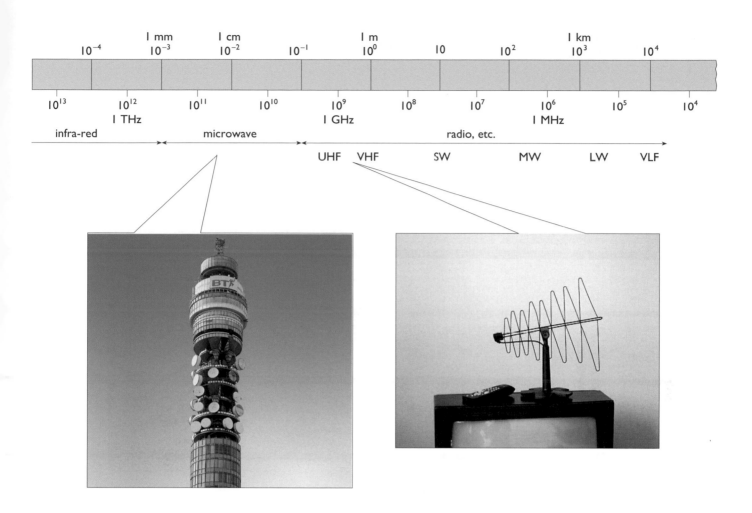

Maxwell's theory, so they could perhaps be described as different aspects of a single form of energy. However, it remains convenient for many practical purposes to treat them as separate forms, as we do here.

Figure 4.17 Max Planck (1858–1947) was born in Kiel but educated in Munich. He was a talented musician – pianist, conductor and composer – but physics took priority, and he became professor in Berlin, winning the Nobel Prize in 1918. He was president of the leading German scientific society, but resigned in 1937 in protest at the treatment of Jewish academics. (He had already crossed swords with the Nazi government in 1934, organising a memorial event for the chemist Fritz Haber.) His eldest son was killed at Verdun in 1916, and his second was executed following the 1944 plot against Hitler. In 1945, escaping on foot from the final battles of the war, Planck, aged 87, and his wife were robbed and beaten. They were found by Americans and taken to safety in Göttingen, where he resumed teaching.

BOX 4.8 Using electromagnetic radiation: solar cells

How much electric power can be expected on a sunny day from an array of photovoltaic cells with an active surface area of one square metre?

A photovoltaic cell exposed to sufficiently bright illumination behaves like a battery. It will maintain a voltage, and if connected in a circuit will deliver a current. The electric power available from the cell depends of course on the power delivered by the radiation reaching it. To answer the question, therefore, we need data on this radiation and on the cell's response.

Figure 4.19 shows the power per square metre delivered by sunlight falling on the cell surface. The power is divided into fractions corresponding to different bands of the solar spectrum. The left-hand band covers the range from ultra-violet to blue-green light, with the rest of the visible spectrum falling in the next band and all the remainder being infra-red (heat) radiation. Adding all these contributions, we see that the total input power reaching one square metre is **810** W.

The reason for dividing the spectrum is that the cell material responds differently to the different wavelengths. If the wavelength is too long the photons are too weak and produce no cell output, and if it is too short the photons have more energy than is needed and the excess is wasted as heat. Table 4.1 shows a notional **conversion efficiency**, i.e. the percentage of the solar power that is transformed into useful electrical output, for each wavelength band.

Table 4.1 Wavelength, conversion efficiency and electric power output

wavelength range /μm	conversion efficiency	contribution to output /W
0.3 – 0.5	8%	12
0.5 – 0.7	10%	23
0.7 – 0.9	25%	45
0.9 – 1.1	50%	40
above 1.1	zero	0

Note: These values should not be regarded as typical because there are many different types of PV cell, each with its own conversion characteristics.

To find the output power, we must calculate the appropriate percentages of the different contributions shown in Figure 4.19. These are given in the table, and their total, **120 W**, is the electric power output. Given that the input is 810 W, the overall **cell efficiency** under these conditions is just under 15%.

For more on PV systems, see *Renewable Energy*.

Figure 4.18 PV array

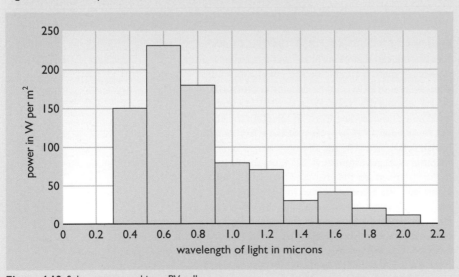

Figure 4.19 Solar power reaching a PV cell

4.6 **Chemical energy**

By the end of the nineteenth century chemistry had become a very well established science. The chemical **elements** were recognised as those substances that consist of just one type of atom, and most of the 92 naturally occurring elements had been discovered. Work by John Dalton at the start of the century had led eventually to a list of the known elements in order of their 'atomic weights' – now called **relative atomic masses**. These values, expressed as multiples of the mass of the lightest atom, hydrogen, were obviously significant: carbon almost exactly 12 times hydrogen, oxygen 16 times, and so on. But some were rather strange: copper, for instance, whose relative atomic mass appeared to be 63.5 times that of hydrogen.

The constituent elements of many **compounds** had also been identified, and **chemical reactions** were understood as rearrangements of atoms into new **molecules**. The molecule of methane (natural gas), for instance, consists of one carbon and four hydrogen atoms: CH_4. When methane burns, its molecules interact with the oxygen molecules of the air. Oxygen is a *diatomic* gas, each of its molecules consisting of a pair of oxygen atoms: O_2. The combustion of methane can therefore be represented by this chemical equation:

$$CH_4 + 2O_2 \rightarrow CO_2 + 2H_2O$$

Each molecule of methane combines with two molecules of oxygen, producing carbon dioxide and water vapour. Once the relative masses of all the atoms were known, it became possible to predict, for instance, the quantity of CO_2 produced per kilogram of methane burned. It was of course also known that some reactions, like that above, release energy whilst others need energy in order to proceed. However, the nature of the 'chemical energy' stored in these molecules could only be established when more was known about the structure of atoms.

^{1}H 1.008																	^{2}He 4.003
^{3}Li 6.939	^{4}Be 9.012											^{5}B 10.81	^{6}C 12.01	^{7}N 14.01	^{8}O 15.999	^{9}F 19.00	^{10}Ne 20.18
^{11}Na 22.99	^{12}Mg 24.31											^{13}Al 26.98	^{14}Si 28.09	^{15}P 30.97	^{16}S 32.06	^{17}Cl 35.45	^{18}Ar 39.95
^{19}K 39.102	^{20}Ca 40.08	^{21}Sc 44.96	^{22}Ti 47.90	^{23}V 50.94	^{24}Cr 52.00	^{25}Mn 54.94	^{26}Fe 55.85	^{27}Co 58.93	^{28}Ni 58.71	^{29}Cu 63.54	^{30}Zn 65.37	^{31}Ga 69.72	^{32}Ge 72.59	^{33}As 74.92	^{34}Se 78.96	^{35}Br 79.91	^{36}Kr 83.80
^{37}Rb 85.47	^{38}Sr 87.62	^{39}Y 88.91	^{40}Zr 91.22	^{41}Nb 92.91	^{42}Mo 95.94	^{43}Tc (97)	^{44}Ru 101.1	^{45}Rh 102.91	^{46}Pd 106.4	^{47}Ag 107.87	^{48}Cd 112.4	^{49}In 114.8	^{50}Sn 118.7	^{51}Sb 121.8	^{52}Te 127.6	^{53}I 126.9	^{54}Xe 131.3
^{55}Cs 132.91	^{56}Ba 137.34	57–71 (see below)	^{72}Hf 178.5	^{73}Ta 180.95	^{74}W 183.85	^{75}Re 186.2	^{76}Os 190.2	^{77}Ir 192.2	^{78}Pt 195.1	^{79}Au 196.97	^{80}Hg 200.6	^{81}Tl 204.4	^{82}Pb 207.2	^{83}Bi 209.0	^{84}Po (209)	^{85}At (210)	^{86}Rn (222)
^{87}Fr (223)	^{88}Ra (226)	89–103 (see below)	^{104}Rf (263)	^{105}Db (262)	^{106}Sg (266)	^{107}Bh (264)	^{108}Hs (269)	^{109}Mt (268)	^{110}Uun (272)	^{111}Uuu (272)	^{112}Uub (285)	^{113}Uut —	^{114}Uuq (289)	^{115}Uup —	^{116}Uuh (289)	^{117}Uus —	^{118}Uuo —

Lanthanides	^{57}La 138.9	^{58}Ce 140.1	^{59}Pr 140.9	^{60}Nd 144.2	^{61}Pm (145)	^{62}Sm 150.4	^{63}Eu 152.0	^{64}Gd 157.3	^{65}Tb 158.9	^{66}Dy 162.5	^{67}Ho 164.9	^{68}Er 167.3	^{69}Tm 168.9	^{70}Yb 173.0	^{71}Lu 175.0
Actinides	^{89}Ac (227)	^{90}Th 232.0	^{91}Pa (231)	^{92}U 238.0	^{93}Np (237)	^{94}Pu (244)	^{95}Am (243)	^{96}Cm (247)	^{97}Bk (247)	^{98}Cf (251)	^{99}Es (254)	^{100}Fm (257)	^{101}Md (258)	^{102}No (255)	^{103}Lr (256)

Figure 4.20 The Chemical Elements. The uppermost figure in each box is the number of electrons in the atom. The figure beneath the symbol represents the mass of the atom, discussed further below and in Chapter 10. One of the most interesting discoveries in chemistry is the repeating periodic pattern shown here in which elements with similar chemical properties fall in the same vertical columns

Table 4.2 List of elements with symbols and atomic numbers

Symbol	Element	Atomic number	Symbol	Element	Atomic number	Symbol	Element	Atomic number
Ac	actinium	89	Ho	holmium	67	Rb	rubidium	37
Al	aluminium	13	H	hydrogen	1	Ru	ruthenium	44
Am	americium	95	In	indium	49	Rf	rutherfordium	104
Sb	antimony	51	I	iodine	53	Sm	samarium	62
Ar	argon	18	Ir	iridium	77	Sc	scandium	21
As	arsenic	33	Fe	iron	26	Sg	seaborgium	106
At	astatine	85	Kr	krypton	36	Se	selenium	34
Ba	barium	56	La	lanthanum	57	Si	silicon	14
Bk	berkelium	97	Lr	lawrencium	103	Ag	silver	47
Be	beryllium	4	Pb	lead	82	Na	sodium	11
Bi	bismuth	83	Li	lithium	3	Sr	strontium	38
Bh	bohrium	107	Lu	lutetium	71	S	sulphur	16
B	boron	5	Mg	magnesium	12	Ta	tantalum	73
Br	bromine	35	Mn	manganese	25	Tc	technetium	43
Cd	cadmium	48	Md	mendelevium	101	Te	tellurium	52
Cs	caesium	55	Hg	mercury	80	Tb	terbium	65
Ca	calcium	20	Mo	molybdenum	42	Tl	thallium	81
Cf	californium	98	Mt	meitnerium	109	Th	thorium	90
C	carbon	6	Nd	neodymium	60	Tm	thulium	69
Ce	cerium	58	Ne	neon	10	Sn	tin	50
Cl	chlorine	17	Np	neptunium	93	Ti	titanium	22
Cr	chromium	24	Ni	nickel	28	W	tungsten	74
Co	cobalt	27	Nb	niobium	41	U	uranium	92
Cu	copper	29	N	nitrogen	7	Uub	ununbium	112
Cm	curium	96	No	nobelium	102	Uuh	ununhexium	116
Db	dubnium	105	Os	osmium	76	Uun	ununnilium	110
Dy	dysprosium	66	O	oxygen	8	Uuo	ununoctium	118
Es	einsteinium	99	Pd	palladium	46	Uup	ununpentium	115
Er	erbium	68	P	phosphorus	15	Uuq	ununquadium	114
Eu	europium	63	Pt	platinum	78	Uus	ununseptium	117
Fm	fermium	100	Pu	plutonium	94	Uut	ununtrium	113
F	fluorine	9	Po	polonium	84	Uuu	unununium	111
Fr	francium	87	K	potassium	19	V	vanadium	23
Gd	gadolinium	64	Pr	praseodymium	59	Xe	xenon	54
Ga	gallium	31	Pm	promethium	61	Yb	ytterbium	70
Ge	germanium	32	Pa	protoactinium	91	Y	yttrium	39
Au	gold	79	Ra	radium	88	Zn	zinc	30
Hf	hafnium	72	Rn	radon	86	Zr	zirconium	40
Hs	hassium	108	Re	rhenium	75			
He	helium	2	Rh	rhodium	45			

The nuclear atom

Unravelling the mysteries of atomic structure was the central scientific problem of the early twentieth century. The first break-through came in 1911, when Ernest Rutherford showed that nearly all the *mass* of any atom is concentrated into a tiny central **nucleus** occupying only a ten-thousandth of the diameter of the atom. This nucleus is surrounded by a cloud of fast-moving electrons, which define the size of the atom. Electrons are known to have negative electric charge, but a complete atom is electrically neutral, so the nucleus must have a corresponding positive charge. The resulting electrical force attracts the electrons towards the nucleus, leading to a stable atom.

It is the electron cloud that determines the chemical properties of the atom, such as the compounds it will form with other elements. Processes such as the combustion of a fuel could now be seen as rearrangements of the electrons when the constituent atoms combine into different molecules. It follows that the energy released must come from changes in the kinetic energy and electrical potential energy of these electrons. 'Chemical energy', like 'heat energy', does not therefore represent a fundamentally new form of energy. However, in a world where the major part of energy demand is met by the combustion of fuels, it is undoubtedly an important form.

Figure 4.21 Ernest Rutherford (1871–1937) was born and educated in New Zealand. He joined the Cavendish laboratory, Cambridge in 1895, and held chairs in Montreal and Manchester before returning to the Cavendish in 1919 as Director. A large exuberant man with a loud voice and brilliant physical intuition, he always attracted a lively group of students and co-workers. The radical change from a picture of matter consisting of many different 'uncuttable' atoms to one in which all atoms are built from just a few fundamental particles was largely his, and he is regarded as the founding father of nuclear physics. (See also Chapter 10.)

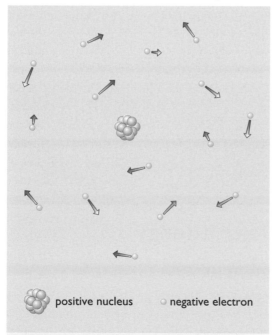

positive nucleus   ◦ negative electron

Figure 4.22 The nuclear atom showing the central nucleus and surrounding electrons

BOX 4.9 Using chemical energy: fuels as energy stores

Table 4.3 shows the energy content of five fuels. How much of each of these would be needed to supply energy equal to that stored by the systems of Boxes 4.3 and 4.4?

Table 4.3 Fuels as energy stores

fuel	energy content /MJ per kg
coal	28
oil	42
natural gas	55
hydrogen	130
wood	15

Table 4.4 Masses needed to store 30 TJ

storage method	tonnes required
hydrogen	230
methane	550
oil	710
coal	1100
wood	2000
lead-acid batteries	69 000
water heated to 80 °C	120 000
pumped storage at 370 metres	8 000 000

Any fuel can be regarded as an energy store. To be precise, the 'stored energy', the energy released in burning the fuel, is the difference between the total chemical energy of the fuel and the oxygen in which it burns and the total chemical energy of the combustion products. However, it is commonly referred as the energy content of the fuel itself.

The stored energy cited in the earlier case studies is 30 TJ, which is 30 million MJ. So the required quantity of coal, for instance, is 30 million divided by 28, or 1.1 million kilograms. As one metric tonne is 1000 kg, this can conveniently be expressed as 1100 tonnes.

Table 4.4 shows the quantity of each of the five fuels needed to produce 30 TJ of heat energy, together with data from the earlier storage systems. (The mass of one cubic metre of water is one tonne, and to complete the picture we take the mass of a lead-acid car battery to be 10 kg.)

Although it is unlikely that all the items in Table 4.4 will in practice be competing in any one situation, the hierarchy does illustrate rather vividly the different 'densities' of the forms of energy: gravitational, thermal, electrical and chemical.

The comparison is of course unfair, for a number of reasons. If electric power is the desired aim, the batteries and the pumped storage plant can store and supply it with relatively little loss, whilst all the others involve appreciable conversion losses. Then we should certainly distinguish between 'storage systems' where the energy can be recycled, or is naturally recycled on a relatively short time-scale, and those such as the fossil fuels where this is not possible. These issues are discussed in more detail in later chapters.

4.7 Nuclear energy

Protons and neutrons

On the Rutherford model, the electric charge of a nucleus must be an exact multiple of a basic *positive* charge in order to balance the total *negative* charge of all the electrons. This led naturally to the idea that all nuclei might consist of different numbers of identical heavy positive particles, the **protons**. Perhaps the nuclei of the successive atoms in the table of elements (Figure 4.20) simply have increasing numbers of protons – one for hydrogen, two for helium, etc. – each nucleus being surrounded by a corresponding number of the much lighter electrons? It is a nice idea, but unfortunately it doesn't explain the *masses* of the atoms. Helium, which should have two protons, has *four* times the mass of hydrogen, and carbon, the sixth element, has *twelve* times the hydrogen mass, and so on. And how can copper have a relative atomic mass of sixty-three and a half?

In 1920 Rutherford offered a solution. He proposed a second nuclear particle, the **neutron**, with almost exactly the proton mass but no electric charge. If the hydrogen nucleus consists of just one proton, the nucleus of helium would have 2 protons and 2 neutrons, carbon would have six of each, and so on. With this idea, the **atomic number** of an element, its position in the table, is simply the number of protons in its nucleus, whilst the relative atomic mass is determined by the total number of **nucleons** (protons plus neutrons), termed the **mass number**. The number of protons must of course be matched by an equal number of surrounding electrons, and it is this number that determines the nature of the element. The consequences do seem remarkable: a mere change from ten electrons to eleven is responsible for the radically different properties of neon (an inert gas) and sodium (a chemically very active metal). The uncuttable atom may have been abandoned, but its purpose had in a sense been achieved, reducing the whole of physics and chemistry – all the physical properties of everything – to the behaviour of just three basic particles: the proton, the neutron and the electron.

Isotopes

Strange relative atomic masses like the 63.5 for copper can also be explained. Natural copper, we now know, consists of two main **isotopes**: atoms whose nuclei have the same number of protons but different numbers of neutrons. Both have 29 protons – essential if the atom is to be copper – but one has 34 neutrons and the other 36. Roughly seven of every ten copper atoms are the lighter **nuclide** ($^{63}_{29}$Cu, or copper-63) and the remaining three are the heavier one ($^{65}_{29}$Cu, or copper-65), giving the 'average' relative atomic mass of 63.5. Such isotopes are not uncommon. More than three-quarters of the stable elements exist as two or more isotopes, and even carbon includes a tiny fraction (about 1%) of the nuclide carbon-13.

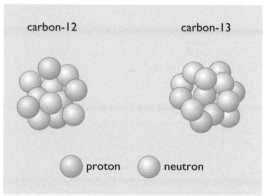

Figure 4.23 Nuclear composition of two isotopes of carbon

The nuclear force

A look at the world around us reveals that most atomic nuclei are extremely *stable* objects. We don't see elements constantly changing into other elements. Oxygen remains oxygen, gold stays as gold, lead as lead. Iron Age tools may rust, but this is a chemical reaction, a rearrangement of the electrons of the iron and oxygen atoms. The iron *nuclei* are the same nuclei that existed when the tools were made, thousands of years ago.

So one important question remains to be answered. What holds a nucleus together? We know that particles with the same type of electric charge repel each other, so an assembly consisting only of positive protons with neutral neutrons should instantly fly apart. Another force is needed to bind the nucleus. (Gravity attracts all objects towards each other, but the gravitational force between nucleons is far too small.) The required force needs to be very strong at the tiny distances within a nucleus and to attract *every* nucleon (proton or neutron) to every other nucleon. In 1935 Yukawa developed a theory of a completely new force to account for the stability of nuclei. There have been changes of detail over the years, but this **strong nuclear force** remains, completing the trio of *fundamental forces* currently known to science: the gravitational, electromagnetic and strong nuclear forces.

Science does not stand still; new experiments are carried out and new theories emerge. During the past half-century two major aims have been to integrate the different forces into one theory and to understand the nature of the particles. In this continuing search for the 'theory of everything', even the distinction between forces and particles begins to disappear, but these developments are beyond the scope of this book.

Radioactivity and fission

The relative atomic masses shown in Figure 4.20 reveal that the numbers of protons and neutrons tend to be equal in the lighter atoms, but an excess of neutrons gradually develops as the atoms become heavier (as for instance in copper, above). We can see why this might be necessary. Because every proton repels every other proton, the energy needed to hold them all together increases steeply as their number rises. Neutrons don't of course experience the electrical force, and *all* nucleons (protons and neutrons) are attracted to each other by the strong nuclear force, so any increase in the number of neutrons helps to maintain stability against the increasing tendency of the protons to fly apart. Eventually, beyond the element bismuth, with 83 protons and 126 neutrons, there are *no more stable nuclei*. All the elements beyond bismuth are **radioactive**: their nuclei spontaneously emit high-energy electrically charged particles, changing into different nuclei in the process. If this is so, how is it that we find uranium (atomic number 92) still existing in the Earth's crust? The answer is that it *is* decaying, but as we'll see in Chapter 10, only very slowly indeed.

The presence of radioactivity shows that nuclei can be unstable, but it is **fission** which offers the prospect of a continuous supply of energy. In fission, unlike radioactivity, the nucleus splits into two *roughly equal parts*. This releases a great deal of energy when the two tightly packed groups of protons fly apart. There is just one naturally occurring **fissile** material: the isotope uranium-235. It is also radioactive, so its nuclei can disintegrate either by emitting a single, much lighter particle or by splitting approximately in half. Both processes release energy, and both occur naturally at very slow rates. The crucial difference is that whilst nothing affects the radioactivity very much, we have discovered how to *control* the rate of fission. It can be increased to billions of times the natural rate, with a corresponding increase in the rate at which energy is released (see Chapters 10 and 11).

The nucleus of an atom, with its closely packed protons, is an extremely compact energy store, and the energy released in the fission of a nucleus is many millions of times greater than that released in combustion of a fuel molecule. As mentioned above, most of it is in fact *electrical energy* released as the two groups of protons separate. In fission as in radioactivity, most of the released energy appears first as the kinetic energy of fast moving particles, but collisions of these with the atoms of surrounding materials rapidly transform it into heat energy.

4.8 Energy and mass

In the summer of 1905 Albert Einstein, then aged 26, published a scientific paper with the title, *On the Electrodynamics of Moving Bodies.* Its subject was a problem that had worried him for some time. (He later told someone that he had 'worked on it for 10 years.') The problem was that the two great theories, Newton's Laws of Motion and Maxwell's Laws of Electromagnetism, led to conflicting results. Einstein's resolution was to replace Newton's laws by a new theory, now generally known as the Special Theory of Relativity. It was a radical proposal, changing the way we look at such fundamental concepts as space and time, but its predictions were confirmed in many different ways throughout the twentieth century.

One consequence of the theory is that we can no longer regard the mass of an object as a fixed quantity. An object has a certain **rest mass** when it is stationary, but when it is moving it behaves in all respects as though it has a greater mass. This increase in mass is the equivalent of kinetic energy in the new theory. Indeed it can be said that the kinetic energy *is* the difference between the mass of the moving object and its rest mass: *mass and energy are the same*. However, the separate concepts have proved too useful to abandon, and the quantity properly called **mass-energy** continues to be given the name and unit that seem most appropriate. The theory specifies how to convert from mass units to energy units:

- the energy (E) in joules is equal to the mass (m) in kilograms multiplied by the square of the speed of light in empty space (c), or $E = mc^2$.

The kinetic energy (in joules) of a moving object is now the difference between mc^2 when it is at rest and when it is in motion.

We don't normally notice the increase in mass as an object accelerates, because the kinetic energy of anything at normal speeds is far too small for the corresponding mass change to be detected. In such cases we can safely continue to use Newton's laws. Only where speeds approach the speed of light is the mass change significant.

The mass-energy equivalence applies equally to cases other than kinetic energy. When a fuel is burned in air in a closed container and the resulting heat energy escapes, the total mass inside the container decreases. Again, we don't normally notice this, because the mass change is very small indeed. Releasing the heat produced in burning a tonne of coal reduces the total mass by about a third of a milligram: one part in three billion. There are however cases where the energy is large enough and the original mass small enough for the change to be detected. When a nucleus undergoes fission, releasing energy, the total mass of all the resulting particles is less than that of the original nucleus. It was the fact that all these masses were known that allowed people to calculate that a very great deal of energy would be released in fission (Box 4.10).

Figure 4.24 Albert Einstein (1879–1955) was born in Ulm but became Swiss while studying in Zürich. Continuing his studies while employed as a patent officer in Bern, he published in 1905 three remarkable theories. One explained the continuous random movement of fine particles (Brownian motion) in terms of molecular bombardment. Another explained the emission of electrons from irradiated metals (the photoelectric effect) in terms of Planck's quanta. The third was the Special Theory of Relativity. After a period in Prague, and a return to Zurich, he became professor in Berlin. Here he turned to gravity – the second part of Newton's great work – and developed the General Theory of Relativity, which became the basis of cosmology. Dismissed by the Nazis, he moved to Princeton, where he spent the rest of his life searching for the theory that would unify gravity and electromagnetism. Despite the failure of this endeavour, he is regarded as the greatest scientific mind of the twentieth century.

BOX 4.10 Using nuclear energy: uranium as an energy store

Calculations show that if all the uranium-235 nuclei in a tonne of natural uranium were to undergo fission, the total mass would decrease by 6.6 grams. How much natural uranium would be needed to supply the 30 TJ of energy considered in Box 4.9?

We start by converting the decrease in mass into a change in energy:

$$E = mc^2$$

The speed of light is 300 million metres per second: $c = 3 \times 10^8$ m s^{-1}, so the energy change in joules is related to the mass change in kilograms by:

$$E = m \times (3 \times 10^8)^2 = m \times 9 \times 10^{16}$$

The mass change in this case is 6.6 grams, which is 0.0066 kg, so the energy released is:

$$E = 0.0066 \times 9 \times 10^{16} = 6 \times 10^{14} \text{ J}$$

which is 600 TJ, or twenty times the required 30 TJ.

The mass of natural uranium needed in principle to supply 30 TJ is therefore one twentieth of a tonne, which is 50 kilograms. Comparison with the other entries in Table 4.4 reveals why the nuclear fuels are regarded as very concentrated energy stores.

4.9 Summary

This chapter has been a journey through time and an exploration of ideas. Its main purpose was to develop an account of the different forms of energy, but in order to do this we found it necessary at the same time to build a picture of the physical world as it appears to present-day science. Ideas about energy are intimately linked with ideas about matter, and as the above account has shown, the two have developed closely in parallel throughout the past three hundred years.

The brief case studies have demonstrated just a few of the energy systems to be discussed elsewhere in this book and in the companion volume, *Renewable Energy*. They also show that each of the forms of energy has direct modern relevance. Kinetic energy, gravitational potential energy, heat, electrical and electromagnetic energy, chemical energy and nuclear energy – all of these will appear in discussions of our present energy systems, and all will have roles to play in potential applications for the future.

References

Bacon, Francis (1620) *Novum Organum.* Excerpts available at http://www.luminarium.org/sevenlit/bacon/baconbib.htm [accessed 24 May 2002].

Boyle, G. (ed.) (1996) *Renewable Energy*, Oxford, Oxford University Press.

Rankine, W. J. M. (1881) *Miscellaneous Scientific Papers* (ed. Millar, W. J.), Griffin, pp. 203, 216

Tait, P. G. (1870) *Nature*, 19th December, 163/2

Further reading

Readers to whom the concepts introduced in this chapter are new may find the more informal and less mathematical approach of one of the following useful additional reading.

Patterson, W. C. (1991) *The Energy Alternative*, London, Macdonald & Co. (Publishers) Ltd.

Ramage, J. (1997) *Energy: A Guidebook*, 2nd ed., Oxford, Oxford University Press.

Chapter 5

Coal

by Janet Ramage

5.1 The fossil fuels

The next four chapters discuss probably the oldest of all technologies: the burning of fuels to provide heat. Already familiar to our pre-historic ancestors, the combustion of fuels still accounts for ninety percent of our present-day use of primary energy. The term *combustion*, however, has a much wider meaning now than in the past. An internal combustion engine or gas turbine may seem very remote from a log fire, but the fuels used in these modern systems have much in common chemically with wood, and their combustion involves essentially the same basic processes. As we saw in Chapter 4, the process of combustion is a chemical reaction between oxygen, normally from the surrounding air, and the constituent elements of the fuel – mainly carbon and hydrogen. The reaction leads to the release of energy in the form of heat which is carried away initially by the combustion products. (The 'burning' of nuclear fuel is an entirely different process, see Chapter 10.)

As their name suggests, all the **fossil fuels** were originally living matter: plants or animals that were alive hundreds of million years ago. With the passage of time, their remains have undergone chemical changes leading to the solid, liquid or gaseous fuels that we extract today.

Oil and **natural gas** have been called the *noble fuels*, by analogy with the noble metals (silver and gold, etc.), and whilst we may have reservations about the results of using them, there is justification for the name. They are amongst the most concentrated natural stores of energy, and being fluids, are relatively easy to move from place to place and very convenient to use. It doesn't require a detailed technical analysis to see why we choose gas rather than coal for household heating and hot water; or why we are reluctant to adopt the electric car when our present vehicles can load hundreds of kilowatt-hours of energy in a minute or so, store it within the volume of a small petrol tank and deliver power at a rate of tens of kilowatts.

In contrast to natural gas or oil, **coal** might be called the *ignoble* fuel. It is indeed a particularly unattractive energy source. Compared with oil and gas, coal is less convenient to transport, store and use. It produces up to twice the amount of carbon dioxide for the same useful heat, and this is by no means its only undesirable environmental effect. A large coal-fired power station can produce enough ash in a year to cover an acre of ground to the height of a six-storey building, while its flue gases may carry several tonnes per day of sulphur dioxide and nitrogen oxides into the atmosphere. The ash, if not carefully isolated, will pollute ground-water with a variety of unpleasant substances such as sulphuric acid and arsenic, and the flue gases are claimed to contaminate lakes and harm trees hundreds of miles downwind. Its extraction leads to land subsidence and spoil heaps or the environmental horrors of strip mining, and is still responsible world-wide for the deaths of hundreds of miners in an average year. There is obviously no difficulty in explaining why coal consumption rose less rapidly than total energy consumption throughout most of the last century, especially in those countries wealthy enough to afford the more attractive alternatives. As Figure 5.1 shows, world coal production in the year 2000 was about three times that of the year 1920, but as a fraction of total commercially traded energy it had fallen from over four-fifths to about a quarter.

There is however a compelling reason why coal is unlikely to fade from the scene completely in the near future. It may be the least desirable fuel, but it has one over-riding advantage: *there is a very great deal of it.*

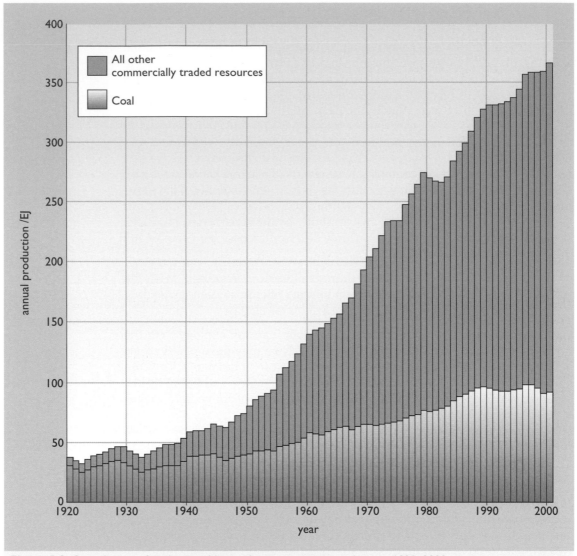

Figure 5.1 Contribution of coal to world annual primary energy production, 1920–2000

5.2 From wood to coal

The early years

Archaeological remains reveal that coal has been used as a fuel for at least three or four thousand years and in many parts of the world. There is evidence of coal-burning in Britain in the Bronze Age, in China and the Middle East over two thousand years ago and in regions of the Roman Empire (including Britain) throughout the period of its supremacy. In these earlier times, as today, fuel was needed for cooking food, for heat and light and for industrial production. (The use of fuel for transport, another present-day requirement, arose only much later, in the nineteenth century.)

During the Middle Ages, the main energy-consuming 'industrial' plants in cities such as London were the lime kilns, burning calcium carbonate (limestone or chalk) as the first stage in the production of lime, primarily for building and agriculture. Other relatively heavy urban consumers included the bakers and the smiths or metal-workers. The initial smelting of the iron ore needed **charcoal** and was generally a rural activity, carried out in the forests. (Charcoal, a secondary fuel produced by the partial combustion of wood or other vegetable matter, is almost pure carbon and burns cleanly and at a high temperature. For a more detailed account, see Boyle, 1996.)

Most citizens of mediaeval London depended on simple open fires for their domestic heating and cooking, and three fuels were available. There was wood, imported initially from the surrounding forests and later from more distant sources. There was charcoal, commonly known simply as 'coal'. And from about the year 1200, there was coal itself, extracted in the northeast of England from surface beds or shallow mines. Carried by sea to London, it was known as 'sea-coal' to distinguish it from charcoal. The varying fortunes of these three fuels during the period from thirteenth to the nineteenth centuries offer interesting examples of the ways in which economic, social, technological and environmental factors influence patterns of energy consumption.

At the start of the period, coal was often carried as ballast by ships returning from the north of England. Delivered directly to wharves on the Thames, close to the potential users, it was a relatively cheap fuel. But coal fires producing sulphurous smoke were ill suited to buildings whose only 'chimney' was a hole in the roof, and wood remained the preferred household fuel. This was also the case for early industrial processes, although there seem to have been occasional policy changes. In 1215 King John insisted on oak for his lime kilns, but by 1264 his successor Henry III was ordering sea-coal for the same purpose. Then in 1306 Henry's successor, Edward I, an early environmentalist, issued a proclamation banning the use of sea-coal. It is said that the penalty for failure to observe the new law was death, but recent historians have expressed reservations about this oft-quoted example of excessive but ineffective punishment (see, for instance, Brimblecombe, 1987). In any case, the penalty became irrelevant when the situation changed again a decade or so later. Diminishing supplies of wood led to a steep price rise, and the cheaper coal became the preferred fuel for London's growing population, with consumption increasing despite the undesirable environmental effects.

The next change had an entirely different origin. The Black Death, reaching England in 1348, killed over a third of the population within a few years. The available wood became sufficient for London's greatly reduced population, and the supply actually increased as abandoned arable farmland reverted to copses and woods. As a result, coal deliveries to London remained almost constant throughout the next 150 years, despite the growing population. But in 1485, at the start of the Tudor period, came a change that proved very important for coal: the re-establishment of brick-making in England. (The great mediaeval brick buildings elsewhere in Europe show that other countries had continued to use bricks throughout the period, but in England brick-making had almost died out with the end of the Roman Empire.)

Figure 5.2 Ornamental brick chimneys: 1. East Barsham Manor House, 2. Hampton Court, 3. Eton College

The new coal-fired brick kilns led to an increasing demand, but equally important was the fact that brick chimneys were resistant to the hot, acidic smoke from coal fires. In a short time, the wealthy were building brick houses with many fireplaces, and coal was their favoured fuel. More modest dwellings gradually began to have chimneys, and in 1585 Elizabethan London was importing an estimated annual 24 000 tons of coal (Nef, 1932). By 1680, in the reign of Charles II, this had risen to 360 000 tons: an average growth rate of nearly 3% per year for 100 years. The environmental effects of this fifteen-fold increase were only too evident, and they attracted the interest of a new group of people: the scientists. The founding of the Royal Society in 1660 had provided a forum for scientific discussion, and several of its early members applied the new investigative methods to problems such as the deleterious health effects of atmospheric pollution, or its corrosive effect on the fabric of buildings. One member, the diarist John Evelyn, wrote a famous book, *Fumifugium*, on the subject (de la Bédoyère, 1995a). He was so concerned about London's smoky atmosphere that he tried to introduce legislation to limit coal burning in the city. The king, Charles II, at first supported him but seems then to have lost interest – or perhaps other more powerful influences prevailed.

The Industrial Revolution

The Industrial Revolution has been the subject of debate almost since Friedrich Engels coined the term in 1844. There has been disagreement about its causes, its effects, even its dates; but for our present purposes, it is worth identifying the starting points of three important strands:

- 1698: Thomas Savery's first steam engine.
- 1709: Abraham Darby's use of coke for smelting iron.
- 1733: John Kay's invention of the flying shuttle.

The significance of Savery's steam engine was that it used *heat* from burning fuel to produce a continuous (or at least repetitive) *mechanical* driving force. However, as we'll see in the next chapter, a century of technical development was needed before steam engines were to play a major role. This was also true of Kay's invention, which should perhaps be regarded as the forerunner rather than the start of an industrialized textile industry. The new spinning and weaving technologies developed by Hargreaves, Arkwright, Cartwright and others appeared only in the second half of the eighteenth century. The great textile mills were initially powered by water rather than coal, but the growing mechanisation of the industry was an important factor in the widespread adoption of steam engines as they became more efficient and reliable.

The work of Abraham Darby was of more immediate relevance to the coal industry. In the late 1600s, a renewed shortage of wood for charcoal was bringing Britain's iron industry to the verge of closure. As mentioned in Chapter 3, Section 3.2, charcoal was essential for the smelting of iron. In this process, the iron ore – essentially iron in chemical combination with oxygen – is heated together with the charcoal, which is almost pure carbon. Under the right circumstances the ore is reduced ('de-oxidized') to metallic iron, the oxygen combining with carbon to produce carbon dioxide.

Over the years, various attempts had been made to use coal instead of charcoal, but its variable composition, inert mineral content and impurities such as sulphur led to very poor quality iron. However, the idea of using the coal to make a new kind of charcoal was already in the air in the 1650s:

> … came home by Greenewich ferry, where I saw Sir Jo Winters new project of Charring Sea-Coale, to burn out the Sulphure & render it Sweete: he did it by burning them in such Earthen-pots, as the glassemen, mealt their Mettal in, so firing the Coales, without consuming them, using a barr of Yron in each crucible or Pot, which barr has an hooke at one end, so that the Coales being mealted in a furnace, with other crude sea Coales, under them, may be drawn out of the potte, sticking to the Yron, whence they beate them off in greate halfe exhausted Cinders, which rekindling they make a cleare pleasant Chamber fires with, depriv'd of their Sulphury & Arsenic malignity: what successe it may have time will discover.
>
> The Diary of John Evelyn, July 11, 1656 (de la Bédoyère, 1995b)

By about 1680 this new, cleaner fuel produced by the partial combustion of coal had acquired a name: **coke**.

Figure 5.3 Coalbrookdale in the early eighteenth century

Abraham Darby was a maker of brass cooking pots in Bristol and a member of an extended Quaker family, some of whom worked in Coalbrookdale, a small Staffordshire village close to one of the largest coal fields of the time. Developing an interest in the potential of coke for iron smelting, Darby moved there, and within a few decades was using coke to produce cast-iron, not only in the form of cooking pots but as large cast-iron cylinders for the new steam engines. (The splendid cast-iron bridge that gives the place its present name was completed by his grandson, also Abraham, in 1779.) By the end of the eighteenth century, Britain's iron industry was flourishing, and the country's coal production had reached an estimated ten million tons per year.

The nineteenth century

At the start of the nineteenth century, steam engines fuelled by coal were already supplying power in mills, mines and factories, coke was in general use for iron smelting, and domestic coal fires and kitchen stoves were common. But Nelson's fleet was still wind-powered; transport and agriculture were still horse-powered; and houses were still lit by oil lamps or candles. A hundred years later, the start of the twentieth century saw a very different world – in those countries wealthy enough to adopt the new technologies. Power from coal had become dominant in nearly all areas of life. (Agriculture, still mainly dependent on the horse, was an exception.) Railways and steamships were providing transport, coal gas was lighting buildings and streets, and electricity, the 'new' form of energy that was beginning to play a role in the industrialized countries, would also depend largely on coal.

The story of the developments that led to this changed world is mainly told elsewhere in this book. In Chapter 4 we followed the evolution of the scientific ideas, whilst the two greatest technological advances remain sufficiently important to warrant separate accounts: the steam turbine in Chapter 6 and the electrical generator in Chapter 9. A third major development, the production of gas from coal, is an interesting example of

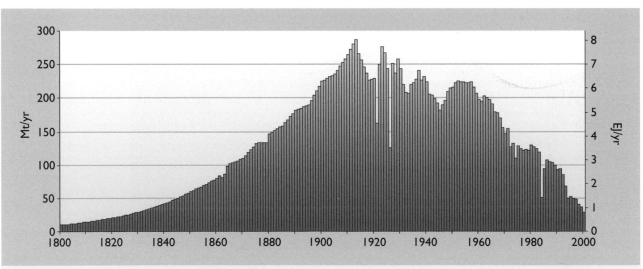

Figure 5.4 UK annual coal production, 1800–2000

a technology-based industry that grew to great importance in many parts of the world, and was then almost entirely superseded (see Chapter 3 and Box 5.3 below).

One significant consequence of the changes described above was the continued growth of coal output throughout the century. Britain's annual production, still rising at about 3% per year, had grown from the ten million tons of the year 1800 to more than two hundred million by 1900 (Figure 5.4). It is an indicator of British dominance during the Industrial Revolution that the United States reached *one* million tons for the first time only in 1840. However, growing demand and an increasing shortage of wood for charcoal led to a remarkable annual growth rate of more than 9% a year, bringing US production to about the same level as the UK by the end of the nineteenth century. The only other country with an output approaching these was Germany, producing a little over a hundred million tons in 1900. World production was an estimated 800 million tons and rising at about 5% per year. In 1905, it would exceed a billion tons.

5.3 The resource and its use

Estimates at different times during the first half of the twentieth century (e.g. Jevons, 1915; Brown, 1950) have suggested that 'total world reserves' of coal were between five and ten thousand billion tonnes. The present annual rate of consumption is about three billion tonnes per year, so the coal should last for well over 1000 years – but such totals must be treated with care, as we shall now see.

Types of coal

Coal seams result from the accretion of layers of plant material, initially protected from atmospheric decay by water and ultimately compressed to a tenth or less of their original thickness. Seams throughout the world that have been mined at different times have ranged in thickness from as little as 300 mm (one foot) to more than 30 m (100 feet), and lie more than a thousand metres below the surface or so close that they almost penetrate it. These features are of considerable economic importance. Coal, like all minerals, is essentially a free resource, waiting to be used. Its cost arises in the processes of extraction, treatment and transport to the user, and the cost of these three stages depends on the form and location of the coal seams.

The nature of the coal itself is also important. The earliest geological strata in which coal has been identified are nearly four hundred million years old (Devonian), and coal is present in all strata down to less than a hundred million years ago (Tertiary). The largest quantities are thought to lie in the Tertiary rocks, but this 'young' coal is less valuable than the deposits of earlier periods. As we shall see below, the different types of coal can be characterized by the extent to which the original plant material has been physically and chemically changed. Obviously the age of the coal is a significant factor in this, but other geological processes, such as subjection to heat and/or pressure are important. Depending on all these factors, the resulting 'coal' can range from substances in which the original plant material can still be identified to those which are almost pure carbon. Not surprisingly, the **heat value** – the heat energy released in combustion – varies considerably, from little more than ten gigajoules per tonne to well over thirty.

In discussing the resource, it is customary to divide this range of types of coal into two main categories. The terminology varies from country to country, but the most common name for the 'upper' end of the range is **hard coal** – although Australia, the world's main exporter, uses the term **black coal**, and in some national or international data the more specific *anthracite and bituminous coal* appears. The 'lower' coals are generally referred to as **brown coal** or **lignite**, with *sub-bituminous coal* sometimes included. A further sub-division of the hard coals, obviously of great practical importance, is based on their potential use: steam coal (or thermal coal) mainly for power stations; coking coal, mainly for the iron and steel industries; and anthracite, for direct use as a natural 'smokeless fuel'. (Section 5.4 below discusses these types of coal and their uses.)

BOX 5.1 Mining the coal

Figure 5.5 Coal cutting in Lilly Drifth Mine, near Newcastle, 1953. The miner lying in a slot only thirteen inches high, hacking out coal with his pick, may seem like a figure from an age long past, but as recently as the 1950s a fifth of Britain's coal was still got by hand.

Figure 5.6 A dragline scraper at work removing overburden in an open cast mine in Decker, Montana. The size of this machine is revealed by the fact that the yellow vehicle just visible at the bottom of the pit is a large caterpillar tractor.

This book is essentially concerned with the nature of our present fuels, the ways in which we use them, and the consequences of these uses. Details of the geological processes which led to their existence and of the technologies involved in locating and extracting them are therefore generally beyond its scope. However, the extraction of coal – or any other mineral resource – is an important determinant of its cost and also has environmental, health and social implications. The following is therefore a brief outline of the main types of coal mine.

Surface mines

The two main types of mine are usually referred to as **surface mines** and **deep mines**, but the depth of the coal below the surface is not necessarily the distinguishing criterion, and some present-day 'surface' mines are extracting coal at greater depths than 'deep' mines in the past or in other parts of the world. The term **open cast**, commonly used in the UK, better describes the essential characteristic of 'surface' mining. The coal seam is accessed by removing the earth and rock or other overburden above it, a process that can involve some of the world's largest machines (Figure 5.6). The coal is then broken up, possibly with the aid of explosives, and removed by mechanical shovels. Various forms of excavation are adopted, depending on the nature, depth and position of the seam. On relatively flat ground, the opened areas can be in the form of long rectangles, with the topsoil and overburden placed to one side for later replacement. In mountainous areas, ledges contouring the slopes may be opened. Or in some cases a single large pit is opened and the overburden is then moved around within it to gain access to different parts of the coal seam. The ultimate depth of excavation can exceed 100 metres.

At the time of writing, just under half of UK coal production comes from open cast mines. For the USA as a whole the proportion is similar, but as we shall see, there are significant variations between different areas of the country. The quantity of coal extracted per worker-day in surface mining can be many times that in deep mines – a factor that is reflected in the generally lower cost of surface-mined coal.

The most obvious deleterious effect of surface mining is on the local landscape. In principle the overburden and topsoil can be replaced and vegetation re-established, but this increases the cost of the coal, and has not always been carried out in the past. Most major coal-producing countries have legislation requiring rehabilitation, or laws prohibiting surface mining on land where rehabilitation would not be possible; but these have not always proved effective – particularly when they conflict with economic or strategic considerations.

Deep mines

The more comprehensive term, **underground mining**, used in the USA, is a better description in this case, as the category includes not only the deep mines reached by vertical shafts, but others with slope or 'drift' entry. Getting the coal is of course less simple than on the surface, and various methods of extraction have been used over the centuries since the first underground mines. The longwall method, already in use in Britain as long ago as the seventeenth century, has proved suitable for mechanisation. Two long parallel tunnels or **roadways** are first driven, and these define the **parcel** of coal to be worked. A tunnel connecting these two roadways then exposes the coal face. In the past, before mechanisation, the coal would be extracted by hand-cutting a deep horizontal slot at the base of this face, along its entire length, and bringing down the coal with the aid of picks, drills and sometimes explosives. The miners would then hand-load the coal for transport back to the mineshaft. The first cutting machines had appeared in the mid-nineteenth century, and a hundred years later, Britain's first cutter-loader was installed; but manual operations, and in particular manual loading, remained common well into the second half of the twentieth century. (In the very early days, children had often hauled the coal underground,

with women carrying it up the shaft to the surface in baskets. In 1842, against some resistance, an Act of Parliament prohibited the employment in mines of children under ten.)

The fully mechanized present-day underground mine is very different. An **armoured face conveyor** perhaps 250 metres long stretches along the length of the face, and the cutting machine works its way along on this. The cutting machine, or **power loader**, has two large rotating drums armed with steel picks. The cut coal falls on to the conveyor and is carried away along the face. In the most recent installations the conveyor is articulated, and advances automatically, together with the roof supports, as the power loader moves along. By the end of each traverse the face is ready for the next run in the opposite direction.

The main adverse environmental effects of deep mining have been land subsidence and waste tips. Tips can leak dangerous chemicals into the ground, and also present other dangers. In 1966, an unstable waste tip from the Merthyr Vale mine in Wales released an avalanche of sludge over the village of Aberfan, engulfing not only houses but the primary school, killing 114 children. Legislation and technical advances have brought improvements in more modern mines, and the closure of many of Britain's older mines appears in some cases to have been followed by successful land rehabilitation.

Accidents and lung diseases have always led to high rates of death or disability amongst miners. In mid-nineteenth century Britain, one in every five underground workers was likely to be killed in an accident before completing a full working life. By 1912, when Britain had about a million miners, the likelihood had fallen to about one in twenty over a working life, although 1913 saw Britain's worst mining disaster, with 439 miners killed in an explosion at the Senghenydd mine in Wales. Towards the end of the twentieth century, the death rate in the USA, with some 160 000 underground workers, was still more than one in fifty over a working lifetime.

It has also been well known for many years that miners, or retired miners, suffer excess rates of respiratory diseases compared with the population at large. Establishing the exact relationship between these rates and exposure to the atmosphere underground is obviously more difficult than counting the deaths and injuries in accidents. Nevertheless, it is perhaps surprising that only in 1950 was pneumoconiosis formally recognised in the USA as a work-related disease – and that another nineteen years passed before the enactment of legislation on dust conditions. (Chapter 13 discusses epidemiological issues in more detail.)

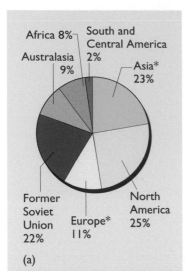

(a)

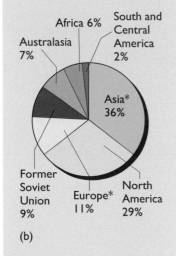

(b)

Figure 5.7
(a) World coal reserves by region, 2000. Total reserves: 745 000 Mtce;
(b) World coal production by region, 2000. Total annual production: 3210 Mtce.
Notes: † Weighed total – see the main text
* Excluding countries of the former Soviet Union
Source: Data adapted from BP, 2001

Reserves and production

Coal is a widespread resource, mined in over 100 countries in every continent except Antarctica. Figure 5.7 shows the regional distribution of reserves and current production for the year 2000. Note that figure for world total reserves is well under a thousand billion tonnes – far below the range given at the start of this section. The main reason for the discrepancy lies in a more restrictive definition of 'reserves'. The reserves of Figure 5.7 are those '*which geological and engineering information indicates with reasonable certainty can be extracted in the future under existing economic and operating conditions*' (EIA 2002). These recent data also take into account different heat values, using the method described in Box 5.2.

BOX 5.2　Assessing the data

In Chapter 2 we discussed the problems that arise in using data from different sources. These problems are particularly acute in the case of coal, for several reasons. The first is the continuing use of different basic units: metric tonnes, tons and short tons (see Box 2.5). This involves tedious conversions, but it isn't the main problem.

The real difficulty arises from the very different heat values of different types of coal, mentioned above. Some data sources ignore these differences and simply add the masses of all types to give one grand total. Others divide coal into the two main categories and assign an average heat value to each. A 'total heat value' is then calculated, expressed usually in million tonnes of coal equivalent (Mtce). The following example shows how the method works.

Data for the year 2000
World hard coal production	= 2860 Mt
World brown coal/lignite production	= 700 Mt

Other values
Average heat value of hard coal	= 28 GJ per tonne
Average heat value of brown coal/lignite	= 14 GJ per tonne
One million tonnes of coal equivalent (1 Mtce)	= 28 PJ

The two heat values translate into 28 PJ and 14 PJ per *million* tonnes respectively, so the total heat value is

$$(2868 \times 28) + (700 \times 14) = 89\,880\ \text{PJ} = 89\,880/28\ \text{Mtce} = 3210\ \text{Mtce}$$

This figure of course differs significantly from the simple total of 3560 million tonnes, but the method is increasingly adopted as more closely reflecting the true energy value of the coal.

Some authorities use a similar method but with more than two categories. Unfortunately it is not always easy to discover which method is being used, or which heat values – or even which conversion from joules to Mtce. And even such reliable sources as the International Energy Agency (IEA), the World Coal Institute (WCI), British Petroleum (BP) and Britain's Department of Trade and Industry (DTI) differ on the most basic information of all: the actual tonnes of coal produced or consumed in a particular country in any given year.

The data given in this chapter and elsewhere in the book, whilst drawing on many sources, are in the main consistent with the BP Statistical Yearbook, 2001, 2002 (BP, 2001; BP, 2002).

The coal resource can vary greatly even on a local level, between adjacent countries or different regions of one country. Table 5.1 the shows the 'known reserves' at the start of the present century for the countries with the largest resource, together with others of interest. As the table shows, even these economically and technically practicable reserves are sufficient to last the world for over two hundred years at current extraction rates.

Table 5.1 Coal production, exports, reserves and R/P ratios for selected countries, 2000

Country	Reserves (thousand Mtce)	Percent of world reserves	Annual production (Mtce)	Percent of world production	R/P ratio (years)	Annual consumption (Mtce)
USA	179	24.0%	856	26.7%	210	846
Russian Federation	103	13.8%	174	5.4%	593	166
China	88	11.8%	747	23.3%	118	720
India	74	9.9%	231	7.2%	318	245
Australia	69	9.2%	233	7.3%	295	70
South Africa	55	7.4%	178	5.6%	310	123
Germany	46	6.1%	85	2.6%	538	124
Kazakhstan	33	4.4%	58	1.8%	564	35
Ukraine	25	3.4%	63	2.0%	402	5
Poland	13	1.8%	102	3.2%	129	86
Brazil	6	0.8%	5	0.1%	1245	18
United Kingdom	1.3	0.2%	29	0.9%	43	57
France	0.1	0.01%	3	0.1%	31	21
WORLD	745	100%	3206	100%	233	3280

Note: The data in this table are weighted values, taking into account the differences in heat value of various types of coal (see Box 5.2).
Source: Adapted from BP, 2001

The USA, with about a quarter of the world's reserves, has a **reserves/production ratio** that is about the same as for the world as a whole. Its coal lies mainly in two broad areas, east and west of the Mississippi (Figure 5.8). The reserves of the two areas are nearly the same in terms of tonnage, but there are significant differences. More than half the eastern coal comes from deep mines. The mining costs are therefore relatively high, but are in part offset by a high energy content per tonne and the proximity to industrial customers. In contrast, about half the western coal is surface mined, in some cases from very large beds. The Wyodak bed in Wyoming and adjacent states, for instance, stretches for hundreds of kilometres and contains an estimated twenty billion tonnes of coal within a depth of 100 metres. Western coal also has the advantage of low sulphur content. However, its average heat value is only about half that of coal from the east, so in energy terms its contribution to coal reserves is little more than a third of the total.

China's reserves and reserves/production ratio are both about half those of the USA, reflecting the fact that their annual coal production is not very different. But coal supplies two-thirds of China's primary energy, compared with less than a quarter for the US. When we also take into account the fact that China has four times the population of the USA, we have good example of the disparities discussed in Chapter 2.

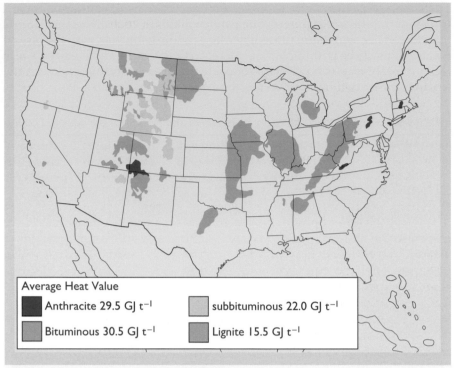

Average Heat Value

- Anthracite 29.5 GJ t^{-1}
- Bituminous 30.5 GJ t^{-1}
- subbituminous 22.0 GJ t^{-1}
- Lignite 15.5 GJ t^{-1}

Figure 5.8 The main coal fields of the USA

The high R/P ratio of the Russian Federation is worth noting. It is a relatively new feature: a consequence of the recent fall in industrial coal consumption following the break-up of the Soviet Union. The rather surprising fact that Brazil has large imports despite a very high R/P ratio is due mainly to the need to import coking coal for the country's steel industry. India is in a similar position, with the world's fourth largest reserves but needing to import about a quarter of her coking coal.

The history of British coal during the twentieth century offers interesting examples of the problems of forecasting – and of the need to interpret resource data with care. In 1915, Jevons predicted that in the year 2001 Britain would consume 350 million tonnes of coal and export a further 400 million tonnes (Jevons, 1915). (For the actual situation in 2000, see Figure 5.4 above and Chapter 2.) He also estimated that Britain's 'remaining coal reserves' in 1915 were about 190 billion tons. A mid-century estimate (Brown, 1950) was slightly lower at 172 billion tons, but some eight billion tons had been mined in the intervening thirty-five years, so the discrepancy was not very great. A more serious problem arises when we look at the year 2000. Britain's total coal production during the period from 1950 to 2000 was about 7 billion tonnes, which should have left more than 160 billion tonnes in place. But Table 5.1 shows the reserves as just 1.3 billion tonnes. Where is all the missing coal? The answer could be that the definition of 'remaining reserves' has undergone a major change, but there is another possible explanation, related to the fate of the British coal industry in the second half of the twentieth century. As Figure 5.4 shows, annual production had risen after World War II, reaching about 220 million tonnes in 1950. But it then fell almost continuously to less than 30 million tonnes by the end of the century. This drop in output reflects the closure of over a thousand mines, leaving just eighteen major mines in 2000. The closed mines may have been economic (or marginally economic) in the past, but would almost certainly be uneconomic to reopen in today's world coal market – and many are already dismantled, flooded or otherwise beyond reopening. In consequence, although most of the coal of 1950 is still in place, the country's 'technically and economically accessible' reserves have fallen to less than a hundredth of their mid-century value.

The Australian situation could hardly be more different. Proven reserves of hard coal increased almost four-fold between 1950 and 2000, and there is obvious potential for similar growth in the future (Figure 5.9). Including all forms of coal, the reserves are just under a tenth of the world total and sufficient to last for almost 300 years at current rates of production. Moreover, the rate of production in 2000 was over *three times* the rate of consumption. Australia may be only the third largest coal producer, but it is by far the world's largest exporter, selling coal to over thirty countries. In 2000, about half the exports were to Japan (the world's largest coal importer). About 30% went to other Asian countries – including 13 million tonnes to India – and most of the rest to Europe and the Middle East. Coal is Australia's largest export, and investment (half of it from foreign investors) has resulted in a modernized system with dedicated trains running from loading terminals near the mines to ports on the coast. The fact that Australia can then send the coal on a sea voyage of some twelve thousand miles and yet sell it at a competitive price in Britain is a striking illustration of the relative importance of two factors in determining the cost of the fuel: the nature of the coal seams obviously matters much more than their distance from the user.

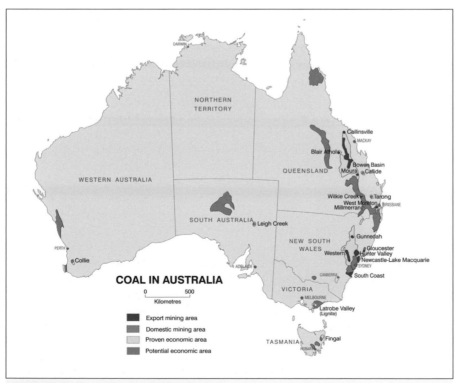

Figure 5.9 Australian coal fields and mines

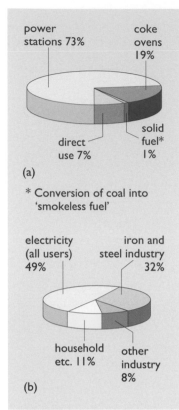

(a)

* Conversion of coal into 'smokeless fuel'

(b)

Figure 5.10 Uses of coal in the UK, 2000
(a) First destinations
Primary energy supplied by coal:
57 Mtce or 1.60 EJ
(b) Final distribution to users
Final energy derived from coal:
29 Mtce or 0.81 EJ

The uses of coal

The major use of coal in the late twentieth and early twenty-first centuries has been for the generation of electricity. In the year 2000, almost two-thirds of world coal output was sold as **steam coal** for use in power stations. An input of some 2100 Mtce of coal fuelled an output of 5400 TWh, about a third of the world electricity total. In energy terms, this represents 19 EJ of output from 59 EJ of input, an overall conversion efficiency of 33%.

The second important user is the iron and steel industry, consuming around 600 Mtce per year – about a fifth of world coal consumption in 2000. We have seen that coke plays an essential role in this industry, and with present-day technology each tonne of steel produced in blast furnaces requires an input of 600–800 kg of coking coal (Box 5.3). About two-thirds of the world's steel is produced in this way, with the rest coming from electric arc furnaces, in which scrap iron and steel are melted together. The electricity for these often comes from coal-fired power plants, so coal is again an important requirement.

In most countries, the proportion of coal used for purposes other than electricity generation or iron and steel production is quite small, often less than ten percent. Figure 5.10(a) shows the initial destinations of the coal consumed in Britain in 2000, with just over 90% going to power stations or coke ovens. In terms of final use, some steam coal goes directly to industrial plants other than power stations, whilst some coke shares with anthracite and 'smokeless fuel' derived from coal a relatively small market for direct combustion in heating boilers in domestic, commercial or public buildings – or even in open fireplaces.

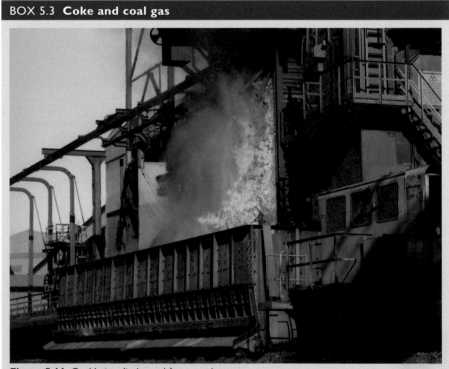

BOX 5.3 **Coke and coal gas**

Figure 5.11 Coal being discharged from a coke oven.

> **Producing coke**
>
> Coke is produced by heating finely divided coal in the absence of air. The coal particles, no more than 3 mm in size, are heated to above 1200 °C over a period of up to twenty hours. During this time the more volatile constituents are driven off as coke oven gas, and the remainder, almost pure carbon, becomes liquefied. As the oven cools, this liquid fuses ('resolidifies') to form the coke. Not all types of coal, when subjected to this process, fuse to produce the required hard porous coke. Types that do are called **coking coal**.
>
> **Producing coal gas**
>
> As we see in the next section, the volatile coke oven gas is itself a fuel. Today it is mainly regarded as a valuable by-product that can be used to heat the coke ovens or for other purposes in the steel works. However, for much of the nineteenth and twentieth centuries in Britain and other coal-rich countries, the situation had been very different. Every town of any size had its gasworks, using the above process to produce the 'town gas' that provided lighting, then heat for cooking, and eventually fuel for the central heating of buildings. In the gasworks, *coke* was the by-product, sold as a fuel for domestic boilers or the central-heating furnaces of larger buildings. (See also WCI, 2002 and see Chapter 7 for more on coal gas and other 'secondary fuels'.)

5.4 **Coal combustion**

The composition of coal

Many of the unattractive features listed in the introduction to this chapter are due to the extremely complex nature of coal. As in all fossil fuels, the constituents that are important for combustion are carbon and hydrogen; but coal, unlike gas or oil, cannot be analysed at the molecular level into relatively straightforward hydrocarbon compounds. Rings of six carbon atoms play an important role in the structure, forming layered arrangements that incorporate not only hydrogen, but significant amounts of oxygen and nitrogen. The structure also includes differing quantities of sulphur and traces of other environmentally undesirable elements, and coal always contains some inert mineral material with no fuel value at all, destined to remain as ash. Finally, all coal incorporates some moisture within its structure.

A full chemical analysis of a sample of coal, called an **ultimate analysis**, lists the proportions by mass of the main elements that are present, usually carbon, hydrogen, oxygen, nitrogen and sulphur. These analyses are often in terms of dry, ash-free samples, excluding any moisture and inert matter. Figure 5.12(a) is an example of such an analysis, for a type of steam coal that might be used in a power station. However, at the most basic level, it is the relative proportions by *atoms* of the various elements that matter, and as we have seen, different atoms can have very different masses. (Table 4.2 in Chapter 4 shows that, for instance, a carbon atom has twelve times the mass of a hydrogen atom.) Figure 5.12(b), taking this into account, shows the percentages of the different atoms – as we shall see, a more informative picture of their significance in the combustion of the coal.

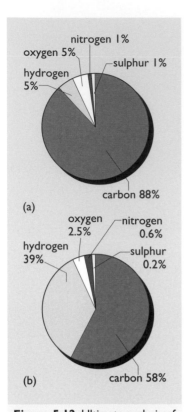

Figure 5.12 Ultimate analysis of a dry, ash-free coal sample
(a) Percentages by mass
(b) Percentages by number of atoms

The ultimate analysis is undoubtedly of interest to the coal scientist, but the technologist needs answers to a different kind of question: Does a particular type of coal ignite easily or only with difficulty? Which types are suitable for use in power stations, and which can be used for the production of coke? In such a complex material, theory cannot yet take us from the ultimate analysis to the answers to these questions. A different analysis is needed; one that reflects the value of the coal to its users. Heating is the essential feature common to all uses of coal, leading either to full combustion or part-combustion, or to the production of other fuels, and it is the sequence from heating to combustion that provides the basis for the most common characterisation of different coal types.

The combustion process

The amounts of inert material and moisture are obviously significant in determining the heat values of different types of coal; but there are other features that play important roles. They are best explained by considering the series of processes leading to the complete combustion of a coal sample.

- In the early stages of combustion, as the coal is heated, any **moisture** in the structure is also heated and evaporates. The moisture content can vary from one or two percent to as much as a tenth of the total mass in hard coals, and twice that in brown coals. The heating and evaporation both use some of the energy of the coal, but as Box 5.4 shows, this 'loss' is only a small fraction of the total energy released in combustion, even for coal with relatively high moisture content.

- As the temperature continues to rise, a range of gases is evolved. These are called the **volatile matter** (**VM**), and arise from the dissociation of the coal structure. They carry most of the hydrogen and oxygen in the coal and some of its carbon, and consist of carbon monoxide (CO), methane (CH_4) and a variety of other hydrocarbons – the 'bitumens' that give the coal type its name. These are fuels, releasing heat as they burn, and as much as half the heat energy from the coal may appear in this form. Anyone who has watched a wood or coal fire will have seen the spurts of intense flame from these little jets of gas.

- The combustible part of the material remaining after the volatile matter has gone is the remainder of the carbon, known as the **fixed carbon** (**FC**). This is effectively charcoal, or coke, and can burn at a high temperature in oxygen from the air:

$$C + O_2 \rightarrow CO_2$$

Depending on the type of coal, the fixed carbon can account for virtually all the heat output or no more than half.

- Finally, with all the fuel burnt, any inert material remains as **ash**. A high ash content is obviously undesirable, and the best coals have less than ten percent ash. However, coals with up to 15% are fairly common, and in some countries as much as 30% is tolerated if the priority is to use local rather than imported coal.

This analysis carries a number of lessons for the designer of furnaces burning solid fuel, a subject to which we return shortly. It is also the basis of the customary specification of different types of coal.

BOX 5.4 Removing the moisture

Energy is needed to heat the moisture in the coal to its boiling point and then to evaporate it. For simplicity, we assume that this 'moisture' has essentially the thermal properties of water, and that it is initially at 20°C. The energy required to heat one kilogram to the boiling point (100°C) can then be calculated using the data and method of Box 4.4 in the previous chapter:

energy needed = $1 \times 4200 \times 80 = 336\,000$ J = 0.336 MJ per kilogram.

The energy needed to evaporate 1 kg of any liquid is its **specific latent heat of vaporization**, and its value for water is 2.258 MJ per kilogram.

The total energy required, per kilogram of 'moisture' is therefore 2.594 MJ, or about 2.6 MJ.

Suppose that the coal has 10% moisture. One tonne (1000 kg) of coal will contain 100 kg of moisture, and the energy needed to heat and evaporate this will be about **260 MJ**.

With 10% moisture, and perhaps 10% inert mineral matter that contributes no energy, the tonne of coal will contain 800 kg of actual fuel. Assuming that this has a heat value of about 35 MJ per kilogram, the total heat energy released when the 800 kg burns will be 28 GJ, or **28 000 MJ**.

Comparison of the above two figures shows that, even for this coal with its high moisture content, the energy 'lost' is less than 1% of the total. (Of course, purchasers of this coal will no doubt wish to take into account the fact that 200 kg of the tonne of 'fuel' they are buying is either dirt or dirty water.)

Proximate analysis

Compounds consisting mainly of carbon, hydrogen and oxygen are characteristic of living materials, and in this sense coal is in closer to its origins than the hydrocarbons of oil and gas. Indeed, the different types or **ranks** of coal, from brown coal and lignite at one extreme to anthracite at the other, can be regarded as members of a sequence which starts with wood and peat. Figure 5.13 shows some members of this series, analysed in terms of the four constituents appearing in the account of combustion above. This **proximate analysis** reveals that the percentage of fixed carbon is a factor in determining the rank. As can be seen, the heat of combustion (given at the foot of each column) also tends to increase with the rank of the coal.

Other properties than those appearing in Figure 5.13 are relevant in the selection of coal for a particular purpose. As we have seen, anthracite is a naturally smokeless fuel, suitable for direct use where other coals would not be permitted. The low-volatile bituminous coal is the steam coal whose ultimate analysis appeared in Figure 5.12. Any of the bituminous coals might be a coking coal, but the proximate analysis alone does not tell us whether the fixed carbon agglomerates when cooled, as required, or appears in a powdery form unsuitable for coke (see Box 5.3 above).

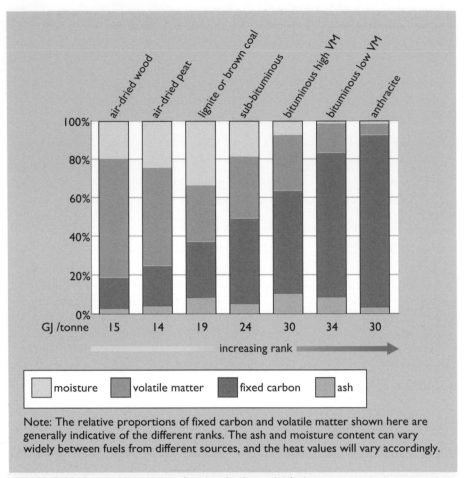

Figure 5.13 Proximate analysis of coal and other solid fuels

Combustion products

In the full combustion of coal, all the constituent elements shown in Figure 5.12, and some of the others present as minor impurities, undergo chemical change in the heat of a coal fire or furnace. Carbon and hydrogen are the main constituents, and their reactions with oxygen lead to the main combustion products: carbon dioxide and water (initially in the form of steam). The quantities of these products can easily be calculated if we know the percentages of carbon and hydrogen in the coal and the appropriate relative atomic masses. As Box 5.5 shows, the combustion of one tonne of coal with the composition of Figure 5.12 releases about three tonnes of carbon dioxide into the atmosphere. In terms of the electrical output of a power station using this coal, this becomes almost one kilogram of CO_2 per kilowatt-hour. (The corresponding figure for a gas-turbine plant is about one third of a kilogram per kilowatt-hour of electrical output.)

BOX 5.5 Carbon dioxide released in coal combustion

How much carbon dioxide is released in the combustion of one tonne of coal, and what are the consequences for the emission of CO_2 from a coal-fired power station?

We assume that all the carbon in the coal reacts with oxygen to form carbon dioxide:

$$C + O_2 \rightarrow CO_2$$
$$12 \quad 2 \times 16 \quad 44$$

The relative atomic masses of carbon and oxygen are 12 and 16 respectively, so the numbers under the chemical equation are the relative masses of the three items, and we see at once that 12 kg of carbon produce 44 kg of carbon dioxide. Each kilogram of carbon in the coal therefore releases 44/12 = **3.67 kg** of CO_2.

We'll consider one tonne of the low-VM bituminous coal in Figure 5.13, which is also the coal of Figure 5.12.

10% of this coal is moisture or ash, so each tonne becomes 900 kg when dry and ash-free, and the carbon content (Figure 5.12) is 88% of this, or **792 kg**.

So the mass of CO_2 released in the combustion of one tonne of this type of coal is

$$792 \times 3.67 = 2904 \text{ kg} = 2.90 \text{ tonnes}$$

The heat value (Figure 5.13) is 35 GJ per tonne. Suppose now that this coal is used in a power station. We have seen in Chapters 2 and 3 that the electrical energy finally reaching the consumer is about one third of the fuel energy input, so for each tonne of coal used, the final electrical energy is about

$$35 \div 3 = 11.7 \text{ GJ} = 11\ 700 \text{ MJ}$$

One kilowatt-hour is the same as 3.6 MJ, so this 11 700 MJ is 3240 kWh.

Burning one tonne of coal therefore releases 2904 kg of carbon dioxide and produces 3240 kWh of electricity: *just under one kilogram of CO_2 per kilowatt-hour.*

There are other combustion products. Nitrogen can combine with oxygen to produce oxides of nitrogen. These compounds (N_2O, NO, NO_2, etc.), known generically as **NO_X** (or NOX), are produced mainly by nitrogen from the coal, but nitrogen from the air may also contribute. Any sulphur present in the coal will readily form sulphur dioxide (SO_2). All coals contain some sulphur, and it can account for as much as 5% of the total mass. In the high-sulphur coals, half or more of the sulphur may be in the inert mineral material, and can be removed by washing the coal. The remainder, however, is incorporated into the structure of the coal, and inevitably forms SO_2 as the coal burns. Australian coals, for instance, have relatively low sulphur content, but even 1% in this form releases 20 kg of sulphur dioxide per tonne of coal burned.

The consequences of these emissions for a large coal-burning power station are described in *Flue Gases* below.

5.5 Fires, furnaces and boilers

The design of furnaces may not seem the most exciting topic in the world, but it is certainly an important one in the world of energy. As we have seen, all but a tiny fraction of our primary energy comes from fuels – fossil fuel or biofuel – and the destination of three-quarters of these fuels is a fire, furnace or boiler. (The destination of the remainder is the internal combustion engine, to which we return in Chapter 8.) The efficiency with which a boiler extracts energy from the fuel is a critical factor in determining the fuel consumption of systems ranging from a domestic heating boiler to a 1000 MW power station. Better design means a lower fuel requirement and therefore lower costs – and lower emission of carbon dioxide and other undesirable products. Or in a very different context, better design means a more efficient cooking stove, lower wood requirement and less destruction of valuable vegetation.

This last example can serve to illustrate the situation facing anyone trying to design a really efficient solid-fuel boiler. How much wood is needed to bring one litre of water to the boil on a fire? Common experience will give a rough answer, but Box 5.6 shows the results of a calculation based on known data. As the concluding remark suggests, the stove shown might have a fuel-to-useful heat efficiency of about 10%, which we can compare with perhaps 70% for a well-run domestic gas water-heater and 90% or more for a modern power station boiler.

BOX 5.6 Boiling a litre of water

How much wood is needed to bring one litre of water to the boil?

Data

Specific heat capacity of water	$= 4200 \, \text{J kg}^{-1} \, \text{K}^{-1}$
Mass of 1 litre of water	$= 1 \, \text{kg}$
Heat value of wood (Figure 5.13)	$= 15 \, \text{MJ kg}^{-1}$
Density of wood	$= 600 \, \text{kg m}^{-3}$
1 cubic centimetre (1 cm³)	$= 10^{-6} \, \text{m}^3$

Calculation

Heat energy needed to heat 1 litre of water from 20°C to 100°C	$= 80 \times 4200 \, \text{J}$
	$= 336 \, \text{kJ}$
Heat energy released in burning 1 cm³ of wood	$= 15 \times 600 \times 10^{-6} \, \text{MJ}$
	$= 9.0 \, \text{kJ}$
Volume of wood required	$= 336 \div 9.0 \, \text{cm}^3$
	$= \textbf{37 cm}^3$.

This suggests that one thin stick about a foot (30 cm) long should be enough. Any reader who has tried boiling water on a simple open wood fire will no doubt find this result surprising, but the well-designed stove of Figure 5.14 would perhaps need only ten such sticks per litre of water.

Figure 5.14 Boiling water

Much research has been devoted to the design of modern household boilers, and many interesting ideas have gone into the development of simple energy-efficient cooking stoves for combustion of biofuels. Unfortunately space doesn't allow detailed discussion of a range of systems, and we concentrate therefore on the major consumers of coal: the large 'boilers' used to produce steam in power stations. (For the other components of a power station, see *Turbines* in Chapter 6 and *Generators* in Chapter 9. For more on the uses of biofuels, see the companion volume, Boyle, 1996.)

Power station boilers provide a useful study for several reasons:

- At their best, they are amongst the most efficient fuel-burning systems we have.
- Their waste products create some of the world's major pollution problems.
- Interesting technological solutions to some of the problems exist or are being developed, but are still to be universally adopted.

Power-station boilers

The starting-point for the furnace designer must be the sequence of events described in the above section on combustion. To extract the maximum energy from the solid fuel, both the fixed carbon and the volatile matter must be fully burnt. As one part is solid material and the other a stream of gases, this is not simple. Modern plants have of course long abandoned the person with a shovel, and operate with a continuous feed of fuel, so the solids and gases must burn completely at about the same rate. The purpose of the plant is to produce steam, so another requirement is for the best possible *heat transfer* from the burning fuel into the circulating water. Finally, the system should minimise the production of undesirable by-products and include a method for dealing with the unavoidable wastes: ash and flue gases.

Coal is of course the main subject of this chapter, but the following brief accounts of solid-fuel boilers include occasional references to other fuels. With the ever-growing problem of disposal of waste materials, the versatility of a power-station furnace becomes increasingly important. The aim is a plant that can deal safely and efficiently with fuels that may have very different physical properties and will certainly have a wide range of heat values.

Grate boilers

None of the boilers to be described here is fed with large lumps of coal. These disappeared with the person with the shovel. In a **grate boiler** (Figure 5.15(a)), the fuel is in pieces a few millimetres across, which are fed in from a hopper or on a conveyor belt, and move across the grate in an upward flow of air. With this arrangement, the fixed carbon burns on the grate and the volatile matter in the space above. Radiant heat from both reaches an array of tubes through which the water circulates, while the hot gases from the combustion pass between another set of tubes. Note that the few tubes shown in the diagram represent the large, much more complex array used in the actual system. Boilers of this type are still used for coal, but mainly for biofuels such as wood chips, processed domestic wastes and other

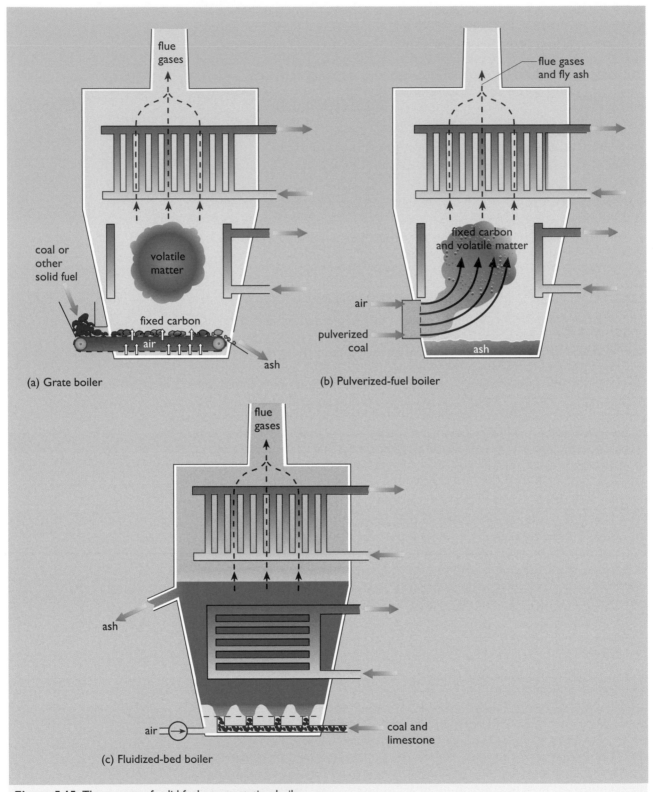

Figure 5.15 Three types of solid fuel power-station boiler

materials that are not suitable for pulverized fuel boilers. Increasingly, however, they are being replaced by the cleaner, more efficient fluidized bed boilers when systems are upgraded.

Pulverized fuel boilers

The plant shown schematically in Figure 5.15(b) is by far the most common boiler type in present-day coal-fired power stations. Plants of this form have been in use for half a century or more, and when well run can transfer over 90% of the energy content of the coal to the circulating water (or steam).

The essential feature of this boiler is that it uses **pulverized fuel (PF)**. The coal enters the furnace in the form of particles less about 100 microns (0.1 mm) in size – a dust so fine that it floats. This pulverized fuel is swept in a controlled flow of air to the burner jets. With such tiny particles, the fixed carbon burns out completely in a very short time, so the volatile matter and fixed carbon burn together in roughly the same part of the furnace, increasing the efficiency of heat transfer. Another advantage of the short time that the fuel spends in the furnace is that it reduces the production of NO_x and other combustion products. However, achieving these ends requires careful control of the air/fuel mixture. Insufficient air for combustion leaves unburnt char in the ash, or carbon monoxide in the flue gases. Too much air promotes the production of undesirable oxides, and also reduces the efficiency by carrying away more heat in the flue gases. Considerable research effort is being devoted to the development of on-line monitoring and control of fuel flow rates and particle size, but such systems have yet to be widely adopted.

A serious disadvantage of PF boilers is that the ash is a fine dust, and without measures to prevent it, this **fly ash** will be carried into the atmosphere with the flue gases. (For more details of this and other pollutants, and the methods used to reduce them, see *Flue Gases*, below.)

Fluidized bed boilers

The third type of plant in Figure 5.15 uses a rather different principle. **Fluidized bed combustion (FBC)** is regarded as offering solutions to some of the pollution problems of coal combustion, and in addition, the possibility of burning other fuels cleanly. FBC plants have been in use for some time in various industrial processes, but investment in the development of coal-burning systems for power was at a relatively low level until the 1970s. Work then accelerated, and the first plants came on stream in the 1980s. By the end of the century a few thousand were in operation, burning not only coal but other fuels including a variety of wastes. The following is a brief account of the principles of the main types of FBC plant. The means by which they reduce pollution are discussed in *Flue Gases*, below.

The essential feature of an FBC plant is a thick layer of inert material – sand or gravel, with particle size in the range 0.3–2.0 mm. This lies on a base plate that has many small apertures through which jets of air are blown. At a certain air speed the whole mass of material expands to a depth of a metre or more and starts to behave like a liquid. Objects in it will float or sink, just as in a liquid. The significant feature for combustion and heat transfer is that as the air flow is further increased it forms bubbles that rise through the bed, and the particles bounce around as if they were indeed in a boiling liquid. In a power-station system, fuel particles are fed into this bed, and because of the constant motion and the air flow, the fixed carbon and the volatile matter both burn quickly and heat the entire bed. The tubes carrying the water to be heated are buried in the bed and/or the containment walls, and the excellent thermal contact with the constantly moving material means that good heat transfer doesn't require the very high temperatures needed in an 'open' furnace. This process, called **bubbling fluidized bed combustion (BFBC)** is used in the majority of present FBC plants, most of them in the small-to-medium range, with electrical outputs below 100 MW.

A different system, **circulating fluidized bed combustion (CFBC)**, was developed a little later, but the number of plants grew relatively rapidly – to more than a thousand world-wide in the early years of the present century. With this system, an increased air flow drives some of the particles out of the fluidized bed, into the space above, where they behave more like a hot gas. A circulating system constantly returns them to the bed, maintaining the high temperature and increasing the time available for combustion, which allows a wider range of coal types – and other fuels – to be used.

The most advanced of the fluidized bed systems uses **pressurized fluidized bed combustion (PFBC)**. The present PFBC plants operating commercially are based on bubbling bed systems, but with the important difference that the pressure in the furnace is increased to about ten times atmospheric pressure. This means that the hot gases from the furnace can be used in a gas turbine as well as raising steam for the steam turbine. (Chapter 8 discusses these combined-cycle systems in more detail.) The largest PFBC

Figure 5.16 Osaki Power Station

plant at the time of writing (2002) is the 250 MW Osaki Power Station in Japan (Figure 5.16). It has two large beds, between 2 m and 4 m high when fluidized, and a furnace pressure of about 1 MPa. (Hitachi, 2001). Pressurized systems using circulating beds are also under development, but have not reached full commercial production at the time of writing. (See also DTI, 2000.)

Flue gases

We have seen that the inevitable result of burning coal is a range of combustion products. Here we look at the quantities of each of these that might be released into the atmosphere in the course of one hour by a modern 660 MW coal-fired power station, in the absence of any pollution reduction system. The figures are approximate and will of course vary with the nature of the coal and the combustion process.

- 2500 tonnes of nitrogen. This is four-fifths of the air drawn in for combustion and has passed largely unchanged through the whole system, except that heating it accounts for about half the energy loss in the boiler. Four-fifths of the air we breathe is in any case nitrogen, so it presumably remains harmless.

- 700 tonnes of carbon dioxide produced in combustion. CO_2 is a minor constituent of normal air and is not thought to be harmful to life in the concentrations produced by power stations. However, the cumulative effect on the climate of the CO_2 from all the world's power stations is a serious issue (see Chapter 13).

- 150 tonnes or more of steam. This is the other combustion product, and some also comes from moisture originally in the coal. Not condensing this steam accounts for the other half of the lost energy, but the flue gases need to stay hot if they are to rise out of the tall chimney.

- A tonne or so of oxides of nitrogen. The higher the furnace temperature, the greater the production of NO_X. These contribute to acid rain and are damaging to health, but at present the world-wide contribution from power stations is considerably less than from internal combustion engines.

- From one to twenty tonnes of sulphur dioxide gas (SO_2). The power station of this example would produce about four tonnes of SO_2 per hour from coal with 1% sulphur. It is this sulphur compound which is held largely responsible for acid rain.

- 10–20 tonnes of fly ash. Otherwise known as **particulates**, this is the fine ash resulting from burning pulverized coal. Its main visible feature is that it is very dirty, but the tiny particles can damage the lungs, and may also contain poisonous impurities.

There are two ways of dealing with these pollutants: remove them, or don't produce them in the first place. The second is technically the more elegant solution.

Fluidized-bed boilers adopt this second approach for sulphur and the oxides of nitrogen. Most of the sulphur compounds are removed at source by introducing limestone particles into the bed. The SO_2 reacts with this to produce calcium sulphate, a hard inert material that can be removed from the bed. The production of NO_X is considerably reduced by maintaining the bed temperature at under 1000 °C, instead of the 2000 °C in the hot gases of conventional plant. Fluidized beds will also release particulates, although in far smaller quantities than the equivalent pulverized fuel plant. The flue gases of FBC boilers that operate at atmospheric pressure are cleaned using the methods described below for PF plants. In the case of PFBC, particulate removal is an integral feature of the plant, as the gases must be cleaned to a high standard before they can be used in a gas turbine.

For the majority of existing power station boilers, the practicable options are more limited. Improved burner design has reduced NO_X emissions to some extent, but fly ash and SO_2 remain, and are dealt with by removing them from the flue gases. Although there are several techniques for doing this, relatively few coal-fired plants world-wide yet reach the best achievable reductions of both pollutants. Before we criticise 'them' for failing to do this, we should perhaps place the problem in context. Three thousand tonnes an hour of gases leave the boiler. They are hotter than the hottest domestic oven, with a particulate level greater than in the worst ever London smog, a concentration of SO_2 a thousand times worse than downtown Los Angeles on a bad day, and enough moisture to cause it to start raining in the gas stream if the temperature falls below that of a moderate oven. Any cleaning system must be able to handle this hot, dirty corrosive mass on a continuous basis. It should remove most of the pollutants (about 90% of the SO_2 and 99% of particulates are current aims), use as little energy as possible, leave environmentally acceptable residues – and be cheap. There is of course no such system.

In Britain, and in Australia with its low-sulphur coal, most effort was initially devoted to reducing the particulates in the flue gases. There are several ways to do this, of which the most common are bag filters, cyclone filters and electrostatic precipitation – or some combination of these. The principles of the first two methods are respectively those of a traditional household vacuum cleaner and of the newer cyclone type where particles are thrown outwards from the fast-spinning air. Both methods are relatively cheap, but they are become inefficient for particles sizes of less than about 10 μm.

In an electrostatic precipitator, the fine particles acquire an electric charge by passing near a high-voltage wire and are then pulled sideways out of the gas stream by an electric field. This method is effective even for very small particles and for most types of coal, but the capital cost is considerably greater.

Flue gas desulphurization (FGD) to remove the sulphur compounds usually involves reacting the SO_2 with finely divided limestone (calcium carbonate, $CaCO_3$). This can be a dry process, but more usually a slurry or spray or jets of water are used to bring the limestone into contact with the flue gases, so that the resulting insoluble calcium sulphate precipitates and can be removed.

All the above processes of course involve costs. Electrostatic precipitators can add over 5% to the capital cost of a new power station, and FSD as much as 15% (IEA, 1999). The processes also use energy and reduce the overall efficiency of the plant, effectively increasing the fuel cost per unit of output. The issues raised when we try to balance these costs against the 'costs' of the deleterious effects are discussed in later chapters.

5.6 **Summary**

This chapter, like the previous one, has involved a journey through time, but with a different emphasis. It has concentrated on technological developments rather than scientific ideas, and the 'history of coal' as been the story of how it has been used. We have seen how the inventions of the eighteenth and nineteenth centuries led to an almost total reliance on coal as the energy source for an increasingly industrialized world. And although 'king coal' is no longer dominant, we have also seen that the mid-twentieth century obituaries were definitely premature. It is always interesting to look at past forecasts. A hundred years ago, coal was going to be the main fuel for centuries to come. Then came oil and gas, and coal would surely follow wood into relative obscurity. Then we had oil crises, and the world's enormous coal reserves were going to rescue us. Now we have global warming.

To the best of our knowledge, however, the oil and gas resources are indeed more limited than the coal resource, and this least attractive of the fossil fuels is likely to remain the one in greatest supply. It *is* possible to burn coal cleanly, reducing if not eliminating the pollutants – even the CO_2 (see the final chapter). Or coal can be converted into more acceptable or more desirable fuels, including alternatives to gasoline (see Chapter 7). The merits and demerits of all these possibilities will be discussed; but unless there are dramatic new discoveries of oil or gas, or dramatic increases in the use of renewable energy sources, it seems that the resource-driven choice in the mid twenty-first century could well be between coal and nuclear power.

References

Boyle, Godfrey (1996) *Renewable Energy*, Oxford, OUP/OU.

BP (2001) *BP Statistical Review of World Energy, 2001* [online], BP, web site http://www.bp.com (accessed November 2001, no longer online).

BP (2002) *BP Statistical Review of World Energy, 2001* [online], BP, web site http://www.bp.com (accessed October 2002, select 'energy reviews').

Brimblecombe, Peter (1987) *The Big Smoke*, London, Methuen & Co. Ltd.

Brown, Frederick, ed., (1950) *Statistical Yearbook of the World Power Conference, No. 5*, London.

de la Bédoyère, G.(1995a) *The Writings of John Evelyn*, Woodbridge, Boydell and Brewer.

de la Bédoyère (ed.) (1995b) *The Diary of John Evelyn*, Woodbridge, The Boydell Press.

DTI (2000) *Fluidised Bed Combustion Systems for Power Generation and other Industrial Applications*, Department of Trade and Industry [online], web site http://www.dti.gov.uk/cct/pub/ (accessed October 2002).

DTI (2001) *Digest of UK Energy Statistics*, Department of Trade and Industry [online], web site http://www.dti.gov.uk/epa/dukes (accessed October 2002).

EIA (2002) *International Energy Outlook 2002*, Energy Information Administration, web site http://www.eia.doe.gov (accessed October 2002).

Hitachi (2001) *A Large-Capacity Pressurized-Fluidized-Bed-Combustion Boiler Combined-Cycle Power Plant*, Hitachi Review vol. 50, no. 3, pp. 105–109 [online], web site http://global.hitachi.com/Sp/TJ-e/2001/revaug01/pdf/r3_109.pdf (accessed October 2002).

IEA (1999) *Greenhouse Gas Emissions from Power Stations*, IEA Greenhouse Gas R & D Programme [online], web site http://www.ieagreen.org.uk (accessed June 2002).

Jevons, H. Stanley (1915) *The British Coal Trade*, republished in 1969, Newton Abbot, David & Charles (Publishers) Ltd.

Nef, J. U. (1932) *The Rise of British Coal Industry*, London, Routledge.

Ramage, J. (1997) *Energy: A Guidebook*, 2ed., Oxford University Press.

WCI (2002) *Coal in the Steel Industry*, World Coal Institute, London, web site http://www.wci-coal.com (accessed October 2002).

Chapter 6

Heat to Motive Power

by Janet Ramage

6.1 Introduction

Fossil fuels and nuclear power together account for about 95% of the world's commercially traded primary energy. The first stage in our use of any fossil fuel is combustion, because this is how we extract the stored chemical energy. Whether the means is a furnace or fire, an internal combustion engine or a gas turbine, the essential product is energy in the form of heat. This is also the case for a nuclear reactor, although the source of heat is different. So it is true to say that we use virtually all the world's primary energy to produce heat. True, but misleading, because what we ultimately want in many cases is not heat at all. We want lighting, a refrigerator, TV, and all the other benefits of electricity; and we want to travel rapidly and in comfort. An essential step in meeting almost all these needs is the conversion of heat into **motive power**: the driving power of machines.

Any system designed to obtain continuous motive power from heat is called a **heat engine**, and this term embraces not only the steam engines that are the main subject of this chapter, but the internal combustion engine and the gas turbine, which are discussed later, in Chapter 8.

BOX 6.1 Prime movers

The term 'prime mover' has a long history. In mediaeval astronomy the *primum mobile* was the outermost crystal sphere that carried the stars on their daily rotation around the earth. Gradually the term came to mean any natural source of energy such as moving water or the wind. Then a final shift led to the present meaning: a machine that converts the energy of any natural source into motive power.

In Section 6.2 of this chapter we start with a look at the early steam engines – the first **prime movers** in the modern sense (Box 6.1). Although the earliest steam engines have long been superseded – in many cases by very different machines – their relatively simple modes of action provide a useful basis for discussion of the general principles governing the performance of *all* heat engines.

The science that embodies these principles came to be called **thermodynamics** (*thermo*: concerning heat; *dynamic*: concerning force, or power), and its laws are the laws that govern all energy technologies. But what do these laws say?

- The First Law of Thermodynamics rules out perpetual motion machines. It says that you can't get something for nothing in the world of energy.

- The Second Law of Thermodynamics imposes more severe constraints on any heat engine. It tells us that when heat is the input you can't break even: there must inevitably be wasted heat energy.

- The Third Law of Thermodynamics says that there is a lowest possible temperature, an absolute zero, below which nothing can be cooled.

Like any other scientific theory, thermodynamics could be superseded by new theories in the future; but for the present, designers of machines are well advised to take note of its laws. The First Law is essentially the law of conservation of energy, already discussed (and in many cases assumed) in earlier chapters. In Section 6.3 we look at the Second Law and its consequences, with a glance at the Third Law.

Following this rather theoretical section, we return in Section 6.4 to the history, with a review of the principal developments during the great period of steam engines in the nineteenth century. Many of the innovations designed to improve the efficiency of these machines, and the increasing understanding of *why* they reduced the losses, remained relevant to the steam turbines that replaced them. These turbines are the subject of the final two sections of the chapter, which finishes with an account of the processes in the steam turbine system of a modern power station.

Looking at the laws of thermodynamics, one might well ask why so many of our energy systems start by producing heat, given the inevitable inefficiencies in its use. The first thing to note is that we don't *always* do this. Processes in which heat does not play an essential part are not subject to the same constraints. A hydroelectric plant, for instance, transforms the potential energy of water in a high reservoir or the kinetic energy of a running river into electrical energy with very little loss. Unfortunately, similar direct conversion of the stored chemical energy of a fossil fuel into mechanical or electrical energy has proved much more difficult to achieve. Fuel cells, to be discussed in Chapter 14, can do this, but the technology is fairly new and these are only now on the verge of being able to make a significant contribution to the production of electricity. (It could be argued that we have in fact achieved one type of large-scale, relatively cheap, direct conversion from chemical to mechanical energy, in the form of the explosives used regularly in mining and civil engineering.)

6.2 Steam engines

The early years

The idea of using steam to drive a machine predates the Industrial Revolution by at least 1700 years. Figure 6.1 shows the *aeolipyle* devised by Hero, an engineer working in Alexandria about 2000 years ago. It is the earliest known example of a **reaction turbine**. Steam shooting out of a jet in one direction produces a reaction force in the opposite direction, causing the sphere to spin. It does not seem to have been developed into a continuously operating machine, and remained just one of a number of 'executive toys' devised by Hero.

Some 1400 years later, a renewed interest in science and technology, together with the enthusiasm for the classical world that characterized Renaissance Italy, led Leonardo da Vinci and others to propose many different prime movers using the power of steam. Amongst those inspired by Hero's devices was Giovanni Branca, who designed probably the first steam-operated **impulse turbine** (Figure 6.2), using the force or impulse exerted by the jet of steam striking the vanes or blades of a wheel. Branca's proposal was that the rotating wheel would drive a machine for crushing ore, but with the technology available at the time it is very unlikely that the force from the steam jet would even have rotated the wheel itself, and it is not surprising that little more was heard of steam turbines for another 500 years.

Figure 6.1 Hero's aeolipyle – the first steam-operated reaction turbine? The name, which means 'gateway of Aeolus', refers to the gusts that supposedly blew out of the mouth of the cave inhabited by the god of the winds

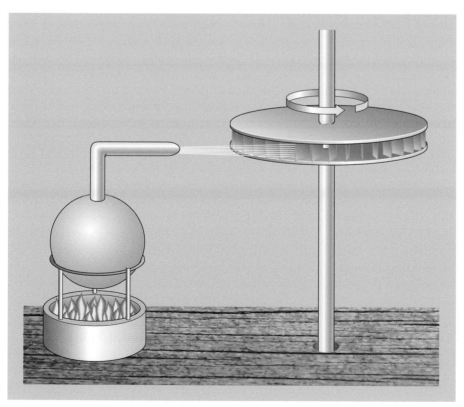

Figure 6.2 Giovanni Branca's 'prime mover' – the first steam-operated impulse turbine?

BOX 6.2 **Producing steam**

The essential first stage in any steam engine is, of course, the production of steam. The energy needed to convert one kilogram of water at an initial temperature of 20 °C into steam at the normal boiling point of 100 °C was calculated in Box 5.4 of the previous chapter:

- to heat 1 kg of water from 20 °C to 100 °C requires 0.336 MJ of heat – roughly a third of a megajoule

- to convert 1 kg of hot water into steam at 100 °C requires 2.258 MJ

- to produce 1 kg of steam at 100 °C from water initially at 20 °C therefore requires 2.594 MJ.

The second quantity, the energy needed for vaporization, may be regarded as the additional energy 'stored' in each kilogram of steam. The figures reveal that this is nearly 90% of the total – very much greater than the energy stored simply by heating the water.

We shall take as an example an early steam engine using 250 kg of steam per hour. The rate of heat input needed to produce this from water at 20 °C would be

$$250 \times 2.594 = 650 \text{ MJ per hour}$$

Suppose the boiler of this engine used coal with an energy content of, say, 30 MJ per kilogram. It should in principle need about 22 kg of coal an hour. However,

the early boilers were very inefficient by present-day standards, with poor heat transfer from fire to water and little insulation, and the necessary coal input could well have been more than three times this, perhaps 70 kg an hour, or in the unit of the time, about 150 lb (pounds) an hour.

The purpose of a steam engine is to convert as much as possible of the energy of the steam into useful work. The steam will finally emerge at a lower temperature and/or a lower pressure, perhaps as **wet steam** containing some water, or even completely condensed into water. The final temperature of the steam and the extent to which it is ultimately condensed, releasing its latent heat, are important factors in the performance of steam engines, as we shall see.

It is worth noting that 70 kg an hour of coal with an energy content of 30 MJ kg^{-1} is supplying 2100 MJ per hour, which translates into an input power of about 580 kW. After losses in the boiler, the 'steam power' carried to the engine (650 MJ an hour) is only 180 kW. But as we shall see later in Box 6.4, the final output power of this early steam engine could in fact have been no more than 4 kW. This means an overall coal-to-output efficiency of only about 0.7% – less than three-quarters of one per cent.

Savery and Newcomen

When the first practical steam-driven machines did appear, they operated on entirely different principles from the very early designs. The English inventor Thomas Savery's 1698 machine is worth a close look, because it makes use of a fact that has been important in many subsequent steam engines: when steam is condensed, it gives up the energy that was used in producing it (Box 6.2). 'The Miner's Friend' (Figure 6.3) was in fact a steam *pump*, developed to meet the problem of flooding in mines. The system is quite simple, with no moving parts. Its essential features are a coal-fired boiler, a pressure vessel and a vertical water pipe in which there two 'flap-valves' (C and D in the diagram) which open only for upward flow. It is convenient to start an account of its operation at the stage where steam from the boiler has filled the pressure vessel. Closing Tap B stops the steam supply, and when A is opened, a spray of cold water over the vessel condenses the steam. The resulting drop in pressure draws water up the lower pipe through valve D and into the vessel (Figure 6.3a). When the vessel is full of water, A is closed. If B is now opened, the steam pressure drives the water through valve C and up the top pipe to the surface (Figure 6.3b). The vessel is now full of steam and the process starts again.

Savery's pumps had in fact two boilers and two steam vessels, working in alternation, one pushing water up through the top pipe as the other was drawing it up through the lower one. Unfortunately, Savery's concept was rather ahead of the technology of the day, and the vessel and pipe-work frequently burst when subjected to the high pressure involved in pumping

(a)   (b)

Figure 6.3 Savery's steam engine, 'The Miner's Friend'

water through an appreciable height. The Frenchman, Denis Papin, working at about the same time, had devised a much more sophisticated pump, with an ingenious arrangement of pistons and cylinders, and even a safety valve. (He had already designed a pressure cooker.) But he met the same problems – and died a pauper. Savery seems to have been a more successful entrepreneur, and sold a number of machines despite their unreliability. It is unclear whether any of these operated successfully in coal mines, but Savery seems to have benefited from the growing fashion for water gardens with fountains. Every large estate had to have one, and Savery's pump could be employed to lift water into a high tank to provide the necessary 'head'.

Thomas Newcomen was a Dartmouth ironmonger, and his assistant John Calley was a plumber. Together they produced the first really successful steam engine design. Their first recorded machine, completed in 1712, was used to pump water from a coal mine near Dudley Castle in Staffordshire, England. Newcomen died in 1729, but during the sixty years following his first engine, hundreds were built. Figure 6.4 overleaf shows the main features, and as can be seen, this was not at all like Savery's steam pump. Newcomen was associated with Savery and knew of Papin, and he *was* using steam to pump water. He was also condensing the steam, but more efficiently, by spraying water directly into it. However, the differences between the two machines involve a great deal more than a few energy-saving improvements.

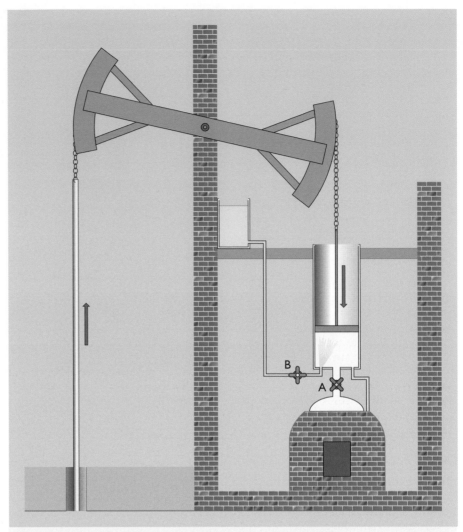

Figure 6.4 Principle of Newcomen's 'Atmospheric Engine', 1712

In Newcomen's engine, it is *atmospheric pressure* (Box 6.3) that does the work of raising the water, leading to its alternative name, the **Atmospheric Engine**. The sequence is as follows:

- Steam at slightly above atmospheric pressure is allowed into the cylinder by opening tap A, shown in Figure 6.4. Aided by the weight of the pump and bucket suspended from the other end of the beam, this raises the piston to the top of the cylinder.
- Tap A is now closed and the steam is condensed by a spray of water admitted by opening tap B. This creates a near-vacuum under the piston, which is pushed down by atmospheric pressure, tilting the beam and raising the water.

High steam pressures were not needed in this machine, and many of the problems that troubled Savery disappeared. Achieving a good seal between the cylinder and piston introduced a whole new range of difficulties, but many of these were gradually resolved over the years. John Smeaton, already famous for his detailed studies of windmills and water-wheels, carried out careful experiments with model steam engines, which led to numerous improvements in efficiency, as shown in Table 6.1 in Box 6.4 below.

BOX 6.3 **Pressure**

The 'atmosphere'

It is not at all obvious to most of us that we live with a force equivalent to a one-kilogram weight pushing on every square centimetre of our bodies (about fifteen pounds on each square inch). So perhaps we should not be surprised that only in the seventeenth century was it understood that air has weight. Galileo's pupil Evangelista Torricelli showed in 1643 that the weight of an imaginary vertical column of air, up through the entire atmosphere, was equal to the weight of a similar column of mercury about 76 cm high (or a column of water about ten metres high). A few years later, Otto van Guericke, the mayor of Magdeburg in Saxony, used his newly invented air pump to extract the air from a copper sphere. In a dramatic demonstration of atmospheric pressure, the sphere collapsed inwards.

In more modern terms, we understand that the pressure acting on any surface in a fluid is the result of constant bombardment by atoms or molecules. Each particle briefly exerts a force as it bounces off the surface, and the result of all these random collisions is a net force that is always perpendicular to the surface – *no matter which way the surface is facing*. The **pressure** is then defined as the net force per unit area: the number of newtons per square metre. As with other units such as the newton and the joule, the SI unit for pressure has been given a name, the **pascal** (**Pa**). One pascal is exactly same as one newton per square metre.

Normal atmospheric pressure at the surface of the earth is slightly more than 100 000 Pa. (It varies of course with the weather, and decreases with height above sea level.) The pressure of the formally defined **standard atmosphere**, at sea level and 15 °C, is 1.01325×10^5 Pa. The common earlier unit for pressure was the pound per square inch (lb/sq in or psi), and one atmosphere was usually taken to be 14.7 psi. A further unit, still seen on weather maps, is the **bar**. One bar is 100 000 Pa, i.e. approximately standard atmospheric pressure.

Finally, it is useful and perhaps more illuminating, particularly in discussing high pressures, to use phrases such as 'a pressure of sixty atmospheres'. We shall often use this form where greater precision is not required.

Fluids under pressure

The interpretation of pressure as the result of molecular collisions leads to useful results, as the following general reasoning shows. The pressure, the net force on a square metre of surface, will depend on the number of particles striking it per second, their mass and their average speed. In other words, the pressure depends on the *density* of the fluid and its *temperature*. In practice, the temperature and pressure are the quantities that we are normally able to control directly, so it makes more sense to reverse the statement and say that the density of a fluid (liquid or gas) depends on its temperature and pressure.

In the case of liquids, with their closely packed molecules, the density is not greatly affected by changes in temperature or pressure. The density of water, for instance, falls by about 4% between 20 °C and its normal boiling point of 100 °C, whilst increasing the pressure to twenty atmospheres changes the density at 20 °C by only one part in a thousand. Even under extreme conditions of temperature and pressure the changes in the densities of liquids are rarely more than a few per cent.

Steam is entirely different. Dry steam is the colourless gas seen just at the spout of the kettle. Its density is about 0.58 kg m^{-3} at 100 °C and normal atmospheric pressure. But if heated to 200 °C, still at atmospheric pressure, the steam will have expanded by a fifth, its density falling to 0.46 kg m^{-3}.

Increasing the pressure has even more startling effects. Most people are familiar with the fact that water boils at a lower temperature at the top of a mountain, where atmospheric pressure is lower. This is one aspect of the more general rule that the temperature at which any liquid boils depends on the pressure. If the water in a boiler, for instance, is held under a pressure of five atmospheres, it will only boil when the temperature reaches just over 150 °C. At twenty atmospheres, this rises to 212 °C, and the density of the steam that forms under this high pressure is 10 kg m^{-3} – nearly twenty times that of ordinary steam from a kettle at normal atmospheric pressure.

Steam heated above the boiling point is referred to as **superheated**. For example, at normal atmospheric pressure, steam at 200 °C is superheated steam. But the data above show that with a pressure of twenty atmospheres it would still be water at 200 °C. At this pressure it would become steam only at 212 °C, and would only be superheated steam at temperatures above that.

Ultimately, as the pressure is further increased, a critical point is reached. For water this occurs when the boiling point reaches 374.14 °C, and above this **critical temperature** the distinction between water and steam disappears. There is no 'boiling point' just one **supercritical** fluid whose density depends on its temperature and pressure. The steam produced by the boilers of some modern power stations is in this form. At a temperature of 660 °C, for instance, and a pressure of 20 MPa (200 atmospheres), it is a supercritical fluid with a density of 50 kg m^{-3}.

However, the steam engines of Savery and Newcomen offer a nice example of the fact that technological advance is not *always* a continuous series of slight improvements. Some of Savery's ideas remain in Newcomen's engine, but many – you might even say most – are completely discarded in favour of new, totally different concepts. We shall see this again with James Watt and the great steam engines of the nineteenth century, and later with Charles Parsons and the steam turbines of the twentieth century.

Newcomen's engine also had one potential use that Savery's completely lacked. The mechanical link to a rocking beam meant that it might be used for purposes other than pumping water. It could even in principle supply a continuous mechanical drive. But this development was to come only after the next major advances in steam engines.

James Watt

James Watt's contributions to steam engine design were so numerous that the space available here allows only a brief list of the most important. It can truly be said that the first two items below were critical for the development of steam engines – and their influence remains in today's machines.

- Watt's first patent, and the improvement for which he is best known, came from his experiments with model steam engines (see Figure 6.5). He realized that the Newcomen engine was wasting a great deal of heat by cooling the whole mass of the cylinder and piston each time the steam was condensed, requiring energy from the next steam input to heat it again. So he proposed a **separate condenser**. This could be kept cool all the time and the cylinder, well insulated, could remain hot. The patent also included a feature that is extremely important for an efficient condenser: a **pump**, driven by the main engine, extracted the water and as much air as possible from the condenser at the end of each cycle.

- Although he didn't use high pressures (see below), Watt realized that it was not necessary to continue adding steam until the piston reached its turning point. If the steam supply was closed off sooner, the existing steam would continue to expand, giving up more of its energy as useful work. Using less steam on each cycle obviously reduced the required coal input and improved the efficiency.

- He introduced a double-acting engine. It had a closed cylinder, not open to the atmosphere. Fresh steam entered on one side of the piston as the steam on the other side was being condensed, and then the process was reversed. (This led to yet another invention. A double-acting piston must be able to push as well as pull, so the chain connecting it to the beam in the Newcomen engine (Figure 6.4 above) could no longer be used. Any rigid rod used in place of it must remain vertical to allow the piston to slide freely, but if the top of the rod is fixed to the beam end, it will move in an arc. To solve this problem, Watt devised the parallelogram linkage – the invention, he said, of which he was most proud!)

- Watt also devised the 'sun and planet' method for converting the up-and-down **reciprocating motion** of the beam into rotary motion.

Figure 6.5 James Watt (1736–1819) was born in Greenock, in Scotland. Contrary to popular opinion, there is no evidence that he was inspired by steam from his mother's tea-kettle. In 1764, employed as an instrument maker by Glasgow University, he was given a model Newcomen engine to repair. The job completed, he began to wonder why the little machine used so much steam. Five years and many experiments later, he took out his first patent. In 1773 he went into partnership with the Birmingham businessman Matthew Bolton, and the patent was so all-embracing that until it finally expired in 1800, their firm had a virtual monopoly on steam engines. Watt then retired to his house in the country, devoting his time to 'mechanical pursuits and inventions'.

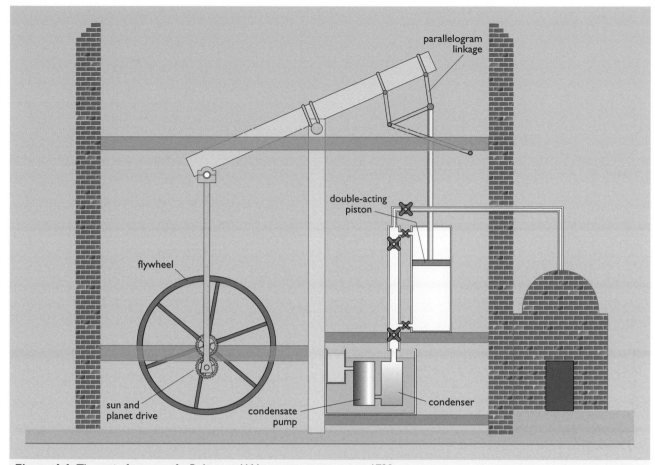

Figure 6.6 The main features of a Bolton and Watt steam engine, circa 1790

Figure 6.6 shows in outline a machine from about 1790 which incorporated all the above features.

Is there anything that Watt did *not* invent? The answer is yes of course, but two instances are rather surprising. He didn't use a crankshaft (see Chapter 8) to produce rotary motion, and he didn't use high pressure steam to improve efficiency. In the first case, his unsuccessful early attempts to obtain rotary motion from steam had discouraged him, and by the time Bolton persuaded him of its importance, it was too late. In 1780, James Pickard had attached a crank to a Newcomen engine and patented the arrangement. The story of high-pressure engines was similar in that Watt was cautious – in this case about the dangers. But metals technology was improving, and other engineers saw the potential. Until 1800, however, they could do nothing without Watt's consent, because his patent covered all steam engines with separate condensers.

Nevertheless people were already experimenting. In Cornwall in England there were many mines extracting tin, copper and other metals, and they needed pumps. But there were no local *coal* mines, and importing coal from elsewhere in the country to fuel the pumps was expensive, so any gain in efficiency would be welcome. There were other factors too. A high-pressure system could be more compact (it might even run without a

BOX 6.4 Horsepower, duty and efficiency

Horsepower

James Watt was responsible for defining one of the earliest units for power. He wanted to specify the power output of his steam engines, and wisely chose a unit with meaning for everybody: the **horsepower**. He based its value on measurements using real horses, but cunningly used 1.5 times the actual 'horse' average, so that his customers would find their machines comparing favourably with any horses they possessed. Rounding his result, he defined the horsepower:

1 HP = 33 000 ft lb per minute

One foot-pound (ft lb) is the energy needed to raise a mass of 1 lb through a height of 1 ft. It is equal to 1.356 joules.

It follows that 1 HP is equal to 745.7 W, or roughly three quarters of a kilowatt.

Duty and efficiency

The **efficiency** of an engine is defined as the output divided by the input (energy or power), usually expressed as a percentage. But this definition would not have been possible in 1800, a few years before the word *energy* was adopted and half a century before Joule identified the different forms. There was however an equivalent term used for the performance of the early steam engines:

The **duty** of an engine was equal to the number of foot-pounds of output obtained by the combustion of one bushel of coal.

There are two problems with this. Firstly, one bushel is 8 gallons, which is a volume, not a mass, so the result depends on the density of the coal. And more significantly, which coal does the engine use? As we have seen in Chapter 5, the heat value of coals can vary over a wide range.

There seems, however, to have been general agreement that the mass of one bushel of coal was 84 lb (38 kg). The heat value of the coal is more of a problem, and various authorities have used figures as low as 24 MJ per kilogram and as high as 35 MJ per kilogram.

Given that one foot-pound is a little over one joule, and that 84 lb of coal will clearly have a heat value of many millions of joules, it is obvious that the duty of any useful engine will be measured in millions.

Assuming a heat value of 30 MJ kg^{-1} for the coal, it follows from the definition of duty, and the data above, that the efficiency, in modern terms, of an engine is related to the duty approximately as follows:

percentage efficiency = 0.12 × duty in millions

or alternatively, 1% efficiency corresponds to a duty of about 8.4 million.

Table 6.1 shows data for a few of the engine types discussed in this section, together with some nineteenth century machines mentioned in Section 6.4 below. (Note that the Newcomen engine is similar to the one discussed in Box 6.2.) It must be stressed that all these figures are very uncertain. Engines could perform under or even over their nominal horsepower, and in the early years, quoted values for duty rarely specified the operating conditions or the type of coal used.

Table 6.1 Data for some early steam engines

Engine type[1]	Approximate date	Horsepower	Coal rate (lb / hour)	Duty (millions)	Percentage efficiency
Savery	1700	0.5	30	2.8	0.33
Newcomen	1720	5.5	150	6.1	0.72
Smeaton	1770	25	440	9.5	1.1
Watt	1790	24	300	13	1.5
Trevithick	1810	16	80	33	3.9
Cornish engine	1830	100	300	55	6.5
Cornish engine	1840	100	180	92	11.0

1 See the main text for an outline of each type.

Sources: Law, 1965 and von Tunzelmann, 1978

condenser), and could perhaps be *portable*. Richard Trevithick, a Cornishman and a brilliant and adventurous engineer (see Figure 6.12), was certainly looking at various options. In 1792, he had built a reciprocating steam pump without a beam: the cylinder was upside-down and the piston rod directly operated the pump. This led to threats of lawsuits from Bolton and Watt, but Trevithick continued his developments, and on Christmas Eve 1801, a startling result emerged into the daylight (Figure 6.7). The age of steam locomotion had begun.

Figure 6.7 Trevithick's steam carriage

6.3 **The principles of heat engines**

As we saw in Chapter 4, the world's *electric* power industry has its roots in fundamental scientific discoveries. No one was trying to build generators in 1800 because before Michael Faraday there was no reason to believe that one coil of wire rotating inside another would produce anything.

The case of *steam* power was completely different. Hero was already trying to produce motion from steam at a time when 'the elements' meant fire, air, earth and water. Scientific knowledge had of course advanced greatly by the time of Savery's first pump in 1700, and the new concept of atmospheric pressure was important for later machines. But steam engine development owed much more to advances in metals technology than to revolutionary scientific ideas such as those of Isaac Newton. The situation in the 1840s illustrates the point. Hundreds of steam engines worldwide were supplying power in mines, factories and even on railways, but James Joule was still trying to convince the scientific world that heat was a form of energy.

Figure 6.8 Sadi Carnot (1796–1832) was born in Paris. His father was a brilliant military engineer and much-admired general, a man of principle whose life in those turbulent times alternated between high political position and prison. At 16 Sadi entered the famous Ecole Polytechnique. Graduating in 1814, he too joined the Corps of Engineers, having already fought in a brief battle under Napoleon. During a visit to his exiled father in Magdeburg in 1821, the two discussed a new steam engine imported by Guericke. Returning to Paris, Sadi immersed himself in the subject and within a year or so came close to the view of heat and mechanical work that Joule was to develop 25 years later. But nothing of Carnot's early work was published in his lifetime, and his famous book appeared only in a small private edition. He died at the age of 36 in the Paris cholera epidemic of 1832.

Carnot's law

A theory of heat engines *had* been developed, twenty years earlier, by Sadi Carnot, a young captain in the French army. In 1824 he published a little book with the title *Réflexions sur la Puissance Motrice du Feu et sur les Machines Propres à Développer cette Puissance*. It would become one of the most famous of scientific texts, probably best known today in its many English translations, usually titled *Reflections on the Motive Power of Fire* (Carnot, 1992). But although it was well received by Carnot's French scientific colleagues, the book was essentially ignored in other countries and initially had little practical effect. Its importance gradually came to be recognized, but was only fully appreciated in the English-speaking world when William Thomson (Lord Kelvin) read it in the late 1840s. Kelvin evidently had a talent for recognizing genius. We met him in Chapter 4 as the person who drew attention to James Joule's ideas, and he seems to have been equally impressed by Carnot:

> Never before in the history of natural science has such a great book been written!

What was the great significance of Carnot's work? He didn't construct any machines or carry out any experiments, and didn't even believe that heat was a form of energy. He held the then-current view that it was a sort of fluid, called **caloric**. But as the title of his book suggests, he *thought* about heat engines. He was aware that even the best steam engines of the time were wasting a great deal of their heat input. Efficiencies were improving, but very slowly, and it seemed that there must be something that made it impossible even to approach the perfect heat engine – in modern terms, a machine that would take in heat energy and convert all of it into mechanical energy. So he proposed a simple law: *It is impossible to have a perfect heat engine*. This can be regarded as a rather informal statement of the **Second Law of Thermodynamics**.

The most remarkable part of Carnot's study was the result that followed from his law. He was able to show by careful mathematical and logical reasoning that this simple statement led to a formula for the maximum possible efficiency of any specified heat engine. Carnot's reasoning was very detailed and precise, and expressed in terms of the caloric theory of heat, so the following account is therefore both simplified and 'modernized'.

The Carnot engine

Carnot took as his basis for discussion an extremely idealized version of a steam engine (Figure 6.9). The **working fluid** of his heat engine passes continuously through a cyclic process, taking in heat from a 'boiler' at an **input temperature**, T_1, producing a mechanical output, rejecting waste heat to a 'condenser' at a lower **exhaust temperature**, T_2, and then returning to its starting point to repeat the procedure. Everything is ideal: no friction, perfect insulation, no sudden changes of any sort.

Suppose the quantity of heat taken in each time from the boiler is Q_1 and the quantity of 'waste heat' rejected to the condenser is Q_2. Because energy is conserved, the mechanical output, the work done, must be equal to the difference:

$$W = Q_1 - Q_2$$

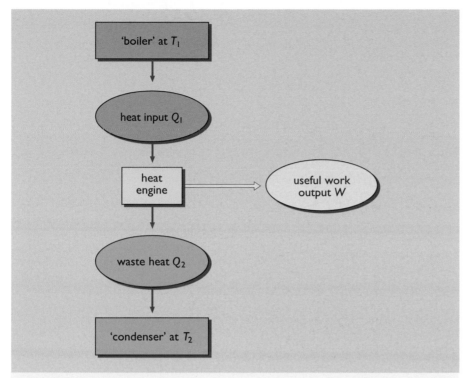

Figure 6.9 Carnot's heat engine

The efficiency of the engine, the output divided by the input, is then

$$\text{efficiency} = \frac{W}{Q_1} = \frac{Q_1 - Q_2}{Q_1} = 1 - \frac{Q_2}{Q_1}$$

There are other details in the full specification of the idealized **Carnot engine**, but for our purposes we need only consider the results.

Firstly, Carnot showed that, for his engine

$$\frac{Q_2}{Q_1} = \frac{T_2}{T_1}$$

i.e. the heat rejected divided by heat input is exactly equal to the condenser temperature divided by the boiler temperature. This leads at once to the following simple formula for the efficiency of the Carnot engine:

$$\text{efficiency} = 1 - \frac{T_2}{T_1} = \frac{T_1 - T_2}{T_1}$$

(The *percentage* efficiency will of course be equal to this multiplied by 100.)

Carnot then used his general law to prove a very important point: any other heat engine operating with the same input and exhaust temperatures must have a lower efficiency than his idealized one. In other words, Carnot's formula tells you the maximum possible efficiency for *any* type of heat engine if you know the temperatures between which it operates. Box 6.5 discusses the temperature scales that must be used in applying Carnot's formula, and includes a sample calculation.

Figure 6.10 William Thomson, Lord Kelvin (1824–1907) was born in Belfast but educated in Glasgow, where his father became Professor of Mathematics. He was an undergraduate at 14, and at 15 wrote a medal-winning essay. The years from 1841 to 1845 he spent in Cambridge, studying mathematics and natural philosophy (physics), and in 1846 he visited Paris to work with the great French mathematical physicists of the time, before his appointment as Professor of Natural Philosophy at Glasgow. He contributed to many areas of physics, including thermodynamics (building on the theories of Joule and Carnot), hydrodynamics, and electricity and magnetism, where his ideas influenced Maxwell. However, his popular fame (and a fortune and a peerage) came from the use of his 'galvanometer' in the first successful transmission of telegraph signals by transatlantic cable.

BOX 6.5   Temperature scales

On closer inspection there is something odd about Carnot's formula. What happens if the condenser is at zero degrees? If $T_2 = 0$ the percentage efficiency becomes 100%. All the heat is converted into mechanical output! But Carnot's Law says that this is impossible.

The resolution to the paradox lies in the definition of the temperature. Carnot's formula obviously can't be used with any randomly chosen scale of temperature. (Fahrenheit and Celsius, for instance, would give two different values for the efficiency.) As Carnot himself knew, his formula is correct only for *absolute* temperature scales. Any scale will do, provided its zero (0 degrees) is set at the absolute zero of temperature: the lowest possible temperature. We shall see later (Box 6.6) that there is such a thing, but for the present a brief account of the most common absolute scale will be sufficient.

The **Kelvin** scale of temperature was defined so that the size of each degree would be the same as in the familiar Celsius scale. The zero would however be at absolute zero, which is 273.15 degrees below zero degrees Celsius (0 °C). So the relationship is

> temperature in kelvins (K) = temperature in °C plus 273.15

(The approximate value, 273, is often used.)

Using Carnot's Formula

We can now apply Carnot's formula to an early steam engine. We'll assume that the steam enters the cylinder at 100 °C, which is about 373 K, and that it is cooled in a condenser to the temperature of warm water, say 40 °C, or 313 K. The percentage efficiency is then

$$\frac{T_1 - T_2}{T_1} \times 100 = \frac{60}{373} \times 100 = 16.1\%$$

Comparison of the data in Box 6.4 with this figure should hold no surprises. As we have seen, the early engines suffered enormous heat losses, and as explained in the main text, there are more fundamental reasons why no steam engine can achieve its notional **Carnot efficiency**.

Nevertheless Carnot's result remains an important guide in the search for improved efficiency. The high input temperature (T_1) of a combined-cycle gas turbine system leads to high efficiency (see Section 6.7 below and Chapter 8). And there are even proposals for a heat engine in space. Exhausting to a temperature T_2 very close to absolute zero, this would potentially have a very high efficiency indeed.

Carnot's result is useful in setting an absolute maximum to the efficiency, but the cycle of processes followed by the working fluid of a particular type of heat engine may be quite different from the Carnot cycle. This is certainly the case for any steam cycle, with the consequence that the maximum possible efficiency of even an 'ideal' steam engine falls well below the Carnot efficiency corresponding to its input and exhaust temperatures. The same is true of the gas turbines and internal combustion engines discussed in Chapter 8. (More detailed accounts of the working fluid cycles of all these heat engines may be found in Sorensen, 2000.)

Atoms in motion

Many people find the Second Law of Thermodynamics rather strange on a first encounter, so it is worth looking at it from a different viewpoint. Its essential feature is the distinction between heat and all the other forms of energy, and the limit it puts on the conversion from heat to these other forms. Common experience shows that there is no similar constraint in the reverse direction. It is only too easy to turn other forms of energy into heat. You bring the car to a skidding halt, converting its kinetic energy into heat mainly in the brakes and tyres; but you can't make use of all this heat to set the vehicle in motion again. The electrical energy supplied to a kettle is all converted into heat; but converting all the heat back into electric power is not possible. (This is why the hot water of Box 4.4 in Chapter 4 is a very poor temporary store for surplus power.)

What is it that distinguishes heat from other forms of energy? We have seen in Chapter 4 that the temperature of anything is a measure of the kinetic energy of its constantly moving atoms and molecules, and that heat is the energy we add or remove to change the average speeds that characterize this thermal motion. With this in mind, we can see how heat differs from all other forms of energy: the characteristic quality of thermal motion is that it is *disorganized* motion. The molecules of the air, for instance, are constantly moving, with a variety of speeds and in all directions, bouncing off each other and the walls of the room, changing both speed and direction all the time. The same picture of random motion holds for the vibrating atoms or molecules of a solid material.

If this is heat, then the principle that all energy tends to become heat is no more than a law that we all know: *things tend to become disorganized.* (Expressed in rather more formal terms, this becomes an alternative version of the Second Law of Thermodynamics.)

Consider, for instance, a stationary car as an assembly of atoms moving at random. In a similar car moving north-east at 50 km per hour all the atoms have a *directed* motion superimposed on their random movement, and this gives the kinetic energy of the moving vehicle. If the car stops this extra energy must go somewhere. It becomes heat – further *disordered* motion of the atoms of the brakes and tyres. We now have a warmer, stationary vehicle. To reverse the process and recoup the heat energy requires part of the random motion to become ordered again, something which is very, very unlikely to happen spontaneously. It is easy to produce chaos. Order presents more of a problem.

A more rigorous approach to thermodynamics introduces the useful concept of **entropy**. In terms of our picture of heat as random motion, the entropy of a system may be regarded as a measure of its degree of disorder. This leads to yet another version of the Second Law: *The entropy of a closed system tends to increase with time.* In other words, if a system is totally isolated from its surroundings (no exchange of heat, no work done on or by the system) then any change within it results in a increase in its entropy. This has led people to refer to entropy as 'the arrow of time'. The direction of increasing entropy is the direction in which time is flowing. We all recognize when a film of a breaking wave is being played backwards because we see disorder turning into order. Entropy is decreasing – and things don't happen that way.

BOX 6.6 **Absolute zero**

The picture of heat as the energy of particles in motion offers an obvious interpretation of the idea of an **absolute zero of temperature**: it is the temperature at which all motion stops. Every particle is frozen in position, and there is no more energy to be extracted. (The modern picture is not quite like this, because quantum mechanics does not allow the energy of a system of particles to become exactly zero, but the concept of the absolute zero of temperature remains.)

In 1906 Walter Nernst formulated a law that says that *the absolute zero of temperature is unattainable.* This is the **Third Law of Thermodynamics**, also known as **Nernst's Heat Theorem**. Nernst was Director of the Institute for Experimental Physics in Berlin, a person of great importance at the University. He was a grave figure, treated with respect by all, and it is said that in his lectures he always carefully referred to the Third Law as 'My Heat Theorem'.

Nernst showed that as the temperature of any object is reduced towards zero, the drop in temperature resulting from the extraction of a given quantity of heat becomes ever smaller and smaller. Effectively, the specific heat capacity becomes greater and greater, with the consequence that although substances can be cooled to temperatures nearer and nearer to absolute zero, nothing will ever reach it.

Heat flow

A one-way process that we meet every day is the flow of heat from a hot object to a cooler one. You pour hot coffee into a cold cup and the result is cooler coffee and a hotter cup. There is a flow of heat energy from the hotter coffee to the colder cup. You would be very surprised indeed if the reverse were to occur, with the coffee becoming even hotter and the cup ice-cold. Yet, provided the heat energy gained by the coffee is the same as that lost by the cup, energy would still be conserved in this bizarre case. Once again we must look to the Second Law of Thermodynamics to tell us the direction in which things happen.

The key point is that *hotter means faster*. The atoms of a hot object have a greater average speed than those of a cold object. To take another domestic example, suppose you were to observe that the handle of a metal fork lying on the table was becoming red-hot while ice crystals appeared at the other end. This would mean that the faster atoms had spontaneously assembled at one end and the slower ones at the other – an extremely improbable event. A simplified version of the UK National Lottery will serve as an example of such improbability. Ten balls, five red and five black, are shaken in a closed box and then allowed to run out one at a time. What are the odds that the five red balls appear first? About 30 000 to one against is the answer, and for ten red balls from a total of twenty, it becomes 700 billion to one against. A typical fork contains a very large number of atoms indeed (about 10^{24}) and the probability of the rearrangement of atoms described above is vanishingly small. We see that the Second Law has something to do with all this when we reflect that once again the issue is order versus chaos. The atoms in the fork don't *spontaneously* adopt the more ordered arrangement: *heat will not of itself flow from a colder to a hotter body*.

The natural way is the reverse, the direction of increasing chaos.

Heat *can* be transferred from a colder to a hotter body. Every domestic refrigerator does it, pumping out heat from the cold interior into the warmer surroundings, but it needs an external energy supply to do it. The refrigeration unit is an example of a **heat pump**, a category of machines discussed in Chapter 9.

6.4 The age of steam

The 'age of steam' had indeed arrived with the nineteenth century, but it didn't end with it. Steam-driven trains and ships remained dominant until the mid-twentieth century, and today, in the early twenty-first, steam engines still drive four-fifths of the world's electricity generators. But these power station machines are steam turbines, not the reciprocating cylinder-and-piston engines we have been discussing. Reciprocating steam engines *are* still in use, particularly in those countries where steam locomotives remain common; but their relative importance today is small, accounting for less than one per cent of the world's steam power output. For this reason, the following accounts of the great age of steam and the development of steam-driven vehicles are rather brief, concentrating on the innovations that remain relevant today.

Improving the efficiency

The data in Table 6.1 above are not very encouraging. The output power of the Savery engine was about that of a fit person running up a flight of stairs; and for each joule of mechanical output, it needed 250 joules of coal energy input. The other 249 joules became waste heat – no doubt making life very uncomfortable for the person fuelling the boiler and operating the taps (Figure 6.3). There was some truth in the saying that you needed an iron mine to build one of these early machines and a coal mine to run it.

The many innovations of the eighteenth century did lead to better performance, but as Table 6.1 shows, real increases in efficiency appeared only in the early decades of the nineteenth century. The expiry of Watt's patent in 1800 freed people to experiment with new ideas, and Trevithick and other Cornish engineers were at the forefront, investigating the improvements brought by high pressures, horizontal cylinders, more compact beam-less engines, compound engines, new boiler designs, new linkages for rotary motion, and so on. Efficiencies increased, and by the 1840s, steam pumps could achieve two hundred times the pumping power of Savery's machine for only six times the coal consumption. The best ever formally recorded performance by a single-cylinder reciprocating steam engine was in 1835, when the pumping engine at Fowey Consols mine achieved a duty of 125 million. This would seem to imply an efficiency, in modern terms, in the region of 16%; but this figure should not be taken too seriously. In addition to the uncertainties mentioned in Box 6.4, the engine was being run in a special 24-hour test under optimum conditions – and it has even been suggested that some wood was packed in beforehand, to supplement the carefully measured quantity of coal!

Some of the improvements listed above are discussed in the next subsection, in the context of machines supplying mobile power. However, two developments in particular are worth a remark here for their continuing importance today.

As mentioned above, it was already recognized before 1800 that **high pressure** steam improves the efficiency of a steam engine. More precisely, the pressure of the steam at input should be as high as possible and the exhaust pressure as low as possible. Why is this the case? A relatively simple answer emerges if we bring together Carnot's rule and the ideas in Box 6.3 above. At very high pressures, the steam must be at a high temperature in order to be steam at all. So the use of high pressure necessarily meant an increase in input temperature and therefore, by Carnot's rule, greater efficiency. (We shall look at high pressures in more detail later, in the context of steam turbines.)

The second innovation with lasting importance was the **compound engine**. The idea, proposed by Jonathan Hornblower as early as 1781, was to have two or more cylinders, each feeding steam to the next. The exhaust steam from the first cylinder, instead of being condensed, would become the input steam to the second, and so on. In terms of Carnot's rule there is no gain in this, as the maximum overall efficiency is still determined by the overall initial and final pressures and temperatures. There is however a gain in *practical* terms, because each cylinder is exposed to a smaller range of temperatures during its operating cycle, and this reduces the heat wasted in re-heating it. For a compound engine to work, each cylinder – in modern terms, each **section** – must be larger than the previous one, because the *volume* of steam that flows through during each cycle increases as the *pressure* falls. There are two main arrangements of the cylinders: side by side or in line, called respectively **cross-compound** and **tandem-compound**, terms that are still used for today's huge power-station turbines (see Figure 6.21).

Hornblower's early ideas eventually came to nothing, however, because his proposed engine used a condenser and air pump, which attracted the attention of Boulton and Watt and their patent lawyers. Later, in the early 1800s, compound reciprocating engines were developed in several countries, including the UK and the US; but at that time they found it difficult to compete with the increasingly efficient single-cylinder engines. The greater capital cost, weight and space-occupation of the compound engine had to be justified by significant efficiency gains which began to emerge only midway through the century, as technologies improved – and then only for some applications, as we will see below.

It is ironic that as the great Cornish machines were reaching their peak, Cornish mining was starting its long decline. Steam engines were, however, finding other uses: in the manufacturing industries, in transport (see below), and in pumping the water supplies needed by the growing populations of towns and cities. They were being exported, too. In 1849, Britain supplied three monster pumping engines to drain the Dutch polders, while a growing demand from new mining regions such as South America and Australia was met by machines from Britain and other European countries – and increasingly from the US. The famous Corliss steam engine of 1876 was the largest of its time. It had a mechanical output power of 1400 HP – just over 1 MW. But it was over 13 metres high and weighed 600 tonnes, giving it a power-to-weight ratio of about 1.7 W per kg. (For comparison, a typical modern car engine might produce 600 W per kg.)

The advances of the nineteenth century were not without penalties, however. During the mid-century decades, steam pressures were rising above five atmospheres, and some of James Watt's caution was proving to be justified. In Britain in the years around 1850, steam explosions were resulting in some 500 deaths a year (von Tunzelmann, 1978). This seems to have been the worst period, and by the end of the century, although pressures in some machines exceeded ten atmospheres the annual death rate from all types of steam explosion was 'only' a hundred or so (Figure 6.11).

Figure 6.11 The path taken by the boiler, circa 1890

Mobile power

In the early days of steam engines, obtaining more useful work from each tonne of coal was a prime objective, and the emphasis throughout the period described above was consequently on improvements in efficiency. (The capital cost of the machine was of course important, but some manufacturers even offered an ingenious arrangement whereby the user, instead of purchasing a replacement machine outright, could pay a fee based on the annual saving in fuel cost.) In the years after 1800, however, other criteria became increasingly significant. Richard Trevithick's first car outing may not have been entirely successful (see Figures 6.12 and 6.17), but he and others remained persuaded of the possibilities of steam as power for a *vehicle*. It was obvious to everyone that in this case the size – both the dimensions and the mass – of the engine would be important. A further consideration was that the complete machine, including the furnace and boiler and all the mechanical linkages and steam pipes, must be able to withstand the accelerations and decelerations (including the jolts and turns) of a moving vehicle.

The space available here does not permit an account of the ingenious mechanical devices in Trevithick's vehicle, but his steam engine itself is important because it embodied new features that are still seen in modern machines. The steam was no longer produced by lighting a fire under the boiler. Instead, the furnace together with the flue pipes that carried away the hot gases were *inside* a horizontal cylindrical boiler. This idea, of maximizing the heat exchange by passing the hot gases in tubes through the water (or alternatively, flowing the water in tubes through the hot gases) has remained in the **tubular boilers** of present-day steam plants. The single cylinder was horizontal – an arrangement that has continued in most steam locomotives of the nineteenth, twentieth and even twenty-first centuries. The major part of the cylinder was also inside the boiler, to keep it hot. It was double-acting and used steam at a pressure of about two atmospheres. There was no separate condenser, another obvious weight-saving measure, and the exhaust steam passed into the smoke stack, producing a drag or **blast** effect that increased the air flow through the furnace.

Some data have been assembled on the successor to Trevithick's first, burnt-out vehicle (Brooklands, 1998). Built in 1803, this second vehicle was very similar to the first. The cylinder had a diameter of 140 mm (less than 6 inches) and the piston stroke was 762 mm. With a steam pressure of about 200 kPa, the engine is said to have delivered 3 HP (2.25 kW) when propelling the vehicle at just under 8 kilometres per hour (5 mph). Its coal-to-motive power efficiency under these conditions appears to have been about 4%.

Trevithick crashed his second machine in London, and seems then to have abandoned the idea of a steam-powered road vehicle. He did however achieve the first commercial use of a **railway train**. Confounding the critics who 'knew' that metal driving wheels on a metal track would be sure to spin, his 1804 locomotive successfully replaced the horses that hauled wagons from an ironworks in Wales. Successfully, but briefly, because the seven-ton locomotive proved too heavy and broke the rails. The next venture was similar: an engine to replace horses drawing a coal train from a mine in Northumberland. This ran on a wooden wagon-way, which unfortunately also proved unable to support the weight of the engine. Then in 1808 came

Figure 6.12 Richard Trevithick (1771–1833), the son of a Cornish mine manager, became an engineer through practical experience. Over six feet tall and very strong, he was as adventurous in his life as in his engineering, and both were sequences of successes and catastrophes. Celebrating the success of his (and the world's) first mechanically driven vehicle, he left it in a shed near the inn, but didn't extinguish its fire, which destroyed machine and shed. After each of his other 'locomotives' came to grief for similarly contingent reasons (see the main text), he returned to stationary machines. Still failing to make money in Britain, he went to Peru as an engineer in the silver mines. There he prospered – until revolution forced him to leave the country, abandoning all his property. Penniless in Colombia, he met Robert Stevenson, who gave him his fare back to England, where he continued to generate ideas but not money. He died a pauper in Dartford.

Trevithick's 'first passenger-carrying train'. Running on a little circular track, about 10 metres in diameter, near the present Euston Square in London, it was more an entertainment than a transport system, but was a success with the public – until it was derailed. For some reason Trevithick was unable to find financial supporters for further experiments with steam vehicles, and he returned to stationary engines.

Trains

Trevithick may not have continued with steam traction, but others did, and were more successful. The second decade of the nineteenth century saw several steam engines hauling trucks on rails, including *Puffing Billy* (Figure 6.13) and George Stephenson's first engine, *Blücher*. The development of these successful engines was obviously an important factor, but Stephenson's role in the growth of the railway system was much wider.

Figure 6.13 *Puffing Billy*

In 1822 he persuaded the directors of the proposed Stockton and Darlington railway to use steam rather than horse-power, and at the opening of the line three years later his engine *Active* drew the world's first public passenger-carrying steam train. (*Active* was later re-named *Locomotion*.) Stephenson soon became involved in the development of other railways, and his belief that a 'locomotive' could draw the trains on the new Liverpool to Manchester route led to the famous Rainhill trials of 1829. The directors of the company had agreed to offer a prize of £500 for the engine that best met their performance specifications. The original ten entrants were reduced to five by the start of the seven-day trials, and at the end there was just one survivor: *The Rocket*, built by George Stephenson and his son Robert. They received a contract for eight locomotives for the new line.

The Rocket incorporated developments of a number of Trevithick's innovations, including a tubular boiler and steam blast. But its power transmission was much simpler, with one cylinder on each side and the pistons connected directly to the driving wheels. With many changes of detail but remarkably few changes in the essential principles, this remained the dominant system in steam locomotives throughout their great period in the nineteenth and twentieth centuries (Figure 6.15). Unfortunately their efficiencies also showed relatively little change. Writing in the late 1920s, an American railway engineer could complain that

> Instead of resorting to improved combustion of fuel, maximum utilization of radiant heat, high rate of heat transfer, efficient convection, higher steam pressures and temperatures and the more economical use of the heat in the steam, the general trend … has been to more extravagant methods of stoking and use of fuel, lower steam pressures in boilers and cylinders and the ejection of a greater percentage of heat from the exhaust nozzle and stack.
>
> Muhlfeld, 1929

Figure 6.15 *The Flying Scotsman* leaving London on her first non-stop run to Edinburgh, 1 May 1928

Figure 6.14
George Stephenson (1781–1848) was born in the pit village of Wylam in Northumberland in England, where his father was a colliery fireman. He started work in the colliery at the age of ten, but learned to read and write in evening classes and eventually, at 31, became enginewright at Killingworth colliery, at a salary of £100 a year. He invented a miners' safety lamp just before Sir Humphrey Davy announced his, and in the resulting controversy was labelled by Davy 'a thief and not a clever thief'. But Stephenson had more ambitious plans, and in 1814 his *Blücher*, the first 'locomotive' to use flanged wheels, started operating between the colliery and the port some five miles away. He built many more engines at Killingworth before his first venture into passenger trains. Following the success of the Liverpool–Manchester line (see main text), George and his son Robert not only continued to develop locomotives but played major roles in the construction of other railways throughout the world.

The author continued with a list of the increasing requirements placed on the steam, which was expected to provide power for labour-saving systems and heat for a variety of purposes – including keeping the passengers warm, or cool. He concluded that the already poor engine efficiency of 6–8% was reduced in average service to an overall system efficiency, from coal input to motive power output, of 4–6%, and contrasted this with the less powerful but more efficient locomotives being developed in Europe.

Ships

The early history of marine steam engines closely mirrors that of rail locomotives. The first steam-driven ship appeared just after Trevithick's road vehicle but before his first train. She was the *Charlotte Dundas*, built by William Symington in 1802 for use as a tug on the Forth and Clyde canal in Scotland. Her single rear paddle wheel worked successfully, but concern about its effect on the canal banks led to her withdrawal from service. In one important respect the engine differed from its contemporaries on dry land: it was a Watt horizontal double-acting *condensing* engine. The less severe constraints on space and weight meant that the engine could take advantage of the better efficiency resulting from the use of a condenser.

The first commercially successful steamboat was the *Clermont*, built in 1807 by Robert Fulton and his partner Robert Livingston to carry passengers on the Hudson River, in New York state. Her Boulton and Watt engines came from England – on a sailing ship, of course. In 1815, towards the end of his life, Fulton constructed the first stream-driven warship, for the United States government. A vessel of 40 tons, she had side paddle wheels, armament, and was called *Fulton*.

The following decades saw many small steam vessels, and increasing numbers of steam-assisted ocean-going sailing ships. Credit for the first fully steam-driven Atlantic crossing probably goes to the *Curaçao*, a 400-ton Dutch paddle steamer who made the passage to the West Indies in 1827, a voyage that took one month. Just over ten years later, in 1838, the steam ship *Sirius*, 700 tons, offered the first transatlantic passenger service, but she was followed into New York within hours by a much more famous ship, Isambard Kingdom Brunel's 1400-ton *Great Western*, which had made the crossing in 15 days, about four days faster than the *Sirius*. Then in 1840 came the first regular Atlantic service, with four almost identical 1100-ton paddle steamers commissioned by Samuel Cunard.

In general, steam had little advantage over sail for speed at this time. A fast sailing ship with a good wind would easily outpace any steam vessel, and this often compensated for the steam ship's ability to proceed in (almost) any wind conditions. In any case, virtually all ocean-going steam vessels of the mid-nineteenth century were still fully rigged for sailing.

The second half of the century saw many advances in marine engine technology. One improvement that soon found its place in this context was the compound engine (Figure 6.21). Ocean liners, cargo vessels and the ships of the world's navies were growing ever larger, and from its introduction in the *Brandon* of 1854, the double-compound (and later triple- and even quadruple-compound) reciprocating steam engine gradually became the standard. By the 1870s, the efficiencies of the best engines had risen to 15%.

Another innovation was the screw propeller. The idea had been considered for some time, and the first operating systems were developed independently but simultaneously by two engineers: the Englishman Francis Pettit Smith, and John Ericsson, a Swede living in London. (In 1829 Ericsson had been involved in the construction of *Novelty*, one of the unsuccessful entrants in the Rainhill steam engine trials.) Both their screw propellers achieved success in 1839, Ericsson's on a naval vessel in the US and Smith's on a steam ship aptly called *Archimedes*. The efficiency advantages of the propeller over paddle wheels were immediately obvious, and led Brunel to change the plan for his next ship, *Great Britain,* completed in 1844. (*Great Britain* represented another marine innovation, as she was not only the largest ship of her time, but one of the first to be made of iron rather than wood.)

The combination of double- or triple-expansion reciprocating steam engines and screw propellers was to have a very long life. By the early twentieth century, with very efficient tubular boilers producing steam at pressures above ten atmospheres, and effective condensers to maintain low exhaust pressure, efficiencies had reached 20% — amongst the best ever achieved by reciprocating steam engines. Long after the advent of more advanced technology, cargo ships were still being built with these propulsion systems, and many remained in service into the 1950s and beyond. But the first of their successors had already appeared more than half a century earlier — a ship driven by **steam turbines**.

6.5 **Steam turbines**

One category of prime movers that had virtually reached their present-day form by the end of the nineteenth century were the turbines. But these were *water* turbines. The name, which comes from the Latin *turbo*, meaning 'something that spins', was coined in about 1830 to describe the new high-speed machines that replaced slow-moving water wheels. The second half of the nineteenth century saw rapid design improvements, and from the 1880s turbines driven by flowing or falling water were providing the motive power for many of the new electrical generators (see Chapter 9). The Niagara Falls generators, commissioned in 1894, with outputs of 5 MW, are an indication of the scale that water power had achieved by the end of the century.

At this time, reciprocating steam engines were also providing power for generators, but many people in the industry were aware that these massive, relatively slow machines were far from ideal for the task. And it had become obvious that they could not compete with the new internal combustion engines as mobile power sources for road vehicles (see Chapter 8). It is not surprising that experts at the time were predicting that the entire steam industry would disappear within a few decades.

Steam, speed and rpm

With the example of water turbines before them, why did the steam engineers fail for so long to follow suit? As we have seen in Section 6.2 above, the idea of a steam turbine was very old indeed, but in the first detailed analysis, in the very early 1800s, James Watt had concluded that a simple turbine driven by steam jets was technically impossible. The essence of the problem is the obvious difference between the densities of water and steam — a factor of several hundred. Suppose we consider a simple impulse turbine — a modernized version of Figure 6.2. The situation is analysed in Box 6.7. To deliver power with reasonable efficiency, a jet of steam needs to travel at extremely high speed — up to 1000 m s^{-1}. To take up most of the kinetic energy of the steam, the turbine blades need to move at about half the speed of the jet, and this means that a turbine with a diameter of about a foot (0.3 m) will need to rotate at over 30 000 rpm! Watt had reached this conclusion and correctly identified two problems. Firstly, to achieve the required jet speed, the boiler must produce steam at high pressure – ten atmospheres or more. And secondly, with the metals technology available at the time, a one-foot turbine rotating at 30 000 rpm would be torn to pieces under the forces involved.

BOX 6.7 Steam jets

To produce a jet of steam through a nozzle there must obviously be a greater pressure on the input side of the nozzle than at the output. The process is effectively converting some of the energy of the hot, high-pressure steam into kinetic energy of the directed jet. So we might expect that the Second Law of Thermodynamics will limit the extent to which this can be achieved. This is indeed the case. With a 'perfect' nozzle, where no energy is lost through turbulence or friction, the jet speed can be calculated from the state of the steam at the input, the pressure at the output and the known thermal properties of steam. Figure 6.16 shows the speeds resulting from different input pressures, assuming an output pressure of one atmosphere (about 100 kPa). The second line on the graph shows the percentage of the energy of the steam that is converted into useful kinetic energy of the jet. The efficiency advantage of using high pressure is obvious.

The problem that then arises is the need to match speeds of 1000 m s^{-1} to a rotating machine – a modern version of Figure 6.2. The aim of course is for the jet to give up all its kinetic energy to the turbine, which means that the steam should effectively come to rest after striking a blade. Consider two extremes:

- If the turbine blade is moving much more slowly than the jet, the steam will 'bounce' off it in the opposite direction, retaining most of its kinetic energy.

- If the blade is travelling at almost the speed of the jet, the steam will continue with little change in speed, again retaining most of its kinetic energy.

Suppose the jet travels at 1000 m s^{-1}. If the blade is moving at just half this speed, i.e. 500 m s^{-1}, the approach speed of the steam *relative to the blade* will also be 500 m s^{-1}. Assuming a 'perfect bounce' the steam then leaves at 500 m s^{-1} backwards, again *relative to the blade*. But the blade is moving forwards at exactly this speed, which means that the steam is not actually moving forward at all. It has been brought to rest by its contact with the blade. In other words, when the blade speed is just half the jet speed, the jet gives up all its kinetic energy – the ideal situation. (In practice, energy losses mean that a speed slightly less than this proves to be the optimum.)

Consider now a 30-cm diameter turbine wheel with blades moving at 500 m s^{-1}. In each complete revolution the blade moves through about 1 metre ($\pi \times 30$ cm). So the wheel must make some 500 rotations per second, which is 30 000 rpm (revolutions per minute), as Watt concluded.

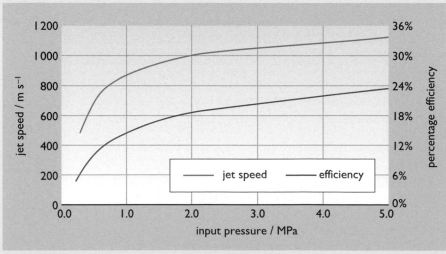

Figure 6.16 Pressure, jet speed and energy conversion efficiency

By the late 1800s, however, high steam pressures were already being used in conventional reciprocating steam engines. And in Sweden, the engineer Gustav de Laval, with the aid of many ingenious design innovations, was showing that turbines could run at extremely high speeds of rotation, up to 40 000 rpm. His impulse turbine, patented in 1887, proved almost twice as efficient as the average steam engine of the time. Its rate of rotation was far too high for direct connection to an electrical generator, but de Laval dealt with this problem by designing suitable high-speed gears. His turbines, with outputs in the range 5–500 HP (roughly 4–400 kW), had a degree of success for some years. But they were fated never to dominate the market, because another brilliant engineer was already designing a very different turbine specifically for power stations.

Parsons' turbo-generator

The Honourable Sir Charles Algernon Parsons was the son of the Third Earl of Rosse. After graduating from Cambridge University in 1877, he worked for various engineering companies, finally leaving in 1889 to set up C. A. Parsons Ltd independently. His first steam turbine had been completed in 1884. To be precise, it was a complete turbo-generator set and it continued in use, generating electric power, until 1900.

Parsons' turbine involved two ideas that were quite new in the context of steam engines. Firstly, the jets of steam did not emerge from conventional nozzles. Instead, they were formed as the steam passed through a ring of fixed blades (or guide vanes) attached to the outer casing. These directed the flow on to the ring of moving blades on the rotor (Figures 6.17 and 6.18), a concept adopted from the increasingly sophisticated water turbines of the time.

Parsons' second innovation was the brilliant idea that really solved the problem of the need for very high speeds. He realized that if many rings of fixed and moving blades were mounted close together in sequence, with each feeding steam to the next, the pressure could drop in relatively small steps along the complete turbine. The speed of the jets resulting from these smaller pressure differences would then be much lower, with the important consequence that the matching rate of rotation of the blades would also be lower. (Box 6.8 describes the action of this type of turbine. See also Parsons, 1911.)

The output of Parsons' first machine was only 10 HP (7.5 kW); much less than the largest reciprocating engines of the time. But the little turbine was remarkably small compared with conventional steam engines of similar horsepower. The complete turbo-generator was under two metres long, and the turbine itself about half a metre long and 15 cm in diameter. The

Figure 6.17 Parsons' first turbo-generator. The upper part of the casing is removed to show the little rotor with its fourteen rings of blades

BOX 6.8 Turbine blades

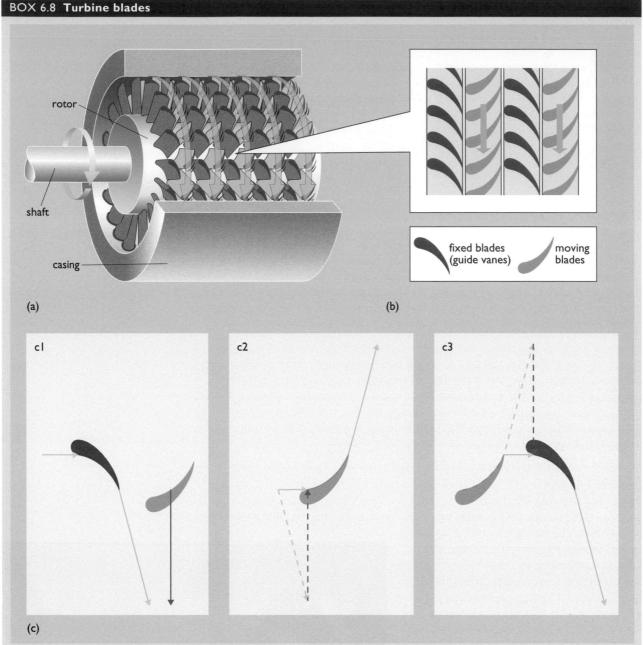

Figure 6.18 A simple steam turbine (a) Rotor and casing (b) Fixed and moving blades (c) The steam flow

Figure 6.18(a) shows the essential structure of a simple turbine, with alternating blades projecting outwards from the central rotor and in from the surrounding heavy casing. The steam flow is axial, from left to right in diagram (b), which shows a few blades in cross-section viewed 'end-on'. In Parsons' original turbine, the blades were simpler, as can be seen in Figure 6.17; but he soon adopted the curved 'aerofoil' shape,

appreciating that the high efficiency of water turbines was the result of careful shaping of their guide vanes and curved turbine blades. In passing between the fixed blades, the steam not only gains speed by the 'jet' effect but is deflected so that it flows smoothly across the faces of the moving blades. The same occurs when the steam meets the next set of fixed blades, and so on along the turbine.

Diagram (c) shows how this smooth flow is achieved. In (c1) the steam enters parallel to the axis, follows the face of the first fixed blade, and emerges in a new direction at greater speed. (The lengths of the arrows represent the flow speed.) This steam leaving the edge of the fixed blade will strike the next moving blade – but not at the angle suggested by (c1), because the moving blade is travelling in the same direction as the steam jet, and at almost the same speed. (c2) shows the situation as it would appear to someone riding on the moving blade. Allowing for the forward motion of the blade we see that the steam actually meets it as shown by the small blue arrow in (c2), i.e. in the original axial direction.

The steam therefore passes smoothly across the face of the moving blade, leaving it as shown in (c2). But to see how it reaches the next fixed blade, we must return to the view from the fixed position (c3). As this shows, adding the blade motion to the steam motion shown in (c2) reveals that the actual direction of the steam is once again axial, ready to pass through the next set of fixed blades.

Diagram (c) also shows that the steam exerts a force on each blade. If the flow direction changes, the blade must exert a force on the flowing steam, which therefore exerts an equal and opposite reaction force on the blade. Ideally, the direction of this force would be in the direction shown by the large arrows in (b), to rotate the turbine, but in practice there is also a significant component along the turbine axis. In Parsons' original turbine, this was balanced by sending a second steam flow in the opposite direction through a complete 'dummy' turbine with the same axis. In today's turbines, as we shall see later, the problem is often overcome by double-flow systems, where the steam flow splits to drive two identical turbines on the same rotor.

As the above analysis shows, the steam leaving each complete stage has same velocity as at the start. The power transmitted to the rotor comes from the fall in pressure as the steam passes through each set of blades, as described in Box 6.7.

'working' section of the turbine consisted of fourteen rings of moving blades mounted on one central rotor, alternating with fourteen rings of fixed blades, or vanes, mounted on the inside of the outer cylindrical casing. The steam entered at a pressure of 6.5 atmospheres (660 kPa) and was exhausted at normal atmospheric pressure (100 kPa). With the fourteen sets of vanes and blades, the average fall in pressure through each stage was only two-fifths of an atmosphere, and the turbine could rotate at a reasonable 18 000 rpm.

How efficient was this new form of steam engine? One answer comes from the basic principles discussed in Section 6.3 above. The input temperature is the boiling point of water at 660 000 Pa which is 163 °C, or 436 K. If we assume that the departing steam exhausts at 100 °C, or 373 K, then Carnot's rule tells us that the maximum possible efficiency is about 14.5%. On average, each of the fourteen little rings of blades would be extracting about 1% of the input energy as useful work. However, no actual heat engine achieves its nominal Carnot efficiency, and it seems unlikely that Parsons' first turbine even achieved an efficiency of 5% – considerably less than the best reciprocating steam engines of the time.

Nevertheless, Parsons continued to develop the new machine, and several hundred were in operation within five years. Although recognition in other countries took longer, it is an indication both of the acceptance of steam turbines and their rapidly increasing size that in 1898 Parsons & Co. were invited to supply a 1 MW turbine for a power station in Elberfield, in Germany.

Parsons' ideas were further developed by engineers such as Charles Curtis in the US, but work on other turbine types also continued. The idea of forming steam jets through nozzles was not abandoned, and work by Auguste Rateau in France and others led to turbines with a ring of fixed nozzles, and jets giving up their energy through impulse forces as described in Box 6.7. Both the reaction and impulse principles are retained in the steam turbines of present-day power stations.

Marine engines

For fairly obvious reasons, the steam turbine never became a serious option as a propulsive system for road vehicles; nor did it displace the reciprocating steam engine for locomotives. It did, however, find one other major application, as a marine engine. Parsons himself foresaw this, and realized that its adoption for naval ships would speed its acceptance in the world of merchant shipping. He also knew that the British Admiralty had a history of resisting all innovation. (They had notoriously insisted on wooden hulls when others were using iron, and had initially rejected steam engines.) Parsons therefore decided on a demonstration, and constructed a remarkable vessel: *Turbinia*. She was 30 metres (100 ft) long, by less than 3 metres beam, by less than 1 metre draught, with three turbines, in cross-compound arrangement, delivering a total power of 2300 HP (1.7 MW) to her screw propellers. Taken secretly to Cowes, she roared out into the 1897 Jubilee Review of the Fleet at Spithead with Parsons at the controls and travelling, it is said, at the unheard-of speed of 33 knots (about 61 kph).

Figure 6.19 *Turbinia*

Parsons achieved his aim, and the Admiralty did adopt turbines, but the early vessels were not entirely successful (mainly for reasons other than the engines), and a few years elapsed before turbine engines appeared in merchant ships. The first was the *King Edward*, entering service as a passenger ship on the Clyde in 1901. Others followed, and in 1907 came one of the most famous of many Cunard 'liners', the *Mauretania*. She was the largest ship of her time, and her 68 000 HP quadruple screw direct-coupled turbines gave her a maximum speed of 29 knots. From 1910 until as late as 1932 she held the 'Blue Riband' for the fastest Atlantic crossing. (For a remarkable photograph of *Mauretania* with *Turbinia* alongside, see Parsons, 1911.)

However, steam turbines were designed to run electrical generators (and are excellently suited to this task, as their continued use shows). As marine engines driving propellers, they suffered two major disadvantages. One was the mismatch between their rate of revolution, typically thousands of rpm, and the hundred or so rpm of a large screw propeller. A gearing system invented in 1910 by Parsons and developed by others helped to solve this problem, but many ships retained direct drive and a consequently lower efficiency. The second problem was that a turbine cannot be put into reverse. A commonly adopted solution to this was to have a separate 'astern' turbine. (The reader will appreciate that ships do not have brakes, so a reliable astern drive is essential.)

The 1920s saw the introduction of **oil-fired** steam turbines as marine engines. Although the cost per unit of energy was greater for oil than coal, the savings in other respects were significant. Fuel could be loaded more quickly and stored more compactly, and a tenfold reduction in engine-room manpower meant appreciably lower running costs. These oil-fired, high pressure steam turbines continued to provide the power for the great ocean liners of the 1930s, including the '*Queens*' (*Queen Mary*, 1934, and *Queen Elizabeth*, 1938) on the North Atlantic route, and many steam turbine ships remained in use into the late twentieth century. (They included, surprisingly, the Royal Yacht *Britannia*, launched in 1953. Powered by two oil-fired steam turbines delivering 12 000 HP, she was decommissioned only in 1997.) But these were a vanishing, if distinguished, minority. As early as 1912 a ship with a strange appearance had been launched in Denmark. She was the *Selandia*, the first **diesel-powered** ocean-going vessel, proudly announcing her new-style engines by having no funnels at all.

6.6 Power station turbines

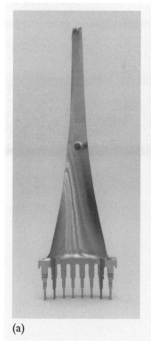

(a)

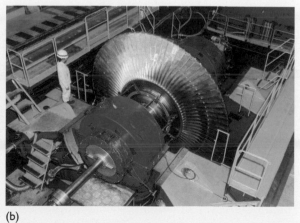

(b)

Figure 6.20 (a) A 43-inch long (about 1.1 metre) LP rotor blade (b) Testing of the 43-inch blades with a full-scale model rotor

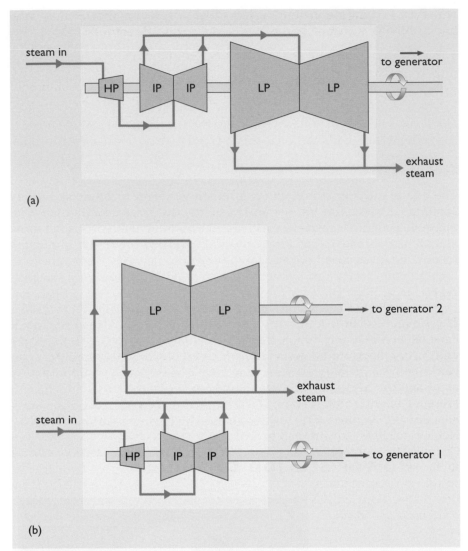

Figure 6.21 (a) Tandem-compound and (b) cross-compound arrangement of turbines

The operating principles of a modern power station turbine are essentially those of Parsons' original machine. So how does the present-day turbine, more than a century later, differ from its ancestor?

The turbines

The modern system will almost certainly use a *compound* arrangement. It will normally have high-pressure (HP), intermediate-pressure (IP) and low-pressure (LP) turbines, increasing in size as the pressure decreases. The tandem-compound arrangement shown schematically in Figure 6.21(a) is common, with all the turbines on a single shaft and driving a single generator. Cross-compound systems (Figure 6.21b) are also used, particularly in plants with large power output, as this arrangement permits two smaller generators on the separate shafts instead of a single very large one. Note also that the IP and LP turbines in Figure 6.21 are **double-flow**, with the incoming steam dividing into flows in opposite directions, driving two turbines — which of course rotate in the *same* direction!

The present-day turbine is undoubtedly very much larger than Parsons' original. Figure 6.20(a) shows a blade from a LP turbine, and (b) shows a complete stage of LP blades under test. The blade lengths in a compound system increase not only from turbine to turbine but from stage to stage within each turbine. As shown, the largest, in the final stage of an LP turbine, can be over a metre long.

Rate of rotation

The rate of rotation of a modern turbine tends to lie in the range 1500–3500 rpm, a fraction of the 18 000 rpm of Parsons' machine. But the size is relevant. The tip of a metre-long blade on a metre-diameter rotor spinning at 3000 rpm is moving at nearly 500 m s^{-1} (about 1000 mph), which is several times the speed of Parsons' little blades.

Steam pressures

The steam pressures used today are of course a great deal higher than Parsons' 660 kPa, and the megapascal (MPa) has become the customary unit for the pressure of steam entering the turbine. The steam condition varies between different types of power plant, depending on the size of plant and the nature of the heat source. For a large fossil-fuel plant, the pressure might be as high as 25 MPa, with temperatures in the range 500–600 °C. At these temperatures, the steam is supercritical (Box 6.3). Nuclear reactors, constrained to work under less extreme conditions (see Chapter 10), commonly supply steam at 4–6 MPa and with temperatures in the region of 300 °C. The boiling point of water at 5 MPa is about 240 °C, so this is superheated steam. At the speeds involved in the turbine, any tiny water drops will damage the blades, so it is important to minimize condensation. Ideally, the steam should exhaust to the condenser as *dry* steam.

The boiler

The design of the boiler that produces the steam is an important factor in the efficiency of the plant. Three centuries of development since the earliest steam engines have led to today's boilers, producing thousands of tonnes of steam per hour, with a fuel-to-steam efficiency as high as 90% (see Chapter 5). A modern water-tube boiler is a complex structure, designed to maximize the heat exchange between the burning fuel or hot gases and the water. It will normally have several separate sets of heat exchangers, designed to increase the overall efficiency. These will include one or more **superheaters**, to raise the final temperature of the steam leaving the boiler, and **reheaters**, in which the steam re-visits the boiler between successive turbines in a compound system. The hot flue gases may pass through an **air preheater** to heat the incoming air, or an **economizer** to heat the incoming water, again reducing the fuel requirement.

There must of course be control systems in place to monitor and govern the rate of steam supply. The purpose of the turbine is to drive an electrical generator, and this brings its own special requirements. The rate of rotation must remain constant within a very narrow permitted range, and the entire system must be able to respond effectively to the variations in demand – sometimes very rapid variations – that are a feature of the life of any power station.

The condenser

Unlike Parsons' original turbine, which exhausted the steam to the surroundings, the modern power-station plant includes a condenser. Carnot's rule shows that the lower the final temperature the greater the efficiency, so a well-designed condenser is essential. The aim, as in the boiler, is to maximize the heat exchange – in this case between the steam and a flow of cooling water. A common arrangement is for the cooling water to flow through many small tubes surrounded by the exhaust steam. The steam condenses on the surfaces of the water tubes, and the resulting **condensate** (water) is pumped away from the condenser, to return eventually to the boiler. The pump is important in maintaining as low a pressure as possible. This pump, and another to circulate the cooling water, will use a per cent or so of the electrical output of the power station, a loss that is usually justified by the resulting improvement in efficiency.

Figure 6.22 Cooling towers

Where a plant is near the sea or a large river, the cooling water can be returned, usually 10–20 °C warmer, to its source. Otherwise a **cooling tower** (Figure 6.22) is needed. In these large, familiar structures the warmed cooling water is sprayed down through a rising current of air. Some of the water evaporates, and the loss of latent heat needed for this process cools the remainder. (A process familiar to anyone who has stepped out of a hot shower into cool dry air.) The plume seen above the cooling tower is produced when the water vapour carried up by the air stream cools and condenses in the colder air outside. The evaporated water, a few per cent of the total, must of course be constantly replaced as the cooled water returns to the condenser.

A 660-MW turbine

Figure 6.23 shows schematically the turbine system of a large power station. In practice, of course, the plant will be much more complex than this, incorporating the features described above and others, designed to maximize both its efficiency and the working life of its component parts. However, this simplified example gives an indication of the flows of materials and energy that characterize a plant of this size, delivering an output of 660 MW to the generator.

As Figure 6.23(a) shows, the working fluid (water/steam) flows in a closed loop at a rate of about 1800 tonnes per hour, or 500 kg s^{-1}. To follow the successive processes, we might start at the point where it leaves the condenser. At this stage it is warm water, maintained at a low pressure by the pump that draws it along. At this pressure – only about a tenth of an atmosphere – the temperature of 45 °C is just a little below the boiling point of the water.

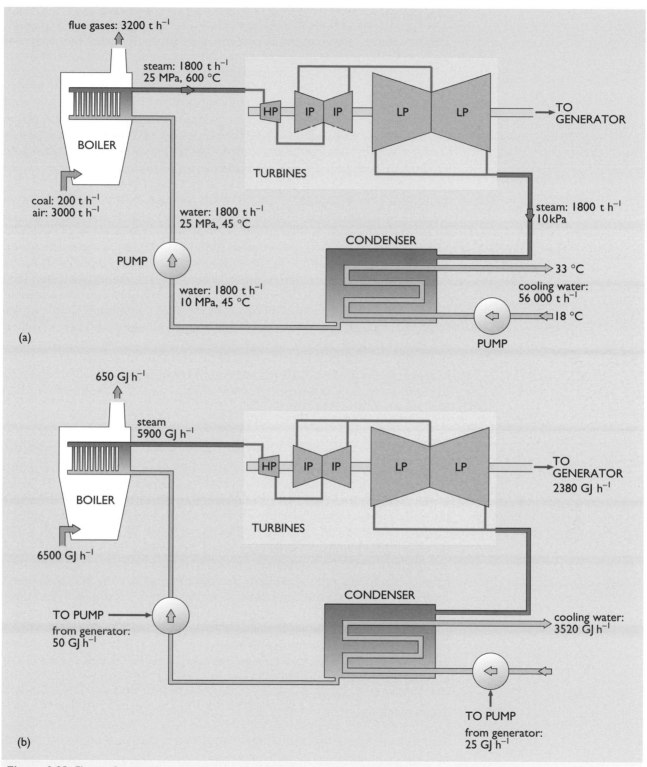

Figure 6.23 Flows of materials and energy in a power station steam turbine system

The pump

The first stage in converting this warm, low-pressure water into very hot high-pressure steam is the pump. In the present example, the water pressure is raised from a tenth of an atmosphere to about 250 atmospheres, and each tonne of water passing through requires just over 25 MJ of energy from the pump for this. With a flow rate of 1800 tonnes per hour, and assuming a pump efficiency of about 90%, the required electrical input to the pump becomes about 50 GJ per hour, or 14 MW. This electric power comes from the final generator output and must of course be taken into account in assessing the overall performance of the system.

The boiler

To increase the temperature of the high-pressure water from 45 °C to the required 600 °C needs a heat input of about 3.25 MJ per kilogram of water. The boiler must therefore supply heat at a rate of 5850 GJ per hour. If the fuel-to-heat efficiency of the coal-fired boiler is 90%, an input of about 200 tonnes of coal an hour will be needed, and at least 3000 tonnes an hour of air to allow full combustion. In a pulverized-fuel boiler, most of this entire mass will exit as flue gases (see *Flue Gases* in Chapter 5). As mentioned above, in practice some of the heat carried by the flue gases may be used to preheat the incoming air or the feedwater, and the boiler may also include reheat systems. If these features serve their intended purpose of increasing the efficiency, they should of course lead to a lower fuel requirement than that shown in the diagram.

The turbines

The turbine system shown in Figure 6.23 above is the tandem-compound arrangement of Figure 6.21, but we shall treat it as a single entity in terms of the input and output flows. The steam enters at 600 °C, which means that it is a supercritical fluid (Box 6.3). At the pressure of 250 atmospheres, its density is about 70 kg m^{-3} (over a hundred times that of ordinary steam). By the time it exhausts to the condenser, at a pressure of one tenth of an atmosphere, the volume occupied by each tonne of steam has increased by a factor of a thousand, reducing its density to about 0.07 kg m^{-3}. If the mass flow rate is to remain the same throughout the system, it follows that volume available in each of the final stages of the low-pressure turbines must be a thousand times that in the first stages of the high-pressure turbine. This is achieved in part by the double-flow system, but it is still necessary for the blade length to increase by a factor of ten or so between the input and exhaust ends of the complete system.

The condenser

The fluid entering the condenser is steam at a pressure of about 10 kPa, and the fluid leaving it is water at approximately the same pressure. As its name suggests, the function of the condenser is to condense the steam, extracting its latent heat and maintaining as low a temperature as possible. The heat released must be carried away by the cooling water, but this water itself must remain appreciably cooler than the condensate. These factors determine the required rate of flow.

In the present case, heat is to be removed at a rate of 3520 GJ h⁻¹. The cooling water enters at 18 °C, and its temperature should not rise above 33 °C. From Chapter 5 we know that the heat energy needed to raise the temperature of 1 tonne of water by 1 °C is 4.2 MJ, so the 15 °C permitted rise allows each tonne of water to carry away 63 MJ. Extracting 3.52 million megajoules an hour therefore needs about 56 000 tonnes an hour of cooling water. (This could be visualized as a flow at about 5 m s⁻¹ through a pipe 2 metres in diameter.)

The pumps to circulate this large volume of cooling water must supply enough power to pump the water through the condenser pipes, and to raise it through any necessary height differences. The total power demand will also depend on whether the water is returned to a river or the sea, or cooled in cooling towers. Again the power for the pumps must come from the generator output. The rather arbitrary figure of 25 GJ h⁻¹ used here is equivalent to a power of just under 7 MW, or about 1% of the gross output.

Efficiency

What is the heat-to-work efficiency of this system?

Using the formula from Section 6.3 above, we can easily calculate the efficiency of a Carnot engine working between our input and output temperatures of 600 °C and 45 °C, respectively.

$$\text{percentage efficiency} = \frac{T_1 - T_2}{T_1} \times 100 = \frac{555}{(600 + 273)} \times 100 = 63.6\%$$

But no real machine would be expected to achieve this, even if there were no heat or frictional losses to take into account, and the steam cycle of our turbine system is very far from the idealized Carnot cycle, as the data in Figure 6.23 reveal.

Suppose we consider just the turbines. The heat supply from the boiler is 5850 GJ h⁻¹, and the diagram shows a work output to the generator of 2380 GJ h⁻¹. The heat-to-work efficiency is therefore

$$\text{percentage efficiency} = \frac{2380}{5850} \times 100 = 40.7\%$$

This is a more realistic figure, taking into account the fact that the turbines are not even 'ideal turbines'.

But we have considered only the turbines, making no allowance for the 650 GJ h⁻¹ heat loss from the boiler or the energy needed to run the pumps and other systems, such as the 75 GJ h⁻¹ for the two pumps shown. These three items alone would reduce the overall efficiency to little more than 35%. But this is now unrealistically low for a modern plant of this size – a result of the over-simplification of our model. Inclusion of reheat, and the other efficiency-improving methods described above, can increase the overall turbine efficiency by as much as 5%, bringing it back above 40%.

Generator losses will reduce this by a few per cent, leading to an overall power station fuel-to-power efficiency of perhaps a little less than 40%. The figure of about 33% quoted throughout this book is lower, in part because not all turbines are designed for the operating conditions described here, but also for the obvious reason that power station turbines – like every other energy conversion system in the world – are not run constantly at optimum efficiency.

6.7 Futures

Is there a future for steam? In Chapter 5, we saw that coal fuelled the industrial revolution, and that it still has a role as the main primary resource for electric power, and in this chapter we have seen that the steam engine has been the enabling technology throughout. In the coming chapters, however, we shall find that the other fossil fuels have followed very different routes, and their associated technologies are different too. When *motive power* is the requirement, the direct use of oil or gas – in the internal combustion engine and the gas turbine – dispenses with the need for steam as an intermediary. And for *heating*, these 'convenience fuels' frequently occupy the place once held by coal.

Does this mean that the futures for steam power and coal are destined to remain inextricably linked? Not necessarily. Coal doesn't *have* to produce steam. As we shall see in the next chapter, it can be converted into the more desirable liquid or solid fuels, as replacements for oil and gas in the direct uses mentioned above. If this route were to be adopted, the coal-fired power station as we know it might eventually disappear from the scene – although there will need to be compelling reasons to replace the well-established existing technology by these more complex processes.

It is even more obvious that steam does not need coal. A sixth of the world's present electricity comes from nuclear power stations, and a smaller fraction from oil-fired plants, all of them using steam turbines. Steam also has a role to play in power generation from renewable resources such as geothermal energy, wood or other biofuels, and even solar energy. (The power plants using these and other renewable sources are discussed in the companion volume to this book, *Renewable Energy*.) And there is the combined-cycle gas turbine (CCGT), the current system of choice for many new power stations in countries where natural gas is available. This compound arrangement of a gas turbine and a steam turbine uses the exhaust gases of the former to produce steam for the latter, and the consequence of the very high input temperature is an overall efficiency that currently approaches 60% (see Chapter 8). It seems unlikely, therefore, that the steam turbine will disappear in the near future.

In the much longer term, the key question is whether we shall continue to use heat engines at all. Shall we eventually avoid the need to convert heat into motive power? The present situation, described in the opening paragraph of this chapter, is very far indeed from this; but there are two possible routes for change. We could find ways to convert the chemical energy of fuels directly into motive power or electricity (as in the fuel cells discussed in Chapter 14). Or we could replace the fuels by primary sources that provide motive power directly: hydro power, wind, wave and tidal power – all, as it happens, renewable resources.

References

Boyle G. A. (ed) (1996, 2003), *Renewable Energy*, Oxford, Oxford University Press in association with the Open University.

Brooklands (1998) 1803 Trevithick London steam carriage, http://www.brooklands.org.uk/Goodwood/g9828.htm [accessed 18 July 2002, no longer online].

Carnot N. L. S. (1992) *Reflections on the Motive Power of Fire*, Peter Smith Publisher

Law R. J. (1965) *The Steam Engine*, London, HMSO.

Muhlfeld J. E. (1929) 'Locomotive', *Encyclopaedia Britannica*, 14th edn, London.

Parsons C. A. (1911) *The Steam Turbine*, Cambridge, Cambridge University Press [also available online at http://www.history.rochester.edu/steam/parsons/ , accessed 22 October 2002].

Sorensen B. (2000) *Renewable Energy*, 2nd edn, Academic Press.

von Tunzelmann G.N. (1978) *Steam Power and British Industrialization to 1860*, Oxford, Oxford University Press.

Chapter 7

Oil and Gas

by David Crabbe

I expect our grandchildren to ask:

You burned it? All those lovely organic molecules; *You just burned it?*

Sorry, we burned it.

Ken Deffeyes, 2001

For my grandson Sean, always asking questions.

David Crabbe, 2002

Part 1 **Oil and gas as primary fuels**

7.1 **Introduction**

The five pound note issued by the Bank of Scotland has, on its reverse side, two roustabouts hard at work on an oil rig. The caption below reads 'Oil and energy' which, for those with a scientific mind, doesn't seem quite correct somehow. The meaning is clear, but its manner of expression inelegant. 'Oil as energy' or 'Oil for energy' would perhaps grate less, but how much neater would 'Oil and gas production' be, reflecting its vital role within the local economy? Still, that banknote does serve the purpose, albeit clumsily, of reminding us that **crude oil** is an essential source of energy for transport, commerce and industry throughout the UK, which in itself mirrors all other developed economies around the globe. Even in less developed countries, oil products are just as essential, if only for rudimentary cooking and lighting.

The increasingly important role of **natural gas** cannot be overlooked either. In the last century natural gas has gone from being seen as little more than an inconvenience that must be disposed of to being a very important commodity. It is set for an expanding world role long after the peak in oil production has passed.

Consequently, Part 1 of this chapter will concern itself with both these vital sources of primary energy. It will present a broad description of exploration and production, derivatives and uses. Part 2 of this chapter will take a look into the future at possible substitutes for primary oil and gas, concluding with an attempt to answer the oft-asked question 'when will it all run out?'.

Nonetheless, it has not been possible to include all the issues relating to the use of oil and gas. The important subject of CO_2 and noxious gas emissions has been deliberately sidestepped for the most part, and for a discussion of these the reader is referred to Chapter 13. Matters relating to international politics and the large oil companies have also largely been omitted as they are not central to the theme of this book and have been much written about with great authority elsewhere. Suffice to say that oil is one of the dominant traded products in the world economy, and the economic, political and indeed, on occasions, military implications of this fact have shaped the world we live in, and more specifically, have shaped our pattern of energy use and its environmental impact. The main aim in this chapter, however, is to look at how we reached this point in technological terms, and to review the various ways these resources have been and can be used.

7.2 The origins and geology of petroleum

The word 'petroleum' has the literal meaning 'rock-oil' and in that sense embraces all hydrocarbon material deposited within the earth's crust. In its narrower, commercial sense, the term 'petroleum' usually refers to liquid deposits (crude oil) although gaseous deposits (natural gas), along with certain solid deposits (bitumens, waxes, etc.) have similar origins.

It is now generally accepted that petroleum deposits arise from the decomposition of aquatic, mainly marine, animals and plants successively buried under layers of mud and silt several hundred million years ago. Essential pre-conditions for petroleum formation include an abundance and diversity of plant and animal life, and a seabed environment which discourages immediate breakdown of dead organic material either by bacterial action or by oxidation. Burial of such hydrocarbon-containing plant and animal material under sea-borne mud and silt 'preserves' the material for the next stage of the petroleum-forming process. Over time, innumerable layers of marine silt are laid down and compressed over this 'source' material, and as the thickness of these deposits increases so do the pressure and temperature regimes therein. It is at this stage that the petroleum forms through a mix of chemical and physical processes that even now are not fully understood. Thus large reservoirs of petroleum will only have been formed in those areas where thick successions of marine strata have been laid down.

But the actual *formation* of petroleum is only part of the picture: it must also be effectively contained. Today, countless million years after they have formed, oil and gas are only found where migration from the original **source rock** to **reservoir rock** has occurred as a result of the high pressures in the source rock, and the hydrocarbon has been trapped in the reservoir. For this to have happened two general conditions apply.

The first is that further *upward* migration has been prevented by an impermeable seal known as **caprock**, and the second condition is that further *lateral* migration has been prevented, or greatly reduced, by the natural occurrence of **geological traps** within the reservoir beds themselves. There are different types of geological trap, but the most important in terms of oil and gas accumulation is the **structural trap.**

The structural trap can be further divided into three subclassifications, the **anticline**, **fault trap** and **salt dome** (Figure 7.1). Amongst these subdivisions it is the anticline which is the most common, accounting for around 80% of the world's oil and gas resources. Anticlines are formed by folds in the earth's geological strata and when occurring in large structures can hold appreciable quantities of petroleum in place. A fault trap, by comparison, is formed when a bed of reservoir rock is brought into contact with impermeable strata by movement along a geological fault within the earth's crust. And a salt dome, as the name suggests, is a dome-shaped formation of rock salt which has been forced upwards through overlying strata until it lies under caprock. These three types of structural trap are illustrated opposite.

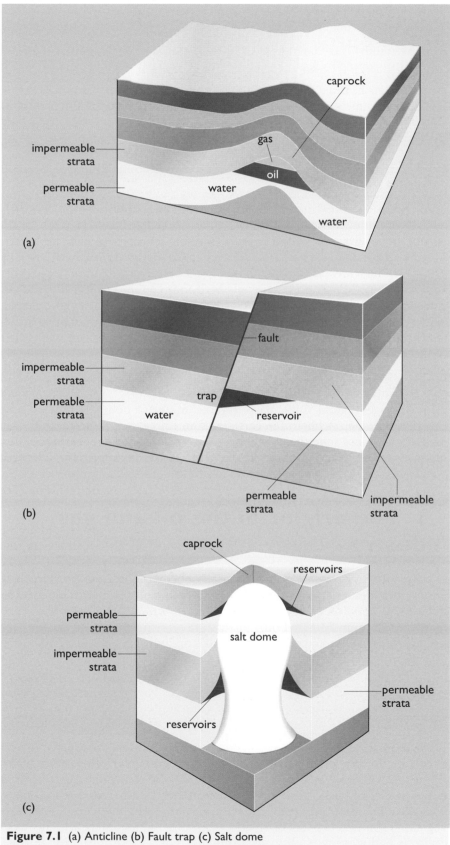

Figure 7.1 (a) Anticline (b) Fault trap (c) Salt dome

So, to summarize, there are four essential prerequisites for accumulations of petroleum to occur:

1　There must have been a suitable source of marine deposits contained within source rocks.

2　These rocks must allow for lateral movement of hydrocarbon deposits via a permeable pathway to layers of porous and permeable reservoir rock.

3　The reservoir must be sealed at the top by an impermeable layer or caprock.

4　A trap must be present to prevent lateral migration of the hydrocarbon away from the reservoir.

And so, in the most general geological terms, where do modern-day petroleum deposits occur? The answer is, in sedimentary basins – areas within the earth's crust where layers of rock and marine sediment are known to have accumulated over time. The global distribution of such areas is actually quite widespread. With the geology of landward sedimentary basins such as the Middle East increasingly well-mapped and explored, much of current oil exploration is now focused on sedimentary formations that extend outward from a continental shelf. Exploration in the Gulf of Mexico led the way in this regard some sixty years ago, to be followed by drilling in the North Sea in the early 1970s.

But wherever petroleum may be found, and in whatever quantities, we do well to remember its primitive and most unlikely origins, stretching back into deep geological time, with organic material being deposited and transformed deep beneath the earth over an immensely long period. This now emerges as a remarkably abundant yet ultimately finite resource. The relentless exploitation of this resource has fired the extraordinary changes enacted in the world economy throughout the twentieth century.

7.3 The origins of the oil and gas industry

Local small-scale exploitation of naturally occurring 'petroleum' goes back many centuries, even to biblical times. Evidence of this is provided by the remains of Babylonian monuments, some of which have been shown to contain bitumen. Marco Polo, on his travels through Asia in the late thirteenth century noted oil springs at Baku on the Caspian Sea, commenting that people travelled great distances to collect this oil, as it was good to burn. He also remarked on a local temple in which a sacred flame burnt continuously, fired by a local occurrence of what we would now call natural gas. Even early travellers to the 'Indies' commented on the phenomenon of the Trinidad Pitch Lake – Sir Walter Raleigh used the lake's pitch to caulk his ships.

However, the true origins of the twenty-first century oil and gas industries lie arguably some two hundred years ago in the quest for superior illumination. Until that time, almost all illumination had been accomplished by burning solid or liquid matter obtained from animal or vegetable sources, e.g. fish oils, tallow candles, etc., much of which was of dubious and varying quality, providing light of poor illuminating power.

Petroleum for illumination

Throughout the eighteenth century experimenters had been hard at work 'cooking coal' to yield a variety of interesting products, but by and large the *gaseous* products were considered to have little more than nuisance value. However, in the 1790s, quite independently of one another, Lord Dundonald in Scotland and William Murdoch, a Scot working at a tin mine in Cornwall, began to exploit the potential of **coal gas** as an illuminant. By the early years of the following century gas distribution for lighting had begun in London and within a generation had become commonplace. However, in rural areas or those without access to a piped supply, substandard lighting remained a problem and a lucrative opportunity awaited anyone who could provide a superior alternative.

Likewise, around the same time, experiments were being conducted into the use of various 'manufactured' oils as illuminants, most notably by Carl Von Reichenbach, who in 1830 made solid and liquid 'paraffins' from beechwood tar. Ultimately however, it was another Scot, James Young, who by dint of observation, good fortune, entrepreneurship and constant vigilance was not only able to produce a superior illuminating oil of consistent quality, but in the process founded what we recognize today as the industry of oil distillation and refining. See Box 7.1 for an account.

BOX 7.1 Beginnings: 'Paraffin' Young

Source: 'Shale oil: Scotland' by David Kerr

The founder of the world's oil industry, James Young, was born in Glasgow in 1811, the son of a carpenter and cabinet maker. His first job was as a joiner with his father but he also studied chemistry in evening classes at the Andersonian Institute (now the University of Strathclyde). In 1831 he obtained a post there as a laboratory assistant to the eminent Scottish chemist Professor Thomas Graham, famous for his laws of gaseous diffusion. Young progressed to become a lecturer at the Andersonian before transferring with Graham to University College, London.

In 1838 Young started a career in industry, ultimately joining the firm of Charles Tennant and Co., the Glasgow heavy-chemical manufacturer, in 1844. While working for them at their Manchester Works, he was sent a sample of oil from a spring in the 'New Deeps' coal mine connected with the Oakes family's Riddings Ironworks near Alfreton in Derbyshire. The thin, treacle-like oil was able to be distilled to a clear liquid. Tennants were not interested in exploiting the oil due to the limited availability (300 gallons per day), but allowed Young to set up and operate a small plant at the ironworks in 1848 with capital provided by Oakes, and employ a chemist, Edward Meldrum, to deal with it. The main product was lubricating oil for the Manchester cotton mills. A solid was also seen to separate out from the oil and this was found to be paraffin wax, which had first been systematically studied by Carl Von Reichenbach in 1830 in tar obtained by dry distillation of wood. (Christison of Edinburgh isolated similar material from Burmese petroleum at about the same time.) However wax was not separated from the Derbyshire oil on a commercial scale. Within two years the operation had become uneconomic due to reduced flow of oil.

Young's patent

Young had the idea that the oil had been distilled out of coal by volcanic activity long ago. He therefore began evaluating coals and cannel coals for oil yield by distillation. ('Cannel' is Scots for candle, as these coals got their name from the bright flame obtained when they burned. Miners used them in braziers to light their homes and they were also used to improve the illuminating properties of coal gas.) An unusual cannel coal known as Boghead 'Parrot' coal – so called because of the noise made when it burned – from a mine near Bathgate, proved to give the most oil: 120 gallons per ton of coal.

In 1850 Young took out a patent in Britain (No. 13292) and in America, for a process for obtaining oil from bituminous coal by low temperature distillation, refining the oil by further distillation and chemical treatment, and obtaining and purifying paraffin wax from it. From then on Young devoted himself entirely to oil production. He was one of the first in the world to produce oil and refine it on a commercial scale and the world's oil industry, which was soon to develop, was based on his refining processes.

Bathgate chemical works

In 1851 with his partners, the chemist Edward Meldrum and Edward W. Binney, a lawyer who helped provide capital, Young opened Britain's first commercial mineral oil refinery at Bathgate in West Lothian. The company was known as E.W. Binney and Co. Ten thousand tons of Boghead coal was bought under contract (at 13/6d [i.e. 67.5p] per ton) from Russel and Sons coal mines at Torbanehill, near Bathgate, and distilled in 21 benches of horizontal retorts (D-shaped cross section with three iron retorts per bench) at the Bathgate Chemical Works. The works were situated at Inchcross, south of Durhamtoun coal miners' rows. Meldrum was works manager, responsible for day to day running, and had a house on the site. At its peak the works employed 700 men. (A railway was built down the west boundary of the site and around 1860 a new enlarged works with retorts, stills and an acid plant, was built to the west of this line.) Meldrum's house is all that remains today.

Initially the works produced around 900 gallons per week of lubricants and naptha solvent for the paint and rubber industries. Naptha was also sold retail for cleaning purposes. At this time the world lacked cheap and reliable light sources, and whale oil, which was one of the better lighting materials, was becoming scarce. So Young found a use for an intermediate boiling fraction of his mineral oil, as burning oil (paraffin) for lighting. The lighting oil he developed was superior to the vegetable and animal oil already used for lighting, and was added to the products made at an early stage. An initial problem of smell had to be overcome, and a suitable lamp had to be developed from a Continental design and was manufactured at the company's own Clissold Lamp Works in Birmingham. Dwellings in Bathgate were the first to benefit from the 'brilliant light' from the new oil. The lamps and the products were sold all over Britain (marketed direct to the retailer from depots to keep prices down) and on the Continent. Before long one quarter of the lamp oil used in London came from the Bathgate Works. In 1856 Young started to use the brand name 'Paraffine Oil' for his burning oil, which remained the main product for many years. By 1854 the amount of the associated lubricating oil product made at Bathgate was 8000 gallons per week. This corresponds to retorting 10 000 tons of Boghead coal per year. Products were exhibited at International Exhibitions from 1851, and won many gold medals.

Conacher says that Young developed the use of steam in distillation and continuous distillation by coupling several stills together. He is recognized as the first in the world to have extracted and purified paraffin wax on a large commercial scale: at first the paraffin wax in the oil was largely left in the lubricating oil fraction and the excess burned. From 1854, the crude paraffin wax was separated and stored at the works and refined wax first marketed in 1858. A key part of Young's wax preparation lay in cooling the oil to allow the wax to crystallize out. Young employed as his chief engineer Alexander C. Kirk, who built the first practical air-compression refrigerator in the world for this purpose. (These were made under licence by the Norman Company of Glasgow and sold all over the world for ice-making.) The Boghead coal gave about 22 pounds weight of wax per ton. By 1861 five tons of wax were being produced each week at Bathgate for sale to candle manufacturers. The impact of the brilliant light from the new candles is hard to imagine now, but at the time the achievement was likened to having coal gas, compressed into a white stick!

Young's works were known as the 'Secret Works' as they were surrounded by a high, massive stone wall, and the workforce were sworn to secrecy. Names given by him to oils at various stages of the process such as 'green oil' and 'blue oil', had a factual basis, but also helped to maintain secrecy. Bremner writing in 1869 describes the Bathgate Works as occupying 25 acres, with a series of broad roadways lined with retorts, stills, boilers and tanks, some covered by iron roofs and some open to the elements. There were 200 retorts projecting down into pits of water. Gas produced by the retorts was used to light the works and also supplied to nearby Bathgate. A correspondent in the *West Lothian Courier* comments on the malodorous vapours emanating from the plant as he walked past!

Figure 7.2 James 'Paraffin' Young

One of the notable aspects of Young's work was that his 'oil' refining was founded not on conventional crude oil, but on bituminous coals of limited availability, and when these ran out he turned in the 1860s to local deposits of oil shale (see Section 7.16) for use as a raw material. Indeed, oil shales were to be exploited in east central Scotland until the 1960s when BP closed the last works and mines, concentrating its activities (as it still does) at the nearby refinery complex at Grangemouth.

Figure 7.3 Grangemouth port and refinery today

By coincidence, it was during the period when James Young's company was at the height of its commercial powers that naturally occurring crude oil was first discovered through drilling at Titusville, Pennsylvania in 1859 by Edwin Drake. This date is generally taken as the commencement of the modern petroleum extraction industry.

As with Young's work in Scotland, the principal outcome of this activity was the isolation of an illuminating oil, for which there was a ready market. It is said that Drake, who had close links with the chemical industry, was actually searching for brine when drilling his famous well – something of an irony when on so many occasions those drilling for oil come across only salt water! Within a few years drilling for oil had spread from Pennsylvania to many other parts of the United States and the petroleum industry was born.

But apart from 'kerosene' for lamps and stoves the only other petroleum product which elicited any demand was lubricating oil for machinery. The remainder of the petroleum, what we would now refer to as the lightest and heaviest fractions, was discarded, including 'gasoline' which was initially deemed too dangerously volatile to use. Within 40 years, however, the oilmen had discovered that it was worthwhile to carefully separate out their hard-won petroleum into a dozen or more fractions, each with their own special uses.

Figure 7.4 Colonel Edwin Laurentine Drake (1819–1880) (foreground right, wearing a top hat) taking to an engineer in front of his oil well at Titusville, Pennsylvania. In the background are 'Uncle Billy' Smith and his two sons who drilled the well. They struck oil at a depth of 69 feet on 27 August 1859

What an extraordinary resource this 'petroleum' was proving to be! It could be found in numerous places, and after some initial effort and expenditure could readily be brought to the surface under its own pressure. Once isolated it could then be subdivided into numerous saleable products with a range of commercial, industrial and domestic uses. As illuminant, lubricant, solvent and raiser of steam, oil products were infinitely superior to what had been available previously. As production rates increased and prices fell, oil-based products rapidly swept the board in a whole range of applications.

Petroleum for transport

But at this time there was no role for oil to play in transport systems. This seems a curious statement given that the period under discussion was little more than an century ago! In the early 1880s the roads were still the province of the horse, and the railways the province of the 'iron horse', the steam locomotive fired by coal. Steam-powered road vehicles were finding a role in local, heavy duty work, but were cumbersome and impractical, and attempts to introduce them for long-distance road transport were defeated by the opposition of the coaching companies and railroad operators. However, the entire situation was about to change.

The cause was the introduction of the petrol-driven internal combustion engine. The story of its development, leading in 1887 to Gottlieb Daimler's first 'motor car', is told in Chapter 8, *Oil and Gas Engines.* Daimler's most notable breakthrough, and a major reason for the success of his vehicle, was that it not only carried its own fuel store, but incorporated a means of feeding the fuel *continuously* to the engine. A generation after petroleum deposits were first exploited, a market had at last been found for one of its most abundant products – in short a marriage made in heaven. And the happy couple, gasoline and the motor car, are still very much entwined today.

But it was more than land-based transport that was to be revolutionized in this way. The Wright brothers' first free flight of a heavier-than-air machine in 1903 was powered by gasoline, and as we see in Chapter 8, oil products have remained the sole fuel for aircraft for 100 years. At sea, the British navy adopted oil-fired steam turbines for their fast, light destroyers as early as 1908, and the decade following the First World War saw oil gradually replacing coal in both naval and merchant vessels. By 1930, about half the world's shipping was powered by oil. Oil-fired steam turbines remained dominant, particularly for large ships, into the 1950s, but the second half of the twentieth century saw their increasing replacement by large diesel engines. (See also Section 6.5 *Marine engines* in the previous chapter and Section 8.3, *Diesel power for ships*, in the next.)

During the twentieth century, oil was to become the mainstay of all developed and developing economies: under its influence warfare would be conducted differently; and oil itself was to become a strategic resource over which international conflicts would break out. In little over a century the oil industry grew from the tiniest beginnings to arguably the world's most important industry, controlled by a handful of extraordinarily powerful companies. Production, which was negligible in 1860, and which still totalled under a million barrels a day in 1900, grew to over 70 million barrels a day by 2000, or around 35% of the world's total energy consumption. UK consumption is currently running at around 1.6 million barrels a day, around one third of our current primary energy needs.

The natural gas industry

Natural gas has a rather different story. To begin with it should be noted that it is not always, strictly speaking, based on petroleum. Natural gas does frequently occur in association with petroleum deposits, so-called **associated gas**, but it also occurs in the absence of petroleum. This **non-associated gas** is believed to have its origins in freshwater rather than marine plant material. Associated gas was once thought of as a downright nuisance, at best being useful in aiding oil lift by re-injection into the wells, only to be flared to waste afterwards. Indeed, in parts of the world where there is either no immediate market or transportation facility, or in small isolated oilfields, flaring at the well-head still occurs, but in general every effort is now made to bring natural gas to market.

In contrast to oil markets, the market for natural gas evolved only slowly, initially in the United States in the 1920s, where it was sold both as a heating fuel substitute for manufactured gas, and also as a power station fuel. In post-war Europe discoveries of natural gas led to its gradual adoption both as a heating fuel and also as a petrochemical feedstock. The boost to

European production and consumption, and indeed the entire North Sea exploration effort, came about from the discovery of the huge Groningen non-associated natural gas field in the Netherlands in 1959. The first similar find off the UK continental shelf was BP's 'West Sole' field in the southern sector of the North Sea in 1965.

As further fields containing appreciable quantities of gas were identified offshore in these shallow waters, the gas industry in the UK was faced with something of a dilemma. By the mid-1960s the manufactured gas mentioned above, generally known as **town gas** (see Section 7.14), was widely distributed throughout towns and cities in the UK. By this time town gas was being made from cheap imports of oil as well as from the coal which had spawned the gas-making industry many years previously. With the discovery of clean natural gas (over 90% methane with a minimum of impurities) the decision facing the gas industry was whether to convert this fuel into a town-gas type mixture or distribute it 'as we get it' and instead convert every gas-using appliance in Great Britain to burn methane rather than the calorifically fifty per cent-inferior town gas. It turned out to be 'no contest' in reality, for it made little sense to convert methane, at great expense, into an inferior fuel, especially as supplies of natural gas seemed assured for the foreseeable future. And so began the North Sea gas conversion process for appliances in homes, commerce and industry, by the end of which provision of gas and the use of gas within the UK economy had been entirely transformed. As if from nowhere, a new primary fuel had entered into the energy equation. In the ten year period from 1967 to 1977, gas supplied within Britain was totally transformed, from being 100% a secondary fuel, to 100% primary, and over the same period consumption increased fivefold. Natural gas now accounts for some 20% of the annual consumption of primary fuel globally, and 35–40% of UK annual consumption. All the signs are that these figures, in both percentage and absolute terms, are likely to grow further.

7.4 Finding petroleum

How do we get it?

Historically, the presence of oil in any particular location was detected by identifying surface seepages. Early prospectors correctly surmised that, in many cases, drilling close to these sources would result in their locating commercially attractive petroleum deposits. The earliest US and Canadian finds came about this way, though like much else in exploration the early indications did not always bear fruit and great patience proved necessary. Some US fields were up and running very quickly, but despite many years of effort the first significant oil discovery was not made in Canada until 1902.

A favourite strategy in western Pennsylvania in the early years was said to be to drill next to the town cemetery, customarily located on the nearest prominent hilltop – what we now know as possible evidence of an anticline, potentially with oil trapped within. Since then, exploration techniques have necessarily become increasingly sophisticated, as more obvious and easily found sources of oil were identified, appraised and exploited early on.

Figure 7.5 Early prospecting – oil workers at Petroleum Springs, Dalaki, near Bushehr in Persia (modern-day Iran)

For some time now drilling for oil has not been used as the primary method for identifying new sources of petroleum. Instead, a range of geophysical techniques are employed to decide whether, prima facie, an area is likely to contain oil-bearing rocks of sufficient thickness. Only if these initial analyses give promising results will the more expensive drilling phase begin.

The initial analysis of a likely area may begin with what is known as a **gravity survey** or **gravimetric survey**. This makes use of the variation in the earth's gravitational field due to the differences in the density of sub-surface rocks. Because oil-bearing, sedimentary rocks are generally far less dense than other types, they effectively reduce the earth's gravitational pull within that area, so a substantial thickness of sediments can be located by measuring the earth's gravitational field and searching for a low reading. Gravimetric surveys are far from infallible – the presence of granites, for instance, produces anomalous results – but they are cheap and speedy to conduct.

In a similar way a **geomagnetic survey** of a region may be carried out. Instead of looking for the gravitational effects brought about by presence of sedimentary rocks, a magnetic survey detects potentially oil-bearing strata by their low, uniform magnetic field. Base rocks frequently contain large amounts of iron-rich minerals, and are consequently highly magnetic, and so can be distinguished from rocks of sedimentary deposition quite readily.

Finally, there is the **seismic survey**. Gravity and magnetic surveys may accurately detect the presence of sedimentary formations but they cannot pick up on the presence (or absence) of oil-bearing traps within the rock strata. The seismic method is based on recording the time taken for sound waves to travel from a source at the earth's surface down into rocks

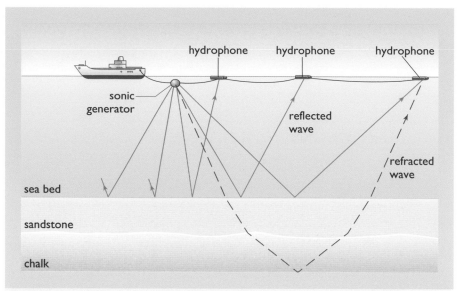

Figure 7.6 Seismic survey – reflection and refraction patterns for sound waves

below, reflect off a rock boundary, and travel back to surface detectors (Figure 7.6). The source of these sound waves was originally a small charge of dynamite, but today various techniques are employed – for example in marine environments compressed air guns are fired just below the sea surface. As a result of sound waves being reflected or refracted back, patterns are built up representing subterranean discontinuities between different types of rock, their arrangement and thicknesses. This can then provide further primary evidence of the existence of a petroleum-holding formation.

In the final analysis, however, the presence of oil can only be confirmed by physical exploration itself – drilling. But at this point, unlike the case for prospectors in Canada a hundred years ago, at least all the clues are in place. Historical data from the US suggests that only one well out of every eight drilled actually bears oil, and some of these have only a low productivity (Shell, 1966). In more recent times, with reserves diminishing and the increasing need to explore less hospitable areas such as the North Sea, commercial imperatives have meant that it is necessary for initial analyses to be more conclusive, and the results of drilling more certain. Despite this, however, not every venture is successful, and exploration is still a very hit-and-miss affair.

Having confirmed the existence of an oil-bearing structure deep below the earth's surface – and the depth can be anything up to 4500 metres, depending exactly when in geological time the local rock strata were laid down – the next stage is to determine the significance of the find in terms of the volume of oil contained, extent of the field, and likely productivity. To achieve this it is necessary to drill further wells, known as **appraisal wells** and, depending on the circumstances, a number of these may be required in order to obtain sufficient information upon which to base a decision regarding the economic viability of a particular find. Should full development then be considered appropriate, a series of **production wells** will be sunk. In the case of the North Sea, where oilfields are largely exploited from a fixed structure, carefully targeted **deviation drilling** away from the platform is carried out in order to maximize the physical (and financial) return from the discovery.

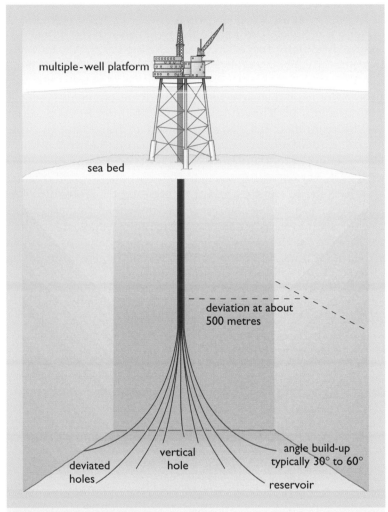

multiple-well platform

sea bed

deviation at about
500 metres

vertical
hole

angle build-up
typically 30° to 60°

deviated
holes

reservoir

Figure 7.7 Deviation drilling

With a large expenditure of capital having been incurred up to this point, without any corresponding return, it is hardly surprising that the oil exploration companies then seek the highest possible levels of initial production in order to boost their income stream. There are, however, a range of economic, political and geological factors to consider at this point. For instance, might the real price of oil rise over the productive lifetime of the field, making oil in the ground worth more than oil in the barrel? What are the likely tax regimes in the foreseeable future? What are the likely production costs, as opposed to the fixed costs already incurred? And for reasons of national security of supply, might governments introduce restrictions on depletion? It is answers to questions such as these which will determine not only whether an oil field will be exploited, but also the rate of exploitation.

Where do we get it?

Oil

Since Drake's gusher in Pennsylvania first produced oil some 150 years ago, the question has always been 'where next?' To begin with, the obvious place to look was 'next door' and so the remainder of the 'lower 48' states and Canada were well prospected in succeeding years; Alaska was to wait until the more advanced searches were undertaken a century later. Russia – then a Tsarist state – and Eastern Europe also figured prominently in early production, and the early years of the twentieth century saw Mexico and Argentina join in. The Middle East first figured in oil production in 1911, a few years after the discovery of oil in Iran. This in turn led to exploration and discoveries in both Iraq and Saudi Arabia, where production for export got under way on a fairly modest scale in the 1930s. By then, the US was out on its own in terms of production and consumption, having sufficient excess to warrant a large export industry. The other principal world exporter at the time was Venezuela. Demand was driven ever upwards by the needs of motorized transport – especially in the United States, where demand for petrol or 'gasoline' was only met by improvements in petroleum refining which maximized the yield of gasoline per barrel of throughput.

The following table shows the figures for twentieth century oil production, highlighting the three main production regions.

Table 7.1 World oil production 1900–2000[1]

Year	World production (millions of barrels per day)	Percentage accounted for by:		
		North and Central America	Middle East	FSU[2]
1900	0.4	43	–	50
1910	0.9	65	–	22
1920	1.9	87	2	6
1930	4.1	66	3	15
1940	6.0	66	5	13
1950	10.9	57	17	8
1960	21.9	40	24	16
1970	48.0	28	29	15
1980	62.7	23	30	20
1990	65.7	21	27	18
2000	74.5	19	39	11

1 Figures include crude oil, shale oil, oil from tar sands, 'heavy' oil and natural gas liquids.

2 Area covered by the former Soviet Union.

Source: Shell, 1966; BP, various years

Figure 7.8 Derricks stand behind traffic on a highway at the Signal Hill oil fields in California

As with so much else, the Second World War proved a watershed for the oil industry. For the first time conflict was vigorously pursued across the globe on land, sea and in the air, and unsurprisingly petroleum became a crucially strategic resource. With post-war reconstruction and continued economic and industrial expansion, demand for oil was to rise even faster, and as early as 1948 the United States became a net oil importer, originally importing just from the nearby Caribbean, but from the early 1970s onwards also from the Middle East, in particular from Saudi Arabia. In the 1940s and 1950s exploration in Iran, Iraq, Saudi Arabia and neighbouring states uncovered vast reserves of crude oil and these were soon to become fuel for the engine which drove a large part of the industrialized world throughout the latter half of the twentieth century. Of course, this is a situation which still applies today.

Other parts of the world have since become significant producers, too – notably China (principally for its own consumption) and Indonesia in Asia; Algeria and Libya in North Africa; and Nigeria in West Africa. International trade in crude oil became increasingly important as, by geological and geographic chance, reserves were discovered away from the main areas of consumption. Table 7.2 overleaf gives some idea as to how the magnitude and pattern of this trade have altered since 1960.

Note the dominant position of the Middle East producers in the export trade; the steady, inexorable rise in imports to the US; and the recent sharp increase in imports to the 'rest of world'. This last is almost completely accounted for by sustained year-on-year increases in consumption by the expanding economies of South East Asia.

Table 7.2 World trade in crude oil, 1960–2000, in millions of barrels per day

Year	Total	Imports to:				Exports from:			
		US	Europe	Japan	Rest of world	Middle East	North and West Africa	Canada and Latin America	Rest of world
1960[1]	10.0	2.0	6.0	1.2	0.8	6.0	1.0	2.5	0.5
1970	25.5	3.2	12.9	4.3	5.1	12.9	5.8	4.1	2.7
1980	31.4	6.2	11.8	5.0	8.4	17.5	5.3	4.3	4.3
1990	30.5	7.1	9.8	4.8	8.8	14.2	4.9	4.7	6.7
2000	38.8	10.2	9.7	5.3	13.6	18.9	6.0	6.5	7.4

1   Figures for 1960 are estimates extrapolated from Shell data.

Source: BP, various years

The discovery of natural gas off the east coast of England in 1965 led to informed exploration in the northern sector of the North Sea and the discovery of a new oil province, much of it straddling the median line between Great Britain and Norway. From being a 100% importer of oil in 1974, within ten years the UK had become, and still is, a net exporter. A combined UK and Norwegian production of 6 million barrels a day accounts for around 8% of world production – a relatively modest amount in global terms but of obvious significance locally. While wider social and political factors have certainly had an influence, without North Sea oil coming into production in the 1980s, the UK's economic performance would not have been nearly so strong in recent years. Self-sufficiency has also served as an insurance against disruptive events in a politically turbulent world, and provides a vital component in the balance of payments. The North Sea continues to provide copious quantities of both oil and gas but, like all resources, the North Sea basin is finite and this fortunate situation cannot be expected to last for ever; indeed, on the basis of published statistics it appears that peak production from the UK sector has already been attained.

Given that US production is in decline, demand from developing countries in particular increasing steadily, and much of the easy-to-find, cheap-to-produce oil already discovered, we can reasonably question the long-term sustainability of such a picture, even taking into account the existence of large untapped reserves still to be exploited in the Middle East and elsewhere. We will return to this important issue in Section 7.17.

Natural gas

Thus far, this section has concentrated on liquid petroleum – oil, rather than its fossil twin, natural gas. Natural gas, however, is set for an expanding world role long after the peak in oil production has passed. This is in sharp contrast to the early days of the oil and gas era, when, as we saw in Section 7.3, discoveries of natural gas were largely associated with oil discoveries, and it was considered to have little more than nuisance value – when not used for well re-injection to enhance oil lift, it was generally flared to waste. However, such was the magnitude of the resource which became available that it was soon piped across the United States to centres of population for local distribution to home, commerce and industry. In areas of plentiful availability and no large local heating or process markets (such as Texas) it soon found a use as a 'steam raiser' in power stations. The only other economies at that time to make use of this clean, easy-to-use and plentiful natural resource in a similar way were Canada and the Soviet Union, as the table below clearly demonstrates.

Table 7.3 World natural gas consumption 1964, millions of tonnes of oil equivalent

US	400
FSU	100
Canada	30
Western Europe	15
Middle East	5
Rest of world	negligible
Total	**550**

Source: Shell, 1966

This situation was to develop quickly, however. Small amounts of liquefied natural gas were already being traded internationally and plans were in hand to expand this trade greatly. Even prior to natural gas coming to market in the UK, liquefied gas from Algeria was being transported to Canvey Island on the east coast of England for reforming into town gas. Shipments to Japan from far afield had started and were due to be expanded further. But it was discoveries of the massive Groningen gas field in the Netherlands in 1959, and the UK offshore a few years later, which proved the spur to building transnational distribution networks across Western Europe. Continental pipelines were laid to the Netherlands and for the first time Britain was to get a national transmission system for the supply and distribution of gas to augment the many systems of local distribution (see Figure 7.9). With trans-European networks in place, large-capacity pipelines would be built from the Soviet Union into Central Europe and from Algeria to Italy on the back of sky-rocketing consumption. Between 1964 and 1974 natural gas consumption in Europe increased tenfold, albeit from a low base, and within a further twenty years had doubled again to around 300 million tonnes of oil equivalent – around half the corresponding figure for oil consumption. In the United Kingdom, the year 2000 saw natural gas consumption reach the equivalent in energy terms of oil consumption, at around 90 million tonnes of oil equivalent per annum, or 40% of total UK primary energy use.

On the world stage, however, and in contrast with oil, widespread penetration of natural gas into Asia, Africa or South America has been really rather modest up to the present day. Leaving aside the more 'mature' markets of the former Soviet Union and North America, the only other area experiencing a large rise in consumption has been Japan, taking advantage of spectacular increases in liquefied natural gas imports from

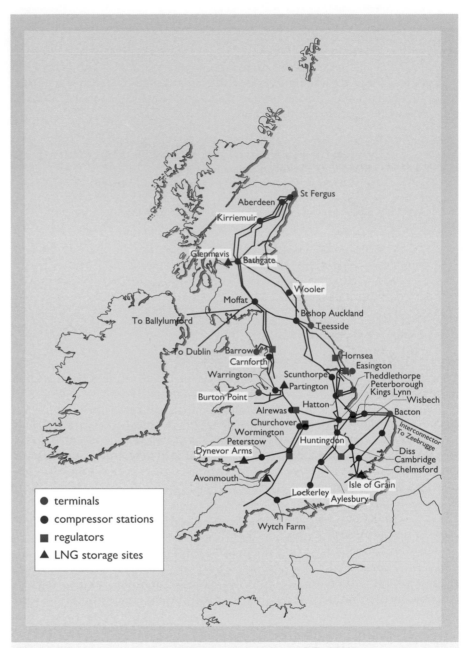

Figure 7.9 UK national gas transmission system (source: DTI, 2001)

its Asian producer neighbours, Australia and the Middle East. Some recent consumption and trade figures illustrate this and are shown in Table 7.4 opposite.

Looking further into the future, there is also the reasonable prospect of continuing, or indeed expanding, imports of gas into Western Europe from huge Russian and Central Asian resources, as well as vast Middle East reserves which could be brought to the market in the future. So there is no immediate likelihood of supply constraints being brought to bear on

Table 7.4 World natural gas consumption 1970–2000, millions of tonnes of oil equivalent

Year	Total	North America	FSU	Japan[1]	Western Europe	Other
1970	955	590	185	negligible	70	110
1980	1270	550	350	25	190	155
1990	1770	540	600	50	230	350
2000	2160	650	490	70	350[2]	600

1 All Japanese consumption is imports.

2 Western Europe currently imports around 30% of its gas from Russia and 5% from Algeria.

Source: BP, various years

consuming nations, and thus no particular reason for the UK to feel especially vulnerable about relinquishing its self-sufficient status in the next few years. But the point bears repetition: despite its apparent abundance, natural gas, like all finite resources, cannot and will not last forever.

7.5 **Refining and products**

Introduction

Since its earliest commercial application, the real technical genius of the oil industry has been to maximize the yield of useful products from a given amount of crude petroleum. Crude oil is essentially a complex mixture of hydrocarbons of varying chemical composition, and its various constituents therefore have different physical and chemical properties, including different boiling points. These different properties make possible the separation of the constituent parts of crude oil by distillation into various oil fractions, a process also known as **fractionation**. This in turn is the basis of **oil refining**, the process of obtaining useful, saleable products from crude oil. There are of course many important variations on this theme, some of which will be described below, but for the time being we should describe the distillation process and the products obtained from it in more detail.

In the early days, refining stills were simply large tanks in which crude oil was heated so that the volatile components boiled and vaporized. This hot vapour would rise, cool and condense, and could thus be collected. This process was repeated at different temperatures to separate different fractions. Following development of the internal combustion engine the need for improved product separation during the distillation process led to the use of simple 'fractionating columns'. These allowed the different boiling point 'cuts' or 'fractions' to be separated out in a single process (see Figure 7.10 overleaf). This distillation was a 'batch' process, whereas modern day refining operation demands a continuous process in which fractions are separated out and then further processed according to the type of crude feedstock available and the types of product in greatest demand.

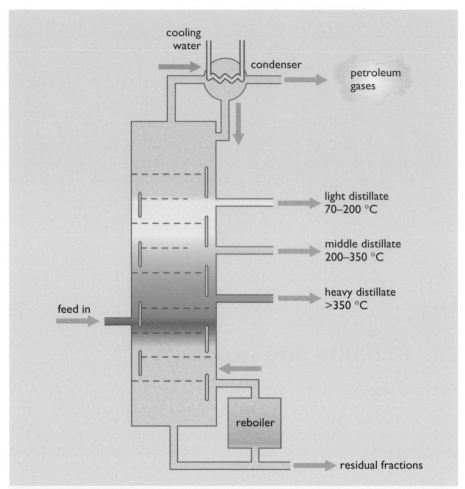

Figure 7.10 Oil distillation

The fractions

There are several ways of classifying different fractions obtained from crude oil, but one of the simplest modes of classification is division into three: 'light', 'middle' and 'heavy' distillate. These can then be further subdivided according to more highly defined physical properties or, perhaps more usefully from our point of view, product types. There are of course two further 'fractions' to consider – gases too volatile to condense under normal conditions of temperature and pressure (e.g. propane, butane) and heavy residual material remaining behind after distillation is complete. It is important to remember, however, that familiar oil products are not single chemical entities but carefully orchestrated mixtures blended for a particular end purpose, and this is reflected in the broad ranges and overlap of their respective boiling points.

Light distillate, that is products with boiling points in the range 70–200 °C, is arguably the most important of the petroleum fractions. It comprises products such as petrol (or motor spirit or gasoline), aviation turbine fuel (or jet fuel), and domestic boiler fuel (known as 'paraffin' or sometimes 'burning oil' in the UK, and as 'kerosene' in the United States). Light distillate is also used as a feedstock in the production of petrochemicals, when midst the baffling array of alternative names which pepper this industry it is frequently referred to by the oft-mispronounced and misspelt name of naphtha.

The **middle distillate** range is defined by the products with boiling points in the temperature range 200–350 °C. The two most recognizable oil products falling within this range are diesel fuel, colloquially referred to as 'DERV' (DiEsel for Road Vehicles), and an overlapping fraction known as 'gas oil', which is mainly used as burning oil for high capacity commercial and industrial heating boilers. Gas oil, as its name suggests, was originally used to provide hydrocarbon enrichment in the manufacture of town gas.

The least volatile of the petroleum fractions is referred to as **heavy distillate**, and is defined as constituents of crude oil with a boiling point in excess of 350 °C. Heavy distillate is either solid or semi-solid under ambient conditions and may require heating to encourage free flow. Diesel fuel for large ship's engines – marine diesel – falls on the boundary of middle and heavy distillate. In volume terms the most significant product falling within this range has been 'fuel oil' used both for power stations and the largest industrial boilers.

Finally, there are the 'residuals', that is the portions of crude oil not amenable to distillation under standard conditions. These residuals are far from useless, however, and indeed some of the most vital of all oil products are obtained from the very bottom of the barrel, usually by steam or vacuum distillation. They include paraffinic waxes and various grades of lubricating oils. Nothing is wasted, and what then remains is used as either heavy fuel oil, bitumen for road-making, or solid 'petroleum coke', useful either for simple steam raising or even as a source of graphite.

Getting more of what you want

It must be appreciated that the description of oil refining given above is very much a simplified version of what happens in reality. Crude oils themselves are very different mixtures, depending on which part of the world, or indeed which particular oilfield, they originate from. North Sea and North African crudes, for example, tend to be 'light' and 'sweet', i.e. the lighter end of the barrel predominates and the sulphur content is low, whereas Venezuelan oil and some Middle East crudes are much heavier and contain greater amounts of unwanted mineral impurities. Historically, UK refineries were geared up for dealing with these heavier feedstocks, whereas over the past twenty years or so much of their throughput has been of a more benign variety. At the same time product demand has changed markedly, a subject we will be returning to in the next section.

So how do the oil companies take into account both changing crude inputs and changing product demand in the refining process? Mention has already been made of **blending**, i.e. familiar oil products may be blends of more than one fraction, or sub-fraction. There may also be blending between different refinery 'runs'. Different crudes may naturally contain different proportions of different types of hydrocarbon but with similar boiling points. (For the chemists among you, the proportions of saturated, unsaturated and aromatic constituents are the distinguishing characteristics.) By careful mixing and blending, for instance, required grades of petrol are obtained. But there are further processes within the oil refinery that can better match refinery output with product demand, and some of these techniques go back to the early days of the internal combustion engine and US demand for gasoline.

Then, as now, supply was skewed to the heavy end of the barrel whilst demand was very much at the lighter end. So refining techniques known as **cracking** were devised to break down the larger (heavy) hydrocarbon molecules into smaller (lighter) ones. Initially this was done by so-called **thermal cracking** – simply by application of heat to heavy distillate and residuals. Later techniques employed the use of catalysts (**catalytic cracking**) to reduce the temperatures required and improve the suitability of the light distillate so obtained in the blending of motor spirit.

A variation of catalytic cracking occasionally employed is **hydrocracking**, where the presence of additional hydrogen under extremes of temperature and pressure further boosts the quantity of gasoline (motor spirit, petrol) obtainable from heavy oil fractions.

Figure 7.11 A catalytic cracker

The many products – a summary

To end this section, Table 7.5 gives a listing and brief description of the types of product available from the various fractions obtained by the distillation of crude oil. This is not intended to be comprehensive, but should convey the extraordinary range of products that can be obtained from this one naturally occurring fossil fuel.

Table 7.5 Products obtained from distillation of crude oil

Product	Description	Uses
methane, ethane	gases	petrochemical feedstock
propane, butane	gases	heavier gases liquefied for use as bottled LPG[1]; petrochemical feedstock
motor spirit (petrol, gasoline)	light distillate blend with well-defined spark-ignition characteristics	motor vehicles (petrol engines)
aviation spirit	light distillate blend with well-defined spark-ignition characteristics	piston-engined aircraft
naphtha	light distillate mixture with b.p.[2] 70–200 °C	petrochemical feedstock
industrial spirit	light distillate mixture with b.p. 100–200 °C	industrial solvent
white spirit	light distillate mixture with b.p. 150–200 °C	paint thinner, dry cleaning
aviation turbine fuel	a kerosene with b.p. within the range 150–250 °C	jet engines
burning oil (paraffin, kerosene)	an oil at the upper end of the light distillate range, b.p. up to 250–300 °C	lamps, stoves, domestic heating installations
DERV	middle distillate blend suitable for compression ignition engines	motor vehicles (diesel engines)
gas oil	middle distillate within b.p. range 200–350 °C	commercial and industrial boilers for heating/process heat; feedstock for 'cracking' at refineries
marine diesel oil	middle/heavy distillate mixture for use in large engines	ship's engines
fuel oil	heavy distillate and residual oils with b.p. in excess of 350 °C, classed typically as 'light', 'medium', and 'heavy'	fuel for power stations; feedstock for 'cracking' at refineries
lubricating oils	oils obtained by vacuum distillation of petroleum residues	to reduce friction and wear at any pair of moving surfaces, use is commonplace and widespread; lubricants are classed according to their viscosity
waxes	hydrocarbons of high molecular weight obtained from primary lubricating oils by crystallization	waterproofing, polishes, insulators, candles
bitumen	solid/semi-solid residue remaining after crude oil distillation	waterproofing, road-making
petroleum coke	solid residue left behind after crude oil distillation	steam raising, source of graphite

1 LPG = liquefied petroleum gas

2 b.p. = boiling point

7.6 UK demand for oil products: past, present and future

Introduction

The oil industry has come a long way in its 150-year history, not only in terms of size and importance but also in the way in which the focus of demand for its different products has changed, from illumination and lubrication, to transport, bulk heating and chemical feedstock. To obtain some insights into how these comparative demands may change in the future it is instructive to examine some recent trends (though over a long enough period that we can get an accurate picture) and seek to draw some conclusions from them.

Here we will focus chiefly on the UK, as this serves as a useful example for which data is readily available.

Time comparison, 1967 and 2000

A look at the *Digest of UK Energy Statistics* (*DUKES*) reveals that total usage of oil products in the UK for all purposes in the year 2000 amounted to some 76.5 million tonnes. This is equivalent to just over 1.5 million barrels per day (accurate comparison of weight with volume is difficult, but a conversion factor of 50 is close enough for our purposes). Let us now look back in time and examine the last occasion on which oil consumption was at this aggregate level, and then look at the constituents of that demand and how they have altered in the interim. The obvious period to consider is the 1960s, the decade in which UK oil consumption grew in a spectacular manner, from under 40 million tonnes to double that figure in 1968, before rising to its peak of over 100 million tonnes in 1973.

The year 1967, with a consumption of 78.1 million tonnes, is about the closest fit we can find to 2000. Table 7.6 opposite gives a comparison of the demand for different products in 1967 and 2000.

Various features stand out:

(i)   Feedstock for manufacture of town gas was a significant component of demand in 1967, at around 5.4 million tonnes, or 7% of the total.

(ii) Demand for motor spirit has nearly doubled and that for aviation fuel and DERV has approximately quadrupled in the intervening period.

(iii) Fuel oil demand has fallen from being by far the most significant single component of demand to just 10% of its former level.

(iv) The drop in fuel oil demand corresponds almost exactly to the aggregate rise in demand for transport fuels.

(v) With the exceptions of (i) to (iii) above, other categories of use/demand have remained remarkably constant.

Two questions emerge, then: how can the trends noted above be explained; and what conclusions, if any, can we draw from them with the next thirty years or so in view?

Table 7.6 UK demand for oil products for 1967 and 2000, in millions of tonnes

	1967	2000
feedstock for gasworks	5.4	–
LPG	0.4	1.2
motor spirit and aviation spirit	12.2	21.4
aviation turbine fuel	2.8	10.7
burning oil	1.8	3.7
DERV	4.3	15.7
gas oil and marine diesel	8.1	6.5
fuel oil	30.6	3.3
Sub-total: petroleum products used for energy	**65.6**	**62.5**
petrochemical feedstock	4.6	5.8
industrial and white spirit	0.3	0.2
lubricating oils, bitumens, waxes, other residuals	2.9	3.1
refinery fuel	4.7	4.9
Sub-total: oil-specific demand	**12.5**	**14.0**
Total	**78.1**	**76.5**

The absence of the use of oil in gas manufacture in 2000 as compared with the 1967 position has an obvious explanation in the discovery of large natural gas resources in the 1960s. Less obvious is the reason behind the huge drop in the use of fuel oil in the intervening period.

Back in 1967 the large UK manufacturing sector had a seemingly insatiable appetite for energy. Historically this had been based on coal, but during the 1960s with the increasing availability of cheap oil, mass substitution took place and fuel oil, much of it only part-refined, the least expensive end of an already cheap barrel, became the fuel of choice for the 'smokestack' industries.

At this time fuel oil was also becoming the fuel of choice for electricity generation. Large oil-fired power stations were constructed close to refineries to take advantage of cheap fuel oil. For a short while this seemed a sound investment, and these stations burned all the excess fuel oil the refineries could provide. But by 1980 the writing was on the wall for fuel oil as a combination of excess generating capacity and the ability of the nationalized coal industry to enter into massive tonnage contracts with the nationalized electricity industry at minimum prices saw coal temporarily resume its market dominance. When eventually it lost that dominance, it was gas rather than oil that was to take its place. In the meantime the oil industry had not been slow in recognizing these trends and in all probability did not regret them. Heavy fuel oil was a bulk commodity which was sold at minimum advantage. It would prove much sounder economically to invest in up-to-date cracking technology at the refineries and break down the heavy end of the barrel into higher value, more refined (literally), lighter products that carried a premium in the market place. Thus we have a situation today where fuel oil has almost

(a)

(b)

Figure 7.12 (a) Fawley oil refinery (foreground) and power station (background); and (b) Fawley oil refinery by night

disappeared; industry requires cleaner and more controllable fuels for process and space heating, such as gas oil or natural gas.

Another aspect of key importance during this period is the large increase in the use of the various transport fuels. In this part of the market, demand was quite assured, and growing all the time. At the same time the UK oil industry 'got lucky', not just because indigenous oil from the North Sea became available but because, fortunately, this crude oil was cleaner and lighter, and was more suited to the growing demands for transport fuel. The market for transport fuels now appears more unassailable than ever. The figures in Table 7.7 below give a flavour of the current dominance of transport fuels within the oil sector.

Table 7.7 Consumption comparisons

	1967	2000
Percentage of UK oil consumption used for transport	25%	62%
Percentage of UK oil consumption used for transport and other oil-specific demand	41%	81%

These figures are quite stark, demonstrating as they do the increasing concentration of oil consumption in areas that to all intents and purposes are oil-specific: refinery operations, waxes and lubricants, solvents and feedstock, but most dominant of all, transport. During the dark (literally) days of the 1973–74 oil crisis the then-Shah of Iran famously opined that oil was 'too noble a fuel just to burn' and it seems that, gradually, history is proving that statement correct. Bulk steam-raising, space heating and cooking are now marginal uses for oil and oil products. Certainly, in the developed Western world, transport demand is crucial and currently non-substitutable. If we wish to look into oil's future, it is transport that needs to be examined in greater detail.

Figure 7.13 shows the changes in demand for the three principal transport fuels in the UK over the past thirty years.

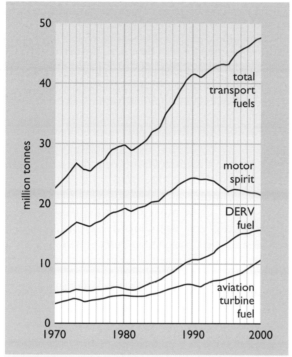

Figure 7.13 Inland deliveries of transport fuels 1970 to 2000 (source: DTI, 2001)

From this diagram we can see that the total for all transport fuels is steadily increasing – but closer examination shows that since 1990 the dominant component of this increasing trend has been air, rather than road, transport. Within the road transport sector sales of petrol have actually *fallen* slightly over the past ten years as engines have become more fuel-efficient and motorists have switched to diesel-engined vehicles. Overall, demand for road transport fuels has thus increased by only 10% over ten years whilst air transport fuel demand has doubled. On this basis, it may be that demand for land-based fuels will approach some peak over the next ten to twenty years. If this were the case, a mix of saturation, demand management, price hikes or efficiency improvements in air transport would be the principal regulators of growing aggregate demand within the transport sector. However, with the growth of low-cost airlines and general deregulation a key feature of the air industry in the early years of the twenty-first century, a low growth future for this sector currently seems unlikely in any 'business-as-usual' future.

7.7 **UK demand for gas: past, present and future**

Measurement

In the previous section, changes in oil consumption over a lengthy period were examined by product type and a trend identified towards an oil-specific demand, chiefly in the transport sector, rather than a demand for energy per se. For gas, however, the situation is rather different, in that instead of numerous products there is just one saleable commodity, namely piped natural gas.

By UK statute, piped gas has to be sold with a **calorific value** or 'the amount of heat contained' declared, and you should be able to find this information on any domestic gas bill. You might recall from Chapter 2 that two sets of units are used for oil, one according to *volume* (barrels, scaled up to million barrels per day) and one according to *weight* (tonnes, scaled up to million tonnes per year). This is complicated and imprecise enough, but simplicity itself compared with all the units of measurements employed in the gas industry. These go back to the days of units such as rods, poles and perches, from which emerged the British Thermal Unit and the therm. The non-standard unit 'thousand cubic feet per day' was frequently employed too. The therm was a nice unit of measurement, in as much as its size was sensible; on a really cold day the central heating system in an average-sized house would soak up four or five therms of gas. For those interested in such things, a tonne of coal would contain about 250 therms-worth of heat and a tonne of crude oil around 400 therms. The therm is now obsolete, however, and the gas supply industry has chosen the kilowatt-hour (kWh) as its standard unit of measurement; about thirty of these make up a therm. However, the *calorific value* is declared in megajoules (MJ) per cubic metre supplied (3.6 MJ = 1 kWh).

This plethora of units gives much room for confusion, and to avoid this, and to facilitate comparison with oil industry figures, we shall use the 'million tonnes oil-equivalent' (Mtoe) measure. The present UK annual gas consumption is about 90 Mtoe. For our purposes this is more convenient than the equivalent 1116 terawatt hours (thermal) figure quoted in the official statistics. Whatever choice we make is essentially arbitrary, but as 'oil and gas' are so frequently discussed together it makes a certain sense to stick to common units of comparison, even though the one chosen (Mtoe) cannot pretend to be scientifically precise.

Demand by sector

Historically, gas was first manufactured for lighting and subsequently for cooking and rudimentary space and water heating in the home. Its use in commerce and industry was largely dictated by where it was produced and was dominated by industries such as coke-making and iron and steel which were involved in both its production and its consumption. As we have already seen, the entire picture changed in the 1960s when natural gas was first brought to market, and within ten years the gas supply industry had been transformed from a 100% secondary to a 100% primary fuel, a remarkable achievement which resulted in natural gas soon making notable inroads into the domestic, industrial and service sectors.

It is worth taking a look at each of these three sectors in turn. By analysing gas usage in these sectors we can perhaps gain some understanding of likely future trends.

Domestic sector

The dominant component of demand in this sector is for space and water heating. It may be hard to imagine in the twenty-first century, but in the post-Second World War years, domestic central heating was something of a rarity in the UK. Only in the 1960s and 1970s did the coal fires and paraffin stoves make way for electric storage heating and gas-fired boilers, as increasingly affluent households invested in convenience and increased comfort. It is estimated that in 1967 only 1 million UK households had gas-fired central heating. By the year 2000 the figure had climbed to 15 million and, for practical purposes, was approaching saturation. Originally the 'gas boiler' would have been a large, free-standing cast-iron construction, sitting in the kitchen corner, garage, or outhouse, with a conventional flue, offering at best 70% efficiency at full load, and part-load efficiencies (e.g. in supplying hot water in summer months, when heating is not required) of only around 20%. Lighter, wall-mounted 'balanced flue' models followed, offering advantages in terms of installation and operating efficiencies, but still with poor part-load performance. Only more recently, with the advent of the gas condensing boiler, have efficiencies of 85–95% under all operating conditions been attained.

Although slightly more expensive to install and maintain, at the time of writing (spring 2002), the gas condensing boiler is due to become the industry standard and offers considerable savings in terms of gas consumption and running costs. The principal operating difference between a conventional gas-fired boiler and the condensing boiler is that instead of high-temperature water vapour being released directly to the atmosphere, in the condensing boiler it is condensed via a second heat exchanger and the heat thereby recovered rendered 'useful'.

Following the huge surge in gas use of the 1970s, the growth in consumption in the domestic sector since then has been comparatively modest at around 2% per annum, and consumption currently stands at around 30 Mtoe per annum. With saturation effects beginning to bite, the gradual penetration into the replacement market of far more efficient boiler equipment, and increasing standards of home energy efficiency, it may be that this growth will be further curtailed in the future.

Industrial sector

In some ways trends in the industrial market follow those in the domestic sector quite closely: the steep increase when natural gas first became widely available and modest increases thereafter. The industrial sector, however, provides a large amount of the flexibility necessary for the gas supply industry to operate economically.

Because so much of overall gas demand is for space heating, peaks and troughs in demand follow the seasons. In winter, sharp fluctuations in temperature cause surges in gas demand which can be up to 50% greater than the daily average. The gas industry, therefore, has storage facilities able to deal with demand peaks (including depleted offshore gas-fields), but in addition to spare capacity on the supply side it seeks to manage consumption on the demand side by entering into 'interruptible' contracts with industrial customers. At times of anticipated high demand – and the gas supply industry is very skilled at accurately predicting overall demand for a given set of meteorological conditions on any given day – industrial consumers will be cut off for a pre-advertised period, at which time they typically revert to gas oil or burning oil as replacement fuels for their process and space heating needs, until the mains gas supply is restored. To put it another way, industrial consumers provide a very useful sink into which excess gas supplies can be dumped at times of low and medium demand. As for the future, given the diminution in the UK industrial sector over recent years (especially heavy industry) and the limited scope for substitution of other fuels, it seems unlikely that UK industrial demand for gas in years to come will change markedly from its current level of some 25 Mtoe per annum.

Figure 7.14 A 'gasometer', London, circa 1930. These huge storage tanks can still be seen in many locations in the UK

The 'service' sector

The 'service sector' is something of a catch-all which includes commercial and retail premises, offices, hospitals and all kinds of public buildings. As in the case of domestic buildings, demand is largely for space heating, but in contrast to the manufacturing industry, this sector of the economy has expanded greatly in recent years and so concomitant increases in electricity and gas consumption are unsurprising. Gas use in this sector has doubled over the past twenty years, increasing at an average rate of 3.5% per annum, and now stands at some 10 Mtoe per annum. With increased economic activity expected to continue in this area, further sustained increases in gas consumption can be reasonably anticipated.

Across these three principal sectors of UK gas demand we have a current annual consumption of 65 Mtoe, with only modest future increases anticipated. However, there is one further sector of consumption to be considered, one that had grown from next to nothing in the early 1990s to some 25 Mtoe by the turn of the century – namely, the combustion of gas in the generation of electricity, an area due for further rapid expansion in the near future. Much of this generation is carried out using the combined-cycle gas turbine (CCGT), a technology to be described in greater detail in the next chapter.

As mentioned in Chapter 5, compared with conventional generation in coal-fired power stations, gas-fired generation has a number of advantages, such as a lower capital cost, less noxious pollution, and higher generating efficiencies. Its greatest virtue, however, is that generating electricity this way results in far less CO_2 being pumped into the earth's atmosphere than with more conventional means of generation. Indeed, it is the substitution of gas for coal in electricity generation these past ten years that has most enabled the UK to meet its CO_2 reduction obligations under the Kyoto agreement.

The full analysis described above is summarized in the following chart.

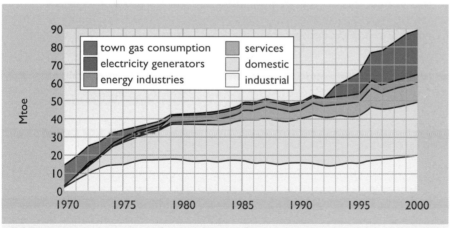

Figure 7.15 Consumption of town gas and natural gas 1970 to 2000 (source: DTI, 2001)

The conclusion to be drawn is that whilst in most sectors gas consumption is reasonably steady, the overall picture is totally skewed by the new entrant, power generation. Any modest changes to domestic or commercial use will be more than outweighed in the future by gas supplies tied up in electricity generation contracts. But despite the advantages outlined above, one might perhaps question whether such large quantities of a fuel so well suited to *direct* use are best employed in *conversion* from one form of energy to another.

7.8 Why so special?

Before leaving the first part of this chapter on oil and gas behind, it is worth briefly considering why exactly we make such a fuss about these two vital sources of primary energy. What *is* so special about them? Some things are obvious, others less so. We will again focus on the UK as a useful example, but it is important to remember that the picture may vary for countries without easy access to supplies. With this in mind, let us take a more detailed look at the key characteristics of oil and gas:

- cheap and readily available
- available from indigenous production, giving security of supply
- convenient and easy to use
- clean to burn (compared to coal)
- easy to distribute/store/carry about
- high energy density.

Cheap and readily available

Natural gas and petroleum products are inexpensive by most standards. Certain oil products, notably DERV and motor spirit, are expensive in the UK because of the taxes applied to them by HM Treasury, but even then, as Figures 7.16(a) and (b) opposite show, the real prices of road fuels are little more today than they were thirty years ago, before the first oil price shock of the 1970s. A crude oil price of $25 per barrel translates to about 10p per litre. This is roughly equivalent in energy terms to the price of natural gas in the market place, and if one looks at, for example, the real price of domestic gas (i.e. allowing for inflation) over the past thirty years, it is easy to see that it has never been cheaper. (See also Chapter 12.)

Apart from occasional strike action or blockades, availability of these fuels in the UK has not been an issue these past twenty years given its self-sufficiency. This is not always true for other countries dependent on imports from politically sensitive areas of the globe.

Indigenous production/security of supply

One of the reasons the UK has remained confident about the availability of gas and petroleum products is that for the past quarter century it has been a substantial producer of oil and gas, and indeed, a self-sufficient one. In relation to oil, that status seems likely to continue for some years, although with further increases in gas consumption for electricity generation the UK is about to lose its self-sufficient status in natural gas. However, such is the anticipated abundance of natural gas routed by pipeline, and under contract, to UK markets that, short of unforeseen circumstances, security of supply seems assured. The counter-argument is, of course, that 'unforeseen circumstances' occur all too frequently around the globe.

Convenience and ease of use

There is not much to expand on or analyse here, save to say that this is more than a mere advertising slogan. There will be very few people who have not experienced first hand the convenience of these fuels. In the

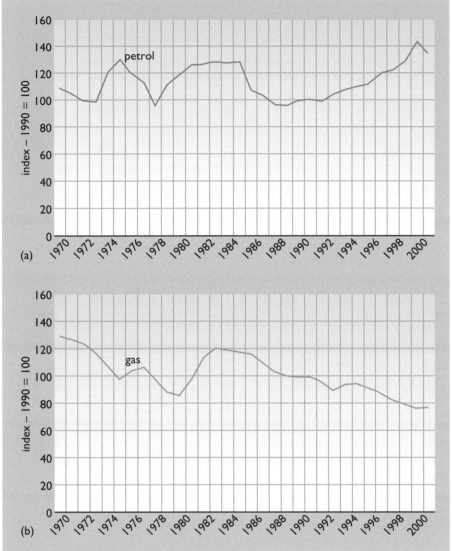

Figure 7.16 (a) Real cost of road fuels, 1970–2001 (b) Real cost of domestic gas, 1970–2001 (source: DTI, 2002)

'developed' Western world, everyone encounters their effects, one way or another, every day, and even in less developed countries oil and gas are used as primary fuels for light, cooking and heating.

Clean to burn

This point is rather more contentious, and its validity does depend on one's definition of 'clean'. Natural gas, as delivered to the consumer, is obviously the cleanest *fossil fuel*. The only products of its combustion are carbon dioxide and water, and per unit of heat output the amount of CO_2 produced is less than for coal or oil (see Table 7.8). Whilst the carbon dioxide that is produced does contribute to the greenhouse effect, it is not a chemical pollutant in the same way as, say, the oxides of sulphur or nitrogen, characteristic by-products of the combustion of fuels such as coal and oil.

Table 7.8 Carbon emissions from combustion of fossil fuels[1]

Fuel	Carbon emissions (kg of carbon per GJ)
coal	24
oil	19
natural gas	14

1 Emissions are expressed in kilograms of carbon per gigajoule. To convert this to kilograms of carbon *dioxide* per gigajoule, multiply by 44/12, the ratio of the relative atomic masses of carbon dioxide and carbon.

Source: DTI, 2000

The main danger that can arise with natural gas occurs when it burns in an insufficient/limited supply of oxygen, resulting in the highly poisonous carbon *mon*oxide being produced, rather than carbon *di*oxide.

Combustion of various oil products, although apparently 'clean' to the user, is invariably polluting to the environment, albeit at a low level in some cases. Generally speaking, the heavier the oil product being burnt, the more likelihood that its combustion will bring in its wake deleterious effects. Hence bottled gas, a propane/butane mixture, should be as clean to burn as natural gas, whereas at the heavy end of the scale, fuel oil is invariably high in sulphur and other impurities. Even the cleanest DERV fuel contains some sulphur, if only at a minimal level. Upon combustion, DERV fuel can emit tiny particles into the atmosphere, with associated health hazards, and even a properly maintained and operating petrol engine will add to the environmental concentration of NO_x (oxides of nitrogen), which contribute to acid rain. This subject is dealt with in more detail in Chapter 13.

Ease of distribution/storage/portability

These aspects are easy to take for granted, but they are some of the key features behind the penetration of oil and gas into various energy markets that might otherwise feature coal and electricity (irrespective of generation method) more strongly. Solid fuels cause handling and storage problems at all points of the supply and distribution chain and only score over some oil products in respect of flammable hazard. Electricity may be instantaneous in its distribution, but it cannot be stored except at great expense and with associated losses. Local storage of oil presents no inherent difficulties but requires stringent safety measures. Natural gas can be shipped, and indeed stored, in a liquefied form, but under normal modes of operation storage is carried out within the transmission system itself.

Energy density

This is where oil and gas, and oil in particular, score highly. It is still something of a marvel, to this writer anyway, that the tank at the back of his house contains sufficient oil to keep it warm through the long Scottish winter; and that the far more modest storage facility in his diesel car could enable him to get from his home in Edinburgh to the Open University campus in Milton Keynes and back without refuelling. Whether by weight, or by volume, oil has a remarkably high energy density, at around 40 MJ per litre in volume terms or 50 GJ (50 000 MJ) per tonne, whereas the best lead-acid battery, for example, cannot achieve within two orders of magnitude of this (see Chapter 4).

For all of the above reasons, as long as they remain available and affordable, oil and gas will usually be the fuels of choice. It may be that generations to come will look back on the twentieth and twenty-first centuries as periods when ready availability of cheap, easy-to-use fuels was not fully appreciated for the easy benefits they brought in their wake. Most things in life are not properly appreciated until they are no longer available or attainable, and in that respect convenient, inexpensive primary energy sources are no exception.

7.9 Substitutes for oil and gas?

Part 2 of this chapter, subsequent chapters and the companion volume to this book, *Renewable Energy*, will examine the possible alternatives to primary fossil fuels. This first part of Chapter 7 has set some benchmarks by which any rivals might be judged. For whether it is hydrogen, solar energy, or power from a tidal barrage, each of these energy sources has the metaphorical mountain to climb to match the properties of usefulness, affordability, adaptability, and above all practicality, offered by the familiar liquid and gaseous fuels of today.

Part 2 **Oil and gas as secondary fuels**

7.10 **Introduction**

From the concluding part of the previous section we can readily see the various reasons why oil and gas have caught on in such a big way, and with the pair of them now providing over half the world's and three-quarters of the UK's primary energy needs, we are completely dependent on them. But the day will inevitably come when these fuels are not so readily available, and so it makes sense to look into the future and seek alternatives to oil and gas. The main theme of Part 2 of this chapter, therefore, is conversion processes whereby a primary fossil fuel (usually coal, but in certain circumstances even oil or gas) is changed into a more desirable secondary fuel. By 'more desirable' we mean more suited to, or more needed for, a given set of circumstances. An example would be conversion of naphtha to town gas in the 1960s, for which this cheap, clean petroleum feedstock was ideal. A more recent example might be the converse of this, i.e. converting natural gas, where there is no obvious or local market, to a liquid gasoline for use as a road fuel. We will look at examples of these conversion processes later on in the chapter.

One key way of satisfying future needs, though, is to do what was done in the past (only better) and once again manufacture transport fuels, piped gas, etc., using coal as a primary feedstock, and this is something that the discussion in this section will focus on in particular. First, though, there are at least five major stumbling blocks that stand in the way of the success of coal conversion processes, each worthy of brief consideration.

7.11 **Obstacles to coal conversion**

Technology

The first point relates to the technology available. The history of coal conversion technology goes back over 200 years, so in many respects there is nothing novel about using coal in this way. What is new, however, is the imperative to maximize output of certain products, those that are the 'most desirable' from today's point of view – e.g. a light or middle distillate-type oil. Scaling up from prototypes or small-scale plants is likely to present real practical difficulties in terms of coal handling and waste disposal, efficiency of operation and economic viability. As we shall see, although **coal liquefaction** technology has been around for the best part of a century, it has never yet had to prove itself as an open market competitor to other primary or secondary fuels.

Environment

Environmental considerations also loom large. The products of coal treatment are many and varied and not all are desirable or saleable. No matter how complete any distillation/combustion process is there will always be an appreciable ash residue to dispose of, something which was for many years a notable problem afflicting coal-fired power stations. Were coal liquefaction technology required at some time in the future then its required scale of operation would be such that proper disposal of waste, scrubbing of noxious vapours and careful treatment of all distillate products would be subject to legislative scrutiny designed to protect the environment. Whilst this in itself would not rule out large-scale coal liquefaction it would dictate higher capital and running costs than might otherwise be the case and this could be the critical factor in deciding whether coal conversion technologies are adopted in the UK or elsewhere.

Conversion efficiencies

The third obstacle to be overcome is that of conversion efficiency – that is, the proportion of energy appearing in the secondary fuels compared with the input energy contained in the primary fuel. As we have seen in previous chapters, in terms of useful product output, all energy conversion processes are necessarily less than 100% efficient, and frequently far less so. For coal conversion in particular, with its myriad different possible products, the exact definition of the term *useful* will clearly have a bearing on any exact measure of *efficiency*. Yet however generous one is in making such a definition, conversion of coal to liquid and/or gaseous fuels will always be significantly less than 100% efficient. Therefore the market price of any fuels made from coal would have to fully reflect the energy losses involved in converting one to the other.

Figure 7.17 (a) A car fitted with a gas-bag and (b) an electric car refuelling – two very different alternatives to conventional motor fuels

Cost and price

The next hurdle to be overcome is the double-sided coin of cost and price. The *cost* of production of conventional oil currently ranges from a few tens of cents per barrel in the most productive oilfields of the Middle East to perhaps an average $10 per barrel for oil from resources that are more difficult to tap, such as those from the North Sea. The market *price*, at the time of writing around $25 per barrel, effectively dictates whether a marginally viable oilfield is developed or not, though it must be remembered that, for much of the world's current production, the cost (of production) and price (in the marketplace) have little to do with each other. Price will always be determined by demand, or more accurately demand for the **marginal barrel**, that is, the *next* barrel, and this is where coal liquefaction technology will need to compete in accordance with its own cost of production. Only when the marginal barrel of conventional crude oil is more expensive to produce than synthetic crude oil from coal will the latter have attained a competitive position. And yet in such a situation, that of increasing energy prices on the world market, the price of coal will surely have risen, too, rendering any calculation of economic viability subject to further examination and correction. Hence there is a situation akin to shooting at a moving target, and one moving always in a contrary direction. The effect is to render the economics of any coal conversion scheme even more problematic. These are complex issues, and are discussed in more detail in Chapter 12.

Capital cost

Finally, we must look at what may prove to be the biggest obstacle of all. Even if the technology is sound, environmental requirements properly satisfied, and coal feedstock contracts entered into at an advantageous price, there is still the crucial matter of capital cost, as illustrated by the simplified comparison in the box below.

BOX 7.2　Cost of electricity versus cost of synthetic oil

Operating at full load, for 90% of the year, and at 30% efficiency, a 2000 MW$_e$ output coal-fired power station produces 16 TWh of electricity. Allowing for losses occurring in generation, this would require an input of approximately 7 million tonnes of coal. Assuming a wholesale price of 2.5p per kWh, the yield from electricity generation is £400 million.

Were this amount of coal instead diverted to a coal-to-oil conversion plant operating at 40% efficiency as defined by 'useful output', i.e. oil products rather than electricity, annual production would be around 1.75 million tonnes, or 35 000 barrels per day. So to match the yield from electricity generation, the oil products obtained would need to command a price of £230 per tonne, in round figures £30 per barrel or about $45 per barrel.

At first glance, therefore, the coal-to-oil plant might appear to be a competitive future option for, in their simplified forms, the above sums appear to subsume both capital and running costs. Yet while the electricity generator is an example taken from 'real life', the coal-to-oil plant is not. Its necessarily complicated technology is likely to incur capital costs substantially in excess of the more mature technology of raising steam to drive a turbine, and therefore a sale price significantly higher than $45 per barrel would in fact need to be realized for commercial viability.

But then, even if coal conversion technology were competitive with electricity generation, if one takes into account end-use efficiencies then 'electric transport' would appear to have a clear advantage over that based on synthetic oil.

Summary

We have examined some of the possibilities and pitfalls awaiting the possible adoption (or re-adoption) of coal conversion technologies in the future. In a world where the production of primary natural gas or crude oil becomes increasingly expensive, manufacturing gas and oil products from a different primary source may become a competitive option. The obstacles standing in the way of large-scale coal conversion are indeed considerable at the present time, but the economics may one day justify taking this approach. Besides this, there are matters beyond mere economics, e.g. increased funding for prototyping, the need for security of supply, and national need to revive manufacturing, etc., which could serve to encourage the adoption of coal conversion technology sooner than might otherwise be the case.

Let us, therefore, take a closer look at some of the technologies involved in conversion processes. Before looking in some detail at the technology of converting coal to other useful products it is instructive to briefly consider gas-to-oil and oil-to-gas conversion processes.

7.12 **Gas from oil**

The making of gas from oil for distribution as a public supply may be seen as something of an historical oddity in that the need for such a process is not obvious. As we have seen, from the early 1800s until the 1960s, in Great Britain the basic technology of gas manufacture from coal changed very little. During the 1960s natural gas began to take over, and by 1980, public supply of gas had changed out of all recognition, from a manufactured product (town gas) delivered locally to a primary source (natural gas) delivered nationally. However, within the short period between 1960 and 1975, the availability and convenience of cheap oil made it a strong competitor to coal as a primary source from which to make a manufactured town gas, especially in areas such as the south of England with no local source of coal.

The process employed to convert light distillate feedstock to gas for mains distribution was one of a general class of processes in the petroleum industry known as **reforming**. As the name implies, this involves the rearrangement of the hydrocarbon molecules making up the oil feedstock and, depending on conditions of temperature, pressure, catalyst and other reagents, a whole range of different products can be synthesized. Like so much else in the petroleum industry, reforming technology was originally developed to improve the yield and ignition characteristics of gasolines (i.e. motor fuels) and was subsequently adapted for other purposes – and one such purpose was the conversion of oil to gas. It was found that treating naphtha with high pressure steam at elevated temperatures (700 °C), the carbon–carbon bonds sheared completely, producing a gaseous mixture containing carbon monoxide and hydrogen (and a little methane). This became known as **synthesis gas** and it will be referred to again in subsequent sections of this chapter, as it is a commonly occurring precursor to many useful products during hydrocarbon conversion processes. Production of town gas – a hydrogen-rich gas with small amounts of carbon monoxide and methane – was achieved by subjecting the synthesis gas to a further steam reforming stage.

This cheap and easily made sulphur-free product, with no difficult by-products associated, was then available for mains distribution. With the advent of natural gas in Great Britain, oil-based town gas had a limited lifespan, but its manufacture continued in Northern Ireland until the 1990s, when piped natural gas took its place there too.

7.13 Oil from gas

As suggested in earlier sections of this chapter, although oil supplies around the world may become more restricted over time, future supplies of gas seem a little more assured. There therefore exists the possibility that it could prove useful to be able to convert these plentiful natural gas (methane) supplies into scarcer liquid fuels. One advantage of this, were it to be carried out at the point of production, would be to obviate the costs and large energy losses involved in natural gas liquefaction and its subsequent transport to distant markets.

We have seen in the previous section that synthesis gas is the staging post between liquid and gaseous hydrocarbon products. In the same way that

Figure 7.18 Mobil's methanol to gasoline plant in New Plymouth, New Zealand

oil can be converted to gas, the reverse process is also possible. For many years, production of the bulk chemical methanol, CH_3OH, has been achieved on an industrial scale by the steam reforming of natural gas (methane, CH_4). Although methanol is a potentially useful energy source for internal combustion engines, its widespread deployment would demand engine modifications, rendering its use uneconomic. However, conversion of methanol to a light distillate akin to petrol has been achieved successfully on a commercial scale in New Zealand, where gas from the large offshore Maui field was piped ashore and made into petrol at Mobil's pioneering Methanol-To-Gasoline (MTG) plant. From the mid-1980s until the mid-1990s it produced some 15 000 barrels daily, a quantity which represented over 10% of New Zealand's domestic oil needs, in a country otherwise totally dependent on petroleum imports. The relative success of this scheme suggests that should natural gas still be abundant at a time when oil reserves have become severely depleted, then the gas-to-methanol-to-petrol route may prove a viable option for the production of secondary fuels. There is also the possibility of converting natural gas directly to petrol or even diesel. This has been demonstrated at pilot plant level, but it is only beginning to be exploited on a commercial scale, e.g. as in Shell's Middle Distillate Synthesis (SMDS) process. Prospects for further developments over the next couple of decades seem promising.

7.14 Gas from coal

A little history

The booming iron industry of eighteenth century Britain was founded on the ready availability of both iron ore and timber. The latter provided the source of charcoal which was the 'reducing agent' employed to convert iron ore to pig iron (that is, crude iron) in the primitive furnaces of the day. But as early as the first decade of the century it had been demonstrated that in the same way that charcoal is produced from wood by the action of heat in the absence of air, so a parallel product, called **coke**, could be made from coal. And although its appearance was different to that of charcoal it had the same reducing effect on iron ore. As the century progressed, coke became a live competitor to charcoal in the iron industry, superseding it in Great Britain almost completely by the turn of the nineteenth century.

During this **carbonization** of coal to produce coke little attempt was made to make use of any of the volatile components of coal, and indeed the coke ovens of the eighteenth century must have been the source of the most dreadful atmospheric pollution. By the late 1700s, though, it was generally recognized that, along with coke, both tar and gas were consequential products of coal carbonization, and attention turned to making use of these (see Box 7.3).

Reference was made in Part 1 of this chapter to Lord Dundonald, one of the pioneers of gas lighting. He was also one of the early pioneers of coal tar exploitation, which in time would spawn a whole new sector of the chemical industry. By the middle of the nineteenth century, after Dundonald's time, the products of coal carbonization were much better understood, a remarkable feat given that so many different products are formed during the process of carbonization.

BOX 7.3 Distillation of coal

To maximize the yields of coke and gas produced, coal is distilled at high temperature in the absence of air or any other reagents. In addition to coke residue and an impure coal gas two other fractions are obtained: an aqueous layer containing ammonia and ammonium salts; and an organic, tarry layer amenable to further fractional distillation in much the same way that a crude petroleum of today might be processed. The ammonia is converted to ammonium sulphate, which finds a ready use as a nitrogenous fertilizer, and the coal tar is further distilled to yield four volatile oily fractions and a residue of pitch, used in road-making and roofing. The four volatile fractions roughly correspond, in terms of their boiling points, to the petroleum fractions described in Part 1 of this chapter, but their compositions are very different, consisting largely of aromatic (carbon ring) type hydrocarbons rather than the aliphatic (open chain) type hydrocarbons characteristic of petroleum. Many of these aromatics find a use in the chemical industry and for many years the lightest of the tar fractions was the chief source of the bulk chemical benzene, which is the basis of all aromatic organic compounds.

The coal carbonization process is summarized in the diagram below.

The yield of liquid organic products is a miniscule 2.5%. Of course, this yield is achieved under conditions best suited to gas and coke production, but it gives some idea of the magnitude of change required for production of economically viable quantities of liquid hydrocarbons, a subject to which we will return in the next section of this chapter.

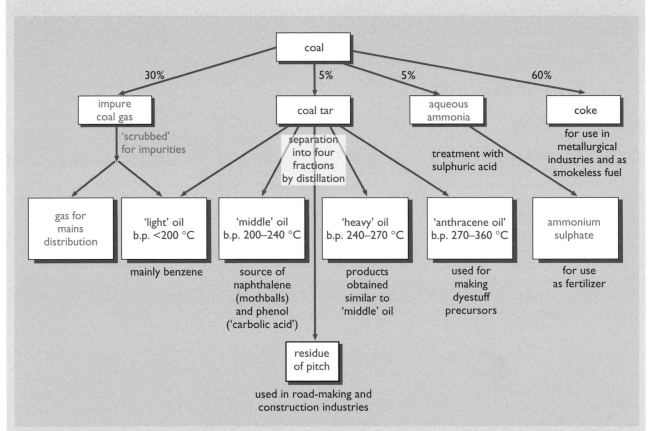

Figure 7.19 Products of coal distillation. The yield of coal tar is just 5%, of which half is pitch. The other half comprises the four oil fractions

Between initial production of coke in the mid-to-late eighteenth century and the well-established coal tar industry of the mid-nineteenth century came the introduction of gas lighting on a wide scale. It is difficult to imagine, or overemphasize, the technological change occasioned by the introduction of gas light and the beneficial effects for the population as a whole. But in the early decades of the nineteenth century, when it was introduced to the towns and cities of Britain, only the rich could afford it and even by the end of the century its cost was a very substantial sum out of the weekly budget of most users. It was familiar to all, however, as civic supply for street lighting, much welcomed by the police and public alike as it accompanied a diminished level of street crime.

Figure 7.20 A nineteenth century lamplighter going about his work on the Thames Embankment in London

Figure 7.21 Large-scale gas works, Southall

The first public supply of gas began in London in 1813, courtesy of a German entrepreneur Frederick Winsor who was the first to promote the idea of gas being manufactured in a central works and distributed by means of underground pipework. The previous year he had formed the Gas Light and Coke company, which was to grow to be the largest company of its kind in the world. Before long the better-off parts of London were able to avail themselves of this novel form of lighting, albeit at a remarkably high cost. Other towns and cities were to follow suit, but this expansion in the production of coal gas resulted in a difficulty. Ready markets were available for as much coke as could be produced, but massive accumulations of coal tar were building up with no apparent uses, for the technology of secondary distillation and separation of products had yet to be developed. Rudimentary uses of coal tar were stumbled across, e.g. as organic solvents, wood preservative or crude heating fuel, but it was not systematically exploited as a valuable resource until halfway through the nineteenth century. By then the coal carbonization industry was producing not just gas for cooking and lighting and coke for the iron and steel industry, but also chemicals (and soon dyestuffs) for the chemical and textile industries, fixed ammonia for fertilizer production and all manner of tars and pitches for preserving, weatherproofing and road-making. And it was to survive in this basic form for a further one hundred years until it was cheaper to make gas, organic chemicals, and even road construction materials from oil. Coke is still manufactured for the UK's contracted iron industry, with coke-oven gas used on-site and coal tar distillation carried out on a somewhat limited scale.

Gas from coal – the future

As we have just seen, past practice featured coal as the starting point in the production of many different saleable products, of which gas was just one. If coal were once again to become a primary source for manufactured gas, because of a scarcity of natural gas, it should not be assumed that there will exist a corresponding demand for coke, tar, or aqueous ammonia. Nor should it be assumed that 'town gas' will be an acceptable product even in times of scarcity; quite the opposite, in fact. The entire system of storage and transmission and all final uses of gas are now suited to methane and not a fuel such as coal gas, which has quite different combustion characteristics and which has roughly half the calorific value of natural gas. So clearly, if coal were once again to be the starting point in gas manufacture then the end point would have to be a synthetic natural gas of comparable burning quality and calorific merit to the natural gas of today: coal will need to be converted to virtually pure methane.

As illustrated in Figure 7.19, when coal gas is manufactured by high temperature distillation of coal the yield obtained is only around 30%; ammonia solution and tar account for some 10%, leaving 60% as coke – not surprising perhaps given that the principal constituent of coal is carbon. So very little methane is actually produced in this process and most of the carbon remains behind.

If the quantity and quality of gaseous products is to be boosted to a level sufficient to meet our needs today, then an additional reagent needs to be introduced, principally to react with the 60% carbon left behind by simple distillation and to convert this carbon to methane, CH_4. Such conversion of coal to gaseous products with the aid of external agents is generally referred to as **gasification**.

What might these 'external agents' be, then? An obvious choice for conversion of coal to methane might appear to be hydrogen – but hydrogen is already a clean combustible fuel, and there is little economic or thermodynamic value in swapping one good fuel for another. As the principal element required is hydrogen, the next obvious choice is water – though the chemistry involved means that this demands rather more than simply dropping a lump of coal into a pail of water! What is required to effect a reaction is steam rather than water, typically accompanied by air or oxygen.

The production of methane in this manner is essentially a three-stage process. The first stage is the reaction of heated coal with steam, to yield a mixture of carbon monoxide and hydrogen known as **water gas**. The second-stage reaction involves taking the water gas produced in the first part, and increasing the hydrogen to carbon monoxide ratio from 1:1 to 3:1. This requires the use of more steam over a cobalt catalyst and results in the conversion of carbon monoxide to carbon dioxide. This step is often referred to as a **shift reaction** and renders the gas stream suitable for the final stage of conversion into methane; this final step is known as **methanation.**

If we take the three steps and add them all together, the entire process can be summed up thus:

$$2C + 2H_2O \rightarrow CH_4 + CO_2$$

In other words, half the original carbon in the coal ends up as methane and the rest as carbon dioxide. The details of how the process works are described in Box 7.4 opposite.

BOX 7.4 Coal gasification: reversible chemical reactions and equilibrium

Chemical reactions frequently occur instantaneously, or at least they appear to be instantaneous. In practice, such reactions are very fast and essentially spontaneous. Lighting a match or the burning of petrol are familiar examples. Very fast reactions usually produce heat as the reaction occurs; such reactions are termed exothermic. Other reactions may absorb heat as the reaction occurs; these are termed endothermic. Endothermic reactions usually need a high temperature in order to supply enough heat for the reaction to occur. Endothermic reactions are often reversible, i.e. the reaction can proceed in either direction, depending on the conditions. In most cases the forward reaction occurs at high temperatures while the reverse reaction occurs at low temperature.

The above concepts are useful in explaining how coal can be converted to more useable secondary products. The first stage of that process, the 'water gas reaction', requires an input of energy not only to start it but also to keep it going. Thus the water gas reaction is an endothermic reaction, and the equation representing that reaction is written as:

$$C + H_2O \rightleftharpoons CO + H_2$$

Note the 'two-way arrow', which signifies that this reaction can take place in both directions. In chemical terms this represents a dynamic equilibrium reaction between reactants (carbon and steam) and products (carbon monoxide and hydrogen). We have stated that the 'forward' reaction i.e. production of water gas is endothermic, requiring an input of heat, so logically the greater the heat applied, the more the reaction will be driven to the right-hand side of the equation. And so it is in practice – the greater the temperature the higher the quantity of gas produced, and it was long ago found out that the optimum temperature for water gas production was around 1000 °C.

The second stage in methane synthesis, the so-called 'shift reaction', can be written as:

$$CO + H_2O \rightleftharpoons CO_2 + H_2$$

This is a mildly exothermic chemical change and so, using the converse reasoning applied to the previous stage, this reaction should be favoured by a low operating temperature. Indeed, that is the case, but for practical purposes it is necessary to keep the temperature up, for the general rule which applies to *all* chemical reactions is 'hotter means faster'. The shift reaction stage must take place at a reasonable speed, for in essence this is part of a production process and the methane factory cannot be kept waiting! So a compromise temperature is arrived at, around 400 °C. In this case the reaction is also assisted by the presence of a metal catalyst – iron, copper or cobalt – which brings the reaction to equilibrium at a much faster rate. The catalyst strongly promotes the chemical reaction but at its conclusion itself remains unchanged. The most familiar example of a catalyst in everyday life is probably the catalytic converter fitted to many modern cars to clean up exhaust emissions; the active metal ingredient there is commonly platinum. Experimentation with, and choice of catalyst, and their mode of deployment, is frequently the key to the many chemical reactions encountered in the fuel conversion industries.

The final stage to consider is methanation.

$$CO + 3H_2 \rightarrow CH_4 + H_2O$$

Note how the 'equilibrium sign' is not deployed here as, in the presence of a nickel catalyst, this exothermic reaction proceeds spontaneously. So rapid is the reaction that care has to be taken to avoid heat damage to the surface of the catalyst. The reverse reaction is possible, but it involves the most rigorous conditions to no useful purpose.

To summarize, the three stages involved in the production of methane from coal, taken together, are roughly 'energy neutral'. The first stage, the water gas reaction to produce the CO/H_2 mixture, is highly endothermic, this being balanced in turn by the shift reaction (mildly exothermic) and the methanation stage (more so). The same general argument also holds true when the Lurgi gasifier substitutes for the traditional water gas route at the first stage (see the main text and Figure 7.22 for a description of the Lurgi gasifier).

However, in practice coal-to-methane conversion efficiencies close on 100% will be impossible as not all the exothermic energy can be usefully employed, and maintaining high temperatures (or high pressures) in the first stage involves expenditure of energy that cannot be recovered.

In short, any commercially viable process for producing a secondary fuel has to satisfy the following general conditions. These are in addition to those already laid down in Section 7.8.

1 All chemical reactions must proceed in the desired direction without recourse to extreme operating conditions.

2 These chemical reactions must take place not only in the desired direction, towards the desired end product(s), but also at due speed.

3 Any energy losses involved in processing/conversion must be at least balanced by the advantages gained in producing liquid or gaseous substitutes.

To date, only the first stage, the water gas reaction, has been employed commercially, the shift conversion and methanation steps having been demonstrated in pilot plants only. The water-gas producing reaction was known about long before further practical developments were applied by the Lurgi Chemical Conglomerate in Germany, some 80 years ago. The breakthrough there came in the form of a high-temperature, high-pressure gasifier suitable for varying grades of coal, which was originally developed for the conversion of brown coal, or lignite. In the Lurgi gasifier, coal is fed automatically into the top of the pressure vessel and distributed evenly through it. A mixture of steam and oxygen is introduced through a rotating grate at the bottom of the gasifier, where ash is also removed. The gases pass upwards through the coal bed, volatile constituents of the coal are drawn off, and the resulting carbonaceous char gasified. Reaction temperatures vary with the grade of coal, but are typically between 650 °C and 850 °C. At the high pressures employed some of the coal is converted directly to methane and as this reaction is exothermic (heat-producing) the oxygen demand is partially reduced. The raw gas, synthesis gas (see

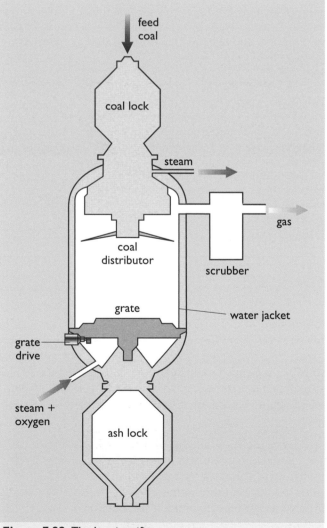

Figure 7.22 The Lurgi gasifier

Section 7.12), can be manipulated and used to produce a variety of liquid or gaseous products, depending on reaction conditions. Here, it exits the gasifier and has all the tar, oil and other by-products of gasification removed before any further upgrading processes are implemented.

With the abundance of primary natural gas available, there has been little incentive to take the process further and develop large-scale commercial 'methane factories'. However, taking into account all the requirements of the three-stage process described above, good thermal efficiencies of up to 70% have been obtained in pilot developments. Assuming similar efficiencies could be achieved on a commercial scale, conditions may, in the future, exist where synthetic natural gas manufacture becomes economically viable: there is no obvious reason why the final two stages in the manufacture of methane should not scale up satisfactorily. But the need for this resource remains some way off and, accordingly, current research and development effort in coal gasification is relatively modest.

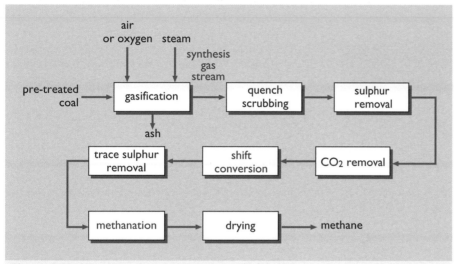

Figure 7.23 Schematic of methane production from coal

7.15 **Oil from coal**

Although conversion of solid fuels to liquid fuels may, at a first glance, seem a novel concept, coal-to-oil technology has a lengthy, if somewhat chequered past. The hydrogen content of coal is typically 5%, whereas that of crude oil is around twice that and for transport fuels it is between 12% and 14%. Hence it is necessary to 'add hydrogen' or to 'remove carbon' from the coal in order to convert it to useful liquid hydrocarbon material.

We came across removal of carbon (or carbonization) in Section 7.14 in the production of town gas by coal distillation at high temperatures. Lower temperature carbonization of bituminous (oil-rich) coals produces greater proportions of liquid products than the miserly 2.5% depicted in Figure 7.19. This was first exploited by James Young in the 1850s and continued to have commercial applications until the 1930s, but was reliant on a high quality feedstock and a ready market for the solid carbonaceous residue left behind after processing. The smokeless fuel still marketed as 'Coalite' was one of the co-products of such a low-temperature carbonization process developed in the 1920s and 1930s. This process also produced good-quality aviation spirit for the RAF as well as a motor spirit (petrol) which was marketed by Carless Capel in south east England as 'Coalene'.

Numerous other small-scale low-temperature processes were in operation at the time, but the basic limitations in terms of low yield of product could not be overcome, and once the Second World War had come and gone this technology was consigned to history. Indeed, had these oil synthetics not been protected by import duties on foreign crude oil they would have been an even less viable proposition.

A more successful method of conversion was based on adding hydrogen, or **hydrogenation**, rather than carbonization, and is generally referred to as a **coal liquefaction** process. The first successful commercial application of this dates back to 1913, in a process developed by the German chemist Bergius, and which carries his name. The **Bergius process** was further developed in Germany in the years between the two World Wars and continued to provide much-needed fuels for the German war effort up until 1945. A variant of this process was piloted by the UK company ICI with government support from 1935 and through the war years. Its production levels were

Figure 7.24 A car being refuelled with Coalene

fairly modest at 3000 barrels of motor fuels per day – compared with German production, which was estimated at some 80 000 barrels per day.

The basis of the Bergius process is the catalytic conversion of coal, slurried with heavy oil using hydrogen at an elevated temperature (450–500 °C) and pressure (200–700 atmospheres). Products are separated out into light oils, middle distillates and residue, and in the second stage of the process the middle distillates are further treated with hydrogen to produce lighter products suitable for use as motor fuels. The thermal efficiency of this coal-to-oil process is 60% at best, but this figure does not account for the energy requirements of hydrogen production, high pressure conditions, or applied heat, which would drive the overall efficiency of conversion down even further.

The Bergius process is an example of **direct liquefaction** in which a solid is converted to a liquid without the intermediate gasification stage. An alternative process which also 'adds hydrogen' is **indirect liquefaction.** The first stage in the indirect process is Lurgi gasification, which we encountered in the previous section. The second stage is the remarkable **Fischer–Tropsch synthesis** route by which a whole panoply of liquid products may be obtained.

Developed by two German chemists in the 1920s, Fischer–Tropsch synthesis involves the use of carbon monoxide/hydrogen mixtures (i.e. synthesis gas) of varying composition to produce hydrocarbons and other organic molecules. The exact nature of the products obtained depends not only on the make-up of the CO/H_2 stream but also on the temperature, pressure, and type of catalyst employed in the reaction chamber. The 'shift conversion' and 'methanation' reactions described in the previous section (to produce methane) may be thought of as the simplest example of a Fischer–Tropsch synthesis; in this section our interest lies more with liquid fuel products.

The principal experiments carried out in Germany during the 1930s focused on the metal catalysts, not so much their exact type or composition (although these factors are important) but their mode of deployment. A stationary or **fixed-bed catalyst** tends to promote the formation of heavier hydrocarbons such as diesel fuel, esters and waxes; whereas dispersing the catalyst within and through the reaction stream (so-called **moving-bed catalysis**) encourages synthesis of lighter hydrocarbons (e.g. naphtha, gasolines, methanol).

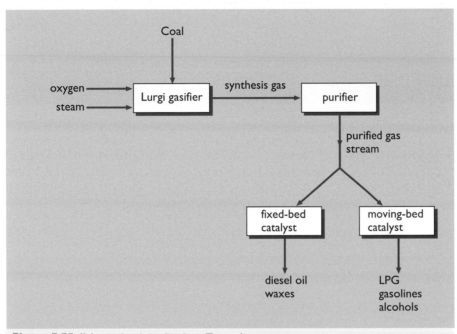

Figure 7.25 Schematic of the Fischer–Tropsch process

Although interest in this technology was maintained in the immediate post-war years, with the advent and availability of cheap oil from the late 1950s onwards the economics of coal liquefaction became disadvantageous and interest in it rather waned. The exception was one country devoid of its own crude oil but where local coal supplies were cheap and plentiful, and whose politics were becoming increasingly isolationist – South Africa. South Africa had ready access to this indirect liquefaction technology and was keen to exploit it for strategic reasons. Consequently, a small Fischer–Tropsch unit, known as Sasol I came into production in the mid-1950s, producing some 10 000 barrels per day of motor fuels, with a panoply of useful by-products. Faced with its increasing isolation in the world, and the alarming prospect of losing access to Iranian imports (a principal source of South African oil) in the immediate aftermath of the 1979 revolution, it was decided to build two further coal liquefaction plants (Sasol II and III) at a capital cost estimated then to be close on $6 billion and with a combined

Figure 7.26 The Sasol plant in South Africa

output of 80 000 barrels per day, or around one-third of South Africa's total oil needs at the time.

It should be remembered that although in a practical and technical sense the new plants were successful, they proved to be no economic bargain, as the following simplified calculation will show.

BOX 7.5 Economics of the Sasol plant in South Africa

Capital and interest

Assumptions: 20 year operating lifetime; 80% plant availability (based on operation 24 hours a day 365 days a year); cost of capital 8% per annum.

Therefore total fixed costs

    = $6 billion capital + $9.6 billion interest = $15.6 billion.

Defraying this cost over 80 000 barrels per day, at 80% plant availability, over 20 years gives fixed costs per barrel of output as

    $15.6 billion $\div$ 80 000 $\times$ 365 $\times$ 20 $\times$ 0.8 = $33.00 per barrel.

Running costs

Assumptions: coal costed at $15 per tonne.

At 30% overall conversion efficiency this translates to $10 per barrel oil output.

Operating costs

Maintenance, labour, ash handling etc. – say $3.00 per barrel.

Total costs

    $33.00 + $10 + $3 = $46 per barrel.

This figure of $46 approximates to the cost estimates produced in South Africa at the time of construction, and when the world oil price briefly touched $40. However, other estimates at that time put the real costs of production of Sasol II and III at virtually double this figure. In any event, it is beyond dispute that the South African apartheid government was operating a siege economy at the time and wished for security of supply at almost any price. In that sense, it proved to be successful, and Sasol continues in operation to this day.

Oil from coal – the future

In the immediate aftermath of the oil price shocks of the 1970s a renewed interest into coal liquefaction technology was born and pilot processes and plants developed around the world, particularly in the United States, with its huge oil import requirement and large domestic reserves of coal. Both the direct and indirect routes ('adding hydrogen') have received attention, rather more than 'carbon removal' via the coal carbonization process. Research and development has focused upon optimizing operating process conditions and reducing the hydrogen requirement in direct liquefaction, and employing more efficient gasifiers and developing superior catalysts for indirect liquefaction processes. Numerous pilot plants have been constructed and tried out with different grades and types of coal – it is worth remembering that 'coal' is not a well-defined, homogenous entity but a mineral with varying physical and chemical composition. For example, a coal containing significant amounts of volatile organic material may be best suited to direct liquefaction, whereas for one with a high ash content (such as in South Africa) indirect liquefaction is more likely to be favoured.

Paradoxically, perhaps, while it is the indirect route which has aroused the lesser research interest, it is the one technology which is operating on an industrial scale, at the Sasol plants in South Africa. The operating history of these three plants, featuring the well-tried processes of Lurgi gasification and Fischer–Tropsch synthesis, suggest that a synthesis gas–methanol–light distillate route could be a viable future option. However, most future hopes seem to be pinned on direct liquefaction or a variation on that theme. The most promising variant appears to be a two-stage catalytic route in which the principal role of the first stage is to dissolve the coal in a suitable solvent and in the second stage treat the coal with hydrogen to yield liquid products. The necessities for high temperatures and high pressures of earlier direct liquefaction plants (the Bergius process and variants) are largely overcome and the hydrogen consumption kept to a minimum. Thus the overall energy efficiency of the coal-to-oil conversion is maximized, with figures of 60–70% being quoted. However, developments employing this technology remain ongoing and a commercial scale plant is yet to be constructed.

To conclude this section we should perhaps return to the Sasol example once more, not to examine the technology employed there, or how it might be improved, but as an illustration of the possible costs associated with large-scale coal liquefaction, and in particular the capital costs. Even allowing for the unique political environment of the time, employment of a well-established, even conservative technology, and the likelihood that the costs incurred were not subject to the full rigours of the marketplace, an estimated fixed cost at 1982 prices of $33 per barrel of oil output over a full operating lifetime is quite striking. Making all the allowances one can, it is difficult to see how the current equivalent capital cost of any coal-to-oil plant could be lower some twenty years later – and indeed, it would probably be considerably higher, as would the costs of the coal feedstock. Indeed, total cost estimates of $80–100 per barrel were calculated some time ago (IEA, 1984). So for the economics of coal liquefaction to look attractive not only will the conversion efficiency need to be maximized, the technology itself will need to be simplified in order to keep the fixed costs at a minimum. That combination seems unlikely, and once primary sources of petroleum become seriously depleted, demands which are currently considered 'oil-specific', such as transport, could well be satisfied by alternatives such as liquefied natural gas, hydrogen or electric power from renewable sources. These remain a subject for another chapter, however, and for the companion volume, *Renewable Energy*.

7.16 **Non-conventional sources of petroleum**

Introduction

Perhaps the most straightforward way of defining 'non-conventional' petroleum is to turn it on its head and define 'conventional' sources instead. These are generally taken to mean crude oils which are produced by natural pressure and contained within the subterranean reservoir, as described at the start of this chapter. The recovery of such products is termed **primary recovery**. Primary recovery is frequently augmented by the injection of natural gas or water into the reservoir; this has the effect of raising reservoir pressure and is referred to as **secondary recovery**. Crude oil produced in this manner is also termed 'conventional'. Primary and secondary recovery typically succeeds in extracting up to 50% of the oil in place within the reservoir.

Crude oils recovered by all other methods and from all other sources are referred to as 'non-conventional' and three such sources will be considered in this section – **shale oil**, **tar sands** and **heavy oil**. There is also the process of **tertiary recovery**, in which high pressure steam or carbon dioxide is used to extract some of the 50% of the resource left behind by primary and secondary recovery of conventional oil. Tertiary recovery has much in common with the production of non-conventional oil sources in that extraction is typically slow and expensive compared with conventional oil recovery, where the actual extraction process is quick and cheap with oil flowing freely and in high volume. Tertiary recovery is therefore frequently classified as 'non-conventional' too.

Oil shale

In the same way that conventional petroleum occurs in sedimentary rocks, oil shale too is sedimentary in origin and contains hydrocarbons, not in gaseous or liquid form, but as a waxy solid known as **kerogen**. If oil shale

BOX 7.6 **Scotland's story**

Should you find yourself a visitor to Scotland's capital city, Edinburgh, arriving by plane on the easterly runway of the city's Turnhouse Airport, take a look out of the port-side window as the aircraft makes its final approach. At first the towns and fields below seem quite ordinary, but with decreasing altitude you will soon pick out several huge iron-red spoil heaps, sparsely covered with wild vegetation, which are the defining features of this part of the West Lothian landscape. Known locally as 'bings' these artificial mountains contain the leftovers of an industry which, during the period from 1862 to 1962, extracted an estimated 180 million tonnes of an oil-bearing rock which some have claimed could yet rescue the world from petroleum-deficient penury – **oil shale**.

Figure 7.27 Shale bings between Broxburn and Winchburgh, West Lothian

is 'retorted', i.e. heated in a closed vessel to drive off volatile material, an oil, analogous to conventional crude oil, may be collected and further refined to yield the full range of petroleum products. Yields of oil per tonne of shale vary, but those worked in West Lothian in Scotland during the industry's heyday averaged between 20 and 40 gallons per tonne of shale worked. Yet even in the boom years of the early part of the twentieth century the total amount of oil produced during any one year was only the equivalent of 5000 barrels per day from 3 million tonnes of shale processed. Today, that would amount to a fraction of a per cent of the UK's annual oil demand, and it would take the mind-boggling figure of 1000 million tonnes of shale to provide the current needs of the UK alone, or five times the quantity of shale processed over an entire century of operations.

One of the many problems associated with shale mining is the volume of spoil left after processing, which is typically double that of the original ore. Hence 'putting it back in the hole' is not a feasible option for disposal and any large-scale future exploitation of this resource has huge environmental implications. For this reason, and others, including depletion of resources, shale mining in the central belt of Scotland is now only an interesting historical diversion. The huge refining and petrochemical complex run by BP at Grangemouth is its legacy.

World distribution

Oil shales are deposited in various parts of the world, but the largest accumulation by far is the Green River formation in the US, where the borders of the states of Utah, Wyoming and Colorado come together. It has been estimated that the amount of oil 'in place' in this one location exceeds by far all the world's conventional crude oil resources. It may be that one day the economics of the oil industry will allow for exploitation of this potentially rich resource, but the capital investment required, environmental effects, and the need for large quantities of water during processing, all suggest that large-scale exploitation of Green River shale is unlikely. Various schemes were commenced in the immediate aftermath of the 1970s oil crises, but have since fallen into decline.

There is a small-scale scheme, the Stuart Project, operating in the state of Queensland in Australia, but even if this proves successful over time, scaling up to significant levels of production is likely to be a lengthy and costly business. Such problems of scaling-up production are typical of non-conventional sources of oil.

Figure 7.28 The Stuart Project in Queensland, Australia

Tar sands

In contrast to oil shale, where the hydrocarbon source kerogen is not, strictly speaking, a fully-formed petroleum, deposits of so-called tar sands consist of loose-grained rock bonded together by heavy bituminous material. Crude oil is obtained either by extraction using steam, or by direct heating to around 80 °C. Thus, compared with shale, oil from tar sands is less difficult to extract from its source rock, and it has therefore received a greater amount of commercial interest over the years.

The world's tar sand deposits are dominated by the Athabasca area of northern Alberta in Canada, where, as in the Green River shale formation, the quantity of oil in place exceeds the world's entire endowment of conventional crude oil. Exploitation of the Athabasca tar sands began as far back as 1967 and continues today despite various setbacks along the way. The environment is an especially harsh one for both workers and machinery, ranging as it does from extreme cold in winter to the sweltering humidity of summer, but these climate-related difficulties have been largely overcome. Mammoth-scale operations involving the processing of over half a million tonnes of material per day, yielding 400 000 barrels per day of oil, have been achieved at a cost claimed to be around $10 per barrel, a figure which suggests economic viability. On this basis, the prospects for the future look good, but the only other location where significant deposits occur is in Russia, and Russian efforts in forthcoming years are likely to concentrate on exploiting their huge untapped reserves of Siberian natural gas.

Figure 7.29 Massive digger, Athabasca, Canada

Heavy oil deposits

There is no strict definition of 'heavy oil', but for the purposes of this discussion it is considered to be a buried petroleum deposit which will not flow to the earth's surface under natural reservoir pressure. Instead, as in tertiary recovery of lighter crude oil, it is necessary to inject steam into the production wells to force the heavy crude oil to the surface, where it can be further processed and refined in the normal way. The world's principal source of this heavy petroleum is the Orinoco basin of Venezuela, where huge deposits exist – not in such massive quantities as the reserves of Green River or Athabasca, but nonetheless potentially significant in world terms and also less difficult to exploit. Current production of around 400 000 barrels per day is set to rise, but the oil is highly sulphurous and attempts in the 1980s and 1990s to market an emulsified secondary product called 'Orimulsion' for the European power station market proved largely unsuccessful, as the advantage of cheapness was more than outweighed by the disadvantage of noxious emissions. Indeed, opponents of Orimulsion successfully labelled it as 'the world's filthiest fuel'.

Leaving that controversy aside, Venezuelan heavy oil has the potential to produce significant amounts of crude oil for many decades to come, but extracting it, cleaning it up and rendering it suitable for premium markets such as transport will be an expensive business – in contrast to the light, sweet, fast-flowing crude oil from the North Sea.

Summary

What role will non-conventional petroleum sources play in the world's future oil supply? We have seen that there are some spectacularly rich deposits of kerogen, bitumen and heavy oil, and that efforts have been and are being made to exploit these sources. Together with tertiary recovery, these sources currently account for around 4 million barrels per day of production out of a global consumption of about 75 million barrels per day. There are problems relating to the effects on the environment, recovery, materials handling, waste and processing associated with each one. Their exploitation has been slow and patchy and past experience suggests that future levels of production may grow, but only relatively slowly, given the fact that these are not clean and easy-to-win energy resources. Current speculation by oil analysts suggests a possible contribution of 10 million barrels per day fifty years hence, by which time world production of conventional crude oil could be only one third of today's level (see Section 7.17 below). The conclusion, therefore, is that if hydrocarbon fuels remain as dominant in the world economy of 2050 as they are in 2000, then heavy oil, tar sands and possibly even oil shale will all have to play a significant role.

7.17 The wider future

Introduction

No modern treatise on oil and gas could be complete without some serious consideration of the oft-asked question: 'how much more is there'? The questioner usually has oil, rather than gas, in mind, but the question is increasingly relevant for gas too. This section will focus upon oil reserves, but we should not forget that gas is increasingly crucial to the UK and world economy, and although the global reserves are quite large, some local reserves are more limited. In time, the issues discussed below for oil are likely to apply also to gas.

Either way, the natural question to ask is exactly how much of the resource is left – as in the oft-heard question 'when will the oil run out?', as though we might wake up with a start one Monday morning and find that the world has, all of a sudden, used the last drop of oil over the weekend. Instead, price signals, brought about by declining production, *should* have long forewarned us that the era of readily available oil was coming to an end. The crucial question, then, is not 'when is the end nigh' (the answer to this is that the 'end' is arguably a century away), but when might we begin to notice that 'business-as-usual' is no longer a viable option? And are there clues to be found in the recent past that can help us in such predictions?

The memory of the oil crises in the 1970s is still with us. First, in 1973 the disruptions to supply brought about by the political situation surrounding the brief Arab–Israeli war gave the OPEC cartel new fame and power, and saw the price of oil increase from $2 to $10 per barrel (see Figure 2.6 in Chapter 2). The queues at petrol stations around the globe were a short-lived phenomenon, though the power of the OPEC cartel was such that oil prices were sustained at this higher level for the following five or six years. In this case, the problem for oil consumers turned out not to be the availability of oil, but where it was located and the ability of OPEC producers to manipulate prices.

Figure 7.30 Petrol stations queues, London, 1979

A second major disruption occurred as a result of the 1979 revolution in Iran, which saw the ousting of the Shah and installation of an Islamic regime largely hostile to the West. Four million barrels per day of production quickly disappeared from the market, and world prices soared briefly to $40 per barrel, before stabilizing at around $30 per barrel – and the world's developed countries were plunged into recession. What saved the day on this occasion was the burgeoning production from new areas such as Mexico, Alaska and the North Sea, so much so that by the mid-1980s the oil industry found itself in a slump where prices dropped to around $10 per barrel for a while.

The only occasions since then when there has been a significant upward pressure on prices were during the Gulf War of 1990, and ten years later in 2000, when world demand had soared to such an extent that the 'swing' OPEC producers (i.e. countries such as Saudi Arabia with spare production capacity) effectively controlled the amount of oil on world markets. It is interesting to note that although, at the time of writing, in the first half of 2002, one reads about 'weak oil markets' the price of crude oil remains in the region of $25–$30 per barrel, and more importantly perhaps, outside the Middle East, world production is running more or less at its full capacity. This suggests, therefore, that maybe the next disruption will be more uncomfortable and long-lasting. Although local politics will be the trigger, supply difficulties brought about by constraints on physical availability could be the legacy.

Lesson from America

To find out what the future may have in store it is often useful to look at the lessons we can learn from the past, from historical data on oil production. If we can then compare these data with rates of oil discovery and estimates of remaining reserves, we can, at least in theory, draw up production profiles for future years based on this information. Such analyses are highly complex and involve the skills of mathematicians, statisticians, petroleum geologists and economists. What is presented here is necessarily a simplified version, but it nevertheless can give us some idea of what the future may hold for oil production.

In 1956, a much-respected geophysicist with strong roots in the oil industry, M. King Hubbert, presented a paper to the American Petroleum Institute in San Antonio, Texas. In this paper, Hubbert predicted that US oil production would peak in the early 1970s. This seemed an extraordinary, even perverse, prediction. US oil production was steadily rising, with prices stable and plenty of spare capacity – the future seemed assured. Apparently, though, according to Hubbert, no one had noticed that the rate of consumption was greater than the rate at which new reserves were being discovered.

In the event, Hubbert was proved correct. US oil production from all sources peaked at 11.3 million barrels per day in 1970, and even allowing for new supplies from Alaska and offshore Gulf of Mexico, has been in irreversible decline since.

BOX 7.7 **US oil production and consumption**

Those with an interest in statistics may wish to examine the figure for US oil production in 1970 a little more closely: it includes production of 'natural gas liquids', which are light hydrocarbons that condense out of producing gas wells. They are valuable blending agents in the production of light petroleum products and are also used in the petrochemical industry. They are not always considered as 'conventional oil' but the figure of 11.3 million barrels per day includes these. Currently, such natural gas liquids account for around one quarter of the US production level. Hubbert's analyses seem to have ignored them, but they do not affect his famous '1970 prediction'.

The industry statistics quoted in this chapter include natural gas liquids and also crude oil obtained by tertiary recovery, for some time now around 10% of the US total. By the year 2000, US production had fallen to 7.7 million barrels per day, but consumption had soared to almost 20 million barrels per day. The corresponding import requirements had increased from about 3 million to over 10 million barrels per day, and are still rising. (See Figure 2.15 in Section 2.7.)

How did Hubbert arrive at his remarkable prediction? Put succinctly, he examined in the minutest detail cumulative production figures from the very beginnings of the oil industry in 1859, right up to 1955, and detailed all the discoveries made over the same period. His view – and the historical record bears this out – was that whatever new reserves remained to be discovered would have little impact, especially in the short term.

Hubbert went further, however, as can be seen from Figure 7.31 above. Historical data on the size and rates of discovery and performance of

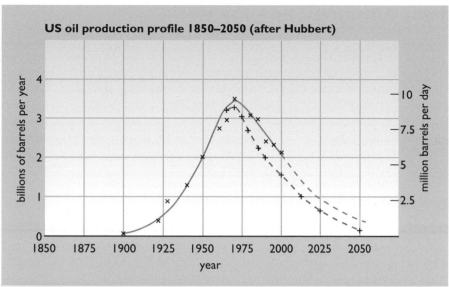

Figure 7.31 Hubbert plot. The dotted red curve represents Hubbert's original projection of 1956, and the area under the curve corresponds to 200 billion barrels of production. The solid green curve represents actual production from 1900 to 2000, with the dotted green section being the projection thereafter. The area under this curve represents 220 billion barrels of production

production wells led him to speculate that after US production peaked, its rate of decline over the ensuing century would roughly mirror the rate of growth over the previous century. In time, the argument went, US oil production would follow a bell-shaped curve, with the year of peak production corresponding to the point where one half of the total endowment of US conventional crude oil would have been used up. On that basis, with cumulative production up to 1970 close to 100 million barrels, the projection is that the US will ultimately produce double that figure, or 200 billion barrels. Calculations by those who have more recently reworked Hubbert's original data, and who had access not just to forty extra years worth of data, but also a far more accurate prognosis of yet-to-be-discovered reserves, have arrived at estimates of 210–220 billion barrels of ultimate recovery – remarkably close to Hubbert's predictions of yesteryear (Campbell, 1997; Deffeyes, 2001). A summary is given in the table below:

Table 7.9 US conventional crude oil production, in billions of barrels[1]

Cumulative production, 1859–2000	180
Reserves identified	20
Yet-to-find reserves	10–20
Total	**210–220**

1 These figures are for conventional crude oil production only and do not include natural gas liquids, nor crude oil obtained by tertiary recovery.

First America... now the world

If we are prepared to accept the foregoing analysis for the United States, it *should* be feasible to carry out similar calculations for other oil-producing regions of the world and thus attempt to calculate when world production of conventional crude oil will peak and subsequently decline. The word 'should' appears in italics for a good reason. The United States is an example of what is referred to as a 'mature' area for oil exploration, having reached the bell-shaped peak – the so-called **Hubbert's peak** – as early as 1970. Few other producers are in this position, though Colombia, Indonesia and the United Kingdom have probably peaked already, with Canada and Norway just behind. Others have experienced irregular patterns of production, largely for political reasons; for example Libya and the former Soviet Union, where, despite current sentiment (many consider that exploration around the Caspian Sea will yield a huge resource in coming years), production may well have reached its maximum in the 1980s. And finally there is the most politically charged area of the oil world, the Middle East, a maturing oil production province due to reach maximum production during the second decade of this century. Here five countries, the so-called 'swing states' – Iran, Iraq, Kuwait, Saudi Arabia, and the United Arab Emirates – possess well over half of the world's remaining endowment of conventional crude oil, and the host governments have the power to exercise significant discretion in the amounts they produce in response to market (and other) conditions.

So, 'modelling the world' is not an easy task, but despite the difficulties involved and the varying assumptions that underlie the different projections, various attempts have been made over the years to assess what the future may hold. At least there seems to be general agreement on one point – that world production of conventional crude oil will more or less follow the United States' experience of bell-shaped depletion. It is the magnitude and timing of the peak, and the rate of depletion which are the source of much discussion and debate.

One factor which suggests that this peak may not be too far off is that, in the same way that the rate of United States discoveries peaked in the 1930s, only for production to reach its maximum thirty-five or forty years afterwards, so it is that *world* oil discoveries peaked in the mid-1960s, suggesting a global peak in production at some time during the period 2000 to 2005. But a stronger argument can only come from a properly detailed look at both production and reserves: production statistics to date, estimates of identified reserves, and estimates of 'yet-to-find' conventional crude oil. Assuming the Hubbert methodology can be applied, then some prediction of the peak production date – i.e. the top of the bell-shaped curve – can be arrived at, giving the point at which approximately half of the world's endowment of conventional crude oil will have been exhausted.

Each of the numbers we need to consider is subject to many uncertainties, including the issue, discussed in Box 7.7, of the types of 'oil' that are counted in the totals. There does however seem to be reasonable agreement on the cumulative production figure for the period from 1859 to 2000, namely 850 billion barrels. The next number we need is an estimate of known reserves. BP's *Statistical Review of World Energy*, generally taken as a reliable source, reproduces summaries of reserve estimates taken from the authoritative *Oil and Gas Journal*. These currently stand at 1050 billion barrels, though some observers (notably Campbell, 1997) have commented both on the arbitrary nature of sudden sharp increases in Middle East reserves reported in the 1980s, and also the apparent year-on-year (gradual) increase in reserves at a time when new discoveries appear to be lagging well behind levels of production. Campbell argues that global reserve estimates are too high by some 140 billion barrels – on this basis there would be 910 billion barrels remaining of known reserves. As an aside, it has been said that reporting on levels of reserves is a political act; we like to appear poor in front of the tax inspector and rich when we meet the banker! So we should not place too much faith in the preciseness of these numbers, and it is instructive to use *both* the above reserve estimates when making projections as to the timing of Hubbert's world production peak.

But there is still the third number to be established, that relating to 'yet-to-find' oil. As one might expect, this is by far the most difficult figure to establish. Campbell's approach is interesting, as rather than establishing the three numbers and adding them together, he establishes first the total, and then subtracts the cumulative production figure to date and the amount of known reserves to give the 'yet-to-find' figure. The 'total' referred to is what Campbell calls 'ultimate production' and he arrives at this figure by bringing together detailed and disaggregated data from all the world's oil provinces, not only in terms of past, present and future production, but also discoveries by size, number and rate of discovery. As an experienced petroleum geologist, with apparently unparalleled access to industry-wide data through his involvement with Petroconsultants SA, Campbell's view regarding future discoveries carries a certain weight, though his methods may necessarily be partially subjective. His calculations suggest that there are 180 billion barrels of conventional crude oil awaiting discovery, and if so, it is likely that within just twenty or thirty more years virtually all of the world's conventional crude oil will have been discovered. The fact is that there are very few prospective oil-bearing areas of the world which have not been appraised. And although Campbell cites Siberia and Central Asia as potential targets for exploration, to date it has been natural gas which has dominated new discoveries in these areas. The South China Sea, offshore West Africa, and even the Falkland Islands all remain possibilities, but outside Asia another new 'North Sea' is hard to envisage realistically. Almost literally, the ends of the earth have been explored for crude oil and most of it has now been found. However, for comparison purposes we shall also examine the situation should Campbell's estimate prove to be 50% too pessimistic, i.e. taking a figure of 270 billion barrels of crude oil awaiting discovery and exploitation, and thus arrive at two different estimates of ultimate production of conventional crude oil.

The two sets of comparison figures are summarized in Table 7.10 overleaf. Note the estimates for the year of peak production. As with Hubbert's analysis of the United States data, these are arrived at by taking the half-way point of ultimate production (970 billion barrels for Estimate 1 in Table 7.10), subtracting the 850 billion barrels consumed up to the year 2000, and proceeding from the year 2000 at the rate of 25 billion barrels per year until the difference is made up. Two things are immediately striking about these two estimates of the peak production year; one is their immediacy and the other is their close proximity to each other despite the

Table 7.10 World production of conventional crude oil, in billions of barrels[1]

	Estimate 1	Estimate 2
Cumulative, 1859–2000	850	850
Known reserves	910	1050
Yet-to-find oil	180	270
Total ('ultimate production')	1940	2170
Peak production year	**2005**	**2009**

1 Estimate 1 is adapted from Campbell (1997). Estimate 2 is based on data from BP (2001) for the estimate of known reserves, and Campbell's figure plus fifty per cent for 'yet-to-find' oil.

apparently substantial difference between the numbers on which they are based. Neither factor should be a surprise; as Table 7.11 below demonstrates, the first twenty years of this new century have for some time been judged as the era of peak production and subsequent decline. The various forecasts for the date of the peak vary by only a few years. And at a rate of approximately 25 billion barrels a year even an extra 100 billion barrels of reserves can, according to the 'half-ultimate' model, only delay peak production by a relatively insignificant twenty-four months. The world is using up its oil endowment *that* quickly.

Table 7.11 Some forecasts of world oil supply

Date of forecast	Source	Forecast date of peak	Ultimate assumed (billion barrels)
1977	Hubbert	1996	2000
1981	World Bank	'Plateau around the turn of the century'	1900
1995	Campbell and Laherrère	2005	1800
1997	Ivanhoe	2010	2000
1998	IEA	2014	2300
2000	Laherrère	2005	2000
2001	Deffeyes	2003–2008	2000
2002	Smith	2011–2016	2200

Source: Bentley, 2002. See also reference list at end of the chapter.

Hubbert's 1977 world projection is interesting; it assumed unrestricted production, reaching 100 million barrels a day (36.5 billion barrels a year) at its peak in 1996, but as we have seen, the price shocks of the 1970s and other political and economic factors have, at times, put the brake on rising levels of demand. They may do the same in the future, too, but the fact remains that crude oil is a finite resource and that whatever the shape of the depletion curve, and timing of the associated peak, the area under the curve (representing the amount of oil produced) is unlikely to change significantly.

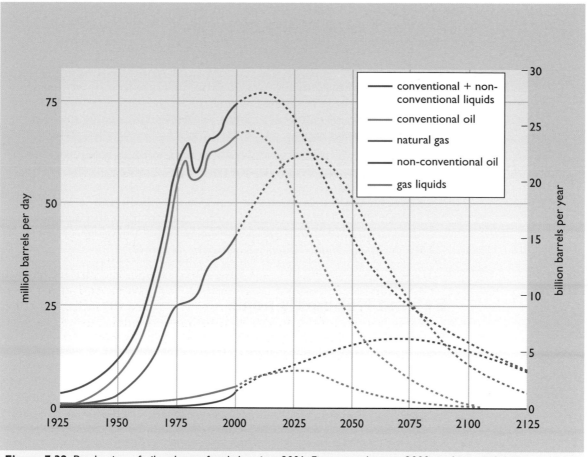

Figure 7.32 Production of oil and gas, after Laherrère, 2001. Figures to the year 2000 are historical data; thereafter the dotted curves represent projections of future supply. These projections are based on ultimate recovery of 2000 billion barrels of conventional oil (and gas liquids), 750 billion barrels of non-conventional oil, and 2000 billion barrels oil equivalent of natural gas.

Figure 7.32 above is due to the geologist and geophysicist Jean Laherrère and depicts his projections of conventional crude oil production along with natural gas liquids and non-conventional oil production. There is also shown the corresponding projection of natural gas production, estimated to peak in 2030. However, future levels of natural gas production are notoriously difficult to estimate, and in any case they rather fall outside the scope of this chapter.

How much reliance are we to put on all this number-crunching? As quoted above, an ultimate production figure of around 2000 billion barrels is not strongly disputed by those who have made a close study of the subject, nor by many oil industry insiders, although there is a recently published estimate (2000) by the United States Geological Survey of 3000 billion barrels, for example. That would amount to the discovery of another Middle East, so unsurprisingly, perhaps, high estimates such as these are very much in the minority, as are views espoused by oil economists such as Peter Odell who famously concluded that 'the world is running into oil, not out of it'. And even if there are positive surprises yet to come in the shape and substance of appreciable accumulations of oil in unlikely places, they will now be simply too late to have any substantial effect on the peak, though they may be able to spin out the tail that little bit longer than would otherwise be the case.

As Deffeyes puts it:

> The finite supply of the world oil is, in my opinion, written in stone. It's written in the reservoir rocks, in the source rocks, and in the cap rocks. No amount of fancy fishing tackle is going to satisfy our appetite for oil...
>
> ...What should we do? Doing nothing is essentially betting against Hubbert. Ignoring the problem is equivalent to wagering that world oil production will continue to increase for ever. My recommendation is for us to bet that the prediction is roughly correct. Planning for increased energy conservation and designing alternative energy sources should begin now to make good use of the few years before the crisis actually happens.
>
> Deffeyes, 2001

Beyond Hubbert's peak

Whether it is 2005, 2008 or even 2012, once the world has climbed Hubbert's Peak, what is it going to do for oil? The first we are likely to know about it is a consumption crisis brought about by a break in supply, unmet growth in demand, or a combination of the two. But this time, the argument goes, unlike the crises of 1973 and 1979, it will *not* go away because it *cannot* go away. Moreover, the five 'swing producers', all situated in a politically fraught area of the globe, will have the power to dictate prices at their will. Theirs will not just be the price-dictating marginal barrel, but one-third of the rest too. The result could be an oil price explosion (perhaps to double or treble the $25 which prevails at the time of writing) and a worldwide recession before the oil supply–demand balance reaches some equilibrium once more at a lower level; with much disruption to society.

Non-conventional oil sources, conversion of natural gas to gasoline and coal to oil synthetics, hydrogen, and fuel cells all begin to look more attractive, and as time goes on and conventional crude oil becomes even scarcer and more expensive, these alternatives become increasingly plausible and economically feasible. In a sensible world we would plan for such a future and be making substantial lead investments in these various technologies, as well as expending more money and effort into energy conservation and higher end-use efficiency.

Perhaps that is why one of the world's largest oil companies, BP (formerly British Petroleum–Amoco), recently ran an advertising campaign branding itself as 'BP – Beyond Petroleum'. Maybe they, too, know that betting against Hubbert could turn out to be a fool's wager.

References

Bentley, R. W. (2002) 'Oil forecasts, past and present', *International Workshop on Oil Depletion*, Uppsala, Sweden.

BP (2002; and various years) *Statistical Review of World Energy for 2001*, BP plc.

Campbell, C. J. (1997) *The Coming Oil Crisis*, Multiscience Publishing and Petroconsultants S.A.

Campbell, C. J. and Laherrère, J. H. (1995) *The World's Supply of Oil, 1930–2050*, Report from Petroconsultants S.A., Geneva, 1995.

Campbell, C. J. and Laherrère, J. H. (1998) 'The end of cheap oil', *Scientific American*, pp. 59–65.

Deffeyes, K. J. (2001) *Hubbert's Peak – The Impending World Oil Shortage*, Princeton University Press.

DTI (1999) *Technology Status Report 010 – Coal Liquefaction*, Department of Trade and Industry, London.

DTI (2002, 2001; and various years) *Digest of UK Energy Statistics (DUKES)*, Department of Trade and Industry.

DTI (2002) *Quarterly Energy Prices*, Department of Trade and Industry, June 2002.

Hubbert, M. K. (1956) 'Nuclear energy and fossil fuels', *Proceedings of the Spring Meeting of the American Petroleum Institute, 1956*, San Antonio, Texas, pp. 7–25.

Hubbert, M. K. (1982) *The Global 2000 Report to the President*, Penguin Books, p.353.

IEA (1984) *The Costs of Liquid Fuels From Coal*, International Energy Agency.

IEA (1998) *World Energy Outlook*, International Energy Agency.

Ivanhoe, L. F. (1996) 'Updated Hubbert curves analyse world oil supply', *World Oil* vol. 217, issue 11, pp. 91–94.

Laherrère, J. H. (2001) 'Forecasting future production for past discovery', *OPEC seminar*, 28 September 2001.

Odell, P. R. (1994) 'World oil resources, reserves and production', *The Energy Journal*, IAEE , vol. 15, pp. 89–113.

Shell (1966) *Shell Petroleum Handbook*, Shell International Petroleum Co. Ltd.

Smith, M. R. (2002) 'Analysis of global oil supply to 2050', *The Energy Network*.

US Geological Survey (2000) *World Petroleum Assessment 2000*, USGS Digital Data Series DDS-60, http://geology.cr.usgs.gov/energy/WorldEnergy/DDS-60 [accessed 15 January 2003].

World Bank (1981) 'Global energy prospects', Working Paper No. 489.

Chapter 8

Oil and Gas Engines

by Bob Everett

8.1 Introduction

In chapter 6 we saw how the reciprocating steam engine developed from its rudimentary beginnings in the late 17th century to a reliable machine capable of powering factories, trains, ships and electric power stations by the end of the 19th century. This required continued effort in improvements in design and the deployment of new materials and progressively higher quality and precision in the manufacturing of components. This brought rewards in progressively higher operating efficiencies. During the 1880s the steam turbine was invented, giving a much improved performance compared with the steam reciprocating engine. The steam turbine remains to this day a key component of practically all fossil-fuelled electric power stations.

But steam technology is that of the *external combustion engine*. The fuel is burnt in a furnace to heat water in a boiler producing steam which drives the engine. Today, about a quarter of the primary energy of the UK is consumed in *internal combustion engines*, where the fuel is actually burnt inside the engine. Development of these had to wait for the widespread availability of suitable fuels, first coal gas for stationary engines and then petrol, kerosene and diesel fuel for transport applications.

This chapter looks at three internal combustion engine designs which have been extensively developed over the twentieth century, the petrol or spark ignition engine, the diesel or compression ignition engine and the gas turbine or turbojet engine. Finally we look at the Stirling engine, an external combustion alternative to the steam engine. Each of these engines is a product of its particular time. Although the petrol engine was invented in the 1880s it did not really start making an impact on transportation until the first years of the 20th century. The diesel engine was invented in the 1890s, yet was not seriously used for road vehicles until the 1930s. The gas turbine in its jet engine form was a product of the late 1930s and World War 2. Its development since then has depended on the availability of new materials capable of withstanding high temperatures. The Stirling engine was used as an alternative to the steam engine right through the nineteenth century. Although it is largely ignored, it could still have an interesting future using high temperature steels and innovative modern designs.

8.2 The petrol or spark ignition engine

Although it was obvious by the early nineteenth century that the steam engine did indeed work, it wasn't clear whether or not the need to use fire to heat water wasn't just an unnecessary complication. Why couldn't fuel be used directly inside a cylinder to drive an 'internal combustion engine'? The newly available fuel town gas was ideal for such a device. The problem was to create a viable engine design. Various inventors produced prototypes, but the first serious working gas engine was a 4 hp unit produced by Etienne Lenoir, a Frenchman born in Luxembourg. He patented his design in 1860 and it went into production in both France and England. A mixture of gas and air was drawn into a cylinder and ignited by an electric spark. The thermal efficiency of Lenoir's engine was low, but it was competitive with the small reciprocating steam engines of the time.

Remember 1 horsepower (hp) = 746 watts

Figure 8.1 Nikolaus August Otto (1832–1891) the German engineer who effectively devised the four-stroke internal combustion engine

Its design interested the German engineer Nikolaus August Otto, who had a couple of gas engines made under licence in Cologne. Otto wasn't happy with their performance, so he sat down and designed his own version, teaming up with an inventive engineer, Eugen Langen. Their first engine was most unusual. Gas and air were compressed in a cylinder and ignited, firing a heavy piston vertically upwards as far as it would go. On the downwards stroke the piston engaged a rack and pinion, turning a flywheel as it descended. A contemporary description of its performance by Franz M. Feldhaus is not encouraging:

> The engine was placed in a dark corner, and the piston rose up with a terrible crash; with a rattling sound it engaged with the balance wheel, and then descended with an anxious whine... The explosions seemed quite arbitrary, and in between there was only sinister silence. The convulsions of this engine inspired real fear.
>
> Feldhaus, 1954

Tests showed that it was much more efficient than competing Lenoir engines and it was awarded a gold medal at the 1867 World Exhibition in Paris. Despite its noise, some 10 000 were produced in the first five years.

Otto and Langen were aware of the problems of their engine design and in 1876 they produced a revolutionary new design, the first commercial four-stroke engine. This was immediately nicknamed the 'Silent Otto' in comparison with the previous model. Although this engine ran on town gas as a fuel, and had a curious ignition device involving a small pilot flame on the outside of the engine, the principles are the same as in a modern car engine (see Box 8.1).

BOX 8.1 Four-stroke and two-stroke engines

Modern petrol engines take two main forms, the four-stroke using the Otto cycle and the two-stroke cycle. Both of these use a piston which is driven up and down inside a cylinder and connected to the drive section by a rotating crankshaft.

At the top of a four-stroke engine there is a cylinder head containing a number of valves controlling the flow of gas in and out. The four 'strokes' are: induction, compression, power and exhaust, illustrated in Figure 8.2 below.

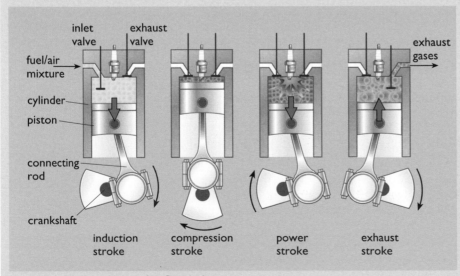

Figure 8.2 The four strokes of an Otto cycle engine

On the induction stroke a small amount of fuel and air is drawn into a cylinder through the open inlet valve, which then closes. On the next stroke this mixture is then compressed into a smaller volume. This reduction in volume is a rather critical factor called the *compression ratio*. In a modern car this is about 9:1, i.e. the fuel/air mixture is squeezed into one ninth of its original volume, creating a highly inflammable mixture. This is then ignited using an electric spark on a sparking plug.

The gases then burn very rapidly reaching a high temperature (750 °C or more) and expand, pushing down the piston on the power stroke. Finally on the exhaust stroke, the burnt gases are pushed out into the exhaust system through the open exhaust valve. The whole cycle then repeats.

In a two-stroke engine, the engine has a sealed crankshaft casing (or crankcase) which allows the bottom side of the piston to function as a pump. The fuel air-mixture travels through the crankcase and up into the cylinder where it is burned. The cycle proceeds as follows as shown in Figure 8.3: in diagram (a) the burning fuel air mixture, ignited by a spark, pushes the piston down, compressing a new batch of fuel and air in the crankcase. When the piston has almost reached the bottom of its travel (diagram (b)), it uncovers an exhaust port in the side of the cylinder. This allows the burned exhaust gases to escape into the exhaust system. As the piston descends right to the bottom of the stroke (diagram (c)), it opens a transfer port. This allows the compressed fuel/air mixture to travel from the crankcase into the working cylinder above. Then the rising piston closes off the ports and compresses the mixture. As it rises more fuel and air is drawn into the crankcase through the non-return valve. When the piston reaches the top of its stroke the mixture is ignited and the cycle repeats.

Most designs of small two-stroke petrol engines need to have oil mixed with the petrol in order to lubricate the bearings of the crankshaft. A small amount does indeed lubricate the bearings, but the rest is burned, leaving an unpleasant trail of white smoke from the exhaust pipe. Tightening emission regulations have meant that this style of engine has largely been abandoned in favour of four-strokes.

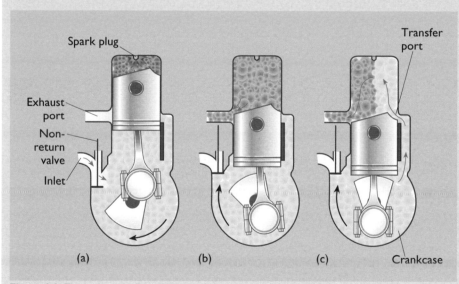

Figure 8.3 The two-stroke Engine (source: Rogers and Mayhew, 1980)

The four-stroke 'silent' engines were about three times as efficient as their predecessors. A Frenchman named Alphonse Beau de Rochas had suggested the four-stroke cycle back in 1862. For a successful design he recommended:

(a) that the mixture should be compressed as much as possible before ignition,

(b) that the maximum expansion of the gases should be achieved after ignition.

Otto's engine was a great success. In the first ten years, more than thirty thousand were manufactured by Otto and Langen's company, now called Gasmotorenfabrik Deutz AG.

Others felt that even this design could be improved on. It was suggested that it was inefficient for the same cylinder to function as a pump on one revolution and as a working cylinder on the next. In 1878, the Scotsman Dugeld Clerk built a two-stroke engine, in which the fuel was fed in by a separate pump cylinder. In later designs the two functions were combined, so that the top side of the piston carried out the working cycle, while the bottom side acted as the pump (see Box 8.1). Although it is difficult to get the two-stroke engine to perform as well as the four-stroke engine (both in terms of efficiency and exhaust emissions) this alternative design remained popular right through the twentieth century, especially for motor bikes, and also for large diesel engines.

The birth of the car engine

As demand for Otto silent engines grew, the Deutz factory took on two new engineers, later to become very famous. One was Gottlieb Daimler and the other Wilhelm Maybach. In 1882 the pair set up their own workshop outside Stuttgart where they carried out the development of a small high-speed engine to run on volatile petroleum fuel rather than town gas. At that time petrol was usually bought at chemists' shops for use as a cleaning fluid. For this engine, Maybach developed a **carburettor**. This is a device which vaporises the petrol and delivers a precise mixture of petrol and air to the engine over a wide range of engine loads.

Figure 8.4 Gottlieb Daimler (1834–1900) received his technical education in Stuttgart and then went to England to study the development of steam cars. He returned to work for Nikolaus Otto. In 1886 he tried his lightweight high-speed petrol engine first on a bicycle, and then a four-wheeled 'car'. In 1890 he founded the Daimler motor company.

Daimler's engine also used a high speed of rotation, 900 rpm rather than the Otto engine's normal 200 rpm. This is important because it is not the *absolute size* of the engine that determines its output power, so much as the *rate of throughput of fuel and air*. The output power depends on the amount of fuel burned in the engine in a given time and its efficiency. A small engine could potentially have a large power output if it could be persuaded to burn fuel efficiently at a high rotational speed.

In 1886 they tried out their engine on a motorized bicycle. Later that year they built their first car, a traditional horse carriage to which Daimler had fitted a version of his single-cylinder engine. By 1889 he had produced a two-cylinder design which he started manufacturing in quantity.

They weren't the only ones working in this direction at this time. Another German, Karl Benz, had set up his own business making gas engines in 1874. In 1886 he patented a three-wheeler car fitted with a petrol engine producing 0.75 hp (about 550 watts). The car had a single wheel at the front steered with a handle that also operated the brake. This design was perhaps unwise in a world of rutted unmade roads. On the first outing he hit a wall. His subsequent attempts at driving weren't much better. Eventually his wife, Berta, 'borrowed' it one morning and set off with their two sons to make the 100 km trip from Mannheim to Pforzheim. It required pushing up hills, but they arrived at their destination, having made the world's first serious petrol car journey. Although Benz exhibited his three-wheeler cars at exhibitions in Paris and Munich, there were few buyers. It took his French sales agent to convince him that a car really should have *four* wheels for stability and comfort. His agent was right, Benz's 'Victoria' of 1893 had a 3 hp engine and a top speed of nearly 18 km h^{-1}. He sold 45 cars in that year. By the end of 1901 his company had sold over 2700 vehicles.

Figure 8.5 Karl Benz driving one of his three-wheeler cars

Meanwhile, in 1890 Daimler and Maybach had set up the Daimler car company. Daimler preferred to concentrate on engine manufacture in particular for buses, ambulances, fire engines and tractors. His company set the lead in motor buses in the UK. In 1904 a 24 hp 34-seater double decker motor bus was displayed at the Crystal Palace Motor show. By the end of 1904 there were 17 motor buses in service in London, and 2000 by 1913. The pattern was repeated in cities across the world. Back in 1880, urban street public transport had consisted of horse drawn buses and trams. By 1913 these had almost totally been replaced by the combined onslaught of a new generation of petrol buses and electric trams.

Gottlieb Daimler retired in 1898, but in 1899, Maybach, who was running the company, received a big order. It was from Émile Jellinek, a wealthy Czech diplomat. He wanted a light car with a powerful engine for touring and racing. He would buy 36 cars if the first was delivered by October 1900 and if the car was named after his daughter, Mercedes. Maybach didn't actually deliver the first car until 1901, but Jellinek kept his promise. The German Daimler factory used the name Mercedes for all its passenger cars from then on.

This new Mercedes was a great success and won its first race at an average speed of 56 kph. It was not cheap, costing £2000 (equivalent to about £130 000 at today's prices). Two years later, came the 60 hp model, capable of over 100 kph with a touring body and over 125 kph when stripped for racing. There was the slight problem of starting the 9 litre engine with the crank-handle at the front which required strong muscles. These were cars for the rich, and concerns about fuel efficiency or the price of petrol were largely irrelevant.

These new cars were much faster than existing speed limits and it was necessary to convince police and local authorities to allow them to be driven at high speeds on the public highway, a battle that continues to the present day.

At the other end of the market, the French firm of de Dion Bouton, which had been making steam cars, became interested in small petrol engines and small cars. Their 3.5 hp 'Petite Voiture' was the world's best-selling car in 1900 and for a while France was the world's leading producer of cars.

BOX 8.2 Power and Speed

The 1903 Mercedes set the performance standards for a modern car. Today a 60 hp (45 kW) engine would be seen as 'small' and a top speed of 125 kilometres per hour as 'slow', despite a speed limit of 70 miles per hour (112 kilometres per hour) on UK roads. Why is all this power needed? Why don't we still have 'Petit Voitures' with 3.5 hp engines?

The key reason is speed. Firstly, the engine must be able to accelerate the car and enable it to climb hills. This is purely a matter of moving its mass. Then it must be able to deal with rolling friction and aerodynamic drag. These involve more subtle aspects of design.

Mass and Kinetic Energy

The words *speed* and *velocity* are usually used interchangeably, but to be precise speed is a **scalar** quantity, it only has magnitude and we are not usually interested whether this is in any particular direction. The more scientific term **velocity** is defined as the rate of change of position. It is a **vector** quantity: it has a magnitude and a direction, say due east. In practice speeds for cars are measured in miles per hour (miles per hour) or kilometres per hour (kilometres per hour). The more scientific unit is metres per second (m s^{-1}).

   50 miles per hour = 80 kilometres per hour
   = 22.2 metres per second

Every time a car is accelerated up to a particular speed, it gains **kinetic energy** due to its motion. This has to be supplied by the engine. When it is slowed down again, this kinetic energy normally has to be dissipated as **heat energy** in the brakes.

As described in Chapter 4, the kinetic energy KE of an object in motion is proportional to its mass m and the square of its velocity, v.

   KE = 0.5 mv^2

A mass of one kilogram travelling at a velocity of 1 m s^{-1} has a kinetic energy:

   KE = 0.5 × 1 × (1 × 1) = 0.5 joules

So a car with a mass of 1000 kg and a velocity of 30 m s^{-1} (approximately 110 kilometres per hour or 70 miles per hour) would have a kinetic energy of:

   KE = 0.5 × 1000 × (30 × 30) = 450 000 J = 450 kJ

The kinetic energy rises in proportion to the mass. Travelling at the same speed a two tonne vehicle would have twice the kinetic energy of a one tonne vehicle. However, since it rises as the *square* of the speed, travelling twice as fast increases the kinetic energy by a factor of four.

If the car has been designed to reach this speed of 30 m s^{-1} in 20 seconds from a standing start, then the engine must deliver an average power to the wheels of:

   450/20 kJ s^{-1} = 22.5 kJ s^{-1} = 22.5 kilowatts or 30 hp.

This figure is critically dependent on the mass and the rate of acceleration. If the car is twice as heavy, then the average power required will rise to 45 kilowatts. This would also be the power required if the mass stayed the same, but the car was only allowed 10 seconds to get up to speed.

Stopping the car poses an equally difficult problem. The kinetic energy has to be disposed of as heat in the brakes, and if anything it is desirable for cars to stop in the absolute minimum of time. Stopping the car in 20 seconds means converting 450 kJ of kinetic energy into heat in that time, requiring a heat dissipation rate of 22.5 kW. Doing so in 10 seconds requires 45 kW and for an emergency stop in 5 seconds the figure becomes 90 kW.

Put simply, the rapid acceleration of a heavy car needs a powerful engine. Stopping it needs very good brakes (something conspicuously lacking on early cars).

Climbing hills

Travelling on the flat is relatively easy, but what if we were to ask our car to climb a hill? Here the engine must increase the **gravitational potential energy** of the car. As described in Box 4.1 of Chapter 4, the energy needed to move a mass m upwards through a height H

against the gravitational pull of the Earth is *mgH*, where *g* is the acceleration due to gravity.

Suppose our 1 tonne car has to climb a 1 in 20 hill, 100 metres high at a modest speed of 60 kph. The energy needed will be

$$m \times g \times H = 1000 \times 9.81 \times 100 = 9.81 \times 10^5 \text{ J} = 981 \text{ kJ}$$

Doing this will involve travelling 2 km along the road. At 60 kph this would be covered in 2 minutes or 120 seconds. Thus the power required, the rate of doing work, will be

$$981 \div 120 = 8.2 \text{ kJ s}^{-1} = 8.2 \text{ kilowatts.}$$

This large power requirement is the reason that low-powered vehicles go up hills slowly (usually much to the annoyance of Mercedes owners)

Once again, mass is important. Doubling the weight of the car will double the required power to climb the hill, as will doubling the speed.

Descending a steep hill is another matter. Now the gravitational potential energy can be used to propel the car forwards, but it is still likely that large amounts of energy will have to be dissipated in the brakes. Disposing of 8 kilowatts of steady heat production is no mean feat, equivalent to the output of two electric fires on each of four wheels. Unlike braking from travelling at high speed, which may only take a matter of seconds, this heat loss has to be sustained for a period of minutes. This is why it is usual to engage a low gear and use the resistance to motion of the engine to dissipate the heat.

Air Resistance and Rolling Friction

Even when travelling on a flat road a car engine has to provide a large amount of power. At low speeds the energy produced by the engine is roughly consumed equally by: the accessories, such as the electric alternator and cooling fan; the drive train, such as the gearbox; aerodynamic drag and the rolling friction of the tyres (see Figure 8.6). At higher speeds, these last two factors dominate the energy consumption. Reducing them becomes very important for determining the engine power required to achieve a given speed, and overall fuel efficiency.

Aerodynamic drag is produced as the car moves forwards pushing against the air in its path. A certain amount of what was stationary air will be trapped by the front of the car, accelerated up to the same speed and pushed out of the way sideways. It will be given

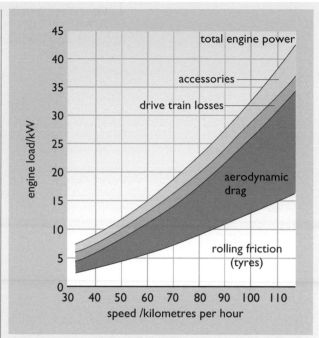

Figure 8.6 Power requirements and speed for a typical small car

kinetic energy as a result, which ultimately has to come from the car engine. Since it is a matter of kinetic energy, aerodynamic drag tends to vary with the square of the velocity and becomes very important at high speeds. It can be reduced by streamlining the body. The degree of success in this is indicated by its **drag coefficient**. A cube has high drag coefficient of over 1. An aerofoil shape like the wing of a plane can have a value of 0.05. A well-designed modern car will achieve a figure of about 0.3. Much of the pioneering work on car body streamlining was carried out by Paul Jaray in the wind tunnel of the German Zeppelin Airship Works between 1914 and 1923. His work led to the rounded shape of the Volkswagen Beetle, first produced as a prototype in the 1930s.

Rolling friction can be reduced by good design of wheel bearings, but mostly by the design of the tyres. Radial-ply tyres, introduced in 1949, have a lower rolling resistance than their cross-ply predecessors.

Overall, a 'Petit Voiture' car engine of 3.5 hp is likely to be able to propel you to the supermarket at a leisurely 20 kilometres per hour. If, however, you want to travel at 120 kilometres per hour *and* do so up hills *and* be able to accelerate to overtake the car in front, you will want something more powerful, more in the 100 hp/75 kW class.

The motorization of the US

It was in the US that sales of the petrol-engined motor car really took off, particularly in New York and California. Initially the US imported large numbers of cars from Europe, particularly from France. Then, in 1908 Henry Ford sold his first Model T for $850. He believed in simple styling for mass sales. 'A customer can have a car in any color as long as it is black', he wrote. In 1913 he installed the first moving production line in one of his factories. Within two years it had cut assembly time per car from 12 hours to 93 minutes. Prices fell and sales soared. By 1923 sales of this model alone reached 1 million per year, but there was no shortage of competition from other manufacturers.

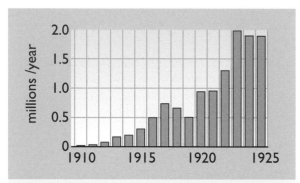

Figure 8.7 Henry Ford's mass-produced Model T was the first car to reach sales of 1 million per year (Source: Model T Ford Club of America (http://www.mtfca.com/)

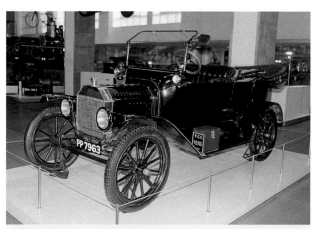

Figure 8.8 The Model T Ford

Figure 8.9 Henry Ford (1863–1947) was born on a farm in Michigan. In 1896 he produced a petrol engined quadricycle and went on to produce racing cars, even driving them himself. In 1903 he founded the Ford Motor Company, producing the Model T car in 1908 and introducing production line manufacture in 1913. The Ford company had a stormy history of labour relations into the 1940s because of his authoritarian management style

The effect on US society was dramatic. By 1923 Los Angeles and Salt Lake City had reached a level of car ownership of one to every three people. By 1925 motor vehicle production had become the largest industry in the US and most cars sold were replacements, not first time purchases. In that year there were 153 cars per 1000 head of US population compared to only 13 in the UK. The 'car culture' of New York and California rapidly spread to the rest of the US over the following decades. The continuing rise in car ownership since 1920 can be seen in Figure 8.10 opposite.

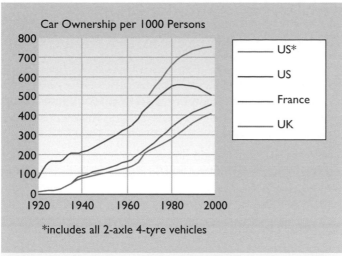

Figure 8.10 Relative car ownership for US, UK and France 1920–1998 (sources: McShane, 1997; EU, 2000)

In the UK and France car ownership has increased following US figures but with a lag of about 35 years. Perhaps fortunately, it took until the 1980s for UK car ownership levels to reach those that those of US in the early 1930s. Recent figures might be interpreted as showing that US car ownership has been declining since about 1980. However this may simply reflect a rather tight definition of a 'car' (something which now has to meet stringent fuel efficiency regulations). Ownership of all vehicles with four tyres and 2 axles, including 'sports utility vehicles' or 'SUVs' continues to rise. In the US there are now three of these vehicles to every four people.

Aircraft petrol engines

The development of heavier-than-air flight required very light and powerful engines. The Wright brothers built their own lightweight 12 hp four-cylinder petrol engine for their 1903 flight. New more powerful designs rapidly followed.

Aircraft piston engines have to deal with a particular problem not shared with cars. The higher they fly, the thinner the air becomes and this reduces the power available. The answer to this was to fit a **supercharger,** an engine-driven pump which compressed the air before feeding it into the engine. In this way engine performance could be maintained up to heights of 10 000 metres or more. An alternative way of compressing the air was to use a small turbine or **turbocharger** driven by the hot exhaust gases.

While supercharging an engine was initially developed for aircraft, it could also be used on the ground, allowing more air and fuel to be pumped through a conventional engine, increasing its power rating. Turbochargers are now commonly fitted to many modern car engines.

Throughout the 1920s and 1930s, aircraft piston engines became larger and more powerful, perhaps reaching the peak of their design performance during World War 2. The 12 cylinder Rolls Royce Merlin engine which powered the Spitfire had a cylinder capacity of 27 litres and a maximum output of 1.48 MW at 3000 rpm. However, this level of performance required the use of high compression ratios and a very high grade of petrol.

Compression ratio and octane number

At the beginning of the 20th century petrol engines used low compression ratios of about 4:1. During the 1920s and 1930s, there was a continuous search for methods to improve their performance. As the years went by it became obvious that higher compression ratios allowed combustion at higher temperatures, producing more work from the same amount of fuel. Typical compression ratios in car engines in the 1930s were about 6:1 rising to 7.5:1 in the late 1940s and up to 9:1 or above today.

However, petrol must not spontaneously ignite in the engine, it must only do so when set off at the right time by the electric spark. If it does so, it will produce a clattering noise known as 'knocking' or 'pinking'. Worse still, it can lead to overheating and damage to the engine. The higher the compression ratio, the higher the grade of fuel required to prevent knocking. This anti-knocking capability is expressed as the 'octane rating' of the fuel. Pure iso-octane (a hydrocarbon with eight carbon atoms) has very good anti-knock capabilities and is given an octane rating of 100. Heptane (which has only seven carbon atoms) has very poor anti-knock properties and is given an octane rating of zero. A mixture of 80% iso-octane and 20% heptane would be assigned an octane rating of 80.

Practical commercial petrol consists of mixtures of a whole range of hydrocarbons, but it is characterized by the anti-knock performance or octane rating. There is practically no difference between the energy content per litre of high-octane and low-octane petrol. What is different is the ability for the fuel to be used in higher-compression engines with higher working temperatures and a higher thermodynamic efficiency. This is well worth doing. For example, raising the compression ratio of an engine from 7.5:1 to 10:1 can improve both output power and fuel efficiency by about 17%. Typically a compression ratio of 7.5 requires 85 octane fuel and one of 10:1 would require 100 octane fuel. Modern 'premium' unleaded petrol has an octane rating of 95 and super unleaded petrol has a rating of 98. In recent years careful design of cylinder heads has produced engines with compression ratios of over 10:1 that can run on 98 octane petrol.

Lead additives

The petrol manufacturers have struggled throughout the 20th century to find ways of making 'high-octane' petrol out of low-grade crude oil. One way was to refine genuinely high grade petrol, and if necessary, use catalytic cracking in the refinery to split long chain molecules into shorter ones, increasing the yield of volatile petrol. The other route was to find a chemical additive.

Tetraethyl lead was first introduced into petrol in 1916 in the USA by Thomas Midgley (who also has the dubious honour of inventing CFCs (chlorofluorocarbons) as refrigerants). He found that adding less than one gram of this compound per litre of petrol improved the octane rating by 10 to 15 points.

It also reduced engine wear, coating the hot exhaust parts in the cylinder head with a protective layer of lead oxide once it had been burned. This allowed cheaper engine components. Ultimately all the lead in the petrol would be discharged from the end of the exhaust pipe as a fine dust of lead oxide.

Although it was appreciated that lead was toxic, investigations had played down the possible environmental effects of it in the urban atmosphere. It was only in the 1960s that concern started to be expressed about possible brain damage effects, especially to children. During the 1980s the proportions of lead were steadily reduced and the octane rating held up by producing a better product at the oil refinery. However, the introduction of totally lead-free petrol has been largely due to the need to remove a different pollutant from the exhaust, NO_x, the various oxides of nitrogen (see below). Ironically, the need to produce 'super' high-octane lead-free petrol has led to another pollution problem. Petrol companies initially included benzene as an octane-enhancer, but because of its toxicity, the use of benzene is now severely restricted.

Leaded petrol has now (2002) been officially withdrawn from normal sale in the UK. Petrol companies now market 'lead replacement petrol' for use in older cars, and oil companies are selling substitute additives using less toxic ingredients. In practice, many companies now offer 'lead-free' conversions at only a modest cost for the cylinder heads of older cars, inserting hardened valve seats capable of resisting the localised high temperatures. However, leaded petrol is still used by piston-engined aircraft.

8.3 The diesel engine

The petrol engine requires the fuel/air mixture to be ignited by a spark (or in the early days with a rapidly-applied flame). The modern diesel engine sucks air into a cylinder, where it is compressed and becomes very hot. At the top of the stroke, a small amount of fuel is injected and it spontaneously ignites, driving the engine through the rest of the cycle in the same manner as a petrol engine. The idea of separately injecting the fuel first appeared in engines designed by Herbert Ackroyd Stuart in 1892. The fuel had to be separately heated in a vaporizer in order to ignite. His engines had an efficiency of about 15%, comparable with the Otto silent engine.

In 1892 Rudolph Diesel patented the idea of compressing air to such an extent that the fuel/air mixture would spontaneously ignite after ignition. More accurately, it 'detonates' as a small explosion, rather than simply burning rapidly, as in a spark ignition engine. This is the reason a diesel engine always produces more engine noise than the equivalent petrol engine. Also the engine components, particularly the pistons, have to be made heavier and stronger in order to resist the rapid rise of pressure in the cylinder. Another early problem was that it was difficult to manufacture exhaust valves that would survive the high temperatures and pressures produced, so most early diesel engines adopted a two-stroke design. Even today, the largest diesel engines are two-strokes.

Diesel did not achieve his original aim of running an engine on powdered coal dust, but the prototype, running on oil, achieved an efficiency of 26%, far higher than contemporary petrol or steam engines. Diesel injected the fuel into the cylinder using a high-pressure air blast. This required an expensive and bulky air pump and storage cylinder, which restricted the engine to stationary and marine use, and absorbed typically about 5% of the engine power.

Figure 8.11 Rudolf Diesel, German engineer (1858–1913) Diesel was a brilliant student, graduating with the highest-ever marks from the Munich Technical University. He patented his compression ignition engine in 1892 which was first exhibited at the Munich Exhibition of 1898. It has been widely used ever since, making him a millionaire. He mysteriously disappeared overboard from a ship in 1913

Figure 8.12 An early British 'oil engine' of the 1920s (source: MAN/ Paxman at www.nelmes.fsnet.co.uk/paxman/paxhfoil.htm)

A key feature of diesel engines was (and still is) that they can be run on almost any grade of fuel as long as it could be pumped. Really thick oil might need preheating, but this could be done using the waste heat from the engine. They could even be adjusted to run on town gas by adjusting the compression ratio. Many modern engines used for power generation run on mixtures of natural gas and diesel fuel.

Diesel power for ships

Large two-stroke diesel engines proved ideal for ships, especially since their efficiency could exceed 40%. They could easily be made more efficient than steam reciprocating engines or even steam turbines. Also diesel fuel was far more convenient to handle than the competing fuel, coal. Diesel could be mechanically pumped. Coal had a lower energy density and so took up more space. It also needed to be shovelled by hand, which required more manpower.

The difference in mechanical efficiency and the savings in wages were so great that it was economic to use diesel engines even though the fuel cost three or even four times as much per tonne. By 1926 over 5% of the world's shipping tonnage was powered by diesel and by 1937 the figure had exceeded 20%. Today it is the dominant power source for commercial shipping.

Although as the 20th century progressed, virtually every other engine has got smaller, lighter and faster, large ship's diesel engines have remained stolidly slow and heavy. Currently, the world's largest production diesel engine is the Wartsila Sulzer RTA96C two-stroke used to power large container ships. It has 12 cylinders each with a piston nearly a metre in diameter and a stroke of 2.5 metres. The maximum continuous power is

89 640 hp (approx. 67 MW), at a stately speed of 100 rpm, typically connected to a large propeller nearly 9 metres in diameter. The engine weighs a mere 2000 tonnes, of which the crankshaft makes up 300 tonnes. Its thermal efficiency can exceed 50%.

Diesel engines for road, rail and air

Small diesel engines for road vehicles did not appear until the 1920s. Even then laws about smoke emissions meant that the two-stroke diesel engine was unacceptable. Road vehicles needed four-stroke engines and a mechanical method of injecting the fuel into the cylinder rather than Rudolph Diesel's complicated air blast system. The **fuel injection pump** has a difficult job to do, forcing a measured amount of fuel into the cylinder against the high pressure of compressed air inside. A basic design had been invented in 1910, but presented many technical problems. Even today it remains an expensive (and sometimes temperamental) part of a diesel engine.

In 1924 the German company MAN introduced a 5 litre diesel engine for road vehicles fitted with a fuel injection pump at the Berlin Motor Show. In 1928 the British manufacturer Gardner introduced diesel engines for marine use and these were immediately experimentally fitted to buses. The Foden lorry company, which had been steadily producing steam lorries, produced their first diesel lorry in 1931 even though they continued to produce steam road vehicles until the 1950s. By 1934 most new trucks and buses in the UK were being ordered with diesel engines.

Diesel even took to the air. In Germany in the 1930s, the Junkers company produced a successful 2-stroke diesel aircraft engine, the Jumo 205, which was used on transatlantic flights because of its good fuel efficiency. After World War 2, the British English Electric company adapted the unusual weight-saving layout of this engine for their successful 'Deltic' railway locomotives produced in the early 1960s. These hauled express passenger trains at up to 160 kilometres per hour.

The replacement of steam locomotives with diesel power had a dramatic effect on the overall fuel efficiency of British Railways. In the mid-1950s the railway network consumed about 5% of UK primary energy use, almost entirely as coal. Although many suburban passenger lines had been electrified, almost every other service used steam locomotives. These were designed for a high power to weight ratio, not fuel efficiency. An express steam locomotive on a long run might reach an efficiency of 11%. However, the figure for those used intermittently could be 5% or less, since they had to be fired up several hours before use and the boiler needed to be kept up to steam temperature even to move the smallest distance. Yet a diesel locomotive could have an efficiency of over 30% when running, and when it wasn't in use the engine was simply turned off. The potential fuel savings seemed enormous.

In 1955 British Railways announced a modernization plan involving the purchase of large numbers of diesel locomotives. This plan was implemented between 1957 and 1967 and included continuing expansion of the electrification of suburban railways. The effect was quite dramatic (see Figure 8.13 overleaf). The total primary energy consumption of British

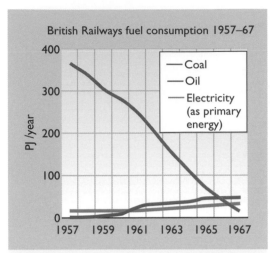

Figure 8.13 Fuel consumed by British Railways, 1957–1967 (sources: Ministry of Power, 1962; DTI, 1973)

Railways fell by almost three quarters. Although there had been a loss of freight and some passenger traffic to roads over this period, the bulk of this energy change is simply due to the substitution of one engine technology by another.

Since then, high speed diesel engine technology has continued to be developed, giving rise to the successful Intercity 125 units introduced in 1975 and now used both in the UK and Australia (see Figure 8.14). These use 2250 hp (1.7 MW) supercharged 12 cylinder four-stroke diesel engines at each end of an eight-coach train. The power is transmitted to the wheels using a diesel-electric transmission. Each engine, which has a thermal efficiency of about 40% drives an electric generator which powers electric motors on the wheels. Although designed to operate at 125 mph (200 kph), they have reached 238 kph. The design has not been without its teething troubles, but careful study put many of the problems down to those of dissipating over 2 MW of waste heat from each engine! More recent UK diesel passenger train designs have turned to having a smaller separate engine under each carriage.

Figure 8.14 Intercity 125 engine unit using a 2250 hp high speed diesel engine

Figure 8.15 The Smart, manufactured by DaimlerChrysler. The diesel version is capable of a fuel consumption of 3.4 litres per 100 km

Back on the road, a small turbocharged diesel engine is now available for the modern Daimler Chrysler 'Petit Voiture', the 2-seater Smart (see Figure 8.15). This 800 cc engine can produce 41 hp (30 kW) at 4200 rpm and, aided by an automatic six-speed gearbox, gives a fuel consumption of 3.4 litres per 100 km (over 80 miles to the gallon). The top speed of the car is electronically limited to 135 kph.

DERV

Although the diesel engine can in theory run on a wide range of possible grades of fuel, concerns about clean combustion and reliability have made it necessary to tightly specify the fuel used for road vehicles, known as DERV (DiEsel for Road Vehicles). DERV is lighter than the heavy diesel oil used in ships, but heavier than the kerosene used in jet aircraft. It cannot be too thick and heavy or it may freeze in the vehicle tanks in winter. Nor can it be too light, otherwise it may not ignite properly under compression. This ability to ignite under pressure is characterized by a *cetane* rating, analogous to the *octane* rating of petrol. In practice modern DERV is very similar to 'gas oil' used for domestic heating, but may contain additives to prevent freezing. Modern 'green diesel' also has the sulphur content refined out, both to stop potential emissions of sulphur dioxide and to allow the use of catalytic converters on engines.

8.4 **Petrol and diesel engines – reducing pollution**

When a fuel such as petrol, diesel or even coal is burned in an excess of air, the main combustion products are carbon dioxide (CO_2) and water vapour (H_2O). If there isn't enough air present, then there are other products such as carbon monoxide (CO) and hydrogen (H_2). Carbon monoxide is quite toxic; it combines with blood and inhibits the absorption of oxygen.

If there is sulphur in the fuel, then this will burn to form sulphur dioxide (SO_2). Diesel fuel, especially heavy fuel oil used in ships and power stations can contain considerable amounts of sulphur, 2% or more. Since sulphur dioxide is a serious contributor to problems of acid rain, it is desirable for refineries to remove as much sulphur from the fuel as possible.

The other serious pollutants are the oxides of nitrogen: nitrogen oxide (NO, commonly known as nitric oxide), nitrogen dioxide (NO_2) and dinitrogen oxide (N_2O, commonly known as nitrous oxide). In the mixture produced by car engines, typically 90% will be in the form of NO. Together they are referred to as NO_x. Like SO_2, these are both an irritant to the respiratory system and a cause of acid rain. NO_x pollution can be particularly unpleasant in cities such as Los Angeles or Athens where there is plenty of sunshine. Under these conditions the NO component can combine with oxygen in the city atmosphere by a photochemical reaction to produce more NO_2 and 'low-level' or 'tropospheric' ozone, O_3, giving a choking brown smog.

NO_x emissions can come about if there is nitrogen actually contained in the fuel. Coal, for example, contains modest amounts in the form of ammonia compounds, which will burn to form nitrogen oxide (NO). However the most important mechanism for the formation of nitrogen oxides in engines is the 'thermal' one. Air is a mixture of gases and contains about 78% nitrogen. If the combustion temperature is hot enough, greater than 1500 °C, then this nitrogen will react with the oxygen in the air to produce nitrogen oxide.

$$N_2 + O_2 \rightarrow 2NO$$

BOX 8.3 More on chemical equilibrium and rates of reaction: **NO formation**

The concept of a chemical **equilibrium** was introduced in Chapter 7. This box takes its study a stage further. At ground level, the air consists of a mixture of gases including 78% nitrogen and 21% oxygen. At low temperatures these exist stably together, but at high temperatures they combine to produce nitrogen monoxide (NO, nitric oxide). This reaction is a problem in all combustion processes but particularly serious for internal combustion engines.

From a chemical point of view we would say that nitrogen, oxygen and nitric oxide exist in some kind of equilibrium together. A system in equilibrium is one in which no further change is apparent. There are two types, a static equilibrium and a dynamic one:

Static equilibrium
A static equilibrium occurs when a system is at rest. A good example is a balance with weights attached (Figure 8.16).

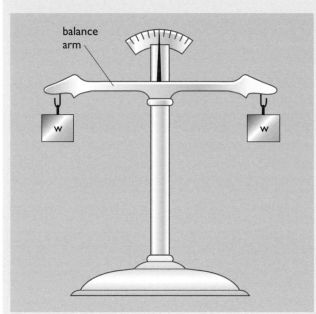

balance arm

Figure 8.16 A weighing balance

Provided the weights are adjusted so that the balance arm is horizontal it will remain in this position indefinitely. It is unmoving or static.

Dynamic equilibrium
A dynamic equilibrium is not at rest. An example is a reversible chemical reaction where a chemical 'A' is

reacting to form a product 'B'. We may write the equation as:

$$A \rightleftharpoons B$$

where the reversible arrows indicate the nature of the reaction. Molecules of A are reacting to form B, while other molecules of B are, simultaneously, reforming to produce molecules of A. When the reaction has reached equilibrium, the forward and backward rates of change will be the same so there is no overall change. At any particular temperature, after sufficient time has elapsed, the reaction reaches a dynamic equilibrium. An observer might assume that the system is static, unless some change is made, such as a change in temperature, to upset the equilibrium.

The reaction to form nitrogen monoxide is such a reversible reaction:

$$N_2 + O_2 \rightleftharpoons 2NO$$

Molecules of nitrogen and oxygen are continuously combining to produce molecules of nitrogen monoxide and these are continually dissociating back into nitrogen and oxygen.

This particular reaction is **endothermic**, i.e. it needs an input of energy to proceed. At room temperature (around 20 °C) the dynamic equilibrium lies so far over to the left-hand side of the equation that there is essentially no nitrogen monoxide in the equilibrium mixture. This is just as well for our health. At higher temperatures >1500 °C, however, there is an appreciable proportion of nitrogen monoxide present in the equilibrium mixture, and this proportion increases rapidly above 1500 °C – hence the problems with internal combustion engines. Also, the higher the temperature the faster the chemical reaction will be so an equilibrium mixture containing plenty of nitrogen monoxide will be reached quite rapidly. It is only necessary for a small part of a flame to reach this temperature to produce a considerable amount of NO.

On leaving a car engine, the exhaust gases containing the NO are rapidly cooled. Although the NO will dissociate back into N_2 and O_2 it will only do so very slowly at low temperature and the pollutant will remain in the air for an appreciable time. However, this reaction can be encouraged by the use of a catalyst, such as the metal rhodium. This is one of the processes that takes place in a catalytic converter (see below).

As the mixture cools, then some of the nitrogen oxide may react with more oxygen to produce nitrogen dioxide (NO_2). 1500°C may seem an extraordinarily high temperature, but it is only necessary for a very small part of a flame to reach this temperature for the reaction to start. The reaction also requires a certain amount of time to take place. If the gases only reach 1500 °C for a few milliseconds, then **equilibrium** will not be reached (see Box 8.3) and production of NO is much reduced.

The formation of NO can be limited by very careful design of *low* NO_x *burners*. In these the combustion of the fuel, be it gas, oil or coal, is carefully regulated to eliminate 'hot-spots' where this reaction can take place. The general principle is to burn the fuel uniformly and keep the flame temperature down to the minimum needed for the particular application. A domestic gas boiler, for example, only needs to heat water to 60 °C, so there is no need for a high flame temperature. A coal-fired power station needs to heat steam to 500–600 °C. This is a little more difficult. A gas-turbine may need to run with a turbine inlet temperature of 1300 °C, which poses serious problems.

When burning fuel for an engine application there is a conflict between the needs of high temperatures to give a high Carnot thermal efficiency (see Chapter 6), and *avoiding high temperatures* to keep the nitrogen oxide emissions down.

The situation is particularly difficult in petrol and diesel engines, where the combustion takes place in a relatively uncontrolled manner inside a cylinder and it has required many years of study and experimentation to reduce NO_x emissions to acceptable levels.

Emissions from petrol engines

Chemically, petrol consists almost entirely of carbon and hydrogen. When it is burned in air it should just burn to produce carbon dioxide and water. Ideally, in a petrol engine the carburettor or fuel injection system should supply 14.7 parts by weight of air to one part of petrol. This is called the *stoichiometric ratio*. It is usually referred to by the Greek letter 'lambda' (λ). This 'perfect' mixture has a $\lambda = 1$.

If, though, there is a 'rich' mixture with too much fuel and not enough air, λ will be less than 1. There will not be enough air for the combustion process to proceed fully. Much of the carbon in the fuel will only be converted to carbon monoxide (CO) and not all the way to carbon dioxide (CO_2). Other products of incomplete combustion are unburned hydrocarbons (HC), ranging from unburned petrol down to a fine sooty carbon dust. Unburned benzene and partial combustion products such as 1,3 butadiene and aldehydes are particularly toxic.

It would seem, then, that the best thing would be to make sure that there was always a surplus of air in the fuel/air mixture, i.e. it would be 'lean'. Unfortunately lean mixtures burn hotter and usually produce more oxides of nitrogen, NO_x.

Since the 1970s, engine designers have wrestled with problem of producing efficient petrol engines that minimize the emissions of the three major pollutants: unburned hydrocarbons, carbon monoxide and NO_x. They have come up with two basic commercial solutions; the lean-burn engine and the three-way catalytic converter.

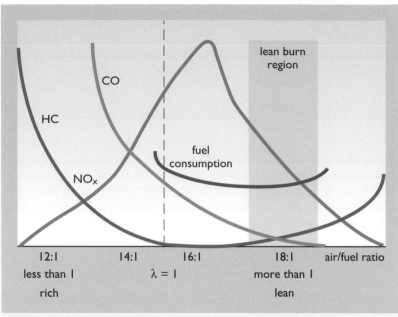

Figure 8.17 Air/Fuel Ratio in a Petrol Engine and Relative Emissions

The lean burn engine

Changing the air/fuel ratio in a petrol engine radically alters the way the combustion process proceeds. As shown in Figure 8.17 above, if the mixture is too rich, there will too much carbon monoxide and unburned hydrocarbons in the exhaust. Also, since the fuel is not all being burnt properly, the fuel efficiency will be poor as well.

If the mixture is made slightly weak (around 16:1) then there will be good fuel efficiency, but plenty of NO_x production. If the mixture is made weaker still (around 20:1), then the combustion can be made to proceed so rapidly that there isn't time for the NO_x to form. This requires extremely careful engine design, but the result is high efficiency, low HC and CO emissions, and NO_x emissions that are far lower than normal engines.

The 3-way catalytic converter

The other alternative is to use a 'normal' engine, but to add what is literally an 'end of pipe' solution. The exhaust gases from the engine are passed through a chamber containing a ceramic substrate coated with a selection of catalysts that can complete the imperfect combustion that has taken place inside the cylinder.

A *catalyst* is a substance that can accelerate a chemical reaction without actually getting changed itself. The metal platinum is widely used in many chemical reactions, including the 'catalytic cracking' of petroleum in refineries. Inside the catalytic converter three catalysts are used, each to convert a different pollutant, hence the term '3-way' catalytic converter. Platinum and palladium are used to help oxidize the unburned hydrocarbons and carbon monoxide to carbon dioxide, and rhodium is used to help convert the nitric oxides back to nitrogen and oxygen.

The main difficulty is that these reactions only work properly if the air/ fuel mixture is very tightly controlled indeed, just slightly rich of $\lambda=1$. The high level of accuracy required has meant that most manufacturers have abandoned the use of the carburettor, an essentially mechanical method of getting the right fuel/air ratio, in favour of fuel injection, usually electronically controlled. The best way to get the precise air/fuel ratio *entering* the engine is to measure the amount of oxygen in the exhaust gases *leaving* the engine and feed the result to a computerized engine management system controlling the fuel injection.

The other difficulty is that catalytic converters only work when they are hot. They only start working once they have been heated by the engine exhaust gases to more than a certain *lightoff temperature*, which can be 150 °C to 300 °C. This means that they aren't likely to be working effectively on short trips. Also, being suspended from the underside of the car, they are very vulnerable to damage, and, rather ironically, especially so from 'traffic calming' measures such as speed humps.

The use of catalytic converters also places more constraints on the petrol being used. Any attempt to burn leaded petrol in the car would result in the catalysts being poisoned with lead oxide and rendered completely useless in a matter of days. Thus the introduction of the catalytic converter has had to be accompanied by the simultaneous introduction of 'lead-free' petrol. Also, levels of sulphur in the petrol have to be minimized, otherwise the catalysts become slowly clogged with deposits of sulphur compounds and the catalytic converter may emit the characteristic 'rotten eggs' smell of hydrogen sulphide on warm-up.

Emissions from diesel engines

Unlike the petrol engine the diesel engine operates with a very wide range of air/fuel mixtures. Normally, at part throttle, the mixture is very weak. There is plenty of air to burn the fuel and emissions of carbon monoxide and unburned hydrocarbons can be very low.

However, at full power, the mixture can easily be far too rich, resulting in clouds of black sooty smoke (called *particulates*), all too familiar from trucks and buses. This pollution can, to a certain extent, be countered by careful engine management controls. Alternatively, various solutions based on exhaust filters are now being fitted to urban buses and trucks.

For diesel cars, 'oxidation' catalytic converters are becoming available to oxidize the remaining unburned hydrocarbons in the exhaust. These are essentially just '1-way' of the '3-way' type fitted to petrol cars. Their introduction has required the removal of sulphur from diesel fuel, since the catalyst would be poisoned by the sulphur dioxide produced in the exhaust. The manufacture of this 'Green' sulphur-free diesel, requires more effort, expense (and energy use) at the refinery.

Generally, NO emissions from diesel-engine cars are higher than those from the equivalent petrol-engine version. The peak temperatures reached in the cylinders of diesel engines are higher than in spark engines, suggesting that they should produce more far more NO. The actual NO production is limited by the fact that the combustion takes place as a rapid detonation, limiting the time for chemical reaction to take place.

Obtaining best efficiency

The other important pollutant from petrol and diesel engines is carbon dioxide. The best way to reduce the emissions of this is to operate the engine at its maximum efficiency. Typically in a petrol-engined car only 24% (about one quarter) of the heat energy in the fuel will be delivered as work to the crankshaft. The figure for a car diesel engine is about 32%. Even these figures can only be achieved if the output of the engine is properly matched to its load.

Car engines are usually sized to meet rather inflated notions of driving performance. An ability to travel at 150 kilometres per hour may be appealing, but in practice the car may spend most of its time on short trips to the local supermarket. Modern petrol car engines are usually designed to operate over a rotational speed range up to about 6000 rpm. For diesel engines this maximum rating is usually slightly lower, about 4000 rpm. However, best engine efficiency usually occurs at about a half of this speed.

The speed of the car is matched to the speed of the engine using a gearbox, which may be manually operated or automatic. In the past, cars were normally only manufactured with three or four-speed gearboxes. The top gear was usually designed to give the absolute maximum speed on a flat road (essential for the advertising). However, considerations of energy efficiency suggest that the car should be able to cruise at a reasonable speed, say 100 kilometres per hour, with the engine turning at its most efficient speed (around 2000 rpm for a diesel and 3000 rpm for a petrol engine). This requires an extra gear 'higher' than top gear. This is the reason that most modern cars are now supplied with five speed (or even six speed) gearboxes or an extra 'overdrive' gearbox. The lower engine speed also has the added benefits of reducing engine noise and extending engine life.

However, to get the absolute best efficiency from a petrol or diesel engine, they should really be operated not just at their most efficient engine speed, but also at close to full power at that speed (i.e. with the accelerator pedal flat on the floor). One solution is to place both the engine and the gearbox under automatic computer control to continually optimise the gear ratio being used and the engine settings for best fuel economy. A further step is to use a *hybrid petrol-electric drive* system. This allows the petrol engine to be used at close to full power when required and also charge up a storage battery. When only small amounts of power are needed, the engine is switched off and the car runs on an electric motor using the stored energy from the battery.

8.5 The gas turbine

Chapter 6 has described how the invention of the steam turbine in the 1880s represented a step change in performance, compared with the steam engine, particularly in power to weight ratio. But why bother with the steam boiler? Why not an internal combustion *turbine engine?*

A Norwegian by the name of Egidius Elling patented a gas turbine as early as 1884, but did not build one capable of actually generating surplus work until 1903. The fundamental problem was that the metals of the day were not capable of withstanding the temperatures of the hot gases produced by a continuously burning flame. His solution was to use the flame to heat a water jacket, projecting a jet of steam and high temperature burned gases into a turbine. This held down the temperature of the mixture to a level that would not melt the turbine blades. In modern parlance this would be described as a **steam injected gas turbine** or STIG. Other developers in France and Germany built experimental prototypes over the next 30 years, but it was difficult to compete with the proven performance of the competing diesel and steam turbines.

The modern gas turbine grew out of the need for a high-performance aircraft engine. In the 1930s various inventors experimented with designs for an engine that could produce a jet of gas to propel an aircraft, rather than relying on propellers driven by piston engines. Two inventors, on opposite sides of World War 2, ended up with working engines, Hans von Ohain and Frank Whittle.

Their designs were basically similar. A continuous stream of air would be compressed, fuel would be injected and burned, and finally the hot gases would be expanded through a turbine. The work derived from the spinning turbine would be used to drive the compressor.

The German jet engine

Hans von Ohain conceived of his jet engine design in 1933 while studying for a doctorate at Göttingen University in Germany. By 1935 he had developed a working model to demonstrate his ideas and approached Ernst Heinkel, an aircraft manufacturer, for support. Heinkel saw it as a means to build the fastest aeroplane in the world and in 1936 offered him backing. By February 1937 von Ohain had tested a turbojet running on hydrogen fuel that produced 250 pounds of thrust, with the turbine spinning at 10 000 rpm. Heinkel was very impressed and urged von Ohain to press on to produce a full-size aircraft engine. The first engine had used a *centrifugal compressor* to compress the air. This is similar to the fan inside a vacuum cleaner, sucking air in at the centre and blowing it out at the rim. Von Ohain's key idea, used in modern turbojet engines was to use an *axial flow compressor*, similar in principle to a turbine, but compressing rather than expanding the gas flow. They also switched to using kerosene, rather than hydrogen as a fuel.

Design work on an aircraft to use this engine, the He-178, began in early 1938 and it made the first jet-powered flight on August 27, 1939. The German Air Ministry were sufficiently interested to give rival contracts for jet aircraft to Heinkel and the Messerschmitt company. The latter subsequently developed to produce the world's first jet fighter, the Messerschmitt Me 262. It made its first fully jet-powered flight powered by two Junkers Jumo 004B turbojet engines on July 18, 1942. A previous attempt made in March was unsuccessful because both of the plane's BMW jet engines failed. Thoughtfully, it had also been equipped with a piston engine and propeller and landed safely.

BOX 8.4 Propulsion and thrust

We all spend time propelling ourselves, on foot or in vehicles, and in both cases we are making use of Newton's Laws of Motion in quite specific ways. But where does the force come from? When you walk, at every step you push backwards on the ground. And this is where **Newton's *Third* Law of Motion** enters. It states that if A pushes against B with a certain force, then B must inevitably be pushing against A with the same force in the opposite direction: *for every action, there is an equal and opposite reaction.*

So when you push backwards on the ground as you step, the ground pushes back, obeying Newton's Third Law, and this gets you moving. Similarly if you drive. The engine forces the wheels to rotate, so they push backwards on the ground where they make contact – and the ground pushes them forward, propelling the vehicle. *Friction* between foot or wheel and the ground is of course essential – as we quickly learn in trying to walk or drive on a sheet of ice. Railway locomotives are often rated by their **tractive effort**, the number of tonnes force they can exert on a train without the wheels slipping.

Thrust

How do we propel ourselves where there are no solid surfaces to provide frictional forces? A rocket can fly through outer space where there is nothing to push against at all. It does by using **Newton's Second Law of Motion**. This is commonly stated as '*the force needed to accelerate a body is proportional to its mass multiplied by its acceleration*'. Its combustion process accelerates a stream of exhaust gases backwards. This is the *action*. The *reaction*, using Newton's Third Law, generates a forward force, the **thrust** of the engine pushing the rocket through empty space.

In the case of aircraft, its engines generate thrust by accelerating a continuous stream of air backwards, be it by using a propeller or a turbojet engine (as in Figure 8.18 opposite).

We have seen in Box 4.1 of Chapter 4 that the force (F) needed to accelerate a mass m with acceleration a is given by

$$F = m \times a$$

To calculate the thrust, therefore, we need to know the mass of air and its acceleration. Let us assume that the engine shown in Figure 8.18 is attached to an aircraft flying at a velocity $v_{aircraft}$. The hot exhaust gases are ejected backwards at a velocity v_{jet}. We'll concentrate on a particular mass, m kilograms of air (and ignore the small mass of the fuel actually burned). Its average acceleration is equal to its change in velocity divided by the time t (in seconds) that it spends in the engine:

$$a = \frac{v_{jet} - v_{aircraft}}{t}$$

and the force F (in newtons) needed to produce this acceleration is

$$F = m \times \frac{v_{jet} - v_{aircraft}}{t} = \frac{m\left(v_{jet} - v_{aircraft}\right)}{t}$$

But m divided by t is the number of kilograms passing through per second: the **mass flow rate**. So we have a simple relationship:

thrust = mass flow rate × change in air velocity

Notice that only the *change* in the velocity of the air is important, so it doesn't matter whether the velocities are measured relative to the ground or the aircraft, provided the same method is used for the initial and final velocities.

For a modern commercial aircraft engine the thrust is in the range 200–400 kN, approximately the weight of a mass of 20–40 tonnes.

The British jet engine

In England a completely parallel development took place. Frank Whittle began experimenting with ideas for a turbojet engine while training as an RAF pilot at the Cranwell Staff College. By 1930 he had designed and patented a jet aircraft engine, but could not get backers in the middle of an industrial depression. He could not even afford to renew his patent. In 1935, while studying for a degree in mechanical engineering at Cambridge University, Whittle was approached by some ex-RAF officers who suggested continuing work on his turbojet. They raised some money to set up Power Jets Ltd. and started work on an experimental engine in Rugby. They tested this in April 1937. Like von Ohain's original engine it used a centrifugal compressor.

The British Air Ministry began to take and interest and in 1939 gave Power Jets Ltd. a contract for a flight engine. The Gloster Aircraft Company was asked to build an experimental aircraft, the E.28/39, which first flew powered by the Whittle jet engine in May 1941. As in Germany, design difficulties led to delays in getting a production engine. A turning point came when the Rolls-Royce company took over the manufacture – they have been making jet engines ever since. A production engine was eventually installed in a twin-engine jet fighter, the Gloster Meteor which first flew in March 1943. Frank Whittle was knighted in 1948 and died in the US in 1998.

BOX 8.5 Principles of the Turbojet

The layout of the modern turbojet engine is little different to von Ohain's design. The key components are an axial compressor, a combustion chamber and a turbine (see Figure 8.18).

Axial Compressor
The compressor carries out the function of the piston in a four-stroke engine, raising the pressure of the incoming air. However it does this by using a succession of banks of turbine blades, progressively squeezing the air into a smaller volume. In a modern engine the compression ratio may be between 16 and 30. It has to be high, because at a jet aircraft's normal cruising altitude of about 10 000 metres, the air pressure is rather low and it is a matter of gathering enough air to burn a reasonable amount of fuel.

Combustion Chamber
The combustion chamber is the heart of the engine where the fuel is burned, fed with highly compressed air. Here, the fuel has to be burned in an even manner in order to minimize NO_x emissions. This must be done all the way from minimum to maximum power output. Strangely, the first task of the combustion chamber is to reduce the velocity of the incoming air, so that the flame does not get blown out. The fuel, kerosene in the case of aircraft engines, is injected, burns and projects a jet of hot gases onwards to the turbine blades. In a modern turbine the gases leaving the combustion chamber can be up to 1300 °C, so the casings have to be made with similar attention to detail as the turbine blades themselves. Industrial gas turbines which have to meet stringent NO_x emission standards are likely to have multiple banks of combustion chambers to cope with changing loads.

Turbine
The turbine extracts kinetic energy from the hot gas stream, converting it into work driving the main shaft through the engine. The hot gases are finally expelled at high speed through the exhaust nozzle, providing the thrust. From the earliest days, finding ways of preventing the turbine blades simply melting has been a major design problem. In the 1950s turbine inlet temperatures were only about 800 °C. Even when using the best nickel-chrome steels, this could only be tolerated by making the turbine blades hollow and pumping cooling air from the compressor through them. Since then, improvements in design and materials have led to a steady increase in inlet temperatures in successive models. The ability to withstand gas temperatures as high as 1300 °C is achieved by additionally coating the blade with a ceramic. Ceramics based on silicon carbide and silicon nitride can withstand higher temperatures than the metal beneath and act as a thin insulating layer.

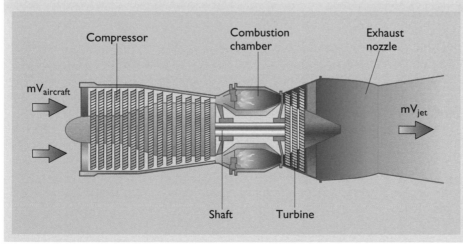

Figure 8.18 Basic components of a turbojet engine (source: Rolls-Royce)

Post-war developments

The jet engine was of little practical use during World War 2. Adolph Hitler did not appreciate the potential of the Me 262 as a fighter plane until it was too late to deploy it. The Whittle engine design was shared with the US (and the Russians under an export agreement) who immediately started their own development programmes. In Germany, Allied troops gathered up the various experimental and production machines as they advanced across the country and spirited them back home to study. Von Ohain ended up in the US.

Figure 8.19 gives a good impression of the sheer power of the jet engine. It revolutionized aircraft design. Even the hastily produced Me262 and Meteor could fly 20% faster than their piston engine equivalents, and on low grade kerosene rather than high-octane petrol. Given the military pressures of the ensuing Cold War both the US and the Soviet Union threw enormous amounts of money into development. By the end of the 1950s their military aircraft were regularly flying at almost three times the speed of sound.

Figure 8.19 Don't try this at home! A primitive gas turbine burning paraffin at the rate of one ton of fuel per hour being tested at Lutterworth Gas-turbine College in 1948 (source: Hulton Archive)

In 1952, Britain unveiled the world's first jet airliner, the de Havilland Comet, which was designed to replace the piston-engined Douglas DC-4 Argonaut. To give some idea of the quantum improvement in performance afforded by the jet-powered Comet, it had a cruising speed of 780 kilometres per hour compared with the Argonaut's 460 kilometres per hour, and a cruising altitude of 11 000 m compared with the Argonaut's 6100 m. The Comet's ability to fly at such high altitudes meant a smoothness of flight above bad weather which the Argonaut, and other piston-engined passenger aircraft of the day, just couldn't match.

The Comet suffered embarrassing setbacks following a series of crashes due to metal fatigue in the bodywork, and in many ways the first commercially successful jet airliner was the Boeing 707 which entered service in 1958. This used four turbojet engines slung under the wings in a manner that made them easily removable. The body of the airliner could keep flying (and earning money) while exchange engines were being repaired or serviced. And even these needed less maintenance than their complex multi-cylindered piston engine predecessors.

Since 1958 the Boeing 707 has been succeeded by ever-larger designs, but essentially of the same layout. These have ushered in a whole new age of cheap mass air travel, with a matching growth in energy demand.

Modern jet engines

Early jet engines had simply concentrated on producing a narrow high-speed jet of gas to push the aircraft forwards. However, for best fuel efficiency, it is better to produce a larger, slower, jet of air.

BOX 8.6 Thrust, Momentum and Kinetic Energy

Which is better – using a jet engine to produce a small amount of high velocity air or a larger amount of lower velocity air?

In order to answer this question, we have to look carefully at the concepts of kinetic energy and momentum. The jet engine turns the *heat energy* of the fuel into the *kinetic energy* of the moving hot exhaust air. As we saw in Box 4.1 in Chapter 4, kinetic energy is proportional to the square of the velocity:

$$KE = \tfrac{1}{2}mv^2.$$

Momentum is the product of mass and velocity. An alternative expression of Newton's Second Law of Motion to that used in Box 8.5 is that the *force is proportional to the rate of change of momentum*.

The thrust force equation can be written as

$$F = mv_{\text{jet}} - mv_{\text{aircraft}}$$

If we first consider a kilogram of air moving at 200 m s^{-1}, this has a kinetic energy of

$$0.5 \times 1 \times 200 \times 200 = 20 \text{ kJ}.$$

Its momentum is $1 \times 200 = 200 \text{ kg m s}^{-1}$.

Next consider a larger amount of slower moving air, 4 kg moving at 100 m s^{-1}. This has the same total kinetic energy, $0.5 \times 4 \times 100 \times 100 = 20 \text{ kJ}$, but its momentum is $4 \times 100 = 400 \text{ kg m s}^{-1}$, twice as large.

The function of the jet engine is to produce the maximum thrust, using the maximum rate of change of momentum, for a given expenditure of energy. The figures above would suggest that it is better to produce a large amount of slower moving air rather than a small amount of high speed air.

This could be done by inserting more blades in the turbine to extract more of the kinetic energy of the gases leaving the engine, delivering it to the shaft running through its centre. In the prop-jet design developed in the 1960s this could then be used to drive a propeller at the front via a gearbox. However, the use of a large propeller limits the aircraft's top speed.

Modern engines for civil aircraft use a large *by-pass fan* which drives air round the outside of the basic jet engine (see Figure 8.20 overleaf). This surround of slow bypass air around the central high speed engine exhaust also has the effect of making the engine much quieter. In a modern by-pass engine design, about 80% of the thrust comes from the fan and only 20% from the hot jet engine exhaust.

Turbofan engines are more efficient than turbojet engines, that is to say that for a given rate of fuel consumption they produce more thrust, and

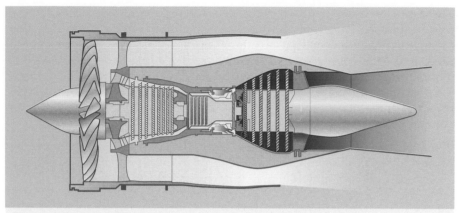

Figure 8.20 Rolls Royce Trent using bypass fan (source: Rolls Royce)

that is what counts to push a plane through the sky. A modern turbofan such as the Rolls-Royce RB211-822 can produce 2.5 times more thrust per kg of fuel burnt than the 1943 German Jumo 004 engine, a product of both the turbofan design and steadily increased turbine inlet temperatures.

Aircraft fuel efficiency is not just a matter of cost, it is critical for the ability to carry out long-haul flights. For example, a Boeing 747-400 flying to New York from Tokyo needs to carry about 145 tonnes of fuel (nearly 40% of the total weight at take-off). 125 tonnes will be burned during the flight, with 20 tonnes kept in reserve fuel. This is completely different to a car. A one-tonne car is only likely to carry 50 kg of fuel at maximum, but of course does have the option to stop and fill up again!

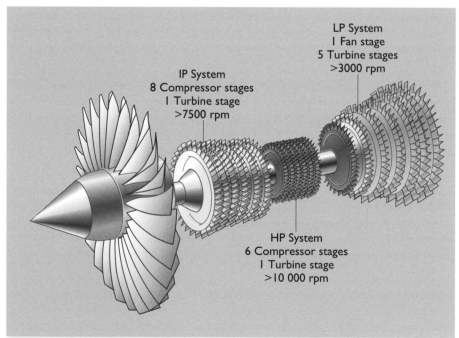

Figure 8.21 Different parts of the jet engine can be connected by concentric shafts running at different speeds. The Rolls Royce Trent has three separate shafts (source: Rolls Royce Education website)

In practice, the engine design can be quite complex. It is desirable that the final stage of the compressor and first exhaust turbine run together at a high speed. This core jet engine then acts as a *gas generator* supplying kinetic energy to the downstream turbines which can turn more slowly on a separate concentric shaft. The Trent engine actually has three concentric shafts linking different sets of fans and turbines. The final five banks of output turbines are used to drive the single bypass fan (see Figure 8.21)

Industrial gas turbines

The ability to extract a large amount of the kinetic energy of the exhaust gases and deliver it to a shaft has meant that gas turbines could be used for purposes which had previously been the province of the diesel and steam turbine. In the 1960s, large gas turbines started to be used in naval ships, where high power to weight ratio is desirable, and then to drive generators in power stations to meet peak loads. The Rolls-Royce Trent engine is also sold in 'marine' and 'industrial' forms. Essentially these are much the same as the aircraft engine, but instead of driving the bypass fan, the shaft drives either a ship's propellers or machinery (usually an electric generator).

Improving power and efficiency

Military aircraft achieve extra thrust by using an afterburner – burning fuel downstream of the turbine. This is extremely noisy and cannot be sustained for long because it is very inefficient.

The simplest way to get more power output of a gas turbine, and make it more efficient, is to go back to the earliest designs and inject water, generating steam to expand through the turbine. This is regularly done in civil aircraft engines to produce the extra thrust for take-off. The increase in performance can be quite considerable. As an example a sample 33.1 MW turbine without steam injection would only have an efficiency of 33%. In a steam injected (STIG) form it could produce 51.4 MW at 40% efficiency. However, to do this continuously requires a continuous supply of high quality water (exactly the same problem as for a steam railway locomotive). This isn't practical for aircraft and is only useful in industry if the plant requires large amounts of process steam.

In practice, the most popular approach in industry has been to use the waste heat from a gas turbine to raise steam in a boiler to drive a steam turbine; this combination was given the name Combined Cycle Gas Turbine or CCGT. This has enabled overall efficiencies of over 50% to be achieved. This is described further in the next chapter.

Gas turbines for cars

After World War 2 it seemed tempting to think that this new gas turbine technology might bring about a whole leap forward in engine design for cars. In 1950, the Rover company produced JET 1, a prototype gas turbine car. Although it showed that such a vehicle was technically feasible the fuel consumption was very poor.

Over the next thirty years, a large number of engine manufacturers experimented with gas turbine designs. Most of these used a very simple gas generator similar to Whittle's original engine. A single centrifugal compressor would compress air to feed the combustion chamber. The hot gases would then drive a single turbine connected back to the compressor by a shaft. The kinetic energy would be extracted by a second single downstream turbine connected via a gearbox to the rear wheels. Most designs featured a heat recovery unit using the hot exhaust gases to preheat the incoming air.

There are a number of fundamental problems that have to be faced. The gas turbine is an excellent device for producing a jet of hot gas. The core gas generator can be made very small and have a high power to weight ratio. However, extracting energy from this to drive the road wheels of a car involves an extraordinary feat of gearing. The turbine may be spinning at 50 000 rpm, but the road wheels may be going at 100 rpm or less. Adding more turbine stages, as in an industrial gas turbine, adds to the bulk of the engine making it difficult to fit in a standard car body. The small gas turbine is also most efficient at the middle to upper end of its power output range. This does not make it very suitable for cars that are only used for short low-power urban trips.

The need to produce car engines with low carbon monoxide and NO_x emissions re-stimulated interest in gas turbines in the 1970s, but this faded away when it became clear that Otto engines fitted with catalytic converters could be made to meet the standards. A 1991 Swedish study (Egnell and Gabrielsson, 1991) concluded that small gas turbines might best be used in hybrid drive vehicles, but they would find competition from Otto engines and Stirling engines.

8.6 The Stirling engine

Figure 8.22 Before the electric fan came the gas-powered fan, driven by a small Stirling engine underneath

The petrol engine, diesel engine and gas turbine are all internal combustion engines, but it would be wrong to think that these are naturally superior to external combustion engines like the steam engine. As was explained in Chapter 6, a high thermodynamic efficiency requires the use of high temperatures. The steam engine is limited in this respect by the physical properties of water. All through the nineteenth century it had a competitor, the Stirling engine, but the thermodynamic performance of this was limited by the properties of the metals of the time. We have seen how the gas turbine in particular has benefited from the development of new metals capable of withstanding high temperatures. The Stirling engine also has considerable potential for development.

The first patent for a practical machine was taken out by the 26-year-old Scot, the Reverend Robert Stirling in 1816. He wanted an engine which avoided the use of high pressure steam, because he was concerned at the number of people killed in boiler explosions. This was a reflection on the poor designs of the time and lax attitudes to safety.

In 1818, he built a full sized pumping engine with a 2.5 metre diameter flywheel and a cylinder 2 metres long, producing just two horsepower. Much of the subsequent development work was carried out by his younger brother James.

Throughout the nineteenth century and early twentieth century, Stirling engines competed with steam engines. The most successful were the smaller ones which could be used by ordinary people without the dangers of high pressure boilers. Thousands of Stirlings were used for farm water pumping and many were in use for 25 years or more. Unlike other engine designs, the Stirling engine can be scaled down to a very small size without serious loss of efficiency. It doesn't have the noise of the internal explosions of the petrol and diesel engines, and a well constructed one can run in almost complete silence. It was thus commonly used for domestic fans, dentist's drills, organ blowers and even fairground popcorn shakers. Many of these applications gave way to the electric motor as widespread electrification spread across the world. Others succumbed to small petrol engine designs developed in the 1920s. This was unfortunate, since new high temperature steels were then being developed which could have increased the Stirling engine's reliability and efficiency.

Principles

The Stirlings' early engines were mainly of the 'concentric piston' type. The essential principle is shown in Figure 8.23.

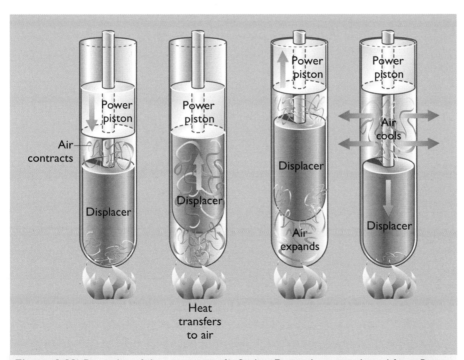

Figure 8.23 Principles of the operation of a Stirling Engine (source: adapted from Ross, 1981)

If a gas is heated, it will expand. If it is then moved and cooled, it will contract. The concentric piston engine uses a cylinder heated at one end and cooled at the other. A loose fitting *displacer piston* moves inside. Its main function is to shift the gas from the hot end to the cold end and back again. When the gas is at the hot end, it will expand, and when it is at the cold end, it will contract. A *power piston* which fits the cylinder tightly moves as the gas expands and contracts, extracting mechanical energy.

The two pistons are driven 90° out of phase by a suitable linkage. A refinement is to make the gas flow through a heat storing *regenerator* usually in the form of a mass of wire gauze or metal plates between the hot end of the cylinder and the cold end. This improves the thermodynamic efficiency of the cycle.

There are a number of alternative forms. The Rinia design, used extensively in modern developments, uses four pistons in an ingenious circular arrangement.

Air was the natural choice of working fluid inside the engine until more refined development took place from the 1940s onwards. For maximum efficiency the working fluid needs to have the right thermodynamic properties. Helium is now commonly used and ideally it should be at high pressures, anything up to 100 atmospheres. As well as working as a heat engine, the Stirling engine can also function as a heat pump if driven by a motor. It can perform as a refrigerator, producing liquid air from a single stage.

The Philips engine

The Dutch Philips company at Eindhoven became interested in portable light-weight generators for valve radios in 1937. They needed to be silent, free from electrical interference and easy to use. They took Robert Stirling's basic concentric piston design and applied new high-temperature steels and auto-engine design.

Amazingly, development went on during the second World War, without the knowledge of the occupying German forces. An almost silent 2.5 hp unit was field tested, hidden under a small cardboard box, powering a small open boat 50 miles around the canals of Holland.

Figure 8.24 An elegant Philips 200 watt Stirling engine-generator set from the 1950s. It ran on either petrol or paraffin and had a string-pull starter. It is about the same size as the valve radios it was intended to power

In 1946, Philips were able to complete the development and produced a prototype engine-generator set of 200 watts output. Compared to late nineteenth-century models they had increased speeds by a factor of ten, efficiency by a factor of 15 and power to weight ratio by a factor of 50. In 1952 they started a small-scale production run (see Figure 8.24 below).

However, by this time Philips realised that the spread of mains electricity and the development of the transistor was undermining much of the need for a small generator. They concentrated instead on manufacturing Stirling refrigerators for producing liquid air. By 1963 a single stage refrigerator had achieved a temperature of 12K, sufficiently low to demonstrate the effects of superconductivity, the complete disappearance of electrical resistance in some metal alloys.

The quest for the Stirling car engine

In 1957, General Motors in the US signed a ten year licensing agreement with Philips. They were interested in developing (a) an outboard motor, (b) a field power unit for the US Army and (c) a solar heated generator for space satellites. At this time the technical development of photovoltaic cells was only in its infancy. Indeed when Arthur C. Clarke first suggested a geostationary satellite in 1945, he saw it as being powered with a solar steam engine!

Over the next ten years General Motors and Philips pressed on with new designs. A 23 kW Philips machine achieved a mechanical efficiency of 38%, making it competitive with diesel engines. However, in 1968, their licensing agreement with Philips ran out and was not renewed, although several large marine Stirlings had been built and shown noise and fuel efficiency advantages over diesels. Shortly afterwards General Motors abandoned all Stirling research, leaving 300 reports unpublished for many years.

Philips continued with some vehicle research. In 1971, they installed a large Stirling into a DAF bus with reasonable success and in 1972 Ford tested a four-cylinder machine in a car. Again, despite promise, this work ceased in 1978. By this time, however, there were over 100 groups world-wide working on Stirling engines; Philips had lost interest and development moved elsewhere, notably to Sweden, where design concentrated on making units suitable for small combined heat and power (CHP) applications.

It must be said that the Stirling is not 100% suited to direct-drive automotive applications. It has an inherent high thermal mass and is thus very sluggish in performance. The best use might be in a hybrid-drive application, where the engine is used to recharge a battery.

The real interest in the Stirling engine is for its ability to out-perform conventional engines on pollution and noise, and, being an external combustion engine, to run on a wide range of fuels. Since the fuel is burned evenly and continuously, the combustion conditions can be optimized for minimum pollution. In 1991 the US Gas Research Institute tested a Stirling engine against the proposed 1995 Californian car emission regulations which were posing considerable difficulties for petrol and diesel engine designers. Yet the Stirling engine achieved the following results: 1/9th of the NO_x emission standard, 1/50th of the carbon monoxide emissions and 1/3000th of the unburned hydrocarbon emissions (indeed in this test there were more unburned hydrocarbons in the 'fresh' intake air than there were in the exhaust!).

Modern developments

The ability of the Stirling engine to act as a low-noise power source for a hybrid drive system has been demonstrated in a rather unusual application. Starting in 1988, the Swedish Kockums shipbuilding company have installed a four-cylinder Stirling engine in a number of Swedish submarines. Although nuclear submarines have no problem running their main engines underwater, conventional diesel submarines have to rely on electric battery power when fully submerged. This Stirling engine can be run underwater and provides enough electric power to keep the batteries charged under normal conditions.

The extraordinary feature of this engine is that it runs on a mixture of diesel fuel and liquid oxygen, which is burned at the pressure of the water on the outside of the submerged submarine. The exhaust gases are then washed out into the sea. The Stirling engine also has the essential quality for a submarine of being a factor of 10 db quieter than a diesel engine.

Another application has shown how, given the right materials, a Stirling engine can be used to produce work from high temperature heat. Experiments in Spain and the US have been made concentrating the sun's rays using a large concave mirror onto an engine driving an electric generator. Although this has demonstrated energy conversion efficiencies of 30% the latest generation of photovoltaic cells have now matched this performance.

There are also intriguing possibilities for very low maintenance engines with virtually no moving parts. The most extreme form of this is the

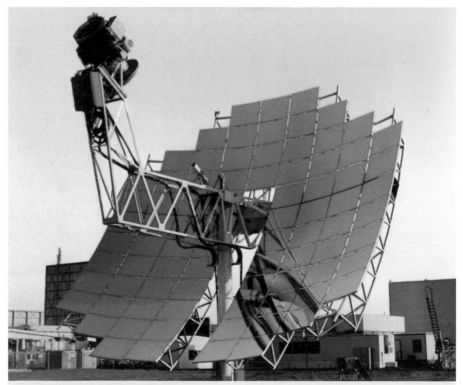

Figure 8.25 Boeing Solar Stirling Engine

thermoacoustic Stirling engine, which could be described as just consisting of a loud noise in a carefully shaped container. The key function of the displacer in a Stirling engine is to move the working fluid from the hot end of the cylinder to the cold end and back again. But this could equally be achieved using a resonant sound wave. A thermoacoustic engine with a thermal efficiency of 30% was described in the magazine Nature in 1999, building on earlier work on thermoacoustic refrigerators, being developed for air conditioning systems. This raises the possibility of a whole range of future devices in which a Stirling engine turning heat into work could be coupled to a Stirling refrigerator to make a heat pump. The only moving parts might be a loudspeaker-like device to start the oscillation and possibly a diaphragm to separate the two halves. It is only by using modern computer modelling techniques that the design of such devices has become possible.

8.7 **Conclusion**

This chapter has looked at a number of different engines developed from the end of the nineteenth century onwards. Chapter 6 has described at the steam engines that preceded them. One might ask 'Which is the best engine?' The answer is that they all have different applications, and one rather than another was used at different times through history.

The reciprocating steam engine, fuelled first by wood and then coal, powered the Industrial Revolution. In railway locomotives, the thermal efficiency rarely exceeded 10%, since they were designed for high power to weight ratio rather than fuel efficiency. The reciprocating engine also has a high starting torque and a reasonable efficiency over a wide range of speeds, which means that it does not need the gearbox necessary for petrol and diesel engines. The best designs of triple expansion steam engines for factories and ships reached an efficiency of 25%. However, there were always considerable practical concerns about boiler explosions, and even today, steam boilers must be thoroughly inspected and even X-rayed to get insurance.

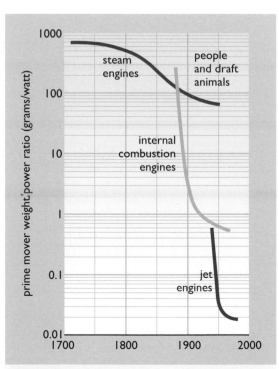

Figure 8.26 Weight to power ratio of different engines (source: Smil, 1994)

The steam turbine, invented in the 1880s had a much better power to weight ratio than its reciprocating engine predecessor (see Figure 8.26). It immediately became the engine of choice for coal and oil-fired steam ships, and for electricity generation, but did not displace the reciprocating steam engine from the railways. It still largely rules in the power station, fuelled by coal, oil or nuclear power. Single machines can be as large as 1 GW output and normal efficiencies can approach 40%.

Stationary internal combustion engines running on gas were introduced in the 1860s and proved popular as power plants for small factories well into the 20th century. Even today small reciprocating engines running on natural gas are used for combined heat and power applications, with electricity generation efficiencies of 25–30%.

The petrol engine as the power unit for the motor car has changed the world since its introduction in the 1880s. There are now about 500 million of them, with typical thermal efficiencies of about 25%. Until the 1930s, the petrol engine was also the main power source for heavy vehicles such as trucks and buses. Until the 1950s it was also the power source for most aircraft. Petrol engines are most useful in the power range 1 kW to about 2 MW.

The diesel engine, introduced in the 1890s, could run on lower-grade fuel than petrol engines. They are useful in the power range 20 kW to 50 MW. Large ship's diesel engines can achieve thermal efficiencies of 40–50%. Even when oil was far more expensive than coal, it would pay to use diesel engines because of their high efficiency and the relative ease of handling the fuel. The development of a practical fuel injection pump in the late 1920s allowed the development of diesel buses, trucks and tractors, with fuel efficiencies of 30–40%, displacing their less efficient petrol predecessors. The replacement of steam locomotives by more efficient diesel ones on British Railways in the 1960s gave a major reduction in overall energy use. Small turbocharged diesel engines are now highly competitive with petrol engines for cars.

The gas turbine, developed in the 1940s to run on kerosene now rules the skies, having totally replaced the piston engine which required high-octane petrol. The modern turbofan design is more fuel-efficient than the original turbojet design used in the 1950s and there is still scope for further development. The industrial gas turbine is now widely used in power stations where natural gas is available, especially when coupled with a steam turbine; the total combined efficiency can reach 50%. Gas turbines are most useful in the power range 100 kW to 500 MW and have a very good power to weight ratio.

The Stirling engine was widely used for small power applications before the widespread availability of mains electricity and small electric motors. As an external combustion engine it can potentially run on a wide range of fuels, or even solar heat. There are also possibilities of very low maintenance engines with an absolute minimum of moving parts. Current devices are being built in the 3–20 kW output range, but efficient devices can be made down to powers of 5 watts.

One of the major questions of the moment is 'which is the best engine for a car?'. In the 1980s this question might have been phrased as 'which will give the least NO_x and CO emissions?' Studies in the 1980s and 1990s commented favourably on steam engines, gas turbines and Stirling engines as alternatives to petrol and diesel engines. Since then, the development of catalytic converters and computer controlled engine management systems has considerably improved the pollution performance of petrol and diesel engines.

Today, we would also be very interested in the overall fuel consumption and consequent CO_2 emissions. The battle is now between petrol and diesel engines (enhanced with computer control and hybrid electric drives) and new technologies such as fuel cells.

References and sources

Anon (1991) 'Stirling Engine Comes Full Cycle', *The Engineer*, 15/22 August 1991.

Backhaus, S. and Swift, G. W. (1999) 'A thermoacoustic Stirling heat engine', *Nature*, vol. 399, 27 May 1999.

Booth, G. (1977) *The British Motor Bus – an illustrated history*, Ian Allan.

Department of Trade and Industry (1973) *UK Energy Statistics – 1973*, HMSO.

European Commission (2000) *EU Transport in Figures – Statistical Pocketbook – 2000*, European Commission: Directorate General for Energy and Transport – in co-operation with Eurostat.

Feldhaus, F. M. (1954) *Die Maschine im Leben der Völker*, Basel/Stuttgart, Verlag Birkhäuser.

GRI (1991) *Stirling Engine Natural Gas Demonstration Program*, Chicago, Gas Research Institute.

McShane, C. (1997) *The Automobile – A chronology of its antecedents, development and impact*, London and Chicago, Fitzroy Dearborn.

Ministry of Power (1962) *Ministry of Power Statistical Digest – 1962*, HMSO.

Monahan, R. (1988) *Development of a Kinematic Stirling/Rankine Commercial Gas-Fired Heat Pump System*, US Gas Research Institute.

Rogers G. F. C. and Mayhew Y. R. (1980) *Engineering Thermodynamics, Work and Heat Transfer*, 3rd edn, London, Longman.

Smil, V. (1994) *Energy in world history*, Boulder CO, Westview.

Stone, R. (1999) *Introduction to Internal Combustion Engines*, Third Edition, Macmillan Press Ltd.

Further Reading

Loftin, L. K. Jnr (1998) *Quest for Performance: The Evolution of Modern Aircraft*, NASA History Office (at http://www.hq.nasa.gov/office/pao/History/SP-468/cover.htm).

Mattingly, J. (2002) Aircraft Engine Design Web Site, http://www.aircraftenginedesign.com/index.html [accessed 6 January 2003].

Nock, O. S. (1980) *Two Miles a Minute - The story behind the conception and operation of Britain's High Speed and Advanced Passenger Trains*, Cambridge, Patrick Stephens Limited.

Paxman, Diesels and Colchester, http://www.nelmes.fsnet.co.uk/paxman/index.htm#introid [accessed 6 January 2003].

Reader, G. T. and Hooper, C. (1983) *Stirling Engines*, E.& F.N.Spon, London.

Rolls-Royce (2002) *How a Gas Turbine Works*, Rolls-Royce Education Web Site, http://www.rolls-royce.com/education/gasturbine/default.htm [accessed 6 January 2003].

Ross, A. (1981) *Stirling Engines*, Solar Engines, Phoenix, Arizona.

Strandh, S. (1989) *The History of the Machine*, Bracken Books, London.

Williams, R. H. and Larson, E. D. (1989) *Expanding Roles for Gas Turbines in Power Generation*, in Johansson, T. B. *et al* (eds) *Electricity*, Lund University Press, Sweden.

Chapter 9

Electricity

by Bob Everett

9.1 Introduction

The previous chapters have looked at wood, coal, oil and natural gas as fuels. These are all *primary* fuels. Chapter 7 has also described 'town gas' made from coal, or more recently from oil. This is a *secondary* fuel, produced because it was more convenient to use. Wood and coal can be burnt to produce heat and drive steam engines. Oil can be used in internal combustion engines. Gas, in its natural or town gas forms, can be put to a whole host of uses. Initially it was used for lighting, but soon found uses for cooking, heating and for driving internal combustion engines. More recently it has been used to power gas turbines.

Electricity is another secondary fuel. Although it may be thought of as something essentially 'modern' most of the basic technology was developed before the end of the nineteenth century. Mains electricity, as we know it today, was commercially unleashed in the 1880s. It was initially produced in coal-fired power stations or hydroelectric plants. (Mains electricity from hydroelectric plants is usually, slightly confusingly, referred to as *primary* electricity.) It could supply lighting with less smell than town gas, drive electric motors that were more convenient than steam, gas or petrol engines and be used in a steadily developing world-wide telegraph network. Émile Zola, writing in 1901, was impressed:

> The day must come when electricity will be for everyone, as the waters of the rivers and the wind of heaven. It should not merely be supplied, but lavished, that men may use it at their will, as the air they breathe. In towns it will flow as the very blood of society. Every home will tap abundant power, heat and light like drawing water from a spring.
>
> Émile Zola, 'Travail', 1901

Given that at the time, it was so expensive that only the very rich could afford to use it in their homes, this must have seemed wishful thinking. As the century progressed electricity was soon seen as not just essential to the running of existing industrial societies, but essential for the creation of new ones.

> Communism is Soviet power plus the electrification of the whole country. Otherwise the country will remain a small-peasant country....
>
> V. I. Lenin, 1920

The development of radio, television and other electrical and electronic equipment since the 1920s has made mains electricity an essential commodity not just in cities but in the furthest corner of the countryside. Once available, electricity slowly took over basic functions of cooking and specialist heating previously catered for by coal, oil and gas. Other uses emerged such as refrigeration and air-conditioning, creating a rising demand to be met by new forms of supply. Zola's words were echoed half a century on with the prospect of cheap nuclear electricity:

> Our children will enjoy in their homes electrical energy too cheap to meter.
>
> Rear Admiral Lewis L. Strauss, chairman of the US Atomic Energy Commission, 1954

Nearly fifty years on, we are still inventing new uses for electricity including computers and the Internet. Consumption is still rising.

This chapter starts by looking back to the nineteenth century, at the basic technology of batteries and mains power generation, at some of the basic physics involved and the early days of electric lighting. Next, we follow the development of the humble electric light bulb and its more energy-efficient relatives through the twentieth century before turning to another key application of electricity, electric traction on road and rail. Then we look at the profusion of different uses for electricity.

This is followed by a description of the expansion of large-scale electricity generation, continuing on from the steam-turbine technology described in Chapter 6 and including a look at the energy-saving possibilities of combined heat and power generation (CHP). Next we look at transmission and distribution, and the development of the National Grid, with a brief account of how the electricity supply industry system is organized and run in the UK.

Then we turn to the generation technology of today, the Combined Cycle Gas Turbine (CCGT) and the expanding use of natural gas in UK power generation, before embarking on a brief tour of the electricity industries of the US, France, Denmark and India.

Electricity is *the* secondary fuel of today, clean and easy to use, and its use has brought many benefits, but is not an unmixed blessing. Fossil-fuelled power stations are one of the leading causes of atmospheric pollution, and the use of nuclear power stations entails other problems – as we shall see in later chapters.

9.2 Making electricity in the nineteenth century

Batteries and chemical electricity

Chapter 4 has described how the battery or 'voltaic cell' was invented right at the end of the eighteenth century. Early nineteenth century scientists found electricity fascinating. They investigated electric sparks and the phenomenon of electric shock, but the main problem was creating a suitable supply on which to experiment. In 1808 the chemist Humphry Davy appealed for money from members of the Royal Institution in London to construct a very powerful experimental battery with 2000 cells. Using this he was able to create a very bright spark which extended over 10 cm when two carbon electrodes were withdrawn from each other. This was the basis of the *carbon arc lamp*. The electric current was flowing through a continuous stream of white-hot particles of carbon. Although technologically interesting, it took another 60 years for electric arc lamps to become a commercial proposition.

Davy also explored the chemical applications of electricity. By passing currents through dilute *aqueous* solutions of metallic salts he was able to produce the metals themselves without the need for smelting. For very reactive metals that react with water, such as sodium, potassium or magnesium, he obtained the metals by passing currents through their *molten*

Figure 9.1 The English chemist Sir Humphry Davy (1778–1829). Most famous for his invention of the miner's safety lamp, he also discovered nitrous oxide, or 'laughing gas', the first chemical anaesthetic used, and explored the use of electrolysis to isolate chemical elements

salts. These processes are known as *electrolysis*. Its widespread use for the production and purification of aluminium and copper did not start until the 1880s.

Inventors explored the electrical and chemical properties of a wide range of materials during the nineteenth century, giving rise to a number of designs of *primary cells*, or non-rechargeable batteries. In 1839 Sir William Grove produced his 'Grove Cell' using zinc and platinum electrodes and a solution of nitric acid between them. It worked well and became a favourite power source for the expanding communications industry – the telegraph. He also demonstrated a 'gas battery' or what would now be called a 'fuel cell' (see Chapter 14), but did not proceed with it because he could see no commercial application. In 1868 the French engineer Georges Leclanché developed a cell based on zinc and carbon. This was an immediate commercial success and 20 years later had been developed into today's familiar zinc-carbon 'dry' battery.

Others were investigating the possibilities of *secondary cells*, i.e. rechargeable batteries, but the commercial need for these did not appear until the birth of mains electricity.

Magnetism and generators

The magnetic properties of electric currents flowing through wires proved equally fascinating. In 1820 Oersted showed that a pivoted permanent magnet would move in the presence of a wire carrying a current. In the same year Ampère founded and named the science of *electromagnetism*. He expounded the theory behind Oersted's discovery. He showed that when a wire carrying a current was placed parallel to a second conducting wire there was a magnetic force between them. Later in the century, the familiar unit of current was named after him. Formally, one ampère (or amp for short) is defined as the current required to produce a force of 2×10^{-7} newtons per metre length between two infinitely long thin wires spaced 1 metre apart in a vacuum. He also went on to experiment with currents flowing through a wire wound around a glass cylinder. Other experimenters found that when iron was used as the inner core it became a powerful magnet, an *electromagnet*, and this could be used to attract or repel other magnets.

By 1831 Michael Faraday had demonstrated that a bar magnet moving through a coil of wire would briefly produce a current. The basic connection between current, magnetism and force was now laid out. A current in a wire could produce a force on a magnet, and a coil moving in a magnetic field could produce a current in a wire.

Figure 9.2 Andre Marie Ampère (1775–1836), French physicist, mathematician and pioneer of electromagnetism

Mechanical energy could be converted directly into electrical energy. By the end of 1831 Faraday had produced a device capable of generating a continuous flow of direct current, the first electric *dynamo* and by 1834 instrument makers were selling the first commercial hand-driven rotary generators. They would have looked very similar to that shown in Figure 9.4, where a small coil of wire is rotated between the poles of a horseshoe permanent magnet made of iron. As the coil is rotated the changing magnetic effect induces a voltage in it. Two metal spring contacts pressing on metal 'slip-rings' on the shaft are connected to the ends of the wire providing an entry and exit path for the current into an external circuit. The version shown will produce an alternating current (AC) flowing first one way and then the other. However, by using a more complicated arrangement of slip-rings, called a commutator, it is possible to produce direct current (DC) that flows in one direction only.

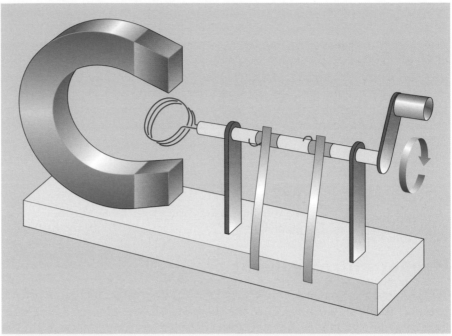

Figure 9.3 A simple generator as might have been sold in the 1830s

Larger generators, with multiple magnets and coils that were turned by waterwheels or steam engines, delivered a more substantial flow of current, and led to the first really practical uses of electricity. One of the first, developed in the 1840s was for the electroplating of metals, in particular silver plating and gilding. This required very large direct currents at low voltages. The 1840s also saw the first demonstrations of powerful electric motors in applications such as a primitive railway locomotive in Scotland, electric boats and electric lathes.

The use that really created a demand for electricity was *lighting*, at this time mainly achieved using inefficient gas flames or oil lamps (see Chapter 3). In our modern world of well-lit streets and buildings, it can be hard to imagine the importance with which this technology was regarded. Practical carbon arc lamps were demonstrated during the 1840s, but it was hopeless trying to run them for any length of time from batteries; they had to be coupled to steam powered generators. Lighthouses were an obvious application. Steam-driven dynamos driving carbon arc lamps were installed at a number of them in England from 1856 onwards.

During the 1860s and 1870s considerable effort was devoted to improving the efficiency of dynamos. Dr. Werner Siemens and Sir Charles Wheatstone looked at using *electromagnets* in generators to replace the heavy horseshoe permanent magnets. These electromagnets could be run from the output of the generator itself. It might seem that such a machine could never be started – no current, no magnetic field. In practice the soft iron pole-pieces held enough residual magnetism to get the process started. Once running they could supply their own magnetic field, i.e. they were *self-exciting*. These machines could be made lighter, cheaper and more efficient than those using the permanent magnet materials of the time. It is perhaps worth pointing out that in the succeeding 130 years, magnetic materials have improved enormously and today small generators and motors using permanent magnets are perfectly viable up to sizes of about 30 kW.

The word **generator** has come to apply to any rotary device for making electricity, often including the power plant to drive it. The word **dynamo** usually describes a small generator using a permanent magnet to supply its field and producing direct current. An **alternator** is a generator that produces alternating current. Its three-phase form, described later, is the normal generator used in modern power stations.

The rise of electric lighting

Carbon arc lamps give off a brilliant white light, excellent for lighthouses, searchlights, and floodlighting. But they were far too bright for most indoor use, and even when used for street lighting there were complaints that they were being 'overlit'.

Nor were they easy to use. A large spark had to be formed by bringing two carbon rods together and then drawing them apart. If they were drawn too far apart, the arc went out. Worse still, the rods slowly burnt away, typically needing replacement after only a few hours. They were no less labour intensive than oil lamps.

Despite all the difficulties, arc lighting created major new possibilities. In 1878 30 000 spectators watched a floodlit football match at the Bramall Lane ground in Sheffield. In the following year Blackpool started its seasonal 'illuminations' when the promenade and pier were lit by six powerful arc lamps, powered by two 12 kW steam generators.

By 1878 the commercial combination of new generators and arc lamp designs was sufficiently attractive to initiate comparative street lighting trials against gas lighting. In many cases the conclusion was that gas lighting was cheaper, but the margin was narrow enough to encourage the design of more efficient and reliable generators. While the town gas companies became worried about losing their street lighting contracts to electricity, they remained confident about hanging on to the domestic and office lighting market, for which the harsh arc lighting was totally unsuitable.

However in December 1878 the Newcastle chemist Joseph Swan demonstrated something far more saleable. He showed that a carbon fibre inside an evacuated glass globe would glow when carrying an electric current. The idea was not new – he had produced an example in 1848 – but it did not last long enough to be useful. The vacuum in the lamp was not good enough. It was only the invention of better vacuum pumps that persuaded him to persevere. Although he had not patented the lamp itself, he did patent a key feature of the use of the mercury vacuum pump that enabled him to get a long life from his filament.

He had produced the incandescent 'light bulb' as we know it, often described as a 'glow lamp' at the time. Unlike gas and arc lamps, they were noiseless, odourless and gave a gentle light comparable to that from an oil lamp.

Figure 9.4 The Newcastle chemist Joseph Swan

Figure 9.5 Thomas Edison (1847–1931) a remarkable inventor and entrepreneur. In the 1870s he worked on automatic telegraph machines before producing a working incandescent light bulb in 1879. He set up an 'invention factory' at Menlo Park, New Jersey, employing a large number of researchers, which is why a large number of inventions are credited to his name. He promoted DC mains electricity supply and a number of US electricity utilities are still named after him

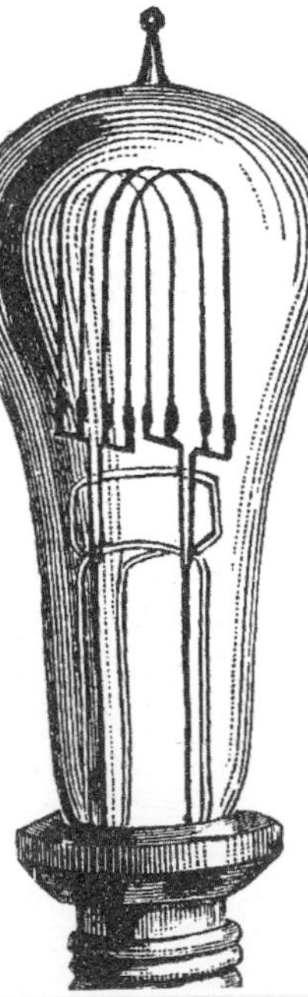

Swan started selling his lamps in 1879 at 25 shillings each (about £80 in modern terms) though by 1881 this had fallen to 5 shillings.

Swan was not the only one with this idea. In 1879 the dynamic American inventor Thomas Edison had produced an almost identical lamp, which he proceeded to patent. A law suit followed. Ultimately, a commercial light bulb required both Edison's and Swan's expertise. In 1883 the two dropped their legal wrangles and joined forces to form the Edison and Swan United Electric Light Co. Ltd.

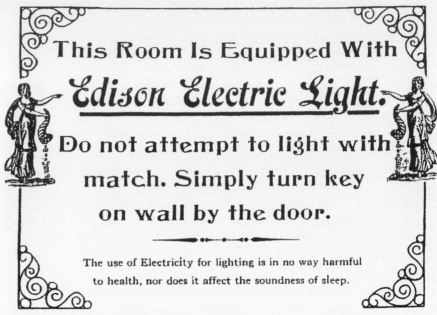

Figure 9.7 Early advertisement for electric light

Indoor electric light with 'glow lamps' immediately became a luxury status symbol. In 1881 the Marquis of Salisbury had electric light installed in Hatfield House, as did the Savoy Theatre in London. Whereas conventional oil lamps (and even electric arc lamps) required continual attention, the incandescent lamps simply ran until their brittle filaments eventually burnt out or broke. They were, and still are, particularly sensitive to the supply voltage. If it is too low, they glow dimly and inefficiently. If it is too high they burn out very quickly. A 5% increase in voltage is sufficient to halve the life expectancy.

Thomas Edison proceeded with entrepreneurial zeal to set up generating systems. The first into service was his demonstration of incandescent electric street lighting on the Holborn Viaduct in London in 1882. This had the distinction of being powered by the first public steam power station in the world. In New York, he formed the Edison Electric Illuminating Company and set up his Pearl Street generating station (see Chapter 1, Figure 1.39), installing six of his new 'Jumbo' dynamos and laying cables in underground conduits. By the end of 1882 192 buildings with over 4000 lamps had been connected. This was the first of a whole series of Edison Supply Companies in the US. In Britain by this time there were over 400 electric lighting installations.

Figure 9.6 An early light bulb

The rush of developers to promote systems led to the same problems that the gas industry had encountered earlier in the century, involving rights to dig up the streets to lay cables. In 1882 the UK Parliament passed The Electric Lighting Act requiring that private electricity suppliers who wished to lay underground cables could only do so with the permission of the local authority.

Figure 9.8 Interior of Brighton Power Station 1882

Initially each lighting system had its own generator, most of which were very small. Brighton Power Station in 1882 (Figure 9.8) had a ramshackle appearance and consisted of a number of 'portable' agricultural steam engines driving separate generators, which were probably each of about 10 kW output.

AC or DC?

The question of whether to use alternating current (AC) or direct current (DC) was a matter of great debate in the early days of mains electricity supply. Thomas Edison was a fervent supporter of DC supply. It could be provided by a steam-driven generator, but augmented with a large battery, carrying out the same storage function that the gasholder did for town gas. This could keep the supply running if the generator had to be shut down for maintenance and could also cope with sudden peak loads.

A DC supply was also convenient for charging other batteries, such as those in the battery-electric vehicles that started to appear in the 1890s. DC was the preferred supply for variable speed electric motors such as those used on trams and early electric trains. Edison was not afraid to use smear tactics to promote DC. In 1889 he allowed the New York prison authorities to use his laboratories to conduct experiments to develop the 'electric chair', but insisted that they should use 'Westinghouse Current', i.e. the AC being promoted by his rival George Westinghouse, hoping that the gory publicity would rub off on him.

Figure 9.9 Nikola Tesla (1856–1943), a Serbian who emigrated to the US, where he worked for the entrepreneur George Westinghouse. Tesla pioneered the use of alternating current electricity to drive motors and developed multi-phase electrical distribution

BOX 9.1 Alternating current

Direct current flows continuously in one direction. In the example in Box 4.5 of Chapter 4, it flowed from the positive to the negative terminals of a battery, through a load such as a torch bulb. An alternating current or voltage is one which goes through a complete cycle of changes periodically in time, reversing direction from positive to negative, as in Figure 9.10.

The time for a complete cycle is called the *period*, normally denoted by the symbol T. The number of cycles per second is called the *frequency*, usually denoted by the symbol f. The unit of frequency is the hertz, named after Heinrich Hertz (see Chapter 4).

Thus

$$f(\text{Hz}) = 1 / T(\text{sec}).$$

The normal frequency of mains electricity in Europe is 50 Hz. In the US it is 60 Hz.

Normal domestic mains electricity in Europe is single phase AC delivered at a nominal 230 volts. Other values for the UK will be found in other books, but in practice it is not important because of wide permissible limits. A figure of 240 volts was chosen in 1946 and later revised down to 234 volts. It has recently been altered to 230 volts −6% +10% (i.e. it can legally be anywhere between 216 V & 253 V).

Obviously the voltage is actually varying all the time, so the figure of 230 volts is an average, a rather special one called the *root-mean-square* or r.m.s. voltage. If the voltage waveform is sinusoidal, as in Figure 9.10, then the r.m.s. voltage is $1/\sqrt{2}$ or 0.70 times the peak voltage.

Chapter 4 described the relationship between power, voltage and current for a DC circuit. As long as the load is a simple resistance (which it is for applications such as immersion heaters or incandescent light bulbs) the same holds for r.m.s. AC voltages and currents.

Power (watts) = voltage (volts) × current (amps)

Voltage (volts) = current (amps) × resistance (ohms)

If the load is a resistance, R, Power = $V^2/R = I^2R$

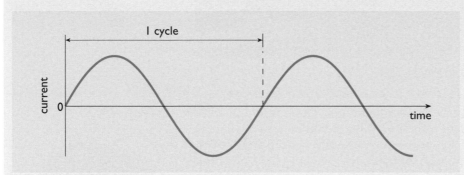

Figure 9.10 An alternating current (or voltage) waveform

But the proponents of AC had good arguments and eventually won the day. AC electricity could be generated at a high voltage for distribution and then easily changed in voltage down to that needed by the consumer using a **transformer**, a set of coils of wire wound round an iron core (see Box 9.2).

Alternating current could be used for running both arc lamps and incandescent lamps, but it was initially not clear how it could be used directly to run electric motors. Then, in 1887, Nikola Tesla (Figure 9.9) showed how two separate AC supplies with the same frequency but displaced in phase could be used to produce a strong rotating magnetic field in a motor, dragging a magnet around with it. Although he started with two-phase electricity, he laid the foundations for the use of three-phase electricity which has subsequently become the normal method of generation and distribution (see Box 9.3). Three-phase motors could be made very powerful and over 90% efficient, eventually leading to their widespread use in factories.

BOX 9.2 **The transformer**

Transformers, as their name implies, are devices for transforming the voltage in an electrical system. They can be used either to 'step up' an input voltage to a higher value, or 'step down' an input voltage to a lower value.

They depend for their operation on the mutual interaction of changing electrical and magnetic fields, so they only work with alternating currents.

A typical transformer consists of two coils of wire, wound in close proximity to each other around a 'core' usually made of thin layers of laminated sheets of a silicon-iron alloy.

The first coil of wire, called the 'primary', is connected to the input source – i.e. the source of electrical energy whose voltage is to be transformed. The second coil of wire, called the 'secondary', is connected to the output – i.e. to the appliance or system requiring a higher or lower voltage for its operation.

When an alternating voltage is applied to the primary coil, it induces an alternating magnetic field in the core. This field in turn induces an alternating voltage in the secondary coil. If the secondary coil has more turns of wire than the primary coil, the voltage induced in it will be higher than in the primary coil; and if it has fewer turns, the voltage will be less. In a 'perfect' transformer, the ratio of voltages is the same as the ratio of number of turns. If the primary coil has N_p turns and the secondary has N_s turns, then the ratio of the primary voltage V_p to the secondary voltage V_s will be:

$$V_p/V_s = N_p/N_s$$

The law of energy conservation still applies, however. The voltage in the secondary coil may be higher than in the primary, but the current will be proportionately less; and if the voltage in the secondary coil is lower, the current in it will be correspondingly higher.

Transformers are very efficient devices. Typically only a few percent of the input energy is lost in the transformation process, mainly in heating the coils and in magnetizing the metal laminations. The energy losses in the laminations can be reduced by using modern silicon-iron alloys incorporating nickel and boron.

Transformers at low frequencies such as 50 Hz can be large and heavy. At higher frequencies they can be made smaller even though they carry the same amount of power. Modern electronic equipment, such as computers, TVs and low energy light bulbs use *switch-mode power supplies*. These convert the 50 Hz to DC using semiconductor rectifiers and then convert this to high frequency AC at about 30 kHz using high voltage transistors. This is then fed to a much smaller transformer to change the voltages to those required. This is one reason why modern TVs are much lighter than their 1970s predecessors.

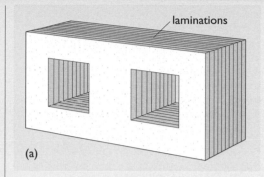

(a)

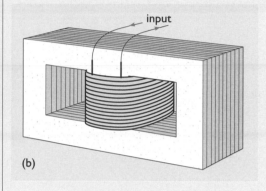

(b)

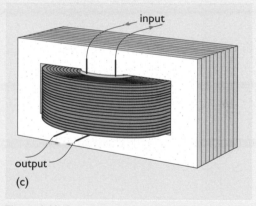

(c)

Figure 9.11 Construction of a transformer

Figure 9.12 A modern 400 kV transformer. The tank on top contains oil for cooling

BOX 9.3 3-phase alternating current

A three-phase AC supply has three main conductors and a central neutral or 'star' wire. The three separately generated voltages between the star wire and each of the three conductors are identical but one third of a cycle out of synchronism with each other (see Figure 9.13). Differences in electrical phase are expressed in degrees, taking a whole cycle as being 360°. Phase 2 is shifted by 120° relative to Phase 1, and Phase 3 is shifted by a further 120°.

This sounds complicated but has immense versatility. A typical 3-phase distribution cable in a residential street has a potential difference of 230 volts r.m.s. between the star wire and each conductor. It is normally described as being rated at 400 volts AC because there is a difference of 400 volts between any one of the phase wires and another. A factory would use all three phases together in electric motors for machines. An individual house would tap off a 230 volt single phase supply between any one of the phase wire and the central neutral wire.

An electricity pylon normally carries two sets of three phase wires hanging on long insulators, one set on each side while the star wire runs along the top. If the currents are perfectly sinusoidal and the loads on all the phases are properly balanced, then theoretically all of the current is carried in the phase wires. The star wire can be made quite thin to simply act as an earth.

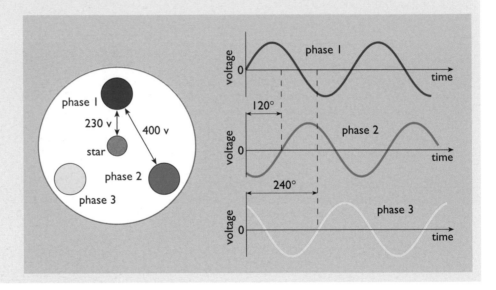

Figure 9.13 Voltage-time waveforms of 3-phase electricity

In 1888, Tesla went on to develop an induction motor which could be used on a single phase AC supply. He showed that these could be made very small, just a fraction of a horsepower, and tolerably efficient (about 50%). This opened the door for a host of new electrical machine tools and domestic appliances, such as an electric table fan, first demonstrated in 1889.

Early power stations had used separate generators to supply different loads. Experiments showed that it was safe to connect all the loads together and supply them from multiple 3-phase AC generators connected together in parallel. This discovery was a key to the development of the modern electricity system. However, whenever a new generator is connected to the system it must carry out a *synchronization* process. It will be run up to exactly the right speed to produce electricity at the right frequency (50 Hz in the UK) and then finely adjusted to that its output will be produce at exactly the right *phase* with the Grid. Only then are the switches thrown to connect its output. Today, this is done automatically under computer control.

BOX 9.4 **The 3-phase alternator**

Figure 9.3 above showed a basic generator, with a coil of wire rotating between the poles of magnet. The modern power station alternator, producing 3-phase electricity is based on a slightly different arrangement.

In its simplest form it can be thought of as a bar magnet (the **rotor**) which is rotated between a set of three fixed coils which together comprise the **stator** (see Figure 9.14). With the wiring connected in what is known as the 'star' connection, the inner end of each coil is connected to the neutral wire of the 3-phase system and the outer ends form the three phase wires. As the magnet is rotated, so a 3-phase voltage is produced as shown in Figure 9.13.

In a more practical alternator, the bar magnet will be replaced by an electromagnet, a coil of wire carrying a large direct current. This is derived from a smaller separate DC generator called the **exciter**. The stator will take the form of a very complex set of windings mounted on an steel frame. The final generated power, which may run to currents of hundreds of amps at voltages of 11 kV or more, emerges from the windings of the stator. Unlike the generator in Figure 9.3 this enormous current doesn't have to pass through a slipring.

Modern power station alternators can be very large indeed, up to a gigawatt in output, and are highly efficient (typically 98.5% at full load). Even so, dissipating the other 1.5% of heat losses may require pumping a coolant (water or even high pressure hydrogen or helium) between the windings.

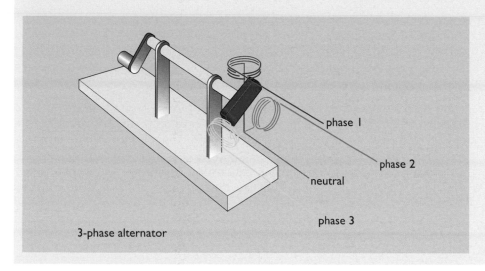

phase 1

phase 2

neutral

phase 3

3-phase alternator

Figure 9.14 The 3-phase alternator

High voltage or low voltage?

By the 1880s, the requirements of arc lamps had already set the normal mains supply voltage at between 100 V and 240 V. This was not ideal for filament lamps, but there were trade-offs to be made when it came to distribution wiring (see Boxes 9.5 and 9.6). Higher voltages allowed thinner wires to carry the same amount of power. They also allowed thinner pins to be used in plugs and sockets.

Early electrical wiring was not particularly safe and electric lighting can only have been slightly less of a fire risk than gas. Distribution voltages of up to 250 volts, although dangerous, were rarely actually lethal. Over the years, safety standards have improved enormously, especially after the introduction of PVC cable insulation in the 1950s. Although an electric circuit only requires two connections, a normal UK mains socket has *three* pins. The live pin carries the high voltage and the neutral pin acts as the current return and will in practice be at a voltage close to earth. The earth pin is what it suggests, a safety connection genuinely at the

BOX 9.5 Choosing a lamp voltage

The current taken by a lamp will be equal to its power divided by the voltage:

$$I = P/V$$

A 100 W lamp designed to run on 230 volts will therefore take 100/230 = 0.435 A

Its resistance when running will be V/I
= 230/0.435 = 529 Ω

Many modern lamps, however, are designed to run on 12 volts. Examples are car headlamps and decorative tungsten-halogen lamps run from the mains with transformers.

At 100 W a tungsten halogen lamp will take 100/12 = 8.3 A

Its resistance = V/I = 12/8.3 = 1.44 Ω

Thus a mains lamp must be designed to have a filament with a resistance (when hot) over 350 times larger than for a 12 volt lamp of equivalent power.

The low voltage lamps have short thick filaments that are quite robust. Mains lamps have filaments that are long, thin and consequently fragile.

However, the 12 volt lamp takes nearly 20 times more current than its mains counterpart. In practice, the supply wiring for 230 volts can be made much thinner than that for an equivalent 12 volt system. It is not just the wire that is affected, but also the size of the pins in plugs and sockets. UK domestic mains wiring is based on plugs and sockets capable of carrying 13 amps or a load of about 3 kW.

---

voltage of the ground. Modern wiring regulations insist that all exposed metal surfaces on electrical equipment, as well as water taps in kitchens and bathrooms are securely earthed. Bare wires at more than 35 volts above earth potential must be kept safely out of reach.

The choice of distribution voltage is much more serious when it comes to transmitting power over long distances. This became important as soon as hydroelectricity was considered for electricity generation since suitable sources were unlikely to be conveniently situated in town centres. Box 9.6 explores a sample calculation on this topic. Attempting to transmit any substantial amount of power at the normal local distribution voltages requires the use of impossibly thick cables. This severely limited the capabilities of early DC systems, as their economic radius of operation was put at about 2 km (see Electricity Council, 1987).

The need to make efficient use of copper for wires was highlighted when in 1887 a French copper syndicate, seeing a boom in demand, cornered much of the world's supplies, forcing up the price. Today, domestic wiring still uses copper, but transmission lines are likely to use aluminium, more abundant but still expensive.

Today, distribution at voltages of 400 kV or more is considered normal. The early development of high voltage distribution seems quite hair-raising by today's standards. One of the pioneers was the brilliant young engineer Sebastian Ziani de Ferranti who started experimenting with electricity while still at school in Kent. In 1885, at the age of 21, he was called in to sort out the Grosvenor Art Gallery system in Bond Street, London.

In 1883 Sir Coutts Lindsay, its owner, had been persuaded to install electric lighting. His neighbours immediately asked if they, too, could have a supply from his generator. Seeing a good thing, he formed an electricity company. By 1885 he had 1 MW of generating capacity with 30 000 lamps connected. The distribution was via bare wires at 2000 volts AC strung on poles across the rooftops right across central London. Every customer had their own transformer to reduce the voltage, but there was considerable confusion as to exactly how it should be connected. The whole system was so unreliable

BOX 9.6 Taking power to the people

Let's consider a town whose normal distribution voltage is 230 V. It needs 1 MW of power to be supplied from a dam 50 km away. If we are prepared to lose another 10% or 100 kW of power in transmission, how thick will the wires have to be?

The necessary calculation has a number of steps:

Step 1   Find the current that must flow to the town to supply 1 MW at 230 V:

amps = watts/volts

1 000 000/230 = 4350 A

Step 2   Find the voltage drop along the cables, if 100 kW of heat are developed when a current of 4350 A flows:

volts = watts/amps

100 000/4350 = 23 V

Step 3   Find the maximum permitted cable resistance, for a voltage of 23 V with a current of 4350 A (you may like to look back to Box 4.5 in Chapter 4):

ohms = volts/amps

23/4350 = 0.0053 Ω or 5.3 mΩ

Step 4   Using data for different sizes of copper wire, choose the one with a sufficiently large diameter:

Wire diameter /mm	Resistance of 100 km of wire /mΩ
20	5500
50	880
100	220
200	55
500	9
700	4
1000	2

Answer We need to choose a cable with wires 700 mm in diameter!

This would require 3500 tonnes of copper!

An alternative would be to use a higher transmission voltage. Transmitting at 23 000 volts (23 kV) instead of 230 volts would only require 43.5 amps instead of 4350. The aim of only losing an extra 100 kW as heat could be achieved by using a cable with conductors a mere 6.5 mm in diameter, and we would only have to buy 300 kg of copper.

that Lindsay had to appoint Ferranti as his chief engineer. Within months he had completely revised the distribution, ensuring that all of the transformers worked in parallel across the mains, supplying a constant voltage, vital for the proper operation of incandescent lamps.

Within another 2 years demand had outstripped supply again. The solution was to build a new steam turbine power station, but the site was 10 km away next to the Thames at Deptford. Ferranti chose to transmit the power to the Grosvenor Gallery at 10 000 volts. This was beyond the manufacturing skills of the cable manufacturers of the day, so he was forced to develop his own, insulated with wax-impregnated paper surrounded by an outer iron casing. It was too thick to be made in coils and was made in 6 metre lengths, the maximum that could be carried through the streets by a horse and cart. Since it had to be buried in the public streets, the Board of Trade were naturally somewhat concerned about public safety. Ferranti assured them that the earthed outer conductor made it perfectly safe and arranged a demonstration. His assistant Harold Kolle volunteered to hold a chisel while it was hammered into the live cable by another assistant (see Figure 9.15). Fortunately the test was deemed successful and permission to lay the cable was granted. It remained in service for 40 years.

Figure 9.15 Sebastian Ziani de Ferranti (1864–1930) testing his 10 000 volt cable in 1890. His assistant (with a look of trepidation) is holding an uninsulated chisel as it is driven through the live cable

Figure 9.16 Deptford Power Station in south-east London: interior (left); exterior (right). Designed by Ferranti and opened in 1889, Deptford Power Station was the first to produce a high-voltage output (10 kV) for transmission

The technology for high voltage transmission developed rapidly. In the US, the Niagara Falls hydro-electric project was started in 1895 with 3-phase transmission at 11 000 volt over the 35 km to Buffalo. By the 1920s overhead lines reached voltages of 220 kV in the US and underground cables were available for use up to 132 kV. This laid the technological basis for the construction of the UK National Grid in the 1930s (see below).

Metering and tariffs

Electricity was originally synonymous with 'electric lighting' and contracts were drawn up in terms of providing lighting to so many streets for a year. Entrepreneurs like Edison provided both the incandescent lamps and the electrical generation plant. If they improved the performance of either, it would be to their commercial benefit in providing the 'energy service' of illumination for a fixed price. This was fine as long as the consumers didn't want to use the electricity for anything else.

Forty years earlier in the 1840s, gas companies had similarly charged for gas on the basis of the number of gas-lamps in a house. As soon as other uses, such as gas cooking, were introduced it became necessary to develop proper gas meters and legislate that gas was to be sold by its energy content.

The same happened with electricity. It became a 'product' metered in Board of Trade 'units' of 1 kWh. Consumers were free to use it for any purpose. Whether they bought the most efficient light bulbs or the cheapest was up to them.

'Rotating disk' meters were introduced in 1889 and the basic design has lasted through to the present day. They are only now being replaced by electronic meters.

The actual way that a charge is made is known as a *tariff*. There are many different forms. Although we might think of 'buying electricity' as just a matter of kilowatt-hours, we are also paying for the transmission wires and the privilege of having a connection. The actual costs of generation only make up about a half of a UK domestic electricity bill today. The connection costs have to be borne whether or not the consumer takes any electricity at all. So, a normal electricity bill consists of a *standing* charge, a fixed quarterly or monthly charge and a *unit rate*, of so many pence per kilowatt hour. More complex tariffs are available, the most common being one that offers cheaper electricity at night. Industrial tariffs can be very complex, featuring not only cheap rates at night and during the summer, when demand is low, but compensating with very high prices during winter evenings, when demand can be very high.

Figure 9.17 The Austrian chemist Carl Auer von Welsbach (1858–1929), who specialized in studying the 'rare earths' using spectroscopy, the fine details of the light given off when chemicals were heated in a flame. He invented the gas mantle, the lighter flint and the first metal filament light bulb

9.3 The continuing development of electric lighting

Gas fights back

Just as the electricity suppliers were beginning to think that they could compete with gas lighting, the Austrian chemist Carl Auer von Welsbach found that thorium and cerium oxides would glow brilliantly in a gas flame. In 1886 he developed the fabric gas mantle, a small fabric sock coated with these chemicals to be fitted over a gas burner. Although the fabric burned away, the residue retained sufficient strength to form a bright glowing mesh. It was extremely fragile, but it improved the lighting efficacy by a factor of about four over the normal gas lamps of the day. By 1895 gas mantles were selling in Britain at the rate of 300 000 per year. This invention allowed domestic gas lighting to last well into the twentieth century, a serious blow to the electricity industry which was forced to diversify into other applications. The gas mantle is still widely used today and appreciated by the 2 billion people world-wide who do not have electric light and are still dependent on lighting using lamps running on kerosene or liquefied petroleum gas (see Figure 9.17).

Although the gas mantle must rate as a really important invention in energy efficiency, it has a pollution price. Thorium is radioactive and US Environmental Protection Agency has recently had to spend several million dollars to remove radioactive residues left on the site of an old gas mantle factory in New Jersey.

Improving the incandescent light bulb

Then in 1904 Welsbach redressed the balance by developing an electric incandescent lamp with a metal filament, using osmium. This could be raised to a higher temperature than carbon filaments, producing a brighter light and improving the lighting efficacy by about 50%.

Figure 9.18 The competition – a modern LPG lamp. The gas mantle is just as important for the 2 billion people world-wide who use fuel-based lighting today as it was in 1886

Today's incandescent lamps have filaments made of the metal tungsten. This was first introduced by Coolidge in the US in 1907. It was capable of running at even higher temperatures and efficacies than osmium. Tungsten is normally very brittle and it required considerable research to be able produce it in a wire form.

The next step was to fill the lamp with an inert gas. Swan and Edison had gone to considerable effort to get a high vacuum inside their lamps, but noticed that the glass bulb slowly became obscured in use by an internal black deposit, a product of the vaporization of the hot filament. In 1913 Langmuir in the US suggested filling the lamp with a gas such as argon, slowing this evaporation and allowing filaments to run hotter and brighter still.

Finally, in 1934 came the 'coiled-coil' lamp, in which the filament is made as a fine coil and then coiled on itself. This reduces convection of heat in the inert gas. The everyday 'light bulb' (also known as the *general lighting service* lamp or GLS) had almost reached its current state of performance. It remains almost unchanged today, over 60 years later, apart from improvements in its mass production.

This is not to say that development has halted. In 1960 the miniature *tungsten-halogen lamp* was introduced. This has a trace of a *halogen* element, such as iodine or bromine, added to the gas filling. This combines with any evaporated tungsten and makes sure it is precipitated back onto the filament allowing it to be run even hotter and brighter, increasing the efficacy by about 20%. In order to work, the glass bulb wall must be very hot, at least 250 °C, so the lamps are much smaller than conventional incandescent bulbs and are made of quartz glass to resist the high temperature.

The fluorescent lamp

Chapter 3 has introduced the terms lumen and efficacy for different light sources. Edison's original incandescent lamps had an efficacy of under 3 lumens per watt of electricity consumed. By 1920 the figure had been improved to almost 12 lumens per watt and they had re-asserted their price advantage over gas for street lighting. They also replaced carbon arc street lights after the First World War interrupted supplies of carbon electrodes from Germany, the main supplier. The dominance of the incandescent lamp for street lighting wasn't challenged until the 1930s brought another contender – the fluorescent lamp.

In 1867 the French physicist Henri Becquerel (the discoverer of radioactivity) showed that an electric current would flow through low pressure mercury vapour inside a long sealed glass tube. In the late 1890s the inventor Peter Cooper Hewitt began developing these to give off a bright bluish-green light. Although useless for domestic lighting, they were widely used by photographic studios. In those days of black and white photography and slow film speeds, they were interested in bright lights, but not too worried about *colour rendering* – the ability to show colours as the eye would see them under natural daylight.

In 1933, the US General Electric Company, which had bought the Cooper Hewitt Company, marketed a new, brighter, mercury vapour lamp, the *high intensity discharge lamp* (HID). This had the high brightness of the arc lamp, but was sealed, did not give off any poisonous fumes, was maintenance-free and had a long life. This was a product to challenge the incandescent lamp for street lighting.

In Europe an alternative, the low pressure sodium vapour lamp, with a more appealing bright orange colour, was being introduced at the same time. The current generation of *high pressure* mercury and sodium discharge lamps was developed in the 1960s, giving light over a wider range of the spectrum (a 'whiter' light) but they still retain their basic blue-green or orange tints. These are now familiar in modern street lights. *Metal halide* lamps are mercury lamps to which other metal vapours are added to modify the colour.

In fact low pressure sodium lamps have the highest lighting efficacies, around 200 lumens per watt, but it is usual to sacrifice efficacy for better colour rendering – a deep red fire engine can appear almost black under low pressure sodium and mercury lamps. This is why modern fire engines carry orange and yellow stripes in order to make sure they can be seen.

The modern 'office' tubular fluorescent lamp also owes its existence to another early twentieth century researcher, the Frenchman Georges Claude. He had found that the newly discovered gas neon would glow a brilliant red in a discharge tube. He also perfected the art of bending the tubes to spell out words. The advertising industry, particularly in the US, loved his red neon and blue-green mercury lamps and was prepared to pay for them. By 1924 there were an estimated 150 000 of them in the US selling for an average of $400 apiece. This specialist application made sure that fluorescent lamps became familiar items.

The problem was to produce a 'white' light. Becquerel's 1867 experiments showed that a mercury vapour arc also gave off ultraviolet light, but if the tube was coated on the inside with a phosphor, such as zinc sulphide, this would be converted to visible light. Different phosphors could be used to adjust to precise colour. It was only in 1938, 60 years after Becquerel's demonstration, that the US General Electric Company, building on Claude's work, produced a commercial tubular fluorescent lamp. This initial version had twice the efficacy of an incandescent lamp and twice the life expectancy.

The basic design has remained much the same since then (see Figure 9.19). It features a sealed glass tube, with metal end-caps. This contains a certain amount of argon gas and a small quantity of mercury, and is coated on the inside with a selection of phosphors. At each end of the tube is a small filament heater. Initially, when the lamp is off and the tube is cold, the mercury will exist as small droplets of liquid. When the lamp is turned on, mains voltage is applied between the two ends of the tube and a special starter circuit energizes the heaters. An arc forms through the argon, and as soon as this happens, the heaters are switched off. The liquid mercury now vaporizes and the mercury arc gives off ultraviolet light, which is converted to visible light by the various phosphors in the coating. A large coil, or *ballast*, limits the current flowing through the tube.

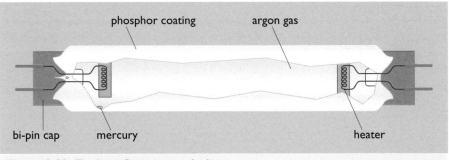

Figure 9.19 The basic fluorescent tube lamp

Since 1938, new design developments and phosphors have enormously improved lighting efficacy, colour rendering and life expectancy of fluorescent lamps. They have become the standard form of office lighting throughout the world.

Even so, by the 1970s the incandescent lamp was still stubbornly popular, though it had remained virtually unchanged in performance for over 30 years. What was needed was a 'compact fluorescent lamp' that could directly replace it. New techniques of bending narrow glass tubing enabled small folded fluorescent lamps to be produced. In 1980, Philips produced its SL series, larger than a normal incandescent bulb, but sufficiently small to fit in most normal light fittings. It was much heavier, with the ballast coil mounted in the base (see Figure 9.20). It used only a quarter of the electricity of an equivalent incandescent lamp, and lasted much longer.

As good as these were, they were almost immediately superseded by electronically ballasted versions. It had been noted that the higher the frequency of the AC applied to a fluorescent lamp, the more the mercury vapour became excited and the brighter the lamp became. But mains electricity only came at frequencies of 50 Hz and 60 Hz. It was not until the development of small, very cheap, high voltage transistors in the late 1970s that this phenomenon could be exploited. The transistors allow the mains power to be converted to high-frequency AC at 35–40 kHz to be applied to the tube. The result has been a lightweight compact fluorescent lamp (CFL) using only 20% of the electricity of a normal incandescent lamp and with 8 times its life expectancy. Phosphors have been carefully chosen to match the colour balance of an incandescent lamp as closely as possible. The fact that these complex lamps can be sold for a price of £10 or less is an amazing feat of mass production.

This development of fluorescent lighting is not without its pollution consequences. Mercury is a toxic chemical and its presence in so many lamps creates concerns about their safe disposal. Since 1980 manufacturers have progressively reduced the amounts in fluorescent tubes. Typically a modern CFL will contain 4–10 mg mercury and a larger high intensity discharge lamp around 200 mg. This is a far cry from the hundreds of grams in Cooper-Hewitt's original tubes. There has been concern that discarded lamps should be specially collected and treated as hazardous waste, but this has to be seen in the context of other large emissions of mercury into the environment, such as from coal-fired power stations.

Figure 9.20 A modern incandescent lamp compared with an electromagnetically ballasted compact fluorescent lamp (centre) and an electronically ballasted CFL (right)

BOX 9.7 Electric lighting today

Today there are a profusion of different electric light sources. Here is a list of the most common, together with typical lighting efficacies. Generally the larger the lamp, the better the performance:

Table 9.1 Lighting efficacies of modern electric lamps

Lamp Type	Details	Lighting Efficacy Lumens per watt[*]
General lighting service lamp	The common incandescent light bulb	9–19
Tungsten halogen	Miniature incandescent lamps, often run on a 12 volt supply from a transformer	17–27
Fluorescent high-pressure mercury discharge lamp	bluish white light, often using for shop lighting	40–60
Metal halide high-pressure mercury discharge lamp	bluish white light, used for street lighting	75–95
Compact fluorescent lamp	fluorescent replacement for normal incandescent light bulbs	70–75
Tri-phosphor tubular fluorescent	Standard type for office and shop lighting	80–100
High-pressure sodium discharge lamp	orange-white light, used for street lighting	75–125
Low-pressure sodium discharge lamp	pure orange light, used for street lighting	100–200

[*]Note: These values are lumens per watt of electricity consumed, i.e. they are not the same as the units in Table 3.1 of Chapter 3 which are lumens per watt of *primary* energy.

Too many street lights may also be a bad thing. Many astronomers have complained of 'light pollution'. Today's city dwellers are no longer able to see and enjoy the stars because of the permanent haze of reflected orange light in the night sky. Émile Zola's 1901 prophecy quoted in the introduction continued '...*and at night it will light another sun in the dark sky, putting out the stars*'. It seems he was right.

The light emitting diode (LED)

Lighting technology continues to develop with the application of semiconductors. Red LEDs were introduced as small indicator lamps in the 1970s. Since then colours have progressed into orange, green, blue and more recently white. Power ratings have increased from milliwatts to watts and the overall brightness and size are now sufficient for use in traffic signals, display boards and bicycle lamps. The luminous efficacies are competitive with incandescent lamps, but LEDs can last 20 times longer, drastically cutting down on maintenance costs for replacements. The challenge at present is to make white LEDs sufficiently powerful and efficient to compete with fluorescent lamps.

9.4 Electric traction

Just as electric lighting can be seen as a clean alternative to its messy oil and gas predecessors, so the use of electric motors to propel cars, buses and trains might seem more desirable than using steam, petrol or diesel engines. Over the past 120 years electric traction has fought it out with its competitors, winning some niche markets and losing others. However, electric drives and controls are increasingly being used in conjunction with diesel and petrol engines, improving their overall capabilities and minimizing pollution.

Electric trams and trains

Relying on lighting as the main outlet for electricity in the 1880s was not good economic sense. The competing gas works could be run steadily throughout the day, storing their output in gas holders. In the evenings, these had no trouble in meeting the peak lighting demands of their customers.

Electricity, in the other hand, had to be made on demand. It was possible to store DC electricity in batteries, but these were extremely expensive. A large amount of electrical generation plant was required to meet only a few hours of actual lighting demand. Today we would say that this was operating with a low **capacity factor**. This is the ratio of the electricity produced over a given period to the maximum it could achieve running flat out continuously. A system with a 100% capacity factor would be one that ran flat out 24 hours a day, 365 days a year. (The term **load factor** is often used in this context though strictly speaking it refers to the demand, rather than the output of the particular generator supplying it.)

Obviously to make the best return on the capital investment in a power station, a high capacity factor is desirable. What was needed were alternative daytime users for electricity, but they had to be willing to pay the extremely high unit prices of the day. The new electric tramways and railways were happy to oblige.

Although Victorian steam powered railways linked almost all the major towns and cities, carrying both freight and passengers, the noisy, smoky, steam engine was not considered suitable for use on urban streets, especially with worries about exploding boilers. So, within towns and cities people either walked or used horse buses or trams.

Electricity had plenty to offer. In 1879 Werner von Siemens demonstrated a miniature electric railway at an exhibition in Berlin. In 1883, Magnus Volk demonstrated his motorized horse tram on the seafront at Brighton, collecting electric current from a 3rd rail laid along the track. Volk's Railway is still running today (see Figure 9.21).

Electric trams systems rapidly spread; Blackpool in 1885, Leeds in 1891 and Bradford in 1892. City corporations were only too glad to get rid of their urban horse populations (see Box 9.8). In the US, Boston, Massachusetts, made 9000 horses redundant

Figure 9.21 Volk's Railway, Brighton – the prototype electric horse tram system of 1883 and still running today

BOX 9.8 What was wrong with the horse?

Given that today many are looking for 'alternative' biofuels for road transport, one might ask why bus and tram companies abandoned a source of motive power that ran on hay and oats in favour of very expensive electric traction.

The first problem was a 'horse shortage'. New railways encouraged the movement of goods and people, but once they arrived in cities, their final delivery involved horse transport. The demand was such that in 1873, a House of Commons Select Committee Enquiry was set up to see whether the country could actually breed enough to keep the cities running.

The second was the logistics of operation. Although a tram or bus normally used two horses, there had to be plenty of spares. An extra 'trace horse' might be used to get it up steep hills, the horses were changed four or five times a day and they were given one day's rest in

every four. By 1898 London's remaining horse tram lines still required 14 000 horses to run their 1451 trams, nearly 10 per tram.

The third was that urban streets were covered in horse droppings. A horse could produce five tonnes of excrement and urine per year, a major health hazard requiring a whole army of 'crossing sweepers' to keep the pedestrian crossings clear.

An electric tram might require complicated wiring with a risk of electrocution, but it was easy to drive, clean, odourless and only required parking space for itself. The electricity generation plant (if actually owned by the tram company) could be placed conveniently out at the end of the line and the stables sold off to property developers. No 'modern' city of the 1890s could afford to be without them.

by electrifying its tram system. Electric traction also increased the speed and capacity of the trams. By the outbreak of the First World War in 1914, the horse tram and bus had virtually disappeared from the major cities of Europe and the US. In their place were the electric tram, the petrol bus introduced from about 1900 onwards, and new electric underground railways.

As the petrol bus developed, the electric tram declined in popularity. They were accused of 'causing congestion'. From the 1920s onwards many tram systems were replaced by buses or rubber-tyred trolleybuses, and even the latter were phased out in the UK. However, trams and trolleybuses survived in many continental cities, such as Amsterdam, Brussels, Vienna and Athens. But time moves on and by the 1990s attitudes to 'congestion' had changed. Towns were prepared to implement far-reaching plans to restrict car access and promote public transport. Could it be a mark of the passing of the 'oil age' that in south London, the tram tracks lifted in central Croydon in 1928 were put back again in 1997 and a new service installed with trams built in Vienna (see Figure 9.22)?

Figure 9.22 A new tram in Croydon in 2002

Electric traction was also key to developing the use of railways in places where the steam engine could not venture, such as long tunnels. The technology of boring long deep tunnels had been perfected by 1870, but at that time the only way of hauling trains through them was with cables driven by steam engines at each end. The combination of reliable electric motors and mains electricity set off an explosion of underground railway building in the major cities of the world from 1900 onwards. It was no longer necessary to fight through city streets choked with traffic; it became possible to move rapidly beneath them. Building underground railways was an extremely expensive activity, but could be financed by property deals. Wherever the lines emerged into the fields on the outskirts of cities, the entrepreneurs were able to buy cheap farmland and immediately sell it to developers for housing, while promising rapid access to the city centres. Not to be outdone, the owners of some existing surface steam railways also adopted electric traction and joined the fray. The result was a growth of suburban sprawl in cities like London and New York that continues to this day. The spread of other cities, such as Los Angeles, was initially due to the availability of rapid electric tram and trolleybus routes before these were replaced by the private motor car.

On main line railways electric traction enabled the use of really long tunnels such as the transalpine Simplon Tunnel in Switzerland, nearly 20 km long and opened in 1906. The Swiss pioneered the use of high voltage AC supplies to trains through overhead wires. This is has become essential to supply the latest generations of high-speed express trains such as the Eurostar (see Figure 9.23) which operates through the Channel Tunnel from London to Brussels and Paris.

As trains become faster, so do their power requirements. The locomotive Mallard, which set the world speed record for a steam train of 202 kph in 1938 had a power rating of 2 MW. The Intercity 125 diesel trains described in Chapter 8 and capable of 238 kph have two 1.7 MW locomotives. The Eurostar, designed for running at 300 kph, requires up to 12 MW of electricity, about the power requirement of a town of 10 000 people. This is delivered to the train through an overhead wire at 25 kV AC. When it has to run over the older 750 volt DC third rail electrification system in the south of England it is limited in speed because it can only draw 3.4 MW from the track.

Figure 9.23 A Eurostar power car

Battery electric vehicles

The battery electric car is often seen as 'the car of the future' and many of the major car manufacturers offer electric versions of their standard models (for example that shown in Figure 4.13 in Chapter 4). Yet electric cars have a long history. In 1899 the land speed record was held by a battery-electric car driven by the Belgian Camille Jenatzy. It was the first road vehicle to exceed 100 kph (see Figure 9.24). It used lead-acid batteries, weighed 1.5 tonnes and was powered by electric motors rated at 50 kW.

Battery electric cars, buses and taxis were widely used during the first 20 years of the twentieth century. They were slow, but in streets still clogged with horse-drawn traffic, this did not matter very much. As the century

Figure 9.24 Camille Jenatzy's 1899 record-breaking electric car, the 'Jamais Contente'

progressed and expectations of speed increased, the battery-electric vehicle was left carrying out the humbler duties of milk float, delivery truck and golf cart, despite repeated attempts to launch new designs of electric car, particularly in response to concerns about urban air pollution.

In the last 30 years there have been advances in magnetic materials enabling the manufacture of lighter permanent magnet electric motors, and modern electronic controls have made the use of **regenerative braking** easier. This is the use of the electric motor during braking to convert some of the kinetic energy of the vehicle back into electrical energy to recharge the battery.

Overall, electric vehicle technology has not advanced in the same dramatic way as that of the internal combustion engine. The tiny diesel Smart car described in Chapter 8 (see Figure 8.15) has to be electronically limited to restrict its speed to 135 kph, yet few modern commercial electric cars are capable of reaching even 100 kph. Typical ranges are around 100 km, yet this is what was being offered by Detroit Electric cars in 1910! This seems disappointing, especially since the basic lead acid and nickel-cadmium battery technologies have been available since the beginning of the twentieth century (see Box 9.9).

There remains a fundamental problem of performance given the limited amount of energy stored the battery. It is possible to travel a long distance slowly, or a short distance quickly, but the ability to travel a long distance at speed remains elusive. Current research is concentrating on rapid charging of batteries. An alternative is to use easily replaceable battery packs. Although modern nickel-cadmium and nickel-metal hydride batteries have longer life expectancies than their lead acid equivalents, purchasing and replacing batteries makes up a significant part of the running costs of any electric vehicle.

BOX 9.9 Rechargeable batteries

The basic form of a rechargeable battery is a number of cells, each with positive and negative plates and with an electrolyte that is either a liquid or gel between them.

The lead-acid battery, invented by Gaston Planté in 1859, uses a positive plate of lead dioxide, a negative plate of lead and sulphuric acid as the electrolyte. The modern form, using an alloy of lead and antimony, was developed in the 1880s. It is now the standard rechargeable battery for starting petrol and diesel vehicles, and has been extensively used for traction applications. Its typical energy storage density is about 30 Wh per kilogram. The principal problem is limited life.

The nickel-cadmium (NiCad) battery, patented in 1900, uses a positive plate of nickel hydroxide, a negative plate of cadmium, and potassium hydroxide as the electrolyte. It has a much longer life than a lead acid battery and is widely used in trains, aircraft and for portable computers. The principal problem is that of disposal, since cadmium is highly toxic.

The nickel-iron (NiFe), battery patented in 1901, is similar to the nickel-cadmium type but uses a negative plate of iron. It was widely promoted by Thomas Edison and has been used for traction batteries.

The nickel-metal hydride (NiMH) battery is similar to the nickel-cadmium type but uses a metal hydride as the negative plate. These are now replacing the nickel-cadmium type.

The lithium ion battery uses metallic lithium as the negative plate. These have a higher energy density and are now being used in portable computers and are being developed for car applications.

The sodium-sulphur battery is a future possibility. This uses liquid sodium and liquid sulphur. Although it potentially has three to four times the energy density of the lead acid battery, it needs to be heated to 300–350 °C before operation.

Electric transmissions and hybrid electric drives

The flexibility and ease of control of the electric motor has led to its use in conjunction with other motive systems. As described in Chapter 8, modern diesel electric railway locomotives use electric transmission. The diesel engine drives a generator, and electric motors drive the wheels. The controls ensure that the system produces the maximum tractive effort and acceleration without the wheels slipping on the rails. There is nothing new in this; an experimental steam-electric railway locomotive was built for the French railways in the 1890s.

A similar system for road vehicles was patented in 1905. Today it would be called a *series hybrid-electric drive*, one where the petrol engine drives the wheels *through* or *in series with* the electric motor. Early petrol-electric buses were popular with drivers graduating from horse buses who could not cope with the complexities of a clutch and gearbox.

Today, the petrol-electric vehicle has reappeared in the form of cars such as the Toyota Prius (see Chapter 1, Figure 1.51). This has a 1.5 litre petrol engine, a 30 kW electric motor and a modest rechargeable nickel-metal hydride battery. The road wheels are driven using the *parallel hybrid drive system*, i.e. *either* directly by the petrol engine *or* by the electric motor *or both*. A computerized control system optimizes the combination. When travelling slowly the petrol engine is cut out and the car travels under its electric motor and battery. At higher speeds, the petrol engine operates and recharges the battery. The motor is also used for regenerative braking when slowing down. In this case the computer-controlled system is designed to minimize fuel consumption and pollution. Toyota has set a target fuel consumption of 5 litres per 100 km (nearly 60 mpg) in European driving conditions.

9.5 **Expanding uses**

Telecommunications and information technology

The use of electricity for communication dates back to the early nineteenth century. Although the railways enabled a good postal service to be set up, there was an urgent need for something faster. Within a few years of Faraday's work with electricity and magnets, the tele-graph – writing at a distance – had been invented by William Cooke and Charles Wheatstone. The railway companies in particular were very interested.

The principle was very simple. A switch was used to open or close a circuit formed with a battery and some wires. Somewhere else on that circuit, perhaps miles away if the wires were long enough, that changing switch position could be detected by deflecting a meter needle. The device shown in Figure 9.25 used a set of five needles to spell out letters of the alphabet and was first tried out in 1837. Its fame was assured after the police used it to arrest a murderer seen boarding a London train at Slough in 1842. They telegraphed ahead to Paddington that he was dressed as a Quaker or rather a 'KWAKER' since the machine had no letter 'Q'. After some confusion the message was understood and the man followed and apprehended. The telegraph soon came to be the key to the efficient scheduling and organization of the railways. Trains ran to timetables under the control of signalmen linked by telegraph.

The five wire system was often unreliable. In 1837 the American Samuel Morse (a painter and sculptor by training) suggested using a single wire telegraph with a buzzer, or an electromagnet which might be connected so it inked a mark on paper. Morse invented a code, which bears his name to this day, consisting of 'dots and dashes' which were formed using short and long switch closures.

Figure 9.25 An early alphabetic signalling device. The diamond shaped display board has the letters of the alphabet written on it. Messages were spelt out using a set of five meter needles mounted across the centre of the diamond

In 1842 the US Congress gave Morse a $30 000 grant for an experimental telegraph to link Baltimore and Washington DC. It was a great success and spurred a rapid spread of Morse telegraphy around the world. It also created a demand for batteries, copper wire and operators. The young Thomas Edison learned about electricity by working as a telegraph operator.

The next major step was the laying of a telegraph cable on the floor of the Atlantic Ocean to join the USA and Europe. A cable laid in 1858 failed after a few weeks operation and another laid in 1865 broke. Finally a successful cable was laid in 1866 and the 1865 one was repaired. Now a message could be sent across the Atlantic in a few minutes, while previously the record was 11 days for the fastest crossing by steamship. By the 1870s there was a world-wide telegraph infrastructure, all powered by batteries.

In 1870 the 'tele-graph' which allowed writing at a distance was joined by the 'telephone' which allowed sounds to be transmitted at a distance. Although its invention is widely credited to a Scottish professor of vocal physiology, Alexander Graham Bell, working in Boston Massachusetts, an Italian, Antonio Meucci, had built a number of working prototypes between 1850 and 1862 but was too poor to patent them.

The telegraph and the telephone relied upon wires for connection. In 1864 Maxwell published his paper predicting the existence of electromagnetic waves that could travel through space without needing wires (see Chapter 4). In 1887 the German scientist Heinrich Hertz made a spark jump across a gap in an electric circuit in response to a powerful discharge in another circuit some metres away. He found that in order to get the best reception it was necessary to 'tune' the crude receiver to the frequency of the transmitter. To this day, his name is commemorated in the unit of frequency, the hertz, or number of cycles per second.

Hertz himself did not believe that long distance radio transmission was possible. Others were more optimistic. The young Italian engineer Guglielmo Marconi successfully established a wireless link across the English Channel in 1899, when he was only 25. In 1901 he demonstrated sending a 2100 mile wireless signal from Cornwall to St. John's Newfoundland and set up the first commercial transatlantic radio link in 1907.

Figure 9.26 Guglielmo Marconi (1874-1937) was born in Italy, the son of an Italian country gentleman and an Irish mother. He began experimenting with radio on his father's country estate in 1895, before taking his equipment to the UK. He opened the first transatlantic commercial radio service in 1907. Later in the 1930s he developed the use of ultra-high-frequency radio and radio navigation beacons

Really practical radio transmitters and receivers had to wait for the development of thermionic valves. Looking somewhat like light bulbs, these allowed the amplification of minute electric signals.

Public broadcasting of news and entertainment did not start for another 20 years. The British Broadcasting Company started regular transmissions in November 1922. Although radio transmissions could be received with simple 'crystal sets', practical receivers required valves. These needed a small amount of high voltage electricity which could be supplied by dry batteries, but a large amount of current at low voltage to supply their heater filaments. If you did not have your own mains supply you would need to use a rechargeable battery that could be taken every week to somewhere that recharged them (usually the corner shop). Understandably this added to the desire for access to mains electricity.

While the introduction of the transistor in the 1950s made the fully battery-powered radio a reality, by then it had been overtaken by an even more power-hungry device, the television receiver. TV transmissions had started in the 1930s, but were interrupted by World War 2. The BBC actually stopped transmission in the middle of a Mickey Mouse cartoon in September 1939. In 1946 they resumed by showing it completely. In 1950 only 350 000 TV licences were sold in the UK. By 1954 the figure had risen to 3 million, largely as a result of the broadcasting of the Coronation in 1953.

A typical valve TV receiver required about 500 watts, while a modern transistorized receiver needs only 100-200 watts. It is not just the receivers that use electricity; a large terrestrial television transmitter may consume a megawatt or more. Terrestrial television transmitters have a maximum range of about 100 km. If you lived further than that from a transmitter, there was no point in buying a receiver.

In 1945 Arthur C. Clarke suggested the possibility of broadcasting TV from a satellite in a **geostationary** orbit. A satellite 36 000 km out in space will rotate around the earth in exactly 24 hours, and if it is above the equator will remain in the same position with respect to the surface. This did not become commercially feasible until the 1980s. A single satellite could direct its transmission to receivers in a 'footprint' up to 1000 km across on the earth's surface below. At the frequencies used, above 10 GHz, electromagnetic waves can be focussed to tight beams using the familiar dishes, allowing the use of very low power signals over long distances. This is essential because satellites are limited in transmission power, being reliant on limited supplies of electricity produced by large arrays of solar photovoltaic panels. (Clarke had originally suggested solar powered steam engines!)

The availability of satellite TV has increased the market for receivers and the desire for access to electricity to the furthest corners of the most rural countries.

Since the 1990s, international communications have been transformed by the use of a combination of point-to-point microwave beams on the earth's surface, and geostationary satellites, and undersea cables using optic fibres rather than wires. A single telegraph wire of the 1890s could carry four simultaneous data channels. A microwave link of the 1990s could carry a thousand or more.

The late 1990s also saw the perfection of the lightweight mobile telephone, working as part of a cellular network of transmitters and receivers all linked together by microwave links. The handset can be small because it only needs to have a range of a few kilometres and a rechargeable battery is sufficient to power it. However, the thousands of cellular phone masts that now litter the country each contain a set of transmitters using 100 watts or more.

The development of the integrated circuit in the 1960s and 1970s enabled the manufacture of practical commercial computers, though still each the size of a large wardrobe. A university or military establishment would have been able to afford two or three. By the 1970s, such institutions began to use long distance telecommunication networks to link computers together.

Figure 9.27 An early IBM PC portable computer. The selling price with a 4.7 MHz processor, 256 kB of memory, one floppy disk drive and no hard disk was $4225 in 1984

These links enabled users to send personal messages (e-mail), and protocols were set up to deal with their handling. Since then, computers have become faster, smaller and cheap enough to become a consumer item and the data links have expanded to become the modern Internet. The early IBM Personal Computer shown in Figure 9.30 is now an 'antique'.

Today, nearly all UK homes have a TV and many have more than one. In many European countries there are now more telephones than there are people. In the past 'landline' telephones were passive instruments that sat unpowered, waiting to ring. Today they are likely to be mains powered containing an answering machine and even a fax machine, continuously consuming two or more watts. Mobile phones have a permanently powered system of transmitters and receivers simply sitting waiting to be used. A more recent phenomenon has been the growth of internet servers, computers permanently switched on, consuming 50 or 100 watts, simply waiting to answer a request. These demands, though modest, all contribute to the rise in electricity consumption of the modern world.

Cooking and heating

Like town gas, electricity quickly advanced from being a lighting fuel to one used for cooking and specialist forms of heating. Electric equivalents of gas appliances appeared very rapidly. By 1890 the General Electric Company in the US were selling electric irons, immersion heaters and an 'electric rapid cooking apparatus, which boiled a pint of water in 12 minutes' – an electric kettle. In 1894 the City of London Electric Lighting Company staged an 'all-electric banquet' for 120 guests to launch their new rental scheme for electric cookers. Although the appliances became available, initially only the rich could afford the cost of the connection, house wiring and the high unit price of the electricity. By 1918 only about 6% of UK homes were wired and that was used almost exclusively for lighting. Prices of electricity fell steeply in real terms during the 1920s and 30s though it remained a far more expensive form of energy than coal or town gas (see Chapter 12 for a comparison of domestic fuel prices through the 20th Century).

Figure 9.28 An early electric kettle produced in 1921

Two decades of promotion had a significant effect. By 1939 about two thirds of UK homes had an electricity supply. In those, almost all would have had electric lighting, 77% an iron, 40% a vacuum cleaner, 27% electric fires, 16% an electric kettle, 14% a cooker and less than 5% an electric water heater (Electricity Council, 1987). Town gas still remained the fuel of choice for cooking, while coal was preferred for heating.

After World War 2 the electricity grid was extended into rural areas of the UK, making mains electricity and electric lighting almost universally available. New manufacturing techniques transformed the heavy cast-iron electric cooker into its white pressed-steel form of today. Today, the competition between gas and electricity for cooking is as strong as ever. Many fitted kitchens are sold with gas rings and an electric oven. They are also likely to include a newcomer, the microwave oven, which first appeared for sale in the US in 1947. It was not until 1972 that a domestic version arrived in the UK. It heats food using a microwave transmitting valve, a cavity magnetron, originally developed for radar. Unlike a conventional cooker which merely applies heat from the outside, this excites the water molecules inside, cooking food right through very quickly. This is turn has created a whole supermarket culture of pre-prepared 'instant' meals.

Although electric fires were widely used as 'secondary' heating in bedrooms, for example, from the earliest days, it was too expensive to tempt consumers away from coal, gas and oil as a main heating fuel in the UK until the 1960s. By this time not only had average unit prices fallen, but the industry was promoting cheaper 'off-peak' tariffs for use with storage heaters. These are essentially large blocks of special heat-resistant brick in an insulated case, which can be heated to a high temperature at night, and then release the heat into the house steadily over the day. These initially sold well, providing a central heating system with a low capital cost, but could not compete against gas central heating once cheap North Sea gas appeared in the mid-1970s. By 2000 90% of the UK housing stock had central heating, but only about 10% used electricity.

Refrigeration

The idea of keeping food fresh by storing it a low temperature in cold stores was introduced in the nineteenth century, initially using bulk ice imported from countries with a plentiful natural supply such as Norway. Mechanical refrigeration was developed during the 1850s. In 1852 William Thomson (Lord Kelvin) observed that when a pressurized gas is allowed to expand into a vacuum it decreases in temperature. This interesting observation subsequently enabled the development of the 'heat pump'. A German schoolmaster, Carl von Linde, successfully used compressed air to reduce the temperature in insulated containers and in 1880, the first cargo of refrigerated meat from Australia arrived in London after a journey of six months. Bulk ice from steam-powered refrigeration plants replaced 'natural' ice in the early years of the 20th century as public health standards tightened.

Shopping was normally done on a daily basis at the beginning of the twentieth century and there was little demand for a domestic refrigerator. Wealthier families might have an insulated 'icebox' to store food, but this was kept cool by ice delivered from a local supplier.

The first electric domestic refrigerators, conventional iceboxes with an added heat pump, appeared in the US in 1912 (see Figure 9.29 for a 1914 example). Initially there was little interest. In 1926 the American General Electric company sold only 2000 refrigerators, but by 1937 sales were up to nearly 3 million. By 1950, 90% of Americans living in towns and 80% in rural areas owned domestic refrigerators (see Weightman, 2001). Other countries have followed suit. By 1975 nearly 80% of UK households owned a refrigerator. These have now been joined by freezers and fridge-freezer combinations. In 2000 domestic 'cold' appliances made up nearly 5% of the UK's electricity demand.

Figure 9.29 An electric refrigerator from 1914 – the large coil of pipes on the top is the condenser

BOX 9.10 **Heat pumps and refrigerators**

A heat pump is a device for pumping heat from a low temperature region and delivering it to a region at a higher temperature. As we saw in Chapter 6, this requires an input of energy from some external source, usually in the form of mechanical work. The best known applications are in the refrigerator (or freezer) and the air conditioner, all used for cooling, though heat pumps can also be used for heating.

The common domestic refrigerator is an insulated box with a heat pump attached. The temperature inside is typically around 5°C, or −10°C to −20°C in a freezer compartment or freezer. The temperature outside is likely to be a normal room temperature of 15–20°C. Since heat will naturally flow from the warm outside through the insulation into the cool interior, the heat pump has to balance this by rejecting an equal amount. It has to move heat from the low temperature interior to a higher temperature on the outside.

The heat pump in a refrigerator uses two properties of liquids. First, when liquids evaporate, they absorb a large amount of energy, their specific latent heat of vaporization (see Chapter 5, Box 5.4). For example it requires about 4.2 kJ of heat to raise 1 kg of water through 1 degree Celsius, but more than 2000 kJ to convert 1 kg of water at 100 °C to vapour at the same

temperature. The second property is that the boiling points of liquids vary according to the pressure. The higher the pressure, the higher the boiling point.

Water has too high a boiling point to be useful in refrigerators – they use special substances called **refrigerants**. Early refrigerators used sulphur dioxide or ammonia, which continues to be used for industrial plant.

Chlorofluorocarbons (CFCs), hydrocarbons which have all their hydrogen atoms substituted by a combination of chlorine and fluorine were developed during the 1930s by Thomas Midgely (who also invented lead additives for petrol). CFCs do not occur naturally but are colourless, odourless, non-flammable, stable, non-corrosive and non-toxic. They were widely used until it was found that the steady build-up of two of them (CFC 11 and CFC 12) in the atmosphere from successive generations of refrigerators was causing damage to the high level ozone layer particularly over the South Pole. They are also powerful agents contributing to global warming. The use of these has now been restricted and they have been replaced by less damaging hydrochlorofluorocarbons (HCFCs) or even simple hydrocarbons such as propane and isobutane.

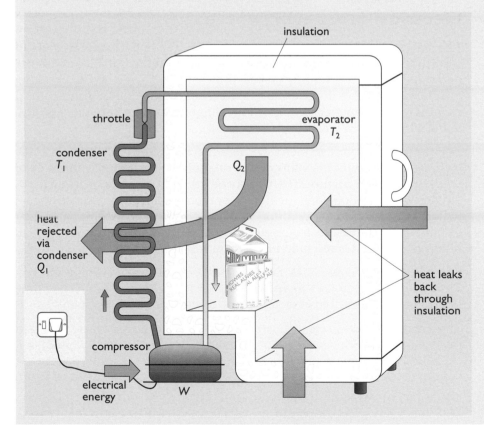

Figure 9.30 Simplified diagram of domestic refrigerator

In a refrigerator such as that shown in Figure 9.30 above, the refrigerant passes continuously through a cycle of processes. Leaving the cold interior as a low pressure vapour, it enters the electrically driven **compressor**. This provides the required energy to increase the pressure. The boiling point of the fluid therefore rises to above room temperature, and it condenses as it passes through the **condenser** – a coil of pipes usually on the back of the appliance. The latent heat of vaporization which it gives up is rejected into the surroundings. It now passes through the **throttle**, which is a fine nozzle, into the **evaporator**, a network of pipes in the walls of the freezer compartment of the fridge. In this low-pressure part of the system the boiling point of the refrigerant falls, and it evaporates, taking up heat from the interior.

This entire sequence, taking in heat from a cooler region (T_2) using an input of mechanical work (W) and rejecting heat at a higher temperature (T_1) is essentially the reverse of the sequence in a heat engine. Reversing all the arrows in Figure 6.9 of Chapter 6 produces a diagram of a heat pump.

The purpose of a heat pump of course differs from that of a heat engine. The aim here is to extract the maximum heat from the evaporator (Q_2) using the minimum work. The **coefficient of performance (COP)** of a refrigeration or air conditioning unit is defined as the ratio of these two quantities: Q_2/W. However, just as for the heat engine, Carnot's Law does limit the performance. Reasoning similar to that in Section 6.3 gives an expression for the maximum possible COP of a heat pump in terms of the Kelvin temperatures[*] of the warm condenser (T_1) and the cool evaporator (T_2):

$$COP_{max} = T_2/(T_1 - T_2)$$

With exterior and interior temperatures of 20 °C and −5 °C respectively, the theoretical maximum COP is about 11. In practice, however, values are very much lower, typically about 2.5, i.e. for every unit of work done by the compressor, 2.5 units of heat are removed from the interior.

A heat pump can also be used to warm a building or to heat water. In this case the condenser is located inside the building and the evaporator is located outside. It can simply be located in the surrounding air, immersed in a nearby pond or stream or buried in the soil. Such ground-source heat pumps are now widely used in the US. As with the refrigerator, one unit of mechanical work can be used to pump two or more units of heat into a building.

[*]Remember: Kelvin temperature = Celsius temperature + 273

Electric motors everywhere

The development of electric motors was a key step in the expanding use of electricity. Small motors allowed the development of labour-saving domestic appliances, such as the electric vacuum-cleaner (1904), washing machine (1908) and dishwasher (1910). These immediately filled a need in middle and upper-class homes brought about by the 'servant shortage' at the end of World War 1 (see Chapter 3).

The three-phase electric motor became the work-horse of large factories from the beginning of the 20th century on. Before then many factories were powered by steam engines. There might be a several separate steam engines, but this would be noisy and inefficient, or more likely, a single steam engine which was linked to many different machines through a complicated set of belts and shafts. Electric motors, though, could be fitted directly to individual machines and individually controlled giving better productivity.

For the electricity companies, factories represented a welcome new market. They often ran day and night with a more or less constant load, thus improving the overall load factor on the electricity supply system.

The trend to increased electrification of industrial processes has continued throughout the 20th century. The extreme is perhaps the fully-automated car production line, where hundreds of electric motors under automatic control now carry out tasks that would once have been carried out by hand.

While electric motors can be extremely efficient on full load, many are only ever used on part load, at poor efficiencies, and often for unnecessary reasons. Some of the worst offenders are the motors used on fans in office air conditioning systems.

Where electricity is used today

Back in 1920 nearly 70% of UK electricity was used in industry and over 10% was used for electric traction. Since then national electricity demand has increased ten-fold. Although it has increased in all sectors of the economy, it has done so faster in the domestic and commercial sectors than in industry. Table 9.2 gives a breakdown of UK electricity use for the year 2000.

Table 9.2 Breakdown of UK Electricity use in 2000

Use	%
Fuel industry use	7.5
Transmission and distribution losses	7.5
Domestic	29.0
Industry	29.0
Commercial and Public Administration	23.5
Transport (including associated buildings)	2.5
Agriculture	1.0

source: DTI, 2001a

Over 100 years on from Edison and Swan, lighting remains an important end use for electricity. In the commercial sector it made up nearly 40% of demand in 1994. In the domestic sector in the same year over a half of the electricity use was for lights and appliances. The increased use of fridges, freezers, washing machines and computers is currently the main driving force behind increasing UK domestic energy use.

9.6 Large scale generation

Thermal power stations

Since the 1880s world electricity consumption has expanded enormously, starting out with a prodigious growth rate of 30% per annum and continuing at a steady 7.5% per annum for 50 years (see Figure 9.31).

In recent decades this world growth rate expressed in percentage terms has declined, but as was pointed out in Chapter 3 demand in the US and France is still surging ahead (see Figure 3.18). The continuing growth in demand has required not just more power stations but also larger ones.

In its early days, the generation of mains electricity was essentially a local affair. Although AC generating had benefits for long-distance transmission, many entrepreneurs favoured DC, for which the economic range of distribution was reckoned to be about 2 km. In the UK DC was initially officially encouraged in preference to AC.

When the First World War broke out in 1914, new munitions factories had to be rapidly constructed and electricity was the natural choice to power them. The military authorities, interested in standardized equipment, were not pleased to find a profusion of different supply voltages and frequencies.

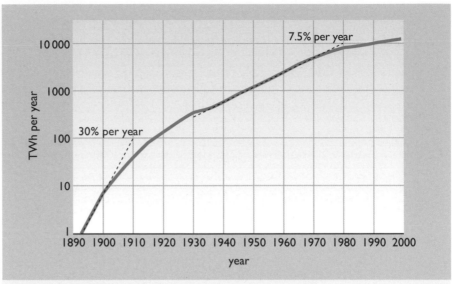

Figure 9.31 Growth in World Electricity Demand (1890–1999)

By 1917, in London alone there were 70 separate companies with fifty different types of system and twenty different voltages. Most of the power stations were small and inefficient. Although the steam turbine had largely replaced reciprocating engines, the average efficiency of UK stations in 1920 was estimated to be below 10%.

This was a time of laissez-faire capitalism in the UK when, like today, competition between electricity suppliers was held to give the maximum benefits to consumers. There was little encouragement for companies to combine to produce economies of scale. They were locked in competition with each other and also the gas companies. This picture was repeated in other countries where cities were peppered with small power stations.

In 1926 a UK government committee set up 'to review the national problem of the supply of electrical energy' recommended that a Central Electricity Board (CEB) be created and given the job of interconnecting the most efficient generating stations in Great Britain by a National Grid of high voltage transmission lines. The story of its construction is described later in this chapter.

Although the stations would remain in private hands, the publicly owned CEB would specify the actual level of generation of the selected stations from a central control room in London so as to achieve the lowest overall production costs. This conflict between ideas of independent electricity generation and enforced centralized control has remained a thorny problem right through to the present day. At the time opponents branded this central control as 'intolerable interference with the management of selected stations'. The press promoted it as implying promises of 'cheap and abundant supplies of electricity' and the Parliamentary Bill setting up the CEB became law at the end of 1926.

This standardized the UK on 50-cycle AC and forced the most inefficient stations to close. By the time the construction of the Grid was completed in 1934 only 140 stations out of a total of 438 were left connected. The net result was that generation costs fell by 24% (see Electricity Council, 1987). Not surprisingly this was seen as a triumph of central planning over piecemeal development.

The nationalization of the industry in 1947 continued the process by combining 560 private and municipal electricity undertakings into a small number of Area Electricity Boards. It was at this time that 240 volts AC was chosen as the 'standard' UK distribution voltage (more recently revised down to 230 volts AC). The coordination of generation continued to be administered centrally. From 1957 onwards the state-owned Central Electricity Generating Board (CEGB) both owned and coordinated electricity generation in England and Wales. In the space of half a century, electricity generation changed from being a local activity to a regional one and then to a state-controlled national one, a situation which continued until the privatization of the industry in 1989.

This centralization process opened the way for the development of larger generating units with higher efficiencies and lower staffing costs. A 300 MW station can produce ten times as much electricity as a 30 MW one, but it doesn't need ten times as many people to run it.

It can be difficult to conceive how large a power station really is in relation to its power output. Battersea power station in London, commissioned in 1934 and in operation until 1982, is a well-known landmark, praised for its architectural style (see Figure 9.32). Its shell, approximately 200 m by 200 m, remains today as a listed building, waiting for a new use. When

Figure 9.32 A mid-twentieth century power station – Battersea in 1971. The 'smoke' is mainly water vapour from the flue gas washing. The station was closed in 1982

first opened it boasted a 105 MW steam turbine and generator, then the largest set in Europe. After World War 2 it was enlarged, bringing its total capacity up to about 500 MW.

During the 1950s, 60 MW was the 'standard size' for a UK power station generator and Battersea was considered a 'large' station. Since then, not only has the demand for electricity grown, but the size of generation plant and stations has also increased. By the 1970s single turbine generator sets of over 600 MW were being used in the UK, and over 1000 MW in the US. A gigawatt generator is enough to supply the average requirements of about 2 million UK households.

These generators were installed in massive coal-fired stations, usually concentrated around the coal mines rather than being local to the consumers (see 'coal by wire' below). This movement of stations away from city centres was also driven by pollution legislation. Urban stations such as Battersea required flue gas cleaning which increased the operating costs. Rural stations could (until the 1990s at least) rely on tall stacks to 'dilute and disperse' the SO_2 and particulates.

As the average size of the plant has risen, so have steam temperatures and pressures and with them the average electricity generation efficiency. In coal-fired stations in England and Wales it increased from about 17% in 1932 to 27% in 1960 and to 36% in 2000. However, this trend has flattened out. Ultimately the efficiency is limited by the chemical properties of steel. High temperature steam is highly corrosive and oxidises conventional steels very quickly. Chapter 6 has described how a modern 660 MW turbine may use steam at 600 °C and 250 atmospheres pressure. This requires the use of expensive high quality stainless steels in the boiler and turbine blades. Most large stations in the UK and US built since the 1960s have limited temperatures to below 550 °C and pressures to 160 atmospheres to give a long station life with cheaper steels. It is a matter of a trade-off of energy efficiency against the capital costs of the plant.

Declining oil prices in the 1950s encouraged the construction of new oil-fired power stations. Hopes that these would be cheaper to operate than coal fired ones were short-lived. Work began on an enormous station at Grain in the Thames estuary in the 1970s precisely at the time when world oil prices soared. It was never fully completed and remains virtually unused today (see Chapter 12).

Nuclear power stations started to appear in the UK during the 1960s; these are the subject of the next two chapters. The early Magnox stations were small compared to the gigawatt-scale coal-fired power stations being built at the time, but the later Advanced Gas Cooled (AGR) reactors and Pressurized Water Reactors (PWR) are of the same scale, powering generators of 500 MW or larger. In the Magnox and PWR types water in the boiler is directly heated by the reactor. Because of the added problems of materials deterioration associated with radioactivity, and the obvious needs to maintain high levels of safety, steam temperatures are limited to below 400 °C (see Chapters 10 and 11).

Natural gas was not really available in quantity in the UK until the 1970s and was little used in UK power stations until the 1990s. Its expanding use since then is described later on in this chapter.

Hydroelectricity

Water power was an obvious candidate for electricity generation from the earliest days of mains electricity. Although capital-intensive, many large scale schemes were brought into use in the early years of the 20th century; almost all of them are still in use. In the UK the bulk of hydroelectric power generation is in Scotland. The Galloway scheme brought 103 MW of generating capacity into operation in 1935. Since then a large number of other schemes have been added bringing the UK's hydroelectric generating capacity to over 1 GW. In addition there is over 2 GW of pumped storage plant (see below).

Developing hydroelectric sources has offered countries the opportunity of freeing themselves from reliance on imported fossil fuels. One of the earliest examples was in the Irish Republic which after gaining its independence from Britain in 1921 did not want to be dependent on British coal. The Ardnacrusha scheme on the River Shannon was built by German contractors for one of the first nationalized electricity boards in the world. The investment represented 20% of the Irish government's budget for the year 1925. The rewards were considerable, its original 70 MW being sufficient to supply 96% of the whole country's electricity demand in 1931. It is a measure of the growth of demand since then that today (even despite some expansion at Ardnacrusha) it only supplies 2% of the Irish demand.

In the US, truly enormous hydroelectricity projects could be undertaken. The Hoover dam project, which was started in 1932 on the border between Arizona and Nevada, has now reached an output capacity of 1.3 GW, about the same as the *total* UK hydroelectric capacity.

Today, the social and environmental effects of large hydroelectricity projects are highly controversial. The Three Gorges Dam project in China, started in 1993 and scheduled for completion in 2009, involves the damming of the country's main river, the Yangtze and the relocation of up to 2 million people. It is planned to produce 18 GW of electricity.

Combined heat and power generation

Traditionally, in the UK, power stations have been seen as making only electricity, yet the generation process produces very large amount of low temperature waste heat. As pointed out in Chapter 3, about a third of the UK's delivered energy is used for space and water heating, i.e. at final use temperatures of less than 60 °C. Combined heat and power (CHP or co-generation) plants produce both electricity and heat at a sufficiently high temperature to be useful. This can enable them to achieve a high overall thermal efficiency.

In the UK at the end of 2001, CHP plant had a total electrical output of almost 5000 MW (usually written as 5000 MW$_e$). Most of this capacity was industrial CHP in a relatively small number of large factories, chemical plants and oil refineries which need both electricity and heat in large quantities (DTI, 2002). These are sufficiently large to operate their own gas and steam turbine power stations, using the waste heat at a range of temperatures. Power plants operated to produce electricity within a large industrial complex and not for external sale are often referred to as **autoproducers**.

Outside the large industrial plant, CHP takes two forms. Small-scale CHP essentially takes the power station to the user, usually in the form of a small reciprocating engine similar to a car or truck engine, but running on natural gas as a fuel. This will drive an electrical generator typically with a power rating between 100 kW and 1 MW output.

In the UK there are hundreds of institutions, especially hospitals, community centres and large hotels, which have a sufficiently large year-round demand for both electricity and heat to warrant investing in their own CHP plant. Although a small-scale CHP unit may only have an electricity generation efficiency of about 30%, less than that of a conventional power station, the ability to use the waste heat makes it more energy efficient overall. This kind of comparison can be expressed in a **Sankey diagram,** which is a flow chart showing the overall energy inputs and outputs of a system. The left half of Figure 9.33 shows the energy flows for a conventional power station such as that used in the example in Chapter 3, Box 3.1. The right half shows the energy flows for a small-scale CHP unit, including the use of waste heat. The actual amount of waste heat that can be used depends on the temperature at which it is required, but for typical heating applications the overall thermal efficiency can be over 80%. Such schemes have been extensively encouraged in the UK since the mid-1980s and are widely used in countries such as the Netherlands and Denmark.

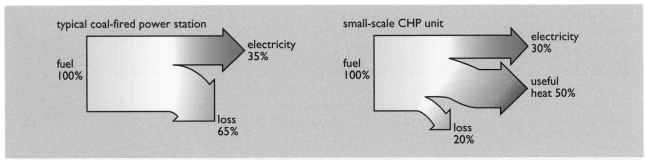

Figure 9.33 Sankey diagram comparing a conventional power station and a small-scale CHP unit

Over the past 20 years there have been repeated attempts to produce really small gas-fuelled CHP units suitable for individual homes. These would need to generate only about 1 kW of electricity. Spark-ignition engines of this size are too small and unreliable, so development has concentrated on Stirling engines (see Chapter 8). Many prototypes have been built, but so far this is a technology still waiting to be commercially proven.

Large scale CHP with community heating

An alternative approach is to distribute waste heat as hot water from existing, or specially-adapted, power stations via thermally insulated pipes to local buildings. Producing water at a sufficiently high temperature to be useful (typically 85–95 °C), does involve some loss in electricity generation efficiency, but this is made up for by the large amounts of heat that can be made available.

The centralized provision of water for heating is known as **district heating** or **community heating**. The heat sources may be central boilers, refuse incinerators or power stations. This has not been very popular in the past in the UK, because there was little tradition of using central heating until the 1970s and since then natural gas has been widely promoted for individual central heating systems. However, it is widely used in Scandinavia. In Helsinki, Finland, 98% of heating is supplied by community heating from a variety of sources.

In the UK only about 2% of homes use community heating. A typical example is the Pimlico District Heating Undertaking, started in 1951 using heat from Battersea power station. 14 MW of heat at 93 °C was distributed via a large insulated heat main to 4000 homes on the opposite side of the river Thames. When the station was closed in 1982 the heat supply was replaced by a large communal boiler (now gas-fired). Today the most visible part of the scheme is a large hot water storage tank, the 'accumulator', built to take up any mismatch between the heat needs of the system and the electricity generation in the power station. It can store enough to keep the whole system running for more than a day.

In 2001 the UK government announced that it aimed to double the amount of CHP capacity to 10 GW$_e$ by 2010. The technical and economic potential far exceeds this, especially through the use of large-scale district heating. A recent report (PB Power Services, 2002) analysed heat loads in a large number of UK cities and the potential for using CHP with district heating. Using a discount rate of 6% (see Chapter 12 for a discussion of what this means) it concluded that the economic potential was 18 GW$_e$ involving the connection of over 5 million homes. The potential for CHP in London alone was nearly 2.5 GW$_e$ – equivalent to the output of five Battersea power stations.

Perhaps the most controversial possibility for CHP is that using the waste heat from nuclear power stations. This has been implemented in many countries including Russia and Switzerland, and the heat mains can extend 20 km or more from the power station (see Kozier, 1999). This technology does, of course, provide heating with virtually no CO_2 emissions, but might face rather severe popularity problems in the UK.

Figure 9.34 The Pimlico Accumulator stores heat for district heating originally supplied from Battersea power station

9.7 **Transmission and distribution**

The National Grid

Generating electricity is one thing, conveying it efficiently to the customer is another. The decision to build the high voltage National Grid in 1926 was a bold one. Only AC stations could be connected and they had to share a common frequency, chosen to be 50 Hz. It was possible to convert from AC to DC and from one frequency of AC to another by using a motor on one system coupled back-to-back to a generator on another. Today this kind of conversion can be done more conveniently using high voltage semiconductors.

It was also perhaps fortunate that the standardization took place so early. The multitude of local distribution systems was very long-lived. It was not until 1962 that the last DC public supply in London was disconnected. Many other countries still suffer from having a variety of supply voltages and frequencies.

Constructing a National Grid is not just a matter of engineering. Obtaining 'way leave' permission for the pylons and overhead cables to be placed on private land required negotiations with thousands of different property owners. These negotiations in the late 1920s were particularly difficult since the landowners themselves may not have actually had a mains supply, or much prospect of getting one in the near future. The CEB employed a number of retired generals and admirals to convince aristocratic rural landowners of the benefits of the Grid. One classic response to a ex-naval negotiator was:

> You mean I'll have to look at that monstrosity passing right in front of my windows, spoiling my best view, interfering with my grazing and ploughing and harvesting, and I won't even get any electricity out of it? What would you have said if I proposed to festoon your battlecruiser or whatever it was with miles of those conductor things and erect a pylon right in the middle of your quarterdeck?
>
> Quoted in Cochrane, 1985

Despite this, by 1933 4800 km of 132 kV transmission lines had been built. As well as enforcing the closure of inefficient generating stations, the Grid gave greater flexibility of operation, especially if there was a failure in any one power station on the network. However, if too much went wrong all at once, there could be progressive collapse of a large portion of the system giving a wholesale black-out over a wide area for many hours. The CEBs response to such a blackout in 1934 was a statement that this had been due to *'a combination of circumstances that is not likely to recur, and there need be no apprehension of any such general failure in the future'*.

Since then blackouts have become larger, more disastrous and politically embarrassing as society has become more dependent on electricity. Widespread grid failures in the UK on Christmas Day 1960 were not very popular. Even less so was the blackout in 1977 that struck New York City for 25 hours during which time there was widespread looting. In 1998 power to the city centre of Auckland, New Zealand was cut off for several *weeks* until large sections of burned out high voltage cable could be replaced.

A major concern today is exactly who is responsible for keeping the lights on, and what to do to put them back on again when they go out. Starting up a collapsed network is known as a **black start** and requires considerable skill in system control.

The initial purpose of the Grid was to interconnect power stations within particular *regions* of Great Britain, and to give local backup when needed. However, experiments in 1936 and 1937 showed that it was possible to run the whole *national* system connected together without disaster. By the following year it was clear that the increasing demand in the south of England could only be met by supplying large amounts of electricity from the north. Starting in October 1938 all seven areas of the Grid were permanently linked together, creating the largest integrated electricity network in the world at that time.

Although by 1950 the Grid had been extended, it was no longer adequate to cope with the 8-fold increase in electricity consumption in the 25 years since its conception. The decision was made to start on a completely new 'Supergrid' initially at 275 kV but upgradeable to 400 kV (though this did not become necessary until 1964). This had pylons whose average height was 42 metres, twice the height of the old 132 kV lines. They also used thicker wires so that at 275 kV it was able to carry six times the power of a 132 kV line.

Once again negotiators had to go in search of permission to lay out another 6400 km of lines. What had been difficult in the late 1920s was now compounded by new planning legislation. New Acts of Parliament had set up National Parks, Areas of Outstanding Natural Beauty and Sites of Special Scientific Interest. The Electricity Act of 1957 also put a duty on the electricity industry to have regard for the preservation of amenity, ranging from the 'natural beauty of the countryside' to 'objects of architectural or historic interest'. It is perhaps fortunate that the 1964 Supergrid crossing of the Thames at Dartford was not in an area noted for its scenic beauty since it required pylons 192 metres high (see Figure 9.35).

Putting cables underground can solve a lot of objections of visual intrusion, but costs can be 20 times those of overhead lines. The 275 kV Supergrid crosses North London almost invisibly under the towpath of the Grand Union Canal, whose water is used to cool the cable (see Figure 9.36).

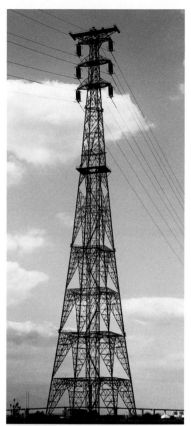

Figure 9.35 The 400 kV National Grid crosses the River Thames at Dartford on pylons each 192m high

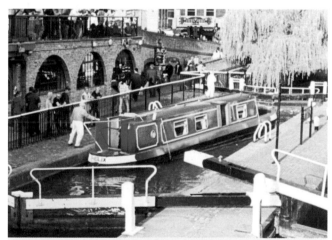

Figure 9.36 The 275 kV Supergrid cables run innocuously through north London under the towpath of the Grand Union Canal

Coal by wire

Between 1950 and 1960 electricity demand in England and Wales doubled again. The problem now was where to locate new coal-fired power stations. The mines were mainly in Wales and Yorkshire, while the growing industrial electricity demand was in the south of England.

A 1 GW coal-fired power station requires about 10 000 tonnes of coal a day to keep it going. That is 10 large, slow, train-loads that were not particularly welcomed by those who wanted to upgrade the railway network for high-speed passenger and freight services. From the 1960s onwards the construction of large new coal-fired power stations has mainly been concentrated close to the mines, and the electricity despatched to the cities over the National Grid. The coal trains only had to make the short journey between the mines and the power station, rather than the length and breadth of the country. The National Grid could thus be thought of as a substitute for an entire railway distributing nothing but coal - 'coal by wire' as it was called.

The Channel link

In 1961 the British and French grids were linked together with a cable under the English Channel capable of carrying 160 MW. This was initially conceived to allow a mutual exchange of electricity since peak demands occurred at different times on the two sides of the Channel. Rather than attempt to synchronize the two grids together the link was made in the form of high voltage DC converted to and from AC at each end. This initial link had to be abandoned in 1982 after repeated damage from ships' anchors and trawler nets but the system benefits had been proven and a new 2000 MW link was completed in 1986. This is laid in underwater trenches cut using a robot digger which crawled across the seabed on caterpillar tracks. Then the cables, each 50 km long and weighing 1700 tonnes, were laid in the trenches from a cable laying ship and buried. Since then the link has largely operated as a conduit for importing French electricity (mostly nuclear) into southern England, rather than on the basis of exchanging power at peak times.

The grid today

In the UK electricity is now mainly generated in very large stations as 3-phase AC at a high voltage of 25 kV or 33 kV. It is then transformed upwards to the level of the Supergrid, 275 or 400 kV (see Figure 9.37). It may then travel hundreds of km to a regional Grid Node, usually visible as a large collection of enormous transformers, where it is reduced in voltage to 132 kV or 33 kV for distribution to local towns and cities, much of it over the original 1930s Grid system. Large industrial users such as steel

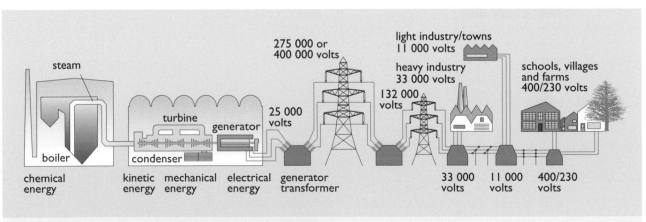

Figure 9.37 A schematic of the UK electricity distribution system

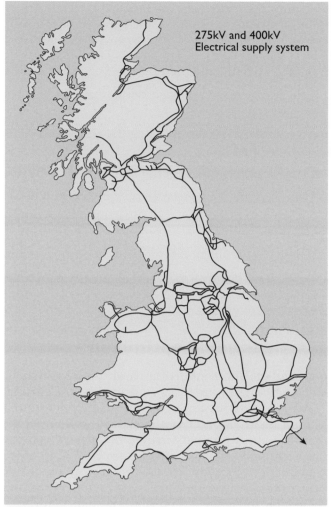

275kV and 400kV
Electrical supply system

Figure 9.38 The National Grid in Great Britain

works are likely to take their supplies at this high voltage. For other users the power is reduced in voltage to 11 kV or 6.6 kV. Small industrial users and others such as large schools and hospitals take their power at this voltage, transforming it down to 400 V 3-phase and 230 V single phase in their own transformers. For domestic consumers the 11 kV supply is likely to be transformed down to 400 volts 3-phase. Individual consumers are given single phase connections into from this 3-phase supply (see Box 9.3). In the UK these cables are usually buried out of sight, but in other countries, the wires very visibly run through the streets.

9.8 **Running the system**

Mains electricity is essentially a commodity that has to be made on demand. Neither the frequency nor the voltage can be allowed to stray outside quite tightly defined limits. In the earliest days a steam turbine driving a generator would be run at a constant speed. Matching the supply and demand would be done by monitoring the mains voltage. If the customers switched on more lights and the demand went up, the supply voltage would fall as a result. This would be sensed (initially by someone continually reading a meter, later by automatic controls) and more steam put through the turbine to hold the voltage up to the required level.

Once large scale grid systems had been set up the problems became more complex. There are four tasks:

1   a technical control problem of keeping the grid voltage and frequency within tight specified limits

2   a more complex problem of keeping supply and demand matched at all times

3   a need to 'keep the lights on' and to carry out a black start in the event of a grid failure.

4   an overall one of optimizing resources in the supply of electricity.

The first three tasks are largely a matter of engineering, but the fourth has been challenging the minds of engineers, economists and politicians since the beginning of the 20th century.

What exactly is being optimized?

To a 'deep green' environmentalist such an optimization might mean supplying adequate 'energy services', such as lighting, heating, etc, using electrical systems, but involving appropriate levels of electricity conservation and distribution of waste heat, as well as actual electricity supply. Even that supply might involve electricity generation at the household or building level using small-scale CHP or photovoltaic cells. Such local generation, provided at low voltage, is often referred to as **embedded generation,** as distinct from supply from conventional power stations via the high voltage grid. The whole system would need to be run to minimize pollution, but doing so at a reasonable financial cost.

At the other extreme, perhaps, a dedicated free-marketeer might see this as all too complicated. Electricity is simply a 'product' and the aim is to maximize sales at minimum financial cost, while obeying any pollution regulations imposed by governments. Conventional economic theory says that the best way to do this is through the use of competitive markets and the best way to optimize such a system is to 'leave it to the market'.

In the middle, there is a long history of state involvement, in the UK and elsewhere. A cheap and reliable supply of electricity is seen as something essential for the growth of an economy. It is something 'too important' to leave to the free market and especially when fuel supplies, be they coal or oil, run short. Also, new technologies such as nuclear power may require large amounts of finance that only the state is likely to be willing to provide. Optimizing the running of such as system requires taking into account not just fuel costs, capital expenditure and the consequences of anti-pollution legislation, but also wider notions of national energy self-sufficiency and expectations of economic growth.

Ownership of the system

In practice, there are many possible mixtures of ownership and control of an electricity supply system. In its earliest days in the UK local private and municipal companies supplied electricity within a few km of their particular power station. After the completion of the National Grid in 1934, the system in England, Wales and Scotland came under central control even though most of the stations and distribution companies were privately owned.

After the 1947 nationalization and the 1957 Electricity Act, electricity supply in England and Wales became what is called a **vertically integrated system**. The generation of electricity was carried out by a state-owned monopoly, the Central Electricity Generating Board (CEGB), which sold it on to state-owned Area Boards, each of whom had a monopoly relationship to its customers. Customers had no choice but to buy electricity from their local Area Board. Although there was some independent electricity generation within large factories, technologies such as small-scale CHP were generally frowned upon.

By the 1980s there was considerable pressure within the European Union to treat electricity as a totally free market commodity, especially since there was considerable cross-border trade between different European countries. In the UK, the Conservative government pushed through the breakup and privatization of the electricity industry in 1989. It was argued that it was bureaucratic and inefficient and that private companies could perform their role more flexibly at a lower cost.

For free marketeers, this privatization was seen as a role model for subsequent similar 'liberalizations' of state-controlled systems elsewhere in the world. Further details of the privatization will be given later in this chapter but for the moment, the key ingredient was that most of the power stations became privately owned and were required to compete with each other. The actual mechanisms of this are complicated but central to the process is the notion of a centrally controlled **power pool**, an hour by hour competitive market in electricity. Electricity is bought in from all the competing power stations and then distributed through the National Grid to consumers. Exactly how the trading is arranged and who gets paid for what is governed by a set of **trading arrangements**. The present power pool in England and Wales is administered by the Office of Gas and Electricity Management (OFGEM). The National Grid Company carries out the day-to-day running of the pool as system operator. Similar power pools exist in other countries and some are international. For example the Nord Pool now covers Sweden, Norway, Denmark and Finland.

Balancing supply and demand

Different power stations have different operating characteristics. A new coal-fired 600 MW plant may have a thermal efficiency of nearly 40%, but, being large, it may take 8 hours or more to reach full power and efficiency when starting from cold. Also, if it is only run at part load, say producing only 300 MW, its thermal efficiency may suffer and only reach 35%. It is obviously best if this kind of station is run continuously at full power.

At the other extreme a small 30 MW 'open cycle' gas-turbine station may have thermal efficiency of under 30%. Its possible fuels, natural gas or kerosene, are likely to be more expensive than the coal used in large stations. However, such a station can be run up to full power in a matter of minutes. This kind of plant is best used for 'peaking' duties and may only be actually run for a few hundred hours a year.

In practice, demand for electricity varies widely, from hour to hour and from season to season. Figure 9.39 shows recent UK system data. On average, over a year, the load on the system is just under 40 GW, but on a summer night it can be under 20 GW and it can rise to over 50 GW on a cold winter evening. The demand can thus be thought of as a continuous 'base load' of around 20 GW, which essentially has to be produced for 365 days a year, with an additional variable demand on top.

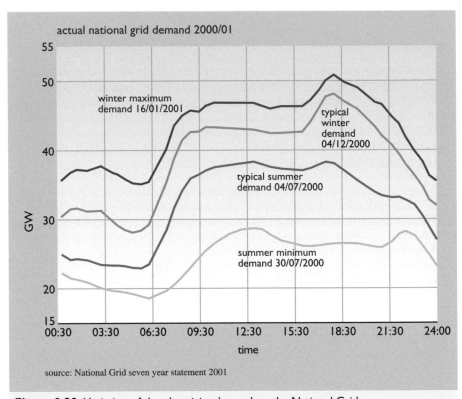

Figure 9.39 Variation of the electricity demand on the National Grid

Under the old nationalized CEGB system, power stations were classified according to a 'merit order'. Those with the lowest running costs, usually the nuclear stations and the largest and most efficient coal stations were assigned to the top of this merit order. It was these that supplied the 'base load'. Next there were a large number of smaller and usually older coal-fired stations with lower thermal efficiencies. These were rated as 'middle merit order' and ran for most of the time in autumn, winter and spring, but shut down in summer. Finally, peaking plant such as simple gas turbines would be brought in to meet the mid-winter peak demands.

Under the current privatized power pool system the decision about which station runs at a given time is governed by a continuous process of competitive bidding under computer control. At any given time a computer model run by the National Grid Company estimates what the national demand will be in the following few hours and invites bids for the supply of electricity. Power station owners reply (or rather their computer programs do), the cheapest offers are accepted and the system is adjusted to bring appropriate stations on line by remote control. The process has required the development of bidding strategies by power station owners to make sure that their particular commercial interests are maximized.

Overall, the system runs in a similar manner to that of pre-privatization days. The stations with the lowest running costs supply the base load, but this is a competitive mix of coal, nuclear and gas-fired stations. The prices bid on summer evenings when demand is low can be very low. As the demand rises, so stations with higher running costs bid into the pool and the electricity price rises. At any time the competitive price is chosen to be sufficient to bring enough stations on line to meet demand.

Peak demands and pumped storage

One particular problem is that of coping with sudden surges in demand. As power station and grid link sizes have grown over the years, so have the potential problems arising from their failure. A 600 MW station might be generating at full power one minute, and the next a vital circuit-breaker could have tripped and it might be completely disconnected. There needs to be a temporary backup, while other stations are brought on-line to cover the deficit.

The growth of radio and television has also produced its own problems by increasingly synchronizing the behaviour of large numbers of people. A mass rush for the electric kettle at the end of a popular TV show can produce an increase in national electricity demand of over 2 GW in a matter of minutes. Figure 9.40 shows the 'demand pickup' when the 1990 World Cup Football Semi-Final was transmitted.

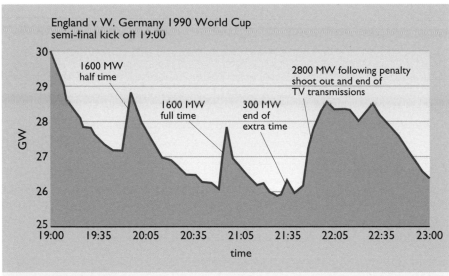

Figure 9.40 Variations in UK electricity demand during the 1990 World Cup Football Match

To cope with these problems a number of **pumped storage** stations have been built. These are hydroelectric plants with very large turbines which can also act as pumps. At off-peak times electricity is used to pump water up to a storage reservoir usually high up on a mountain. When a surge in electricity demand occurs the water is rapidly released back through the turbines. Although the output of these stations can be hundreds of megawatts, they are only designed to operate for short periods to cope while other more conventional plant is brought on-line. The UK currently has three pumped storage plants, two in Wales and one in Scotland (see Chapter 4, Box 4.3 and in the companion volume *Renewable Energy*). Their combined peak output power is over 2 GW or about 5% of the UK's typical winter electricity demand.

The privatized UK system

Prior to privatization, the nationalized UK electricity supply system consisted of three key parts:

- The CEGB power stations which supplied the electricity
- The National Grid, also part of the CEGB, which distributed it at high voltage
- The local Area Boards which distributed it at low voltage and were responsible for selling it to customers and billing them. Domestic customers bought electricity from their local area board.

When the privatization was carried out, these parts were split up (or 'unbundled') into a large number of separate companies. The government's intention was initially to privatize all electricity generation, but it rapidly became clear that some of the older nuclear stations were unsaleable and so they had to remain in state hands. Eventually the existing conventional power stations of the CEGB passed to the hands of two large new private companies, National Power and Powergen. The newer nuclear stations, the Pressurized Water Reactors and Advanced Gas Cooled Reactors, were taken over by British Energy. The older Magnox reactors remained under state control as part of British Nuclear Fuels (BNFL) which owns the fuel reprocessing plant at Sellafield.

The National Grid became the private National Grid Company.

The Area Boards in England and Wales became Regional Electricity Companies (RECs). They kept control of their local medium- and low-voltage distribution systems. They were also allowed to build their own power stations and many took the opportunity to invest in CCGTs.

The electricity system in Scotland and Northern Ireland remained as smaller vertically integrated systems outside the competitive power pool.

Since 1989 there have been a large number of mergers and take-overs of the various companies. Many UK companies have invested in the US. Others have been taken over by foreign companies. London Electricity is now owned by the state-owned French company Electricité de France.

Under the current UK system, any customer in England and Wales is free to buy electricity from any licensed supplier. Domestic customers in Yorkshire, for example, might buy their electricity direct from a company such as Powergen, originally set up simply to generate electricity. They will receive a bill from that company. In practice a portion of the money will be paid out in transmission charges; a small amount to the National Grid Company for transmission at high voltage and a large amount to the local Regional Electricity Company, Yorkshire Electricity, for the use of their low voltage distribution system.

The operation of the UK electricity trading system has not been without its own problems. Fierce competition has undoubtedly achieved its aim of driving down prices. However, there were allegations that some generating companies were manipulating the original trading arrangements (simply known as 'The Electricity Pool') in order to force up the price. In 2001 these were replaced with a new set of rules, the New Electricity Trading Arrangements (NETA). These have been criticised for being to unfavourable to CHP and wind power operators who may find it difficult to guarantee a precise level of generation several hours in advance. Even these 'new' trading rules are due to be shortly replaced by another set, the British Electricity Trading and Transmission Arrangements (BETTA) whose main function is to bring Scotland into the competitive electricity market.

It has become obvious that running a competitive electricity market while avoiding commercial manipulation and simultaneously encouraging CHP and renewables is not a simple task and the same problems are faced by Pool systems in other countries as well (see below).

9.9 The dash for gas

The use of natural gas in power stations in the UK (or Europe for that matter), is a very recent phenomenon. When North Sea gas was introduced from 1970 onwards it was treated as a premium fuel. It was used to replace first the existing town gas supplies and then coal and oil-fired heating in urban areas to reduce SO_2 and smoke emissions. At that time it was not clear how large the reserves of North Sea gas would be. It seemed to make little sense to use it for electricity generation when there were plentiful supplies of coal and enthusiastic investment in nuclear power. This policy was formalized in a 1975 European Union Large Combustion Plant Directive which restricted the use of gas for large scale electricity generation.

In the US the picture was different. Large amounts of natural gas were available and there was a long history of using it for power generation with conventional boilers and steam turbines. In 1960 over 20% of US electricity was being generated from natural gas.

Following the development of jet airliners such as the Boeing 707 in the late 1950s, gas turbines became mass-produced standardized items. When used to drive a shaft to run a generator, the thermal efficiency of a simple 'open cycle' gas turbine is poor, under 30%. However, multi-megawatt engines were small, relatively cheap and could be run up to speed very quickly as 'peaking plant' to assist hydroelectric pumped storage schemes.

The fact that the exhaust gases emerged at a temperature of 500 °C or more, sufficiently hot to raise steam in a boiler, suggested to US engineers that they might be used with a conventional power plant as a **topping cycle**. This is an extra stage added at the 'hot' or 'top' end of a thermodynamic system. To this end in 1963 a 27 MW gas turbine driving a generator was

BOX 9.11 The Combined Cycle Gas Turbine (CCGT) power station

Chapter 6 has described the importance of using the highest possible temperatures in heat engines in order to obtain the maximum thermal efficiency. This is a consequence of Carnot's equation:

Carnot efficiency $= 1 - T_{out}/T_{in}$

where T_{in} and T_{out} are the inlet and outlet temperatures respectively.

A conventional steam turbine can operate with outlet temperatures of 25 °C or less and inlet temperatures as high as 550–600 °C, yet fuels such as natural gas can be burnt to produce far higher temperatures. As described in Chapter 8, combustion temperatures in gas turbines can reach 1300 °C, limited only by the properties of the latest metal alloys and ceramics. Industrial turbines tend to use slightly lower temperatures to give a longer life. Designing a single heat engine that could operate between a very wide range of temperatures is not easy, yet it can be done by using a **combined cycle**, with two separate stages.

Figure 9.41 21st Century Power – Shoreham CCGT power station near Brighton. Pollution regulations do not normally require such a high chimney for a gas-fired power station – this one acts as a navigation landmark for coastal shipping

Such as system is used at Shoreham power station in Sussex, built in 1999 (see Figure 9.41).

The system is shown in Figure 9.42. Natural gas is burned in an industrial gas turbine at 1140 °C. The drive shaft, which spins at 3000 rpm, is connected both to the compressor and to one end of the generator, producing up to 240 MW of electricity. The exhaust gases from the gas turbine, which leave at about 630 °C are fed to a boiler, more formally known as a heat recovery steam generator (HRSG), and used to raise steam at 540 °C. This is then fed to a conventional steam turbine connected to the opposite end of the generator via a clutch, producing a further 140 MW of electricity. Using the terminology described in Chapter 6, this is a **tandem compound** system, with two turbines driving one generator. The station draws cooling water from a river estuary on one side of its site and discharges it into the sea on the other. This gives an average working temperature in the condenser of 22 °C.

The whole system is computer controlled and can be operated by two staff. When starting from cold, the steam turbine is initially left disconnected from the generator. The gas turbine is run up to 3000 rpm using the generator as a motor and the gas burners are lit. It takes about 30 minutes to get the gas turbine stage fully operational and running at low power. Over the next hour steam starts to be produced in the boiler. This is initially used to clean out condensed water in the various stages of the steam turbine. Then the high pressure steam is used to run this turbine up to 3000 rpm and the clutch connects it to the generator. It takes about 3 hours for the whole system to reach full power of 380 MW. The overall electricity generation efficiency is then over 50%.

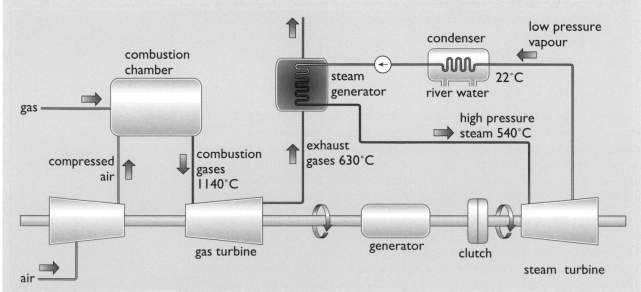

Figure 9.42 A schematic of the combined cycle gas turbine at Shoreham

added to a conventional gas-fired plant using a 220 MW steam turbine in Oklahoma. The hot exhaust of the turbine was directed to the existing steam boiler, supplementing the heat from the gas being burned there.

Although there have been many such 'add-on' schemes to existing steam stations built since then, the combined cycle gas turbine (CCGT) station has been refined to a carefully integrated and optimized form with a high overall efficiency (see Box 9.11).

The first CCGT plant to excite international attention was the Japanese Futtsu plant of Tokyo Electric Power, fired by imported liquefied natural gas (LNG) and started in the mid 1980s. This broke the trend to larger and larger steam turbines. Instead of a single gigawatt-sized boiler and steam-turbine generating set, the Futtsu plant was built as a series of identical modules, gas turbines combined with steam turbines; each with an output of only 165 megawatts. As each module was completed it came into operation and began generating both electricity and cashflow in under three years from the start of construction. Replicating the design for successive modules cut the costs significantly. By 1997, the Futtsu plant had an output of 2000 MW, from 14 identical modules.

In the UK by the end of the 1970s, most electricity was produced from coal, oil and nuclear power stations, with 70% being generated from coal. Only a small number of gas turbines had been introduced as peaking plant. The heavy reliance on coal led to a bitter confrontation between the Thatcher government and the UK miners in 1984. Following this the UK government resolved to reduce the reliance on coal and set about having the EU Large Combustion Directive withdrawn to allow the use of natural gas in power stations. By the time the UK industry was privatized in 1989 the technology of CCGTs was well-developed. Not only were they cheaper and quicker to build than coal or nuclear stations (a subject to be discussed in Chapter 12), but it had become possible to negotiate a fixed-price contract for 15 years future supply of gas. This kind of contracting had never been done before.

Under the terms of the 1989 privatization, the new Regional Electricity Companies (RECs) were allowed to invest in their own power stations. Although now any customer can buy electricity from any supplier, this right was only established after a number of years. Initially the RECs were monopoly suppliers to a sufficient number of small customers to guarantee sales of electricity from any moderate-sized power station that they might choose to build. Armed with a long-term gas supply contract and captive customers all they had to do was approach merchant banks to supply the finance. Seeing an attractive and relatively risk-free investment, they were happy to oblige. The so-called 'Dash for Gas' was on. Between 1990 and 2000 the amount of electricity produced from gas in the UK increased from less than 1% to 39%, largely at the expense of coal-fired generation (see Figure 9.43).

This switch to gas has also had environmental benefits. In generating a kilowatt of electricity, a CCGT produces only about half as much CO_2 as a coal fired power station and virtually no sulphur dioxide. As will be pointed out in later chapters, gas-fired generation has enabled the UK to meet SO_2 emission targets and cut national CO_2 emissions without undue expense. The price paid was a 50% increase in UK natural gas consumption between 1990 and 2000, just to supply power stations.

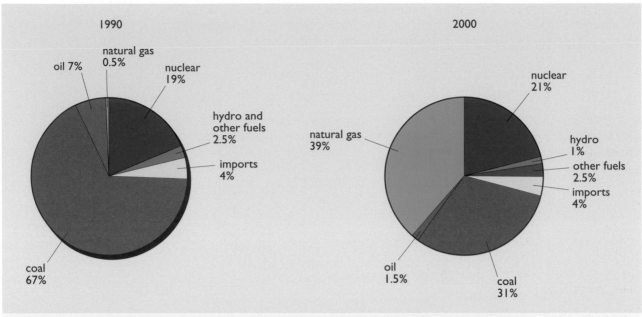

Figure 9.43 The Dash for Gas – UK Electricity generation by fuel 1990 and 2000 (source: DTI, 2001b)

9.10 Electricity around the world

Mains electricity is the convenience fuel of today. Conventional economists and free-marketeers would like to treat a 'kilowatt-hour of electricity' as if it were a uniform entity like a litre of unleaded petrol. Yet the manner of the generation of electricity and its use varies widely from country to country. Each of our sample countries in this book uses different quantities of electricity and obtains it from different mixes of sources, as shown in Table 9.3, below. These have different pollution consequences and implications for sustainability.

Table 9.3 Per Capita Electricity Consumption and percentages of Electricity Generation by Fuel – 2000

	Per Capita Consumption /MWh yr⁻¹	Coal	Oil	Gas	Nuclear	Hydro & Other
UK	5.7	31%	2%	39%	25%*	4%
US	12.8	52%	3%	16%	20%	9%
Denmark	6.0	45%	12%	25%	0%	16%
France	6.7	5%	0%	2%	77%	16%
India	0.5	74%	0%	8%	3%	15%

*Includes imports (mainly nuclear) from France

Sources: DTI, 2001b; Energy Information Administration, 2001; Danish Energy Authority, 2001; MINEFI, 2001; TERI, 2002

Also, the patterns of ownership and control are different in each country, with implications for investment in new generation technologies and in electricity conservation.

We look at each country in turn in the remainder of this section.

United Kingdom: a summary

As can be seen from Table 9.3 per capita electricity consumption in the UK is similar to that in Denmark and France, but far lower than that in the US. The UK figure did not increase at all between 1995 and 2000 (see Chapter 3, Figure 3.18). This is most likely due to campaigns to promote energy efficiency in appliances and lighting, and also strong competition from cheap natural gas for direct heating uses.

Electricity is generated by a mix of coal, gas and nuclear plant, with a small amount of hydro power and other renewable sources. The use of gas has been growing at the expense of coal. This increase in gas demand has drained the UK's North Sea gas reserves all the faster, forced up gas prices and increased concerns about the long-term supply of gas to the country. At the time of writing (early 2003), enthusiasm for the construction of new gas-fired power stations has been dampened, and some existing ones have been 'mothballed'.

There are a large number of nuclear power stations on the system, and an appreciable amount of electricity is imported from French nuclear stations via the Channel link. Although the total output and reliability of UK nuclear stations has been increased since 1990, many of them are fast approaching retirement. Even the more modern ones are in financial difficulties (see below), casting a shadow over the future of the whole nuclear industry. This, in turn, poses problems about how the country will meet its commitments to cut CO_2 emissions. On the brighter side, there is plenty of technical potential for development of gas-fired CHP and renewable energy, particularly wind and wave power. The government has set targets for 2010 that the electricity industry should include 10 GW of CHP generation and that 10% of electricity should come from renewable energy. However, the main electricity demand remains in the south and east of the UK, and much of the prime renewable energy resource is in the west of England and Scotland, so there may have to be a further phase of construction of the National Grid with all the wayleave and visual intrusion problems that this will entail.

As has been described above, the electricity supply industry is privatized, but subject to a government-appointed regulator. At the time of writing, early 2003, the industry is in a state of financial turmoil. The fierce competition between electricity suppliers has led to low electricity prices. While this may be good news for consumers, it has led to financial problems for the suppliers themselves. British Energy, which owns the most modern nuclear stations has had to resort to emergency government loans. Other suppliers facing difficulties include large coal-fired power stations such as Drax in Yorkshire, and many smaller combined heat and power and renewable energy system operators.

United States of America

On a per capita basis, US electricity demand is about twice that of the UK and still growing. Over a half of US electricity is generated in coal-fired plant. Unlike the UK, where the use of coal has been falling, much of the growth in electricity demand in the US has been met using coal, further contributing to the country's high CO_2 emissions. Although 20% of the country's electricity comes from nuclear stations, many are approaching retirement, and there has not been much enthusiasm for constructing new ones. The US has many large hydro-electricity schemes and is linked to others in Canada. Between them, the US and Canada produce about a quarter of the world's hydroelectric power. The US has considerable technical potential for renewably-generated electricity from other sources, particular from solar energy in the south and wind power in the central and northern states.

The US electricity industry is mainly organized on a state by state basis and consists largely of private utilities and generators, but with publicly appointed regulators with considerable powers. The high dependence on electricity has created concerns about reliability of supply, both from a technical and a commercial point of view. The grid failure in July 1977 that blacked out the whole of New York City was due to technical failure. However, in January 2001, the State of California experienced power cuts for commercial reasons. The local electricity supply had been extensively privatized in 1996 and a power pool set up. The legislation, though, did not allow for any increases in prices to be passed on to normal consumers until March 2002. In practice the market price of electricity turned out to be very volatile and in December 2000 the price rose dramatically. The local utility companies Southern California Edison and Pacific Gas and Electric, responsible for distributing electricity to small consumers, lost so much money that they eventually filed for bankruptcy.

Initially this collapse was simply put down to badly thought out regulation (see Beggs, 2002), however there were suspicions that these high prices were the result of manipulation of the trading system by suppliers. One such supplier was the US corporation Enron which had built up a strong position in the US energy infrastructure during the 1990s, and at its peak controlled a quarter of all US gas sales. It was involved in power station projects worldwide, including the UK and India. In the summer of 2001 it claimed to have assets of £62 billion. By the end of the year it was bankrupt.

In October 2002, one of their former directors was charged in a San Francisco court with submitting false trading information and bids to the officials who ran the California power market and electricity grid. He pleaded guilty to conspiracy to commit fraud and agreed to pay back $2.1 million in compensation. He told the judge that he did it 'to maximize profit for Enron'. Overall it is possible that the state was overcharged by $9 billion for its electricity by various suppliers (see the articles in *The Guardian* (Anon, 2002) and *USA Today* (Iwata, 2002).

France

In per capita terms, French electricity use is very similar to that in the UK and Denmark, yet demand is growing strongly because it is being promoted as a heating fuel. France does not have the supplies of natural gas available in the UK, or the extensive district heating schemes promoted in Denmark.

Nuclear power and a modest amount of hydroelectricity are the principle sources. In environmental terms an average kilowatt-hour of French electricity involves very little CO_2 production but raises questions about nuclear safety and the generation of nuclear waste. Although there has been some recent interest in expanding the use of wind power, it seems likely that the policy of heavy reliance on nuclear power will continue.

The electricity industry still consists largely of a vertically integrated monopoly system, Electricite de France (EdF), as set up in 1946. The French government has been very resistant to any privatization and it was only in 2000 that the French parliament separated EdF into three parts, one for generation, one for grid management and one for distribution.

Distributing electricity in dense urban areas is relatively easy and cheap but supplying it to widely spaced rural communities involves a considerable expenditure on constructing and maintaining transmission lines. There is always a worry that privatization might result in cheap electricity for industry and city dwellers but higher prices and reduced reliability of supply to rural consumers. France is a large country, with a politically vociferous farming community. It has been suggested that a further expansion of liberalization might challenge the basic republican principles that define the philosophy of the French public service, in this case, equity among French citizens regarding access to electricity (Guovello, 2002). Similar concerns surfaced in the UK in 1989, when the House of Lords sought protection for farmers from any undue price rises resulting from electricity privatization.

At the time of writing (early 2003), tensions between the EU in Brussels, pressing for increased competition in electricity supply, and the French government's protectionist policies towards EdF are as strong as ever.

Denmark

Danish electricity demand on a per capita basis is similar to that in the UK or France. Danish national electricity consumption doubled between 1975 and 2000, yet, as pointed out in Chapter 3, national primary energy consumption did not increase at all. The pursuit of energy efficiency in electricity generation has been a key element in this achievement.

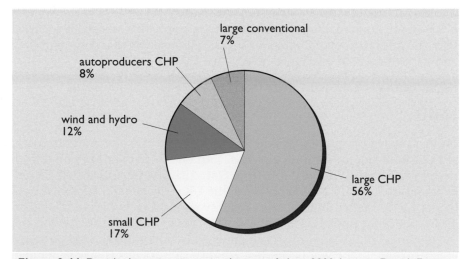

Figure 9.44 Danish electricity generation by type of plant, 2000 (source: Danish Energy Authority, 2001)

In the early 1970s, the Danish electricity industry was almost totally dependent on oil for generation. Denmark's response to the price rises of 1973 was to adopt policies of security of supply, economic efficiency and environmental protection. Power stations under construction designed for oil were quickly switched to coal firing and there was an expansion in the use of combined heat and power generation and district heating. By 2000 about 80% of Danish electricity was generated in CHP plants and the further reaches of the Copenhagen district heating system extended 35 km out from the centre of the city.

The availability of natural gas from Denmark's North Sea sector from 1985 onwards has allowed the replacement of some coal-fired stations by combined cycle gas turbines so that by 2000 gas provided 25% of the country's demand. The use of nuclear power had been considered in the 1970s, but proved politically unpopular and was ruled out in 1985.

Denmark is, of course, famous for its pioneering use of wind power. In 2000, wind supplied 13% of electricity demand and other renewables a further 3%.

In other countries the extensive use of wind and CHP might be seen as excessively restrictive to the efficient operation of an electricity supply system. The Danish system avoids this through strong grid links with Sweden and Norway which have very flexible hydroelectric power sources. There are benefits on both sides. In the wet years of 1989 and 1990, imports accounted for some 40% of Danish electricity consumption. In the dry year of 1996, gross exports from Denmark exceeded 50% of national electricity demand.

Organizationally, the Danish electricity industry has been traditionally based on small urban municipal systems and consumer co-operatives. Although the country only has a population of 8 million, in 1995 there were 103 electricity distribution utilities and eight generating companies. It is these small organizations that have promoted many renewable and CHP schemes.

The electricity market in Denmark is currently being liberalized as part of a common Scandinavian power pool, the Nord Pool, shared with Sweden, Norway and Finland. Marrying the needs of a competitive electricity market with the preservation of Denmark's subsidies to renewable energy and CHP and its traditions of consumer involvement will pose problems for future system regulators (see Gronheit, 2002).

India

India is a developing country and the difficulties of its electricity industry are shared by many similar countries. India has a population of a billion people and this is rapidly growing. Currently its per capita electricity consumption is only a tenth of that of the UK – the general level of access to electricity can be thought of as roughly equivalent to that in the UK in 1930. Reliable supplies of electricity are seen as essential for the development of Indian industry in cities and also for the general policy of electrification of the whole country. About a third of electricity demand is for agricultural use, much of it for irrigation, and this supply is very heavily subsided. This contrasts with the UK where only 1% of electricity demand is used in agriculture.

There is an extensive grid system, particularly from the coal-mining areas in the east of the country to the major loads in the North East. By 1998 almost 90% of villages had some access to mains electricity. Transmission and distribution losses are very high, put at 23% on average, compared to about 7.5% in the UK. However, much of this may be due to theft from local overhead distribution cables.

In 2000 the Indian electricity supply industry had about 100 GW of generation capacity, but even this could not cope with demand and there are frequent power cuts in major cities. It is estimated that there is at least an 18% shortfall in generating capacity to supply the peak demand, or put another way, the country immediately needs another 20 GW of power plants to cope. This may sound a difficult enough problem, but if the country was to aim to reach current European per capita electricity use figures, then it would need to build at least another 1000 GW of generating plant, that is equivalent to five hundred 2 GW stations like the one at Didcot in the UK (see Figure 1.44 in Chapter 1).

Coal is the major fuel used for electricity generation, obtained either from Indian mines or imported from Australia. Most of the recent expansion has been through the use of coal or indigenous supplies of natural gas. The modest use of nuclear power is politically sensitive because of links to the production of nuclear weapons and recent tensions with neighbouring Pakistan. There is plenty of hydroelectric power in the mountainous northern states and there has been a rapid expansion in use of wind power in recent years. India boasted nearly 1.3 GW of installed wind power as of December 2000.

The Indian industry is predominantly nationalized and controlled by power boards in each individual state. Even though domestic and agricultural electricity is heavily subsidized, most state utilities are loss-making and heavily in debt (see TERI (2002)). The national government has encouraged private investment in new power plants, but this has proved fraught with difficulty. One of the largest schemes, a 2 GW plant, was to be financed by the US Enron corporation and this project may well have failed with its financial collapse in 2001.

The Indian government, like those of other developing countries, has difficult choices to make. They have many technical options. Given a national policy of increasing access to mains electricity, do they:

- Invest in new conventional power plant themselves?
- Seek foreign private investment for conventional plant, from sources that may not turn out to be reliable, and may place the country in debt in the future?
- Use cheap coal, increasing the country's CO_2 emissions, or look to lower carbon alternatives such as natural gas?
- Continue to develop nuclear power, risking suspicions that it is mainly for military purposes?
- Cut demand by tackling transmission and distribution losses?
- Cut demand by investment in energy efficient appliances and lighting?
- Invest in renewable energy sources?

Then there are financial questions. Is it right to subsidize electricity to agriculture and domestic users? What incentive is there for these consumers to save electricity if they are not charged a fair price? What is a fair price for a convenience fuel so essential to modern life? The subject of relative energy prices and some of their history are discussed in Chapter 12.

9.11 Conclusion

In this chapter we have reviewed the historical development of electricity and described the relentless growth of the electricity industry during the 19th and 20th centuries, to the point where at the start of the 21st century it has become an essential feature of modern civilization. We have looked at the ways in which electricity is generated, transmitted, distributed and used in present-day electricity systems, and the ways in which the organization and management of the UK electricity system has changed in recent years.

The generation of electricity from fossil fuels gives rise to a variety of deleterious environmental and social effects, many of which will be described in Chapter 13. And although substantial progress has been made in ameliorating many of these effects, there remains an urgent need to achieve further major improvements, particularly in reducing emissions of greenhouse gases. The use of nuclear energy, the subject of the next two chapters, entails virtually no greenhouse gas emissions, but as we shall see, it brings with it other environmental and social concerns.

Growth in electricity demand for the world as a whole seems set to continue in coming decades. Although demand for electricity in some developed countries could be nearing saturation, rising demand in developing countries seems inevitable, driven mainly by rising population levels and aspirations for economic growth. One of the main challenges facing the electricity industry in the 21st century is therefore to meet these growing demands using new and more environmentally friendly technologies. Some of these will be described in Chapter 14 and others in the companion volume *Renewable Energy*.

References and sources

Anon (1976) *A Brief History of the Sign Industry*, Signs of the Times magazine, September, 1976, http://www.signmuseum.com (accessed January 2003).

Danish Energy Authority (2001), *Energy Statistics 1972-2000*, downloadable from http://www.ens.dk (accessed January 2003).

Department of Trade and Industry (1997) *Energy Consumption in the United Kingdom, Energy Paper 66,* Government Statistical Service.

Department of Trade and Industry (2001a) *Digest of UK Energy Statistics (DUKES).*

Department of Trade and Industry (2001b) *Energy in Brief, July 2001.*

Department of Trade and Industry (2001c) *UK Energy Sector Indicators, 2001.*

Department of Trade and Industry (2002) *Digest of UK Energy Statistics (DUKES)*.

EIA (2000) *Annual Energy Review 2000*, US Energy Information Administration.

Embassy of the People's Republic of China in the USA (1997) *Some Facts about the Three Gorges Project*, http://www.china-embassy.org/eng/ 6893.html [accessed January 2003].

Eyre, N. J. (1990) *Gaseous Emissions due to Electricity Fuel Cycles in the UK,* Energy and Environment Paper No. 1, Energy Technology Support Unit, Harwell.

Gronheit, P. E. *Denmark: long-term planning with different objectives*, in Vrolijk, C. (ed.), *Climate Change and Power*, Royal Institute of International Affairs, Earthscan, 2002.

Anon (2002) 'Finance, City Briefing – Enron trader's admission', *The Guardian*, 18/10/02, Manchester, Guardian Newspapers Limited.

de Guovello, C. (2002) *France: focus on non-fossil fuels*, in Vrolijk, C. (ed.), *Climate Change and Power*, Royal Institute of International Affairs, Earthscan.

Kozier, K.S. (1999) Nuclear and Process Heating, presented at the Canadian Nuclear Society Climate Change and Energy Options Symposium, Ottawa, November 1999. Available at http://www.cns-snc.ca/events/CCEO/ nuclearenergyprocess.pdf [accessed 2 June 2003].

Lenin, V. I. (1965), *Collected Works Vol. 30*, London, Lawrence & Wishart, first published in 1920.

Mackenzie, G. (1999) *Pimlico District Heating Undertaking – 48 years of operating experience*, City of Westminster.

Mayhew, H. (1984) *Mayhew's London*, edited by Peter Quennel, London, Bracken Books.

MINEFI (2001) *L'Energie en France - Repères*, Ministère de l'Economie, des Finances et de l'Industrie.

Ministry of Power (1962), *Statistical Digest*, HMSO.

PB Power Energy Services (2002) – *The Potential for Community Heating in the UK*. Downloadable from the Energy Savings Trust web site at http:/ /www.est.org.uk [accessed January 2003].

Porter, R. (1994) *London, A Social History*, Penguin.

Smith, G. (1980) *Storage Batteries*, Pitman, London.

Tata Energy Research Institute, (2002), *Teri Energy Data Directory and Yearbook, 2001/2002*, New Delhi.

Iwata, E. (2002) 'Enron energy trader pleads guilty', *USA Today*, 17/10/ 2002, Gannet Co. Inc. Online article at http://www.usatoday.com/money/ industries/energy/2002-10-17-enron-belden_x.htm, [accessed January 2003].

US Energy Information Administration (2002) *International Energy Outlook 2002*, downloadable from http://www.eia.doe.gov [accessed January 2003].

Weightman, G. and Humphries, S. (1983) *The Making of Modern London, 1815–1914*, London, Sidgwick and Jackson.

Weightman, G. (2001) *The Frozen Water Trade*, HarperCollins.

Williams, R. H. and Larson, E. D. (1989) *Expanding Roles for Gas Turbines in Power Generation,* in Johansson, T. B. *et al* (eds), *Electricity – Efficient End-Use and New Generation Technologies, and Their Planning Implications.*

Zola, E. (1901) *Travail,* Paris, Charpentier.

Further reading

BBC (2002) *The Enron Affair*, http://news.bbc.co.uk/1/hi/in_depth/business/2002/enron/default.stm [accessed January 2003].

Beggs, C. (2002) *Energy: Management, Supply and Conservation*, Butterworth-Heinemann, pp. 37–44.

Byers, A. (1981) *Centenary of Service – A History of Electricity in the Home*, London, Electricity Council.

Cochrane, R. (1985) *Power to the People – The Story of the National Grid*, Newnes Books.

Dettmer, R. (2002) *'Living with NETA'* IEE Review, July, pp. 32–36.

Dunsheath, P. (1962) *A History of Electrical Engineering*, London, Faber and Faber.

Electricity Association (1996) *The UK Electricity System*, London, Electricity Association, p. 24.

Electricity Council (1987) *Electricity Supply in the UK – A Chronology,* London, Electricity Council.

Georgano, N. (1996) *Electric Vehicles,* Shire Publications, Princes Risborough.

Patterson, W. (1999) *Transforming Electricity - The Coming Generation of Change,* Royal Institute of International Affairs, Earthscan.

Tyler, D. W. (1997) *Electrical Power Technology*, Oxford, Newnes.

Vincent, C.A. *et al* (1984) *Modern Batteries: an introduction to electrochemical power sources*, London, Edward Arnold.

Chapter 10

Nuclear Power

by Janet Ramage

10.1 Introduction

In the year 2000, nuclear power stations were providing about a sixth of the world's total electricity. This proportion had remained fairly constant for most of the final decade of the twentieth century, with the total world output and nuclear output both rising at about 2.5% a year. However, the period from 1998 to 2001 has seen the nuclear output, whilst still increasing each year, failing to keep pace with the recent 3–4% growth rate of the world electrical industry as a whole.

At the end of 2001, there were over 400 reactors in commercial operation in 31 different countries, with a total output capacity of about 360 GW_e and an annual output of just under 2500 TWh. About a quarter of all the reactors are in the United States. France and Japan, each with over 50 reactors, together account for another quarter. The UK and Russia have about 30 each, and no other country has more than 20. Almost four-fifths of the world's reactors are of the light-water type (see Section 10.5 below).

Nuclear power has been the subject of controversy since its inception. At a time when very little had yet been revealed to the general public, the scientists themselves were divided, with many who had worked on the atomic bombs having serious reservations about further development of the technology. The post-war period saw the introduction of more weapons of mass destruction and of nuclear power stations, and in the 1950s and 60s the nuclear debate attracted the degree of public attention that today we see devoted to genetic engineering. Nevertheless, as the above data show, the nuclear power industry grew despite its opponents, and it remains a significant part of the world's generating capacity today, despite the effects of structural changes in the electricity industries and the disastrous accident at Chernobyl, both of which severely reduced further investment. The nuclear debate is by no means over today. Fears of global warming have strengthened the case of the supporters of nuclear power, but vigorous opposition remains. These issues, and the question of the future of nuclear power, are the subjects of the next chapter of this book.

The present chapter is concerned with today's nuclear power systems and the developments that have led to them. The first part adopts the approach of earlier chapters, establishing the basic scientific ideas in the context of their history. Radioactivity is probably the feature of nuclear power that gives rise to most public concern, so we rejoin the historical account of Chapter 4 at the point when this first sub-nuclear effect was observed. Moving forward several decades then takes us to the discovery of nuclear fission and the wartime development of the first nuclear reactors and nuclear weapons. A brief account of the approaches to nuclear power adopted by different countries in the post-war period leads finally to an account of present-day types of fission reactor. The rest of the chapter is mainly devoted to the features of these reactors, and associated issues such as safety and nuclear waste. The chapter concludes with a brief introduction to nuclear fusion and its potential as a power source.

We start, however, with an introduction to two new units that are normally used in discussing nuclear processes (Box 10.1).

BOX 10.1   Units for mass and energy

Units such as kilograms and joules are inconveniently large for discussing events at the atomic or nuclear level, so the following more appropriate units are commonly used.

Mass

The unit adopted for mass is the **unified mass unit (u)**. This is slightly less than the mass of a proton or neutron and is equal to 1.660×10^{-27} kg. The mass of an electron is roughly one two-thousandth of this.

Energy

Energies on the scale of single atoms or nuclei are usually expressed in **electronvolts (eV)**. One electronvolt is equal to the kinetic energy gained by an electron in falling through a potential difference of one volt (see Box 4.6). A single electron has, of course, a tiny electric charge, and 1 eV is only 1.602×10^{-19} J, but this is suitable for discussion of single atomic events. (The energy released in the combustion of one natural gas molecule, for instance, is about 9 eV.) For nuclear events, however, millions of electronvolts (**MeV**) are more appropriate, and 1 MeV is 1.602×10^{-13} J.

Mass-energy

We saw in Chapter 4 that mass and energy are essentially two aspects of the same quantity, and that energy (E) in joules is related to mass (m) in kilograms by:

$$E \text{ (J)} = m \text{ (kg)} \times 9 \times 10^{16}$$

If the energy is measured in MeV and the mass in u, this relationship becomes:

$$E \text{ (MeV)} = m \text{ (u)} \times 931$$

In other words, the mass of a proton or neutron in energy terms is very roughly 1000 MeV, and the mass of an electron is about 0.5 MeV.

10.2 **Radioactivity**

Radioactivity is the spontaneous emission of particles from the nuclei of atoms, and as we have seen in Chapter 4, every isotope of every element beyond bismuth in the periodic table is unstable and therefore radioactive. Henri Becquerel discovered the effect in 1896, and within a decade several important features had been recognized.

- Only some chemical elements are radioactive, and the particles they produce and the rates of emission are characteristic of the element.

- In many cases the rate of emission decreases with time, over periods varying from seconds to years. Others seem to show no detectable change over years of observation.

- Physical or chemical changes of the radioactive substances – heating or cooling, compressing, even combining them in different chemical compounds – have *no effect at all* on the radioactivity.

However, the origins of radioactivity were not established for another 15 years. Only after Rutherford had developed the idea of the atomic nucleus was he able to identify radioactivity as a *nuclear* process. There are three main types of radioactivity, easily distinguished by their different penetrating powers.

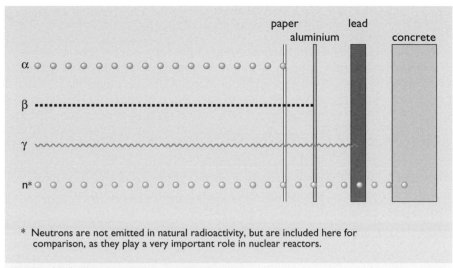

* Neutrons are not emitted in natural radioactivity, but are included here for comparison, as they play a very important role in nuclear reactors.

Figure 10.1 The penetrating powers of radiation

Alpha particles

Alpha particles (α-particles) are the least penetrating, being stopped by a sheet of paper, a couple of inches of air, or human skin. Rutherford identified them as helium nuclei, consisting of two protons and two neutrons. As Figure 10.2 shows, a nucleus emitting an α-particle undergoes a reduction of two in its atomic number and four in its mass number and thus becomes an isotope of the element that lies two places lower down in the periodic table (Figure 4.20 in Chapter 4).

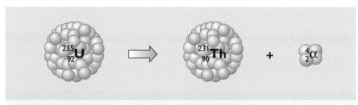

Figure 10.2 Alpha emission from uranium-235

The resulting 'daughter' atom is often radioactive in turn, so that the initially pure sample eventually contains a whole series of different elements, each with its characteristic radioactive emissions. This multiplicity posed a problem for early workers trying to understand the processes; and the fact that some products can be radioactive *gases* remains a safety issue for those dealing with radioactive substances today. Because α-particles are so easily stopped they only present a serious health hazard if the radioactive substance itself enters the body – by inhalation or by ingestion in contaminated food or drink.

Beta particles

Beta particles (β-particles) are more penetrating than α-particles, but nevertheless are stopped by a thin sheet of metal or a few millimetres of almost any material. Historically, they were quickly identified as electrons, exactly the same as those already discovered a few years earlier (see Section 4.4 on electrons) but with a great deal more energy.

It may seem surprising that a positive nucleus can emit a negative particle, but we see that it can happen if the process is viewed as the disintegration of one neutron into a proton and an electron. This is possible because a neutron is slightly heavier than a proton plus an electron. The surplus mass then appears as the energy of the electron (Box 10.2). The consequence is also easy to see (Figure 10.3). The number of neutrons decreases by one and the number of protons increases by one, so the element changes into the next *highest* one in the periodic table but with no change in the mass number (the total number of neutrons plus protons). As we'll see later, one important result of β-emission is the production of plutonium.

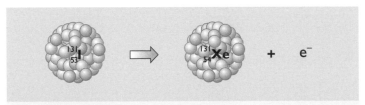

Figure 10.3 Beta emission from iodine-131

Beta-radiation can penetrate into the skin, causing unpleasant burns, but like alpha particles, will only produce internal damage if the source is inhaled or ingested.

BOX 10.2 The decay of a neutron

A neutron is very slightly heavier than a proton, and a free neutron spontaneously disintegrates into a proton and an electron. The net electric charge remains zero, of course.

How much energy is released when a neutron decays into a proton and an electron?

Data

It is convenient to use the units introduced in Box 10.1. The proton and neutron masses are both about one atomic mass unit, and the electron mass is about 1/2000th of this. More precisely, the masses are:

neutron mass	m_n	=	1.00867 u
proton mass	m_p	=	1.00728 u
electron mass	m_e	=	0.00055 u

Calculating the energy

The difference between the neutron and proton masses is 0.00139 u, well over twice the mass of the electron. The spare mass, 0.00084 u, appears in the form of energy. Using the relationship from Box 10.1, we have:

$$E = 931 \times 0.00084 = 0.78 \text{ MeV}.$$

If all of this became kinetic energy of the electron, it would be ejected as a penetrating β-particle travelling at three-quarters of the speed of light.

Gamma particles

Gamma particles (γ-particles) were discovered a few years after the other two types, mainly because they are much more penetrating and thus difficult to detect. Originally called **γ-rays** in the style of the time, they became **γ-radiation** when they were identified as very short wavelength electromagnetic waves. This name is still used, but the photon view of radiation has led to the adoption of the name '**γ-particles**', in line with the other types. Gamma emission is usually a settling-down process by which a nucleus loses surplus energy, often following α- or β-emission or fission. It does not of course change the numbers of protons or neutrons, so the two nuclei before and after γ-emission differ only in their energy. They are referred to as **nuclear isomers**.

Gamma-radiation is essentially the same as high-energy X-rays, and has similar effects on living matter. Gamma-particles are stopped only by several centimetres of lead or steel, or a few feet of concrete, and together with neutrons they are the chief radiation hazard associated with nuclear technology.

Radioactive decay and half-life

The theory of radioactivity developed by Rutherford and his younger colleague Frederick Soddy in 1903 is based on a conclusion drawn from their observations.

The rate at which particles are being emitted depends only on the number of radioactive atoms present at the time, and is proportional to the number of these atoms.

Now take into account the fact that emitting a particle changes the nucleus and therefore reduces the number of radioactive atoms of that type. We then have a situation where the **rate of decay**, the rate at which the number of radioactive atoms decreases, is proportional to the number present at any moment. As the number of atoms falls, the loss rate also decreases. This situation leads to a falling exponential (Figure 10.4).

In the case of radioactivity, the rate of decay is usually measured not by an annual percentage change but by specifying the **half-life**: the time taken for half of any sample to decay. The more rapid the decay, the shorter the half-life. Inspection of Figure 10.4 shows that *the half-life is the same no matter where you start*. Iodine-131 is a β-emitter with a half-life of 8.1 days, so half of any sample will decay in about eight days; half of the remainder will decay in the next eight, half of the new remainder in the next eight, and so on. It follows, of course, that the rate of emission of particles, the radioactivity of the sample, also falls at this rate – although, as mentioned above, in many cases there are radioactive daughter products that start to contribute, affecting the overall activity.

Naturally occurring uranium consists of three radioactive isotopes. The main one is U-238, but about one atom in 140 is U-235, and a very tiny proportion is U-234. All three isotopes are radioactive, but with very long half-lives, which explains why we still find them on earth, and why the early researchers could detect no change in their activity over a period of a year or so. Starting with a piece of U-238 today, you would need to wait

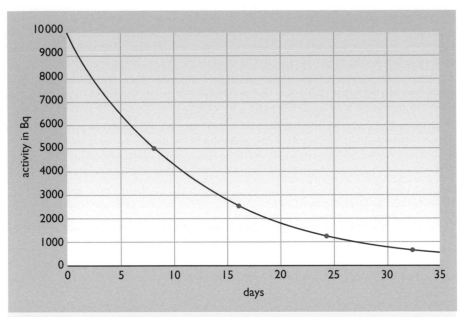

Figure 10.4 The radioactive decay curve of iodine-131

4500 million years before half the nuclei had emitted particles and changed into thorium. For U-235 the half-life is shorter – only 700 million years – so there is less of it left today; and similarly for U-234 with a half-life of a quarter of a million years.

A fact of great practical importance is that the decay of any of these uranium nuclei produces a new nucleus that is again radioactive, and this in turn produces a radioactive 'daughter', and so on, through a long series of radioisotopes, stopping only when the number of protons and neutrons becomes small enough to form a stable nucleus. (In each case the final product is an isotope of lead: Pb-206 or Pb-207.) The significance of this is that uranium ore, buried underground for millions of years, will contain all these isotopes, and all of them will be decaying steadily at the same rate. The result is a total radioactivity many times that of the uranium alone. Moreover, one intermediate product is a radioactive gas, radon, and this has very important consequences, as we shall see in Chapter 13.

An effect without a cause

We have seen that radioactive nuclei are unstable, and that each type has a characteristic half-life. But what causes a particular radioactive nucleus to emit a particle at any particular time? The extraordinary answer is that there is no immediate cause. As Rutherford and Soddy recognized, radioactivity, uniquely, is a *truly random process*, governed by nothing other than the laws of chance. If you start with a sample containing, say, a billion iodine-131 atoms, you can predict with confidence that about half a billion will be left after eight days. But there is absolutely no way to predict *which* atoms will be left, nor when nor whether any particular nucleus will decay. This is not a matter of inadequate apparatus, or insufficient knowledge. The radioactive decay of an individual nucleus is believed to be truly an effect without a cause, and the consequence is that we ourselves can neither cause nor prevent it. Unless a means is devised to convert undesirable radioactive nuclei individually into stable forms, we can do nothing but keep them secure until enough half-lives have elapsed to render them harmless.

BOX 10.3 **Measuring radioactivity**

The question, *'How much radioactivity?'* can be interpreted in two ways. It could be asking about the **strength of a source** – about the emissions from a particular piece of radioactive material. Or it could be about the radioactive **dose** received by a person or group of people, or by structural materials, etc. Here we are concerned with the first meaning.

A radioactive source can be characterized by the number of particles it emits per second. This is its strength in **becquerels** (Bq). A piece of material with an activity of 1 MBq emits one million particles a second, and so on. The becquerel replaced the earlier unit for the strength of a source, the curie (Ci). A one-curie source would emit 3.7×10^{10} particles a second, so the conversion is very roughly $1 \text{ Ci} \cong 40 \text{ GBq}$ or $1 \text{ TBq} \cong 30 \text{ Ci}$.

Table 10.1 shows the half-lives and activities of a few radioisotopes. Comparison of the two uranium isotopes shows, as one would expect, that greater activity leads to a shorter half-life.

To put the data in context, we can note that:

- one gram is the mass of half a tea bag, and one microgram is a millionth of this – a speck of dust;

- an activity of 10 000 Bq per cubic metre (undesirably high for continuous exposure) requires no more than two parts per thousand-million-million of iodine-131 in air;

- the encapsulated sources considered safe for school laboratories have activities of perhaps 10 000 Bq.

So is 10 000 Bq a safe level of radioactivity or not? The first answer is that it isn't a 'level' but the strength of a particular source. And secondly, safety depends on *what* the source is and *where* it is. An encapsulated source used correctly in the laboratory adds a negligible amount to the natural radioactivity that we all receive continuously; but 10 000 Bq of plutonium in soluble form entering the bloodstream would be a very different matter. In assessing safety – or potential danger – we do need data of the type in Table 10.1, but we also need information on the resulting doses and their effects. These topics are treated in more detail in Chapter 13.

Table 10.1 Half-lives and activities

Isotope	U-238	U-235	Pu-239	Sr-90	I-131
type of particle[1]	α	α	α	β	β
half-life	4.5×10^9 years	7.0×10^8 years	24 000 years	28 years	8.1 days
activity of 1 g[2,3]	12 000 Bq	79 000 Bq	2300 MBq	5.3 TBq	4600 TBq
mass for 10 000 Bq[3]	0.81 g	0.13 g	4.3 μg	0.0019 μg	2.2 pg

1 γ-particles are also emitted in all cases except Sr-90.

2 This refers to one gram of the single isotope, and is the activity before it has had time to produce 'daughters' that would change the sample.

3 Notice the different units. 1 μg (microgram) is one millionth of a gram, and 1 pg (picogram) is one millionth of a microgram, or 10^{-12} g.

10.3 Nuclear fission

Experiments with neutrons

The discovery of radioactivity provided a completely new tool for the study of matter. Firing the particles into materials and observing their interactions with atoms or nuclei became one of the most fruitful experimental techniques of the twentieth century, and bombarding ultra-thin gold foil with α-particles led Rutherford to the concept of the **atomic nucleus**. A few years later he detected the first **'free' protons** emerging from the collisions of α-particles with nuclei, and it was his ex-student and collaborator James Chadwick who in 1932 first identified **free neutrons**.

These proved to be by far the most effective 'projectiles' for studying nuclei. Being relatively heavy, they plough through the electrons surrounding the nucleus, and being electrically neutral they are not deflected away by the positive nuclear charge. Enrico Fermi in Rome was amongst the first to use neutrons in this way to study matter. Working his way rapidly through some sixty different elements, he discovered that firing neutrons at target atoms almost always caused the target to become radioactive, emitting β-particles. This led him to believe that elements one place higher in the periodic table were being produced (see the subsection on beta particles, above). Continuing the process, Fermi eventually reached the ultimate target material: uranium, the heaviest naturally occurring element. When he again found radioactivity in the bombarded target, the same reasoning led him to the conclusion that he had produced elements with atomic numbers greater than 92, the so-called **transuranic elements** or **actinides**. He had indeed done this, but closer study of his results showed an even more significant achievement – one that could justly be called world-shattering.

Fission

Fermi's experiments were soon repeated by others, including the German chemist Otto Hahn and his physicist colleague Lise Meitner. In the period from 1933 to 1938, Hahn and Meitner, joined later by Fritz Strassmann, continued the neutron experiments, as did Fermi in Italy and others elsewhere. Identifying the products of these neutron interactions was not easy, and some strange results began to appear.

In Paris, Frederic and Irène Joliot-Curie, using delicate techniques for chemical analysis of tiny amounts of material, reported in 1938 that they seemed to have found *lanthanum* in the products of irradiated uranium. But a lanthanum nucleus is little more than half the size of a uranium nucleus, and nuclei do not fall to pieces. They may release small particles, as in radioactivity, but otherwise they are the most stable objects known (see the subsection on nuclear forces in Chapter 4). The authors concluded, doubtfully, that they must have found some transuranic element with similar properties. On hearing this, Hahn and Strassmann set to work and in a brilliant piece of chemical analysis established that one product was definitely *barium*, atomic number 56. This was so amazing that although they trusted their chemistry they could not at first bring themselves 'to take such a drastic step which goes against all previous experience in nuclear physics' (Hahn and Strassmann, 1939).

Figure 10.5 Otto Hahn (1879–1968) was born in Frankfurt-am-Main and studied chemistry at Marburg. After a year in Montreal working on radiochemistry with Rutherford, he joined the Berlin University Chemical Institute. His role in the discovery of fission led to a feeling of responsibility for its use or misuse, and during the Second World War he distanced himself from the weapons programme. In 1945 he was held with other German scientists in a country house in England, where their conversations were recorded. The group's initial reaction to the news of Hiroshima and Nagasaki was disbelief, but as they became more convinced they kept careful watch on Hahn, seriously afraid that he might kill himself. In the post-war years, Hahn continued to write on the peaceful applications of radioisotopes and fission – including, in 1950, a detailed analysis of the safety measures that would be needed in a nuclear power station.

Figure 10.6 Lise Meitner (1878–1968), daughter of a Viennese lawyer, entered the University of Vienna in 1901 to study physics. With Boltzmann as professor, and others active in the new field of radioactivity, the university was world-famous – an intimidating place for a quiet young woman educated privately at home. But Meitner was an able student and in 1907, her doctorate completed, she moved on to Planck's department in Berlin. There she joined Otto Hahn in a working partnership that was to last for thirty years. She was initially barred from the main rooms on the upper floors of the Institute, but was allowed into Hahn's radioactivity area on the ground floor – as a special concession to him. She eventually became professor and Head of Physics, and as an Austrian citizen, was able to continue after 1933 despite her Jewish background. This changed after the Anschluss, and in 1938 she left Berlin – travelling illegally, escorted by a Dutch friend of Bohr. She accepted a post in Stockholm, declined an invitation to join the Manhattan Project, and continued working in Sweden until retiring to Cambridge, UK, in 1960.

Hahn wrote to Lise Meitner, now in Stockholm, saying 'Perhaps you can suggest some fantastic explanation'. When her nephew Otto Frisch arrived from Copenhagen to spend Christmas with her, he found her reading Hahn's letter. In the next couple of days, sitting in a small hotel, they worked out an explanation. The resulting paper (Meitner and Frisch, 1939) proposed a mechanism by which a heavy nucleus absorbing a neutron might become unstable and split into two lighter nuclei (Figure 10.7). They called the process **nuclear fission** (*kernspaltung*) and observed that it should release a great deal of energy (see Box 10.4). It didn't take long for the news to spread. Niels Bohr, head of the Copenhagen Institute, left for a planned visit to Princeton in January 1939. He was met in New York by Fermi, who had arrived a few days earlier as an émigré from fascist Italy. (His wife Laura was Jewish.) A couple of weeks later Bohr and Fermi described the fission results at a theoretical physics conference in Washington, and Fermi made the startling suggestion that free neutrons might be released in the process. That there would be spare neutrons was no surprise, but it did not necessarily follow that *free* neutrons would appear (see Box 10.5).

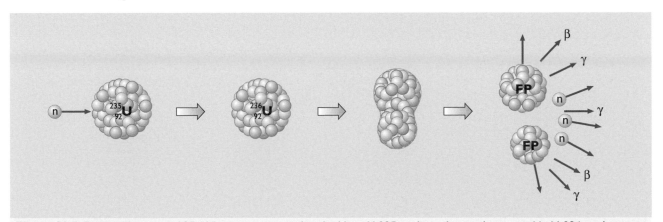

Figure 10.7 Fission of uranium-235. When a neutron is absorbed by a U-235 nucleus, the resulting unstable U-236 nucleus can take up a dumb-bell shape as shown. Electrical repulsion between the two positively charged parts of this can then lead to fission, producing two new radioactive nuclei, the fission products (FP). Two or three free neutrons (n) and gamma radiation (γ) are also emitted in the process

BOX 10.4 Energy from fission

It was not necessary for Meitner and Frisch to wait for measurements of the energy released in fission. The masses of individual atoms, and of particles such as the neutron, were known quite accurately by the 1930s, and the energy could be calculated by the method shown in Box 10.2.

- The *initial mass* is the sum of the masses of the U-235 nucleus and the initiating neutron.

- The *final mass* is the sum of the masses of the fission product nuclei and any spare free neutrons.

- If the final mass is less than the initial mass, the difference will appear as energy released in the process.

The fission of U-235 can have many different outcomes: different fission products and different numbers of free neutrons, but on average the final mass is less than the initial mass by about 0.21 u, a little over one fifth of an atomic mass unit.

Converting this to energy units (Box 10.1) shows that the energy released will be about **200 MeV per fission**.

This is only about 3.2×10^{-11} joules, which may seem a very small amount, but the picture changes if we consider not one atom but one kilogram of U-235. This contains about 2.56×10^{24} atoms, and the energy released in the complete fission of **one kilogram of pure U-235** becomes

$$2.56 \times 10^{24} \times 3.2 \times 10^{-11} = 82 \times 10^{12} \text{ J} = \textbf{82 TJ}.$$

This is equal to the energy released in burning about 3000 tonnes of coal.

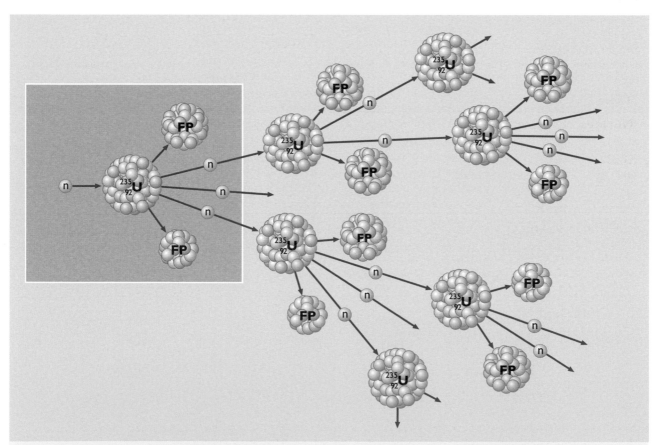

Figure 10.8 An explosive chain reaction. The highlighted area shows the fission process of Figure 10.7 (omitting the intermediate stages). Two of the free neutrons from this fission initiate further fissions, from each of which two more continue the process, and so on. To achieve this rate of multiplication a high concentration of U-235 would be required

The significance of Fermi's remark was immediately obvious to his audience. Neutrons cause fission, so a free neutron could cause a further fission, and more than one free neutron from each fission event could produce a **chain reaction** (Figure 10.8). With only thousandths of a second between events, the result would be a great deal of energy in a very short time – a nuclear explosion.

> This suggestion threw the meeting into an uproar while physicists who had facilities initiated calls to their laboratories to start the search for fission neutrons.
>
> Manley, 1962

Within a day, fission appeared for the first time in the news headlines. Within a year a hundred papers had been published in scientific journals – and in the course of that year the Second World War had broken out.

BOX 10.5 **The surplus neutrons**

A uranium-235 nucleus undergoes fission into two new nuclei. If one of these products is barium, what is the other, and how many surplus neutrons should result from this process?

The atomic numbers of uranium and barium are 92 and 56 respectively. So to keep the same number of protons, the atomic number of the other element must be 92 minus 56, i.e. 36. This is the gas krypton.

Table 10.2 Spare neutrons in fission

Nucleus	Number of protons	Number of neutrons
U-235	92	143
Ba-138	56	82
Kr-86	36	50

Suppose now that the products are the heaviest stable isotopes of these two elements. The first column in Table 10.2 includes the mass numbers (protons plus neutrons) for these, and the third column gives just the numbers of neutrons. Inspection shows that the original U-235 plus the initiating neutron provide twelve more neutrons than are needed by the two product elements.

This calculation is misleading, however, in assuming that the product nuclei are stable, with the correct ratio of neutrons to protons. In practice this is by no means the case, and they are almost always produced with too many neutrons. This has two extremely important consequences. Firstly, it means that the number of *free* neutrons available for the chain reaction is not twelve but perhaps two or three. And secondly, it means that the new 'neutron-heavy' nuclei are unstable. We have seen in Section 10.2 that emission of a β-particle reduces the neutron–proton ratio, so the new nuclei are often β-emitters. As they usually start with excess energy, γ-emission is also common.

Barium and krypton are not the only products of the fission of U-235. Several dozen different elements have been detected, but this example is typical in that the split is usually asymmetrical, with the atomic number of one product in the mid-fifties and the other in the mid-thirties. It is also typical in that the fission products are usually highly radioactive.

10.4 **1939–1945: reactors and bombs**

Explosions, whether chemical or nuclear, are not the subject of this book, which is concerned with the *controlled* use of energy. However, the sequence of events leading to the first nuclear weapons included the development of concepts and technologies that fall within our remit, and the following short account of these events is therefore included.

Figure 10.9 Niels Bohr (1885–1962) was born into a prominent Copenhagen family of academics and bankers. He studied at Copenhagen University, where he later established the world-famous Institute for Theoretical Physics. In his first major work, in 1913, he applied the quantum ideas of Planck and Einstein to Rutherford's nuclear atom. His model, the 'Bohr atom', was a remarkable imaginative leap, for which he received the Nobel Prize. He was in fact a remarkable man: lugubrious in photographs but a great teller of funny stories; an appallingly bad lecturer who was revered and loved by students and colleagues alike. For many theoretical physicists in the mid-twentieth century, Bohr was a greater influence even than Einstein, with whom he shared mutual respect and affection.

The theoretical possibility of a fission weapon was evident to everyone after the Washington conference, but many scientists, including Bohr and Fermi, remained sceptical about its practicability. Bohr had established that only the U-235 isotope readily underwent fission, and as we've seen, each U-235 atom is 'diluted' in natural uranium by about 140 atoms of U-238. On average about 2.5 free neutrons are produced in each fission of a U-235 nucleus, which should be enough for an explosive chain reaction. But experiments showed that many of these high-energy **fast neutrons** are absorbed by the majority U-238 isotope or lost in other processes, with the result that in natural uranium on average *less* than one neutron actually causes another fission, and no chain reaction occurs. A possible solution would be to increase the proportion of the U-235 isotope; but this process, known as **enrichment**, would not be simple. Whilst different *elements* can be separated by standard chemical means, the separation of different *isotopes* of the same element is far more difficult and had never been achieved in large quantities.

Even with highly enriched uranium (if it could be achieved) there would still be neutron losses in processes that did not lead to fission, and neutrons escaping out through the surface. This second fact was recognized as very important, because the proportion that escapes depends on the total volume: the larger the block of uranium, the smaller the *percentage* loss through its surface. There would therefore be a critical size, a **critical mass**, at which the total neutron loss became low enough for a chain reaction to be maintained. Everyone knew that no mass of natural uranium, however large, could maintain a chain reaction, and everyone realized that the greater the enrichment, the smaller the critical mass. But the experimental data needed to calculate critical masses were not yet available – and a bomb requiring tonnes of highly enriched uranium would certainly not be practicable.

Despite the reservations and doubts, work on fission continued. In the summer of 1939, a group of scientists approached President Roosevelt for funding, and in February 1940 he made available the sum of $6000 (!) for fission research. In England at about the same time, Otto Frisch and Rudolf Peierls at Birmingham University were tackling the question of critical mass. Using all available data and some inspired guesses, they concluded that for pure U-235, about *one kilogram* would be enough.

> At this point we stared at each other and realized that an atomic bomb might after all be possible.
>
> Frisch, 1979

They presented their results in the Frisch–Peierls Memorandum (reproduced in Gowing, 1964), which included the suggestion that an explosion could be achieved by rapidly bringing together two sub-critical masses. Their value for the critical mass proved rather optimistic, but even ten times this mass of U-235 (which was eventually the case) would still be practicable.

The first reactor

More experimental data were needed, and by 1941 groups in several countries had projects for the construction of a controllable **nuclear reactor** (called at the time an *atomic pile* in the UK, *uranium-pile* in the US and

BOX 10.6 **Moderators**

The material that slows down the neutrons in order to maintain the chain reaction is called the **moderator**[1].

The main requirement for an efficient moderator is that it should rapidly reduce the speed of the neutrons without absorbing them. The best moderators have light atoms – for a simple reason. If a fast neutron collides with a stationary object of similar mass (a light nucleus) it will be brought almost to a stop; but if it collides with a much heavier nucleus it will be deflected without much change in its speed. (Compare a moving ball bearing colliding with a stationary ball bearing or with a stationary cannonball.)

The moderator atom is never completely stationary at the time a neutron collides with it. Like all atoms it will always have thermal energy (see Chapter 4), and the resulting **slow neutrons** will also carry this residual energy. They are often referred to as **thermal neutrons**.

Hydrogen should be the ideal moderator. Its nucleus, a single proton, has almost the same mass as a neutron, and it has the further advantage of being readily available in the form of ordinary water. Unfortunately, however, the neutron tends to combine with the single proton, and the resulting loss means that ordinary water cannot be used as moderator with natural uranium.

There is, however, another hydrogen isotope: hydrogen-2, or **deuterium**. (Only in the case of hydrogen do isotopes have individual names.) The nucleus, called a **deuteron**, consists of a proton plus a neutron. It is of course heavier, but it does not absorb the colliding neutron and can be used with natural uranium. In the form of **heavy water** (D_2O), it could be an effective *liquid* moderator. Deuterium occurs as 0.015% of natural hydrogen (about 1 atom in 7000). Ordinary water (H_2O) has a similar proportion so there is no absolute shortage of deuterium. However, extracting it involves separation of isotopes, and several tonnes of water must be processed for every kilogram of heavy water obtained.

The most easily available effective moderator is carbon. It can be obtained in very pure form as **graphite**, and although its nucleus is considerably heavier than a deuteron, its neutron absorption is low, so it can be used with natural uranium. This was Fermi's choice for the world's first reactor, but as we shall see, all the moderators described here were in use within a few years.

[1] A common misunderstanding arising from this name is that the moderator reduces the fission rate. Quite the contrary: by moderating the neutron speed, it *increases* the fission rate.

uranbrenner in Germany). Fermi had shown in the 1930s that the chances of a neutron being absorbed by a U-235 nucleus, thus inducing another fission, were greatly increased if the fast neutrons were slowed down by allowing them to collide with relatively light nuclei (Box 10.6). So everyone knew in principle that a sustained chain reaction in natural uranium might be possible. And everyone knew that the others knew.

The US had set up a Uranium Committee, and Fermi, now at the University of Chicago, was already working on reactor development. In Britain, a report based on the Frisch–Peierls Memorandum expressed confidence that a bomb was possible. This was passed to the scientists in America, who at once arranged another meeting with President Roosevelt. On 9 October 1941 he made the decision to throw the huge resources of the US into the development of an 'atomic bomb'. On 7 December, the Japanese attacked Pearl Harbor and the United States entered the Second World War.

In July 1942, Winston Churchill agreed that the bomb programme, now called the Manhattan Project, should become a joint allied effort and most of the people then working on it in the UK, including refugee scientists from Germany and occupied countries such as France and Poland, should move across the Atlantic. The Chicago group and a British–French– Canadian team in Montreal would work on experimental reactors, whilst the main site for weapons development would be Los Alamos in New Mexico, with reactors for plutonium production (see below) located elsewhere in the US.

On 2 December 1942, carefully piling up lumps of uranium metal and graphite, with strips of cadmium (a neutron absorber) to control the process, the Fermi team achieved criticality – the first controlled chain reaction (Figure 10.10). The Montreal reactor and plutonium-producing reactors elsewhere in the US followed. The German project included a heavy water plant in occupied Norway, but still had no operational reactor at the end of the war. Unknown to the allies, it had been decided in 1942 to continue the nuclear work on a small scale only. The decision was made by the politicians, but the extent to which the German scientists, led by Werner Heisenberg, were actively pursuing the development of a weapon has remained the subject of debate for half a century (Pais, 1991).

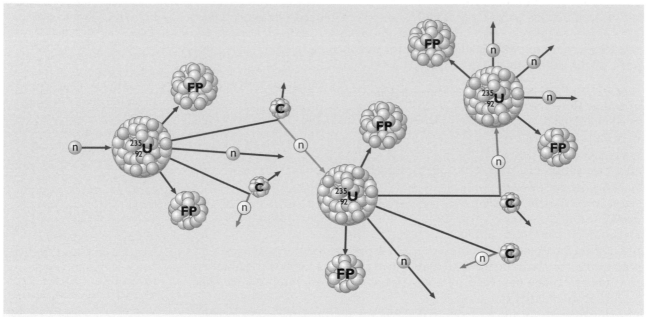

Figure 10.10 A controlled chain reaction. Some of the fast neutrons (n, purple tracks) released in fission are slowed down in collisions with light carbon nuclei (C). The resulting slow neutrons (n, lilac tracks) are able to maintain a chain reaction in natural uranium. In the steady state just one neutron from each fission leads to another fission

New elements

Fermi was correct when in the 1930s he claimed to have produced transuranic elements by neutron bombardment of natural uranium. The quantities were far too small for direct identification, but Hahn and Meitner confirmed the result when they established the presence of a uranium isotope that emitted β-particles. As we have seen, β-emission increases the atomic number by one, so β-emission from uranium must lead to the element with atomic number 93. When this in turn was found to emit β-particles, it was evident that element 94 was also being produced.

Production of these two new elements in quantities just large enough for chemical analysis was eventually achieved in 1941, and they were given the names **neptunium** and **plutonium** (after the planets beyond Uranus).

The sequence leading to their production by neutron bombardment can be written as a nuclear equation:

$$^{238}_{92}U + ^{1}_{0}n \longrightarrow \,^{239}_{92}U \xrightarrow[\text{23 minutes}]{\beta^-} \,^{239}_{93}Np \xrightarrow[\text{2.3 days}]{\beta^-} \,^{239}_{94}Pu$$

The symbol β^- indicates that an electron is emitted in the transformation, and the time below each arrow is the half-life. Plutonium-239 is also radioactive but its half-life is much longer, about 24 000 years, so steady bombardment of U-238 with neutrons will result in a continuous build-up of Pu-239.

The identification of plutonium was probably the earliest major event in this history not to be announced to the world at large. To establish priority the authors at once wrote a paper and sent it to the journal *Physical Review*, but at their request it was withheld from publication until the end of the war (Seaborg *et al.*, 1946). The reason for the secrecy is obvious. As had been predicted on theoretical grounds, *plutonium-239 is fissile*.

A **fissile** nucleus is one that undergoes fission relatively easily – absorption of a *slow* (low energy) neutron is sufficient. The only naturally occurring fissile nucleus is U-235. Plutonium does not exist in nature, except perhaps in minute quantities (Box 10.7), but it is produced in uranium reactors by the reaction shown above. Isotopes like U-238 that can lead to fissile isotopes after absorption of a neutron are called **fertile**, and any reactor containing U-238 will necessarily produce some plutonium. But this process is selective, requiring fast neutrons. The early plutonium-producing reactors had therefore to maintain a fine balance between the slow neutrons required for a chain reaction in natural uranium and the fast neutrons needed for plutonium production. In the later **fast breeder reactors**, designed to produce more fissile material than they consume, the chain reaction is maintained by fast neutrons – but the fuel in these is not natural uranium (see Section 10.7 below).

Figure 10.11 Enrico Fermi (1901–1954) and his brother Giulio were science prodigies, inseparable until Giulio tragically died at the age of 15. Enrico took refuge in study, teaching himself from books found in the local flea market in Rome, and at 17 winning a fellowship at Pisa. When he was appointed professor at Rome in 1927 he had already made his first major contribution to physics, the 'Fermi statistics' that were the basis for future theories of metals and semiconductors. Turning to nuclear physics, he developed a theory of β-decay, and revealing his equal brilliance as an experimentalist, carried out the neutron experiments that led ultimately to fission. Fermi was truly unique. He rarely used advanced textbooks, preferring to derive everything himself, and his approach to newly published work was to read the introduction, solve the problem and see if he agreed with the author. His lectures were dazzling, with new insights instead of standard approaches to problems.

BOX 10.7 A natural reactor

Some 2000 million years ago several natural nuclear reactors spontaneously achieved criticality at Oklo in Gabon, West Africa. The evidence is still there in the ground, in the form of stable isotopes resulting from millions of years of decay of the reactor products.

How could a sustained chain reaction have happened in natural uranium? The first important point is that U-235 decays much faster than U-238. Today, U-235 accounts for only one in every 140 atoms of natural uranium, but two billion years ago it was about one in 30 – enriched uranium, in today's terms. Secondly, the local ore was particularly rich in uranium; and thirdly, water was present in the area to act as moderator.

It is estimated that these reactors operated for about 2 million years and may have produced about 7 tonnes of radioactive wastes, including plutonium. These of course decayed in turn, and after another 2 million years the remaining plutonium would perhaps have amounted to one or two atoms.

Source: Curtin University, 2002

Atomic bombs

With the discovery of plutonium, two fissile materials were available, and with them two potential routes to a nuclear weapon. Neither would be easy. The uranium method required isotope separation on a scale many times greater than had ever been attempted, and the plutonium route required the construction of nuclear reactors on a scale many times greater than the only one in existence. In both cases there were still many unknowns. Nobody had yet attempted industrial-scale chemical and physical processing of intensely radioactive materials. And no one had ever worked with fissile material in quantities approaching the critical mass.

The decision was taken to proceed on both routes, and both reached the intended destination. On 16 July 1945 the first bomb, a plutonium device, was exploded at Alamogordo, New Mexico; on 6 August the first uranium device exploded over Hiroshima; and on 9 August a second plutonium bomb was dropped on Nagasaki.

'Swords into ploughshares'

Not long after the discovery of fission, and some years before the first reactor, it was suggested that controlled nuclear power might eventually fuel the boilers in power stations, and this theme was taken up again in the post-war years (Figure 10.12). All reactors generate large amounts of heat energy – about 25 million kilowatt-hours for each new kilogram of plutonium in the production plants, for instance – and schemes to make use of this heat were gradually introduced. The first nuclear power stations began to appear in the 1950s, and the Soviet Union claimed the world's first commercial plant in 1954, producing 30 MW of heat and 5 MW of electric power.

However, as item '*n*' in Figure 10.12 indicates, most reactor development was still directed towards military ends. The US, while continuing plutonium production at its Hanford site in Washington State, also developed compact uranium reactors using enriched uranium as marine propulsion systems. Needing no oxygen for combustion and with refuelling intervals of a year or more, these reactors were particularly suitable for submarines cruising under the world's oceans. Their suitability as heat sources for power stations is perhaps less obvious, as re-fuelling involves closing the plant for some days, but nevertheless the successors to these reactors came to dominate the world market during the later years of the twentieth century.

The UK, without enrichment facilities until 1953, initially produced plutonium for its weapons in reactors of a different design, using natural uranium. The Calder Hall power station, inaugurated in 1956, was sited near two existing plutonium-producing reactors at Windscale (now Sellafield), in Cumbria, and its reactor was a development of their design. Operated by the United Kingdom Atomic Energy Authority (UKAEA), its main purpose was still plutonium production, but the UK was able to claim the world's first *large-scale* nuclear power station, supplying 90 MW of power to the grid.

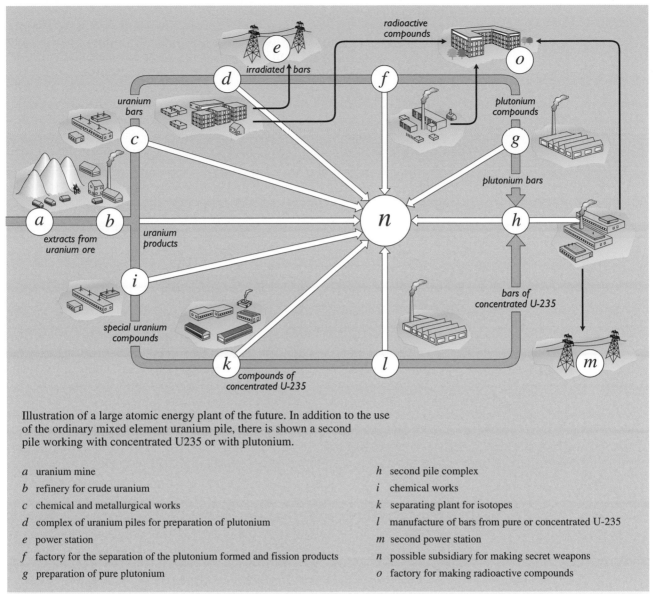

Illustration of a large atomic energy plant of the future. In addition to the use of the ordinary mixed element uranium pile, there is shown a second pile working with concentrated U235 or with plutonium.

a uranium mine

b refinery for crude uranium

c chemical and metallurgical works

d complex of uranium piles for preparation of plutonium

e power station

f factory for the separation of the plutonium formed and fission products

g preparation of pure plutonium

h second pile complex

i chemical works

k separating plant for isotopes

l manufacture of bars from pure or concentrated U-235

m second power station

n possible subsidiary for making secret weapons

o factory for making radioactive compounds

Figure 10.12 'A large atomic energy plant of the future' as foreseen in 1946, based on an illustration in a United Nations public information document (UN, 1946), reproduced in Hahn, 1950

10.5 Thermal fission reactors

A nuclear power station is in many respects similar to a fossil fuel plant, with the nuclear reactor as the boiler producing steam for the turbines. In the half-century since the first nuclear power plants were developed, many different designs of reactor have been tested, of which half a dozen or so have come into use. All of course depend on fission, and most are **thermal reactors**, in which slow neutrons are used to maintain the chain reaction.

It is important to distinguish two uses of 'thermal' in the context of power stations. All nuclear power stations use heat to generate steam or hot gas, and are therefore 'thermal' power stations in the sense introduced in

Chapter 2. However, as we shall see later, not all of them use the *thermal reactors* that are the subject of this section. Here, we look first at the features that all thermal fission reactors have in common, and then see how these are implemented in a few specific reactor types.

The reactor core

The basic requirement for a controlled chain reaction is that precisely one neutron from each fission should interact with another fissile nucleus to induce a further fission. Less than one neutron and the reaction will fizzle out; more than one and it can quickly become an explosion. There are therefore four essential components in the core of a reactor using slow neutrons.

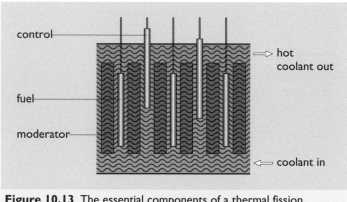

Figure 10.13 The essential components of a thermal fission reactor core

Fuel

A minority of present-day reactors use natural uranium fuel, either metallic uranium or uranium oxide, but most use enriched uranium, with U-235 content increased to between 2 and 5% (Box 10.8). A greater neutron loss can then be tolerated, with the important consequence that ordinary water can be used as a moderator. The fuel must of course be replaced as it becomes exhausted, and in the normal arrangement oxide pellets are enclosed in cylindrical casings, forming rods that can be inserted or withdrawn.

Moderator

Reactors may be identified by the moderator they use. **Light water reactors** use ordinary water – by far the most common today of the three moderators discussed in Box 10.6. **Heavy water reactors**, accounting for a small proportion of existing plants and a few under construction, are the only type still to use natural uranium. The third moderator, **graphite**, is used in most existing UK reactors, with natural uranium in the older type and enriched fuel in later designs, as is the case in one Russian type. Many graphite-moderated reactors are therefore still operating, but none has been built for some time. However, the proposed pebble-bed reactor (see Chapter 11) may change this situation.

BOX 10.8 **Uranium**

Uranium passes through a number of different stages between the mine and the reactor. The following brief account describes the processes involved, and introduces some of the terminology used in this context.

In the context of reactor fuel, the term **natural uranium** always means uranium with the naturally occurring isotopic ratio: 99.27% U-238, 0.72% U-235 and 0.005% U-234. **Enriched uranium** has a higher proportion of the fissile U-235 and **depleted uranium** a lower proportion. These terms are used regardless of the chemical state of the uranium.

Uranium is a fairly common element, found in many types of rock and also in the oceans. As an element, it is a heavy metal, nearly twice as dense as lead; but in nature it almost always occurs as a mixture of uranium oxides, represented chemically as U_3O_8.

In the customary mining terminology, a body of rock is defined as **uranium ore** if it contains an economically recoverable concentration of uranium. Currently this quantity varies from a maximum of a few per cent of U_3O_8 in the rock for a **high-grade ore**, down to a lower limit of about one part in a thousand. After extraction, the U_3O_8 is in the form of a compressed powder called 'yellowcake', which must be further processed to produce the type of fuel required.

In some older reactors, the fuel used is simply uranium metal, but the usual present-day form is another oxide, UO_2, which is a hard ceramic, chemically resistant and suitable for forming into pellets.

If the above chemical and physical processes are the only ones involved in **fuel fabrication**, the reactor is said to use *natural uranium* fuel, whether this is uranium metal, the oxide or some other chemical form.

Most present-day reactors, however, require enriched uranium, in which case the yellowcake goes first to an enrichment plant. Enrichment doesn't change the chemical behaviour of the uranium, so the final fuel fabrication uses normal chemical and physical processes to produce UO_2 pellets, as with natural uranium.

More details of the processes mentioned here appear in Section 10.6 below, and their environmental and other consequences are also discussed in Chapters 11 and 13.

Coolant

The **coolant** is the heat exchange medium: the fluid (liquid or gas) that carries heat from the reactor core. The name reflects its function in the earliest reactors, which was to carry away unwanted heat; but in a power plant reactor, whose whole purpose is to produce heat, the alternative term 'working fluid' seems more appropriate. However, the older term is still commonly used. An advantage of a liquid moderator is that it can also act as a coolant, and this is the case in most water-moderated reactors. Graphite is of course a solid, so a graphite-moderated reactor requires a separate coolant. The UK reactors use carbon dioxide gas, whilst the Russian type, which has more highly enriched fuel, uses water as its coolant.

Control

Routine control of the rate of fission in a reactor, and therefore its heat output, is achieved by designing for slightly *more* than one new fission each time and then incorporating a neutron-absorbing material in the system. Cadmium and boron are good absorbers, and the reaction can be controlled by increasing or decreasing the amount of these present in the reactor core. This can be done by adding an absorber to the moderator or by means of **control rods** that can be moved in or out – or frequently by some combination of these and other methods. Another aspect of control is the ability to deal with events such as an unplanned increase in the rate of fission, and all reactors must include systems for rapid response to such emergencies.

Structures

Certain basic requirements determine the general form of all the reactors discussed here. The fuel must be distributed at the right density, surrounded by the moderator and in good thermal contact with the coolant. The coolant must stream freely past the hot fuel, at the highest possible temperature and pressure for maximum efficiency. This has a major effect on the design, because steel or other structural materials will absorb neutrons. So only those reactors with very low neutron loss in the moderator and coolant can afford to have individual pressure tubes carrying the coolant. In the majority of cases, the whole core is submerged in the flowing coolant in a single large pressure vessel.

It is worth considering for a moment the interior of the reactor core. The coolant must not occupy too much space, which means that it must flow fast in order to carry away the heat. So a very hot fluid streams at high speed and high pressure through narrow gaps and channels past the fuel rods, subjecting materials and monitoring instruments to mechanical stress as well as the combined effects of temperature and pressure, chemical attack and bombardment by sub-nuclear particles. The designer must ensure that even under these conditions fuel rods do not distort, control rods can move freely and nothing impedes the flow of coolant.

Safety

Shielding

With radioactive fission products, and radioactivity induced in all components by neutron bombardment, the radiation in an operating reactor can reach about one trillion (10^{12}) times the level that a person can tolerate for even a short period of time. Without shielding around the core, the 'safe' distance for working would be about eight kilometres, so this is obviously an important aspect of reactor design. As we have seen, the two most penetrating components of the radiation are neutrons and gamma radiation, so the shielding needs to be both a good neutron absorber and thick enough to stop the energetic gamma particles. The radiation shield usually consists of a thick layer of concrete, and if the system also has a steel pressure vessel, this will absorb much of the gamma radiation.

Containment

The safe containment of the radioactive material itself is obviously important. The coolant circulates through the reactor under high pressure, requiring physically strong containment, and as mentioned above, this is often achieved by surrounding the entire core by a strong pressure vessel. The fission products will include radioactive gases, so containing these and monitoring their emission from the plant is an important aspect of routine operation. Finally, the containment system should be designed to safeguard personnel in the plant and people in the surroundings in the case of 'non-routine events'.

Accidents

The main source of public concern about the safety of nuclear reactors is the potential for the release of very large amounts of radioactive substances, which can occur only if the containment fails. Major structural failure during normal operation is possible, but has never occurred, and the more likely event is that the pressure vessel is unable to withstand a sudden *unplanned* rise in core temperature and pressure, or possibly an explosion resulting from this. Any explosion – nuclear or chemical – is the result of a runaway energy-producing chain reaction, multiplying so fast that the energy doesn't have time to escape by the normal 'peaceful' means, as heat and light. Instead, the energy density rises until the bonds holding the material together are broken and it blows apart. In the words of the unfortunate spokesperson after one accident, it is an 'energetic disassembly'. The necessary conditions for a chemical explosion might arise in a nuclear reactor as a result of either a sudden loss of coolant or a 'runaway' chain reaction, and here we shall look briefly at each of these. The specific accidents mentioned below, and their consequences, are discussed in more detail in Chapters 11 and 13.

A major loss of coolant can have particularly serious consequences in a nuclear reactor. In a fossil-fuel plant, such failure would be dangerous, but at least the generation of heat energy could be stopped by cutting off the supply of fuel or air. Not so in a nuclear reactor. Even if the chain reaction stops instantly, radioactive decay of the fission products continues, and the rate of energy release immediately after shut-down can be almost a tenth of the full normal power output – enough to cause the temperature to rise at more than a hundred degrees a minute in the absence of a sufficient quantity of coolant. The half-lives of many of the products are very short, and within an hour the rate will have fallen to a tenth of this level; but the heat produced during this time is enough to melt the core and perhaps the floor of the reactor building as well: the scenario for the China Syndrome (the fictional idea that the contents might continue to sink towards the diametrically opposite point on the globe). The most serious known loss-of-coolant accident was at Three Mile Island in Pennsylvania, in 1979. There was a partial core melt-down and a chemical explosion did occur – the 'energetic disassembly' mentioned above – but the pressure vessel remained intact.

The other event, an uncontrolled divergent chain reaction, should be virtually impossible in a reactor with only slightly enriched fuel. If the coolant is also the moderator, its loss means that no chain reaction at all can be maintained; and in any case all reactors have emergency systems for the injection of neutron absorbers. However, human error can lead even to this improbable occurrence, as was shown by events at Chernobyl in the Ukraine in 1986. A runaway chain reaction led to a chemical explosion that breached the containment, exposing the core. The design of the Chernobyl reactor (see below and Chapters 11 and 13) contributed to the accident, and it is claimed that the sequence of events that led to it would not be possible in the reactors favoured by the US, Canada and the UK. One design feature did however have unexpected merit. When the base of the vessel was blown downwards, molten fuel flowed out, leading to concerns that a self-sustaining chain reaction might re-start. This did not happen, and it was eventually discovered that the fissile material had been diluted by a simultaneous flow of large quantities of sand that had formed part of the screening around the core; an interesting but somewhat unplanned 'safety system'.

It must be stressed that the events discussed here involved *chemical* explosions. The fear that the fuel of a thermal reactor will somehow turn into an 'atomic bomb' is probably not well-founded. A *nuclear* explosion requires a high density of fissile nuclei in a quantity large enough to make the neutron loss negligible. The critical mass is about 10 kg for pure U-235 and about 5 kg for Pu-239; but as we have seen, a much greater mass may be needed if the fissile material is diluted. The only way in which a thermal fission reactor with a few per cent of U-235 could become a nuclear bomb would be for an appreciable proportion of the fissile nuclei distributed throughout 100 tonnes of core to bring themselves miraculously into one small region.

Types of thermal fission reactor

The reactors described in the following brief accounts are the main types currently in operation, including some that are still in use, although superseded. The sizes of the plants vary, but typically a reactor using enriched fuel might produce about 3000 MW (3 GW) of heat energy, providing steam for turbo-generators that supply a gigawatt of electric power. The typical fuel load for this type is about 100 tonnes, a third of which is replaced in the course of each year (Box 10.9).

Light water reactors

Four-fifths of the reactors in today's nuclear power stations are **Light Water Reactors** (LWRs), and three-quarters of these are **Pressurized Water Reactors** (PWRs), direct descendants of the plants developed by the United States in the post-war period as propulsion units for submarines. Typically, a PWR produces about 3000 MW of thermal power, enough for two 500-MW_e turbo-generators. It uses uranium enriched to 3.5% U-235 and holds about 75 tonnes of fuel.

Figure 10.14 shows the structure of the core of a PWR. It consists of little more than the fuel elements, some 50 000 long, thin metal tubes containing

---

BOX 10.9 Power output of a thermal fission reactor

We can calculate the rate at which heat energy should be produced by the reactor described in the text, as follows.

- The quantity of fuel consumed in a year is one third of 100 tonnes, i.e. 33 tonnes.

- If the fuel is uranium oxide, the uranium content of this will be about 29 tonnes.

- Suppose that the fresh fuel has 3.5% U-235. About 0.5% is likely to be lost in non-fission events, and perhaps 0.7% will remain in the spent fuel removed after a year.

- The mass of U-235 undergoing fission in a year will therefore be 2.3% of 29 tonnes, which is 0.67 tonnes, or 670 kg. This is equivalent to rather less than 2 kg a day, or 0.076 kg an hour.

- We have seen (Box 10.4) that the complete fission of 1 kg of pure uranium-235 produces 82 TJ of energy.

- We know that one kilowatt-hour is 3.6 MJ, so one gigawatt-hour (1 GWh) is 3.6 million MJ. The above 82 TJ is therefore approximately 23 GWh.

- Finally, then, we see that the 0.076 kg an hour of U-235 produces 1.7 GWh of heat an hour, i.e. it produces heat at an average rate of 1.7 GW.

But according to the main text, this quantity of fuel supplies about *three* gigawatts of heat. What is the source of the discrepancy?

The answer is that we have not included one important factor: plutonium. Any uranium reactor inevitably produces plutonium, as we have seen, and in a power-station reactor an appreciable fraction of this plutonium will have undergone fission before the fuel is removed. This is the source of more than one third of the heat output of the reactor.

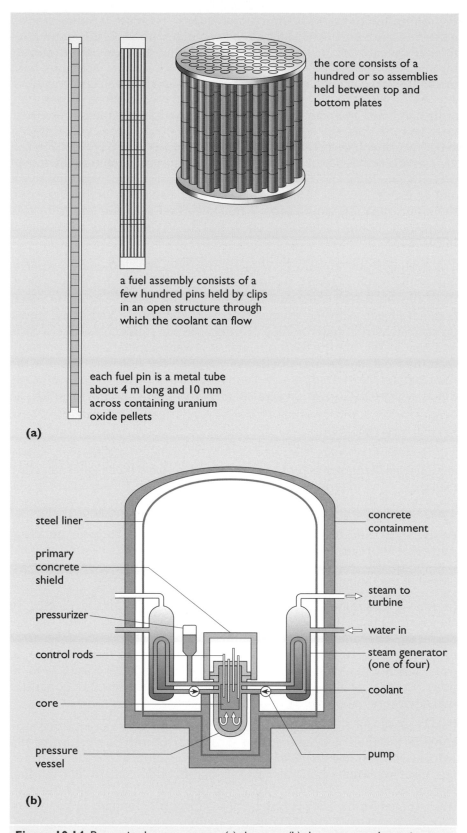

the core consists of a hundred or so assemblies held between top and bottom plates

a fuel assembly consists of a few hundred pins held by clips in an open structure through which the coolant can flow

each fuel pin is a metal tube about 4 m long and 10 mm across containing uranium oxide pellets

(a)

steel liner

primary concrete shield

pressurizer

control rods

core

pressure vessel

concrete containment

steam to turbine

water in

steam generator (one of four)

coolant

pump

(b)

Figure 10.14 Pressurized water reactor: (a) the core; (b) the reactor and containment

the fuel in the form of uranium oxide pellets. The core is submerged in water in a pressure vessel some 12 metres high with walls 20 centimetres thick (Figure 10.14). Insertion and withdrawal of fuel and control rods is through the top of the steel pressure vessel. The water, acting as both moderator and coolant, transmits heat from the core to the steam generators. A pressure of over 100 atmospheres prevents the water from boiling even at its maximum temperature of over 300 °C. Maintaining this pressure is very important, as the cooling requires water – not steam, which would carry away much less heat. The danger of a massive loss of coolant is taken very seriously, because if one of the large pipes were to fracture the loss of pressure would mean that the water would almost instantly become steam and be lost in seconds. Any loss of coolant is also a loss of moderator, so the chain reaction would stop – but nevertheless a fast and effective emergency cooling system is needed, and the PWRs usually have several in place.

The **Boiling Water Reactor** (BWR) is similar to the PWR in many respects, but as the name suggests, the water acting as coolant and moderator is itself allowed to boil to become steam for the turbines. The pressure vessel is much longer, with the submerged core at the bottom and steam under pressure above the water surface. Having no steam generators reduces costs and heat losses, but the BWR operates at lower temperature and pressure than the PWR, so the overall efficiency is not very different. There is the disadvantage that the radioactive steam flows to the turbine, but nevertheless there are more BWRs worldwide than any other reactor except the PWR.

Gas-cooled reactors

The four 23-MW generating sets of the UK's original Calder Hall power station used **Magnox** reactors, the name deriving from the magnesium alloy cladding of the natural uranium metal fuel elements. The moderator was graphite and the coolant carbon dioxide. In 1959, Calder Hall was joined by the similar Chapelcross plant, also designed by the UK Atomic Energy Authority (UKAEA) primarily for plutonium production for military purposes. Both these plants, subsequently owned by British Nuclear Fuels Limited (BNFL), continued in operation into the present century, but are closed at the time of writing and may not reopen.

In 1955 a programme was announced by the UK's Central Electricity Generating Board (the CEGB) for the construction of nine civil Magnox power stations, and these were commissioned over the period from 1962 to 1971 (Figure 10.15). Their electrical outputs ranged from 300 MW to over 1000 MW. In 1984 their operating lives were extended from 25 to 30 years. At the time of writing (2002) four of these have been shut down and two others are not presently operating.

Even before the first Magnox plants came into use, a committee was considering their successors, and in 1962 a prototype version of the **Advanced Gas-Cooled Reactor** (AGR) became operational. Like the Magnox, this used a graphite moderator and carbon-dioxide coolant, but the fuel was enriched to 2.3% and in the form of uranium oxide. In 1964 the AGR was adopted for the UK's future nuclear power programme, and a total of seven twin-reactor (2×660 MW) power stations were eventually completed, all of which are still in operation.

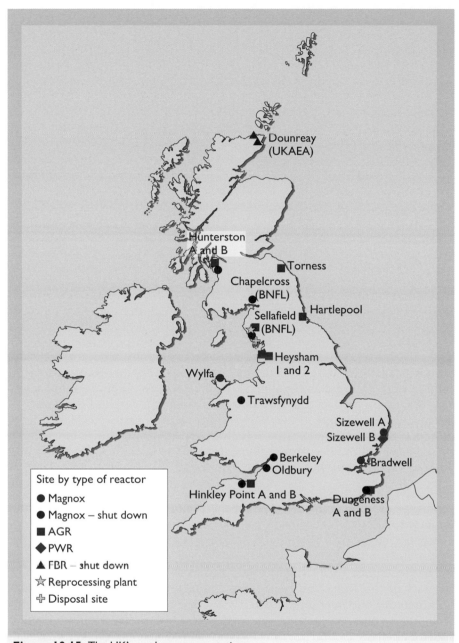

Figure 10.15 The UK's nuclear power stations

With its solid moderator, the AGR is structurally completely different from the PWR. Instead of an open lattice of thin tubes, the core is made of solid graphite. It is about nine metres across and pierced by several hundred vertical channels into which the fuel clusters slide (Figure 10.16). There are 2000 or so of these clusters, which are short and chunky, not long as in the PWR. The carbon dioxide gas coolant is pumped through the core at a pressure of about forty atmospheres, and gas temperatures as high as 600 °C were originally planned, though there were difficulties and this has not been achieved.

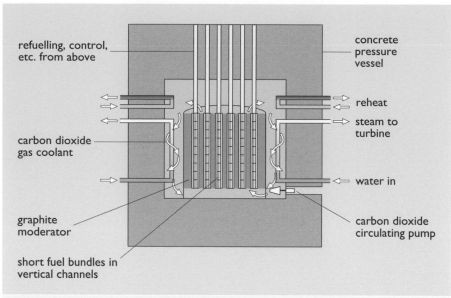

Figure 10.16 Advanced gas-cooled reactor

The pressure vessel is a massive reinforced concrete structure, a system which is claimed by advocates of the AGR to be less likely to suffer catastrophic failure than the steel vessel of the PWR. Other claimed safety advantages are the fact that the coolant is already a gas, and the ability of the heavy graphite core to absorb heat if the cooling system did fail. On the other hand, the planned online-refuelling (without needing to close down the system), which should have given the AGR an advantage over the PWR, has not proved to be very successful in practice.

The AGR had a much earlier potential rival than the PWR. In 1968 the prototype of another British design, the **Steam-Generating Heavy Water Reactor** (SGHWR), had come into operation, and the following decade saw a lengthy debate over the relative merits of the AGR and SGHWR for the next generation of UK nuclear power stations. The SGHWR eventually lost the battle, and in 1978 two more AGRs were ordered. But these were to be the last of the line. The AGR never managed to break into the LWR-dominated world market, and in 1981, long before the final two plants were operational, a planning application was submitted for the UK's first PWR. The site was one previously proposed for the SGHWR, at Sizewell on the Suffolk coast, where there was already a Magnox plant. After a long and bitterly fought public inquiry, the plan was approved, and the Sizewell PWR came online in 1995. At the time of writing the UK has no plans for new nuclear power stations.

The CANDU reactor

The **CANDU** ('Canadian-Deuterium–Uranium') reactors, with fourteen operating in Canada and nineteen overseas, are the only other type of reactor to have made any impression on the world dominance of the LWRs. The CANDU uses heavy water as its moderator and coolant, with natural uranium fuel. Taking advantage of the very low neutron absorption in

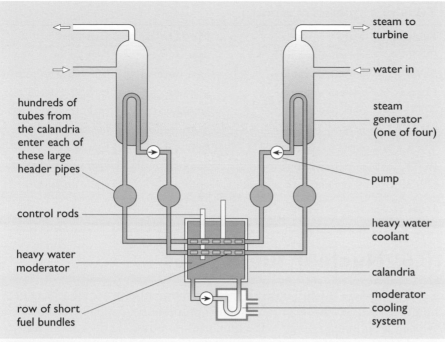

Figure 10.17 CANDU reactor

deuterium, it can afford to have the additional neutron-absorbing metal needed for a separate closed circuit carrying the coolant. The short fuel bundles lie inside hundreds of horizontal double-walled tubes, through which the coolant flows under pressure. These are surrounded by the heavy water moderator, which does not need to be pressurized as it remains below 100 °C. The whole system is inside a horizontal steel cylinder about eight metres long and eight metres in diameter, called the calandria, which is surrounded by a concrete radiation shield. Refuelling can be carried out online, without closing down the reactor. With a steam temperature of about 300 °C, the CANDUs have a relatively low thermal efficiency, but their efficient use of uranium compensates for this. Overall, they produce about 50 million kilowatt hours of electrical output per tonne of uranium, compared with 35 million for the LWRs (and about 3000 kWh per tonne of fuel in a coal-fired plant).

Russian reactors

The USSR developed a number of reactor types in a programme that resulted in about 70 operating power station reactors by the 1990s. Thirty of the present reactors are in Russia, thirteen in the Ukraine and the rest are distributed throughout other countries of the former-USSR. There are two main types of thermal reactor. One, the VVER, is a form of pressurized water reactor, but with significant differences from the American PWR, having, for example, several separate coolant circuits rather than a single pressure vessel. The other type, the RBMK, is a development from early Russian military reactors and uses a graphite moderator, light water coolant and uranium enriched to about 5% U-235. The Chernobyl reactors are of this latter type.

Table 10.3 Types of fission reactor

Reactor	Fuel	Moderator	Coolant
PWR	enriched uranium	light water	light water
BWR	enriched uranium	light water	light water/steam
Magnox	natural uranium	graphite	carbon dioxide gas
AGR	enriched uranium	graphite	carbon dioxide gas
CANDU	natural uranium	heavy water	heavy water
VVER	enriched uranium	light water	light water
RBMK	enriched uranium	graphite	light water/steam
LMFBR[1]	highly enriched uranium	none	liquid sodium

1 See Section 10.7 for details.

10.6 **Nuclear fuel cycles**

In order to assess the full costs or the environmental or social effects of an energy system, we need to look at the complete sequence of events from the original primary energy to the final useful output. In the case of nuclear power, this means the complete fuel cycle (Figure 10.18). We have already studied the reactors in some detail, and this section briefly describes the processes that make up the 'front end' and 'back end' of the fuel cycle. The effects of these processes are discussed later, in Chapters 11 and 13.

Mining and extraction

As described in Box 10.8, uranium occurs naturally in the form of oxides (U_3O_8), and the concentrations that currently justify mining range from about a tenth of a per cent up to a few per cent. Uranium itself is only mildly radioactive, but as we have seen (Section 10.2) it is accompanied in the ore by many radioactive 'daughter' products, including the gas radon. The unhappy consequences of this for miners in the past are mentioned in Chapter 13, but the implications for today are that good ventilation is an essential feature in underground mines.

After mining, the rock is crushed and treated to extract the uranium, usually by dissolving it in sulphuric acid to separate it from other material, and then recovering it again as 'yellowcake'. A potential environmental problem arises from the *tailings*, the residues after extraction of the uranium from the ore. Depending on the grade of ore, there can be as much as 1000 tonnes of these residues for each tonne of uranium extracted. Initially they are in the form of a slurry that may include chemically or biologically undesirable material and will contain all the radioisotopes from the ore (except the uranium, of course). Their total radioactivity can initially be twenty times that of the uranium itself. Appropriate safe storage of the tailings is obviously essential.

The yellowcake, with only the radioactivity of the uranium, does not present a major hazard, and this is the form in which the uranium is normally transported to the next stage. The world's present nuclear power stations require an annual supply of about 75 000 tonnes.

Enrichment and fuel fabrication

Any enrichment method must rely on the very small mass difference between the two main uranium isotopes, as they are essentially identical in all other respects. In both the current methods, the U_3O_8 is first converted chemically into uranium hexafluoride, UF_6, known as 'hex', which becomes a gas when slightly heated. **Gaseous diffusion**, originally developed to produce highly enriched weapons-grade uranium, remains the most common enrichment method. It uses the fact that the gas molecules containing U-235 are lighter than those containing U-238, and thus diffuse through a porous membrane very slightly faster. The difference is extremely small, and a thousand or more repetitions of the diffusion process are needed to increase the U-235 proportion to the required 3% or so. The newer **centrifugal** methods depend on the tendency of the heavier molecules to move outward when the gas is spun at a high velocity, leaving the inner layer richer in U-235. This is much more efficient, needing fewer repeated stages, and is therefore cheaper and less energy-consuming.

The relative amounts of enriched and depleted uranium resulting from the enrichment process depend on the degrees of enrichment and depletion. The U-235 content in the depleted stream is usually in the range 0.2–0.3% – roughly a third of the concentration in natural uranium. With 3.5% enrichment and 0.25% depletion, each tonne of enriched uranium results in about 6 tonnes of depleted uranium. The latter is at present mainly treated as waste, although some depleted uranium metal is extracted for (non-nuclear) military uses, and a small proportion is used in breeder reactors (Section 10.7 below).

As described in Box 10.8, the final fuel fabrication process converts the enriched UF_6 into pellets of the ceramic oxide UO_2.

Spent fuel

Typically, each fuel element spends 3 years in the reactor, and Table 10.4 shows details of the spent fuel removed at the end of this period.

Table 10.4 Spent fuel from a thermal fission reactor. The table shows the main constituents of 1 tonne (1000 kg) of spent fuel removed from a reactor after 3 years. The quantities are approximate and will vary with the enrichment of the original fuel, the type of reactor and the mode of operation. These data might be typical for a PWR using fuel enriched to 3.5% U-235.

Content	Quantity / kg	Notes
U-235	7	Fission of U-235 will have contributed about two-thirds of the power output. Its concentration in the spent fuel is about the same as in natural uranium.
U-238	940	The original U-238 content will have been reduced by neutron absorption leading to plutonium and other actinides.
Plutonium	9	More than half the plutonium produced from U-238 will already have undergone fission, contributing about a third of the total power output.
Fission products	38	These lighter radioactive isotopes, mainly with half-lives from fractions of a second to a few years, contribute over 99% of the initial radioactivity of the spent fuel.
Actinides, U-236, etc.	6	The heavy radioactive isotopes, many with very long half-lives, contribute most of the radioactivity after a few hundred years.

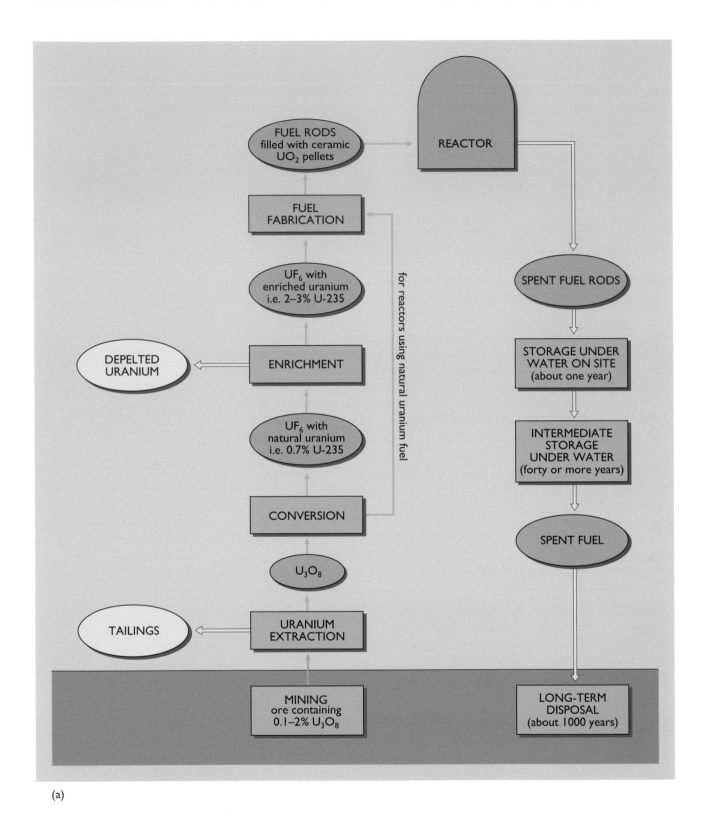

(a)

Figure 10.18 (a) The open fuel cycle; (b) the closed fuel cycle. Note that these diagrams do not include the intermediate- and low-level wastes resulting from each process

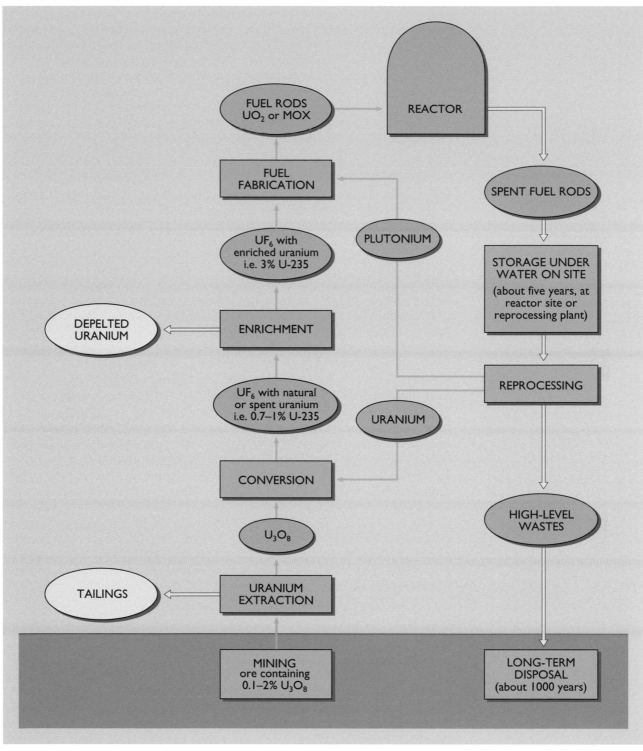

(b)

How the spent fuel should subsequently be treated has been the subject of debate for many years. It is highly radioactive, and the energy transferred by the emitted particles to the surrounding atoms as heat is sufficient to melt the solid material in a few minutes. Any storage system must therefore provide both a radiation shield and an efficient means of heat extraction. After removal from the reactor, spent fuel is normally submerged under a few metres of water in a tank equipped with a cooling system, where it remains for at least a year. Its subsequent treatment falls into two main categories: direct disposal and reprocessing.

The **direct disposal** route involves leaving the spent fuel in tanks for several decades, by which the time the radioactivity and heat production will have fallen to levels that allow other, more compact forms of storage (Box 10.10). Figure 10.18(a) shows the resulting **open fuel cycle**, from ore to final disposal. Of the 'nuclear' countries, Sweden, and for several decades the US, have chosen this option. Others, including Canada, whose CANDU reactor can produce spent fuel with very low fissile content, have effectively adopted it by remaining undecided.

Reprocessing involves separating the spent fuel into three components: uranium, plutonium, and the wastes (the last two items of Table 10.4). The basic chemical separation process has been used since the earliest reactors, being the means by which plutonium was extracted for the first nuclear weapons. In present-day civil reprocessing, the spent fuel is stored at the power station site for at least a year, after which the activity will have fallen to perhaps 10^{10} Bq per gram of material – still almost a million times the activity of natural uranium (see Box 10.3). At this stage it can, if necessary, be moved to the reprocessing and fuel fabrication site, but will remain in cooling tanks until a total of about 5 years has elapsed. The three components are then chemically separated: the uranium to be re-enriched for re-use and the plutonium to be either stored or used as the minor constituent in **mixed oxide fuel** (MOX). This potential alternative to the usual enriched uranium fuel makes use of depleted uranium to which is added 5–8% of plutonium as the fissile material, both constituents being in the form of oxides. The third component, the high-level waste, carries almost all the radioactivity of the spent fuel, and its safe disposal is a critical issue (Box 10.10). This complete sequence is referred to as the **closed fuel cycle** (Figure 10.18b).

The total world reprocessing capacity for fuel from civil reactors is sufficient to treat about 5000 tonnes of spent fuel a year. (A 1000 MW plant produces 25–30 tonnes of spent fuel each year, and the world annual total is more than 10 000 tonnes.) There are currently four major reprocessing centres: in the UK at Sellafield (two plants with capacities of 1500 tonnes per year and 850 tonnes per year), in France at La Hague on the Cherbourg Peninsula (1600 tonnes per year) and at Marcoule (400 tonnes per year), and in Russia at Mayak near Orenburg in the Urals (400 tonnes per year). India has three relatively small facilities, and Japan, which has one small plant at present, is currently building a large complex at Rokkasho, on the north coast of the main island, Honshu. The UK and France currently reprocess spent fuel from other countries including Japan, Switzerland and Germany. The US has not reprocessed fuel from civil reactors for several decades and has no current plans to do so. Belgium recently stopped sending fuel for reprocessing, and Germany intends to do so in 2005.

BOX 10.10 Radioactive wastes

The radioactive wastes arising from the nuclear power industry or other civil uses of radioactive materials are usually classified as high-, intermediate- or low-level, depending on their activity.

High-level waste is responsible for all but a few per cent of the total radioactivity from wastes. It is mainly spent reactor fuel or the separated fission products and actinides that result from reprocessing (Table 10.4). The latter two are initially mainly in liquid form, occupying roughly the same volume as the original spent fuel, and with almost the same activity. The two materials therefore initially require similar shielding and cooling treatment. However, unlike the spent fuel, which remains in tanks for some 40 years, the liquid waste from reprocessing can be concentrated after a few years to about a twentieth of its original volume and **vitrified**. In vitrification, the now dry waste is incorporated into about four times its own mass of molten glass, which is then poured into cylindrical steel containers. Still intensely radioactive and still generating heat, these must be kept in monitored storage for perhaps 50 years. The proposed method for final disposal of both spent fuel and vitrified wastes is to place them in deep underground stores, but at present no country has agreed detailed plans for their type and location, and it remains an issue for debate (see Chapter 11).

Intermediate-level wastes also come mainly from reactors, and account for rather more than a tenth of the total volume and rather less than a twentieth of the total radioactivity of all nuclear waste materials. Any substance in or near the reactor core is subjected to constant neutron bombardment, and many stable nuclei become radioactive after absorbing a neutron. This **induced radioactivity** means that components from the reactor core, and in particular items such as the fuel cladding removed during reprocessing, will be sufficiently radioactive to require shielding and safe storage. They are normally enclosed in concrete and placed underground.

Low-level wastes, mainly materials and equipment that have become contaminated by radioisotopes, are an outcome of medical and research uses of radioactive materials as well as the nuclear power industry. They account for almost nine-tenths of the volume of radioactive material to be disposed of annually, but contribute no more than 1% of the total activity. Nevertheless, they are usually regarded as materials that should be buried or otherwise disposed of at special sites.

The world total capacity for MOX fuel fabrication is about 200 tonnes per year, mainly in Belgium and France, with small plants in the UK and Japan. The three European countries also have vitrification facilities (see Box 10.10) whose combined capacity can treat the high-level wastes from about 5000 tonnes of spent fuel a year.

The supporters of reprocessing argue that it is technically the best solution to the problem of the disposal of spent fuel. It reduces the demand for 'new' uranium, uses up the plutonium as a component of MOX fuel, and reduces the highly active wastes to a twentieth or less of the original volume. Opponents argue that there is no shortage of uranium, that separating out the plutonium makes it more easily available for possible misuse, and that whilst the volume of wastes may be reduced, their activity is virtually unchanged. (See Chapter 11 for further discussion of these issues.)

10.7 Fast reactors

As the name suggests, in a **fast reactor** the chain reaction is maintained by fast neutrons. Consequently, fast reactors differ in two essential respects from thermal reactors: they need no moderator to slow down the neutrons, and they need more highly enriched fuel – as much as 20% U-235. This second requirement follows from the fact that fast neutrons are much less efficient at inducing fission in U-235. Fast neutrons are, however, much *more* efficient at producing plutonium from U-238, and the prospect of 'breeding' fissile material from the growing amounts of depleted uranium was a main motivation for the development of the **Fast Breeder Reactors** (FBRs). They do of course need fissile material to maintain the chain reaction at the start, but provided there are enough free neutrons, the quantity of plutonium will gradually build up, and a cycle can be achieved in which more fissile material is extracted than is loaded as fresh fuel. Like all reactors, an FBR produces heat, and can be used as a power station reactor. Indeed, the first ever 'nuclear' generation of electric power used heat from an experimental breeder in the US in the 1950s. Since then there have been about twenty operational FBRs in eight different countries, but as we'll see in Chapter 11, few breeder programmes remain active at the time of writing.

The liquid-metal fast breeder reactor

The concentrated fuel of an FBR generates more heat per cubic metre of core than a thermal reactor. The coolant must be able to carry away this extra heat, should have reasonably heavy atoms so that it does *not* act as a moderator, and should not of course absorb neutrons. The majority of FBRs have used sodium, a metal that melts at about 100 °C and boils at about 900 °C. It is therefore liquid at the core temperature of about 600 °C without the need for high pressure. Figure 10.19 shows in outline one form of **Liquid-Metal Fast Breeder Reactor** (LMFBR). The vertical fuel elements are steel tubes three metres long and six millimetres in diameter. The central part of each tube contains the highly enriched fissile fuel, but above and below this are sections with the U-238 isotope. Effectively, the central core where fission is maintained is surrounded by a 'blanket' of U-238 in which breeding occurs. Liquid sodium coolant circulates through the core and transfers heat to an intermediate circuit of the same liquid metal. Sodium is chemically extremely active, igniting on contact with air and reacting violently with water, so it is essential that both these sodium flows are very securely contained.

However, the main concerns about a breeder programme centre on its fuel, for three reasons: the concentration of fissile material, the necessity for

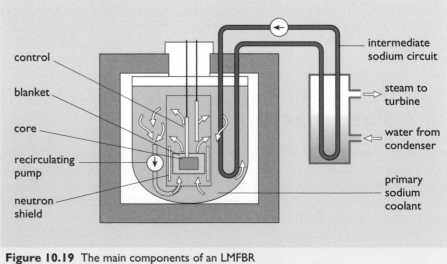

Figure 10.19 The main components of an LMFBR

BOX 10.11 **Plutonium**

To see why plutonium causes concern we need to look more closely at its properties. One central fact is that with enough of it a bomb can be made. Not, it should be emphasized, easily. Plutonium is radioactive, fissile and toxic, and in any but skilled hands is more likely to lead to an unpleasant death than to unlimited power. (Indeed, its extreme toxicity and the consequent blackmail power of a threat to distribute it is perhaps a greater reason for concern.) It could be argued that bombs can also be made from fissile U-235, but the important difference is that in natural uranium or spent fuel this is considerably diluted by non-fissile U-238, and separating the two isotopes requires complex and costly plant. The fissile Pu-239 produced in reactors does have 'diluting' isotopes, and true 'weapons-grade' material should have less than 7% of the non-fissile Pu-240; but the plutonium from breeders, containing 80–90% Pu-239, could still be used in a weapon, if a rather unreliable one.

With a growing world surplus of plutonium, the concept has developed of using a fast reactor as a **burner**, consuming more plutonium than it produces (as opposed to a breeder, which does the reverse). Between 50 and 100 tonnes of weapons-grade plutonium are becoming 'available' following the decommissioning of nuclear weapons by Russia and the US. (To put this quantity in context, about eight kilograms of weapons-grade plutonium are required

for a bomb, and the total quantity of plutonium in existence is thought to be over 1000 tonnes – a million kilograms.)

Several groups of concerned scientists have attempted to reach agreement on a means to make the surplus plutonium safe. It could be used as reactor fuel, which would reduce the quantity and also 'poison' any remaining fuel with radioactive fission products. However, not all ordinary thermal reactors are suitable, and those that are could consume plutonium only at rates well under a tonne a year. (France's Superphénix 1240 MW breeder could in principle have converted up to 2 tonnes per year, but after a number of problems it was closed down in 1998.)

Other proposed options include mixing the plutonium with high-level wastes and then burying it. The United States' 'can-in-canister' plan would embed the plutonium in small ceramic disks to be stacked in tall canisters. Molten glass containing high-level waste (Box 10.10) would then be poured over them, filling the canisters. A similar scheme would incorporate the plutonium into ceramic 'storage MOX' pellets. Packed into tubes like fuel rods, these could be buried together with spent fuel rods, whose high-level radioactivity would prevent access to the plutonium (Macfarlane *et al.*, 2001). Both these methods of course depend on decisions about the final disposal of high-level wastes.

reprocessing, and the central role of plutonium. The concentrated fuel raises the question whether a runaway nuclear reaction could occur. A fully efficient 'atomic bomb' is not possible, and even a 'nuclear fizzle' would require a number of improbable conditions simultaneously. Nevertheless, there is some disagreement about probabilities, and the unlikelihood does seem to be of a different order than the virtual impossibility of a nuclear explosion in a thermal reactor. Similarly with reprocessing, where the significant difference is that whilst this is an option in a thermal reactor programme (see Section 10.6), it is a necessity in a breeder programme, where the extraction of plutonium is an essential feature of the fuel cycle.

10.8 **Power from fusion**

Nuclear fusion is the process that powers the stars, including our sun. It is therefore the original source of almost all the energy that maintains the earth's climate and its living matter. As the name suggests, **nuclear fusion** is the coming together of two lighter nuclei to form one heavier one. As this is the reverse of fission, we might well expect it to consume energy rather than produce it. If we attempted to fuse barium and krypton to create uranium, this would indeed be the case, but as we'll see, the result is quite different for the lightest nuclei.

BOX 10.12 Energy from fusion

The method described in Box 10.4 can be used to calculate the energy released in the fusion of a deuteron and a triton, as the masses of all the particles are known.

Particle masses

deuteron	2.0136 u
triton	3.0160 u
alpha particle	4.0015 u
neutron	1.0087 u

Calculating the energy

Simple arithmetic shows that the total mass will have decreased by 0.0194 u. If all this surplus mass appears as kinetic energy, the α-particle and neutron will share a total of about 18 MeV, or 2.9×10^{-12} joules.

There are 6×10^{26} deuterium atoms in two kilograms of deuterium, and the same number of tritium atoms in three kilograms of tritium. So the energy released in the complete fusion of *five* kilograms of the 'mixture' becomes:

$$6 \times 10^{26} \times 2.9 \times 10^{-12} = 1.7 \times 10^{15} \text{ joules}$$

which is about **350 TJ per kilogram**.

Comparison with the corresponding figure in Box 10.4 shows that fusion should produce over four times as much energy as fission, per kilogram of 'fuel'.

The process at the centre of attempts to achieve controlled fusion power is the fusion of the nuclei of two isotopes of hydrogen. One of these is **deuterium** (hydrogen-2), which we have already encountered in the form of heavy water used as a moderator in fission reactors. The other is hydrogen-3, which is again given a separate name: **tritium**. The fusion of the two nuclei, a deuteron and a triton, is as follows.

$$\text{}^2_1\text{H} + \text{}^3_1\text{H} \rightarrow \text{}^4_2\text{He} + \text{n} \tag{1}$$

As can be seen, the result is a helium nucleus (an α-particle) and a neutron, and as Box 10.12 shows, this process should release considerably more energy than fission, from the same mass of material.

There is, however, the question of obtaining the tritium. Unlike deuterium, it is not present in natural hydrogen because it is radioactive: it is a β-emitter with a half-life of only 12 years. This fact has important consequences. Firstly, it means that the tritium for the above fusion reaction must be produced independently. In principle, this can be achieved by another nuclear reaction:

$$\text{}^2_1\text{H} + \text{}^2_1\text{H} \rightarrow \text{}^1_1\text{H} + \text{}^3_1\text{H} \tag{2}$$

As can be seen, a neutron is effectively transferred from one deuteron to the other, producing a proton and a triton. The process releases energy, about 4 MeV, and a little consideration shows that combining it with the first reaction should lead to the ideal system. Suppose that the spare neutron produced in (1) decays into a proton and an electron, as in Box 10.2. We then have a process whose input is entirely deuterium and whose output consists only of the useful gases hydrogen and helium – and about 380 TJ of energy for each kilogram of deuterium.

As mentioned in Box 10.6, there is plenty of deuterium in the world's rivers and seas, but in a very dilute form. Its extraction is a large-scale endeavour, requiring the processing of up to 100 tonnes of water for each kilogram of pure deuterium. Nevertheless, 380 TJ is about 100 GWh, which is more than the daily fuel requirement of a 1 GW power station operating at 30% efficiency. So if the above reasoning is correct, the entire fuel requirement for such a plant would be met by less than 100 tonnes of water a day. But is this a realistic proposition?

Approaches to a fusion reactor

Nuclear fusion has been the subject of research for more than half a century, but the only device to produce net output power has been the hydrogen bomb. The *controlled* production of fusion power has proved much more elusive. It is easy to see in general why there are problems. Two nuclei can only fuse if they come close enough for the short-range nuclear force of attraction to overcome their intense electrical repulsion (see Section 4.7). To overcome this electrical barrier the nuclei must approach each other at extremely high speeds, and several different methods have been investigated in the attempts to achieve this.

One way to achieve very high speeds is to raise the particles to a very high temperature. This is the principle of the **thermonuclear** approach. Calculations show that the deuterium–tritium reaction shown above, called **D–T fusion**, needs a high particle density at a temperature of several million degrees. The **D–D** reaction, needing conditions a factor of ten more extreme, has not so far been considered practicable, and tritium is currently produced separately, by neutron bombardment of lithium. (Chapter 11 briefly discusses this and other aspects of the use of tritium.) Even for the 'easier' D–T reaction, there are major technical problems in containing the dense mass of particles at these extremely high temperatures. At a temperature of several million degrees all atoms are stripped apart, and the gas becomes a **plasma** of hot nuclei and electrons. No material container is possible, and the method used by stars, where enormous gravitational forces contain the hot plasma at high density, is not available on earth.

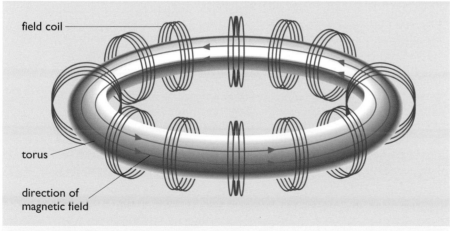

Figure 10.20 The tokamak principle

Much of the fusion research effort of the past half-century has been concentrated on **magnetic containment** – using the force that acts on a charged particle moving in a magnetic field to keep the plasma compressed. In the **tokamak** arrangement, devised in the 1950s by the Russians Andrei Sakharov and Igor Tamm, the particles move in a doughnut-shaped ring (the torus) surrounded by large coils carrying an electric current (Figure 10.20). These coils produce a magnetic field along the torus, and the electrically charged particles spiral around the direction of this field. A current flowing in the plasma heats it, and additional energy may be

Figure 10.21 Andrei Sakharov (1921–1989) was born in Moscow, the son of a physics lecturer. He studied physics at Moscow University, graduating in 1942, and spent the rest of the war in a munitions factory. Declining invitations to join the fission weapon team, he returned to Moscow to work in theoretical physics. Eventually he was drawn into work on the hydrogen bomb, to which he made major contributions. By the 1960s he occupied a privileged position in Soviet society, but was already expressing reservations about nuclear testing. His criticisms gradually became more general, leading in 1968 to the publication in the US of an essay on intellectual freedom, urging an end to the Cold War. Dismissed from his work, he continued to write on human rights. The father of the H-bomb had become, in the citation of the 1975 Nobel Peace Prize, 'the conscience of mankind'. Condemnation of the invasion of Afghanistan in 1980 led to banishment from Moscow, and for five years his friends heard of protests and hunger strikes, or more disturbingly, heard nothing. Then in 1986 he was invited back by Gorbachev to participate in drawing up a new Soviet constitution. He completed a draft a few days before he died.

supplied by fast-moving deuterons and tritons injected into the system. The tokamak principle is used in the Joint European Torus (JET) at Culham in England (see Figure 11.7 in Chapter 11) and the Tokamak Fusion Test Reactor (TFTR) at Princeton in the US. Both of these have achieved fusion for extremely short periods of time, but neither has yet produced more energy from fusion than the energy input needed to run the system.

An alternative approach has been a micro version of the method used in the hydrogen bomb. In the bomb, the D–T core is essentially surrounded by an independent *fission* bomb. When this explodes, it compresses the core, achieving the conditions necessary for fusion. The aim of the **inertial confinement** method is to achieve this implosion in a more controllable way. The Lawrence Livermore National Laboratory in California has been a leader in this approach, using an intense blast of radiation from an array of powerful lasers to explode the surface coating of a sphere containing the D–T fuel. As with the tokamak systems, inertial confinement has yet to produce more energy than it consumes.

All three of the groups mentioned above have plans for new machines that, it is claimed, should be able to achieve a net energy output; but there is general agreement that large-scale fusion power is probably still 50 years away.

BOX 10.13   Hot, cold and wet fusion

As we have seen, a viable fusion power plant using either magnetic containment or inertial confinement is usually thought to be several decades away. There might however be a breakthrough, perhaps even from a novel direction. Recent years have seen much speculation about the possibility of fusion occurring at room temperature, under suitable conditions, and some scientists even claim to have achieved this **cold fusion** on a small scale (see for instance Goodstein, 1994). However, these claims remain very controversial and have not been confirmed by others (Merriman, 1996). This is also the case for more recent claims concerning **bubble fusion**: the generation of very high temperatures, and possibly fusion reactions, in bubbles irradiated with neutrons and ultrasound, making use of a phenomenon called sonoluminescence (see for example Becchetti, 2002; Levy 2002). For the moment, however, as far as large-scale nuclear power production is concerned, it seems likely that the focus will remain on fission, with 'conventional' fusion becoming a significant option only in the second half of the present century.

References

Becchetti, F. (2002) 'Evidence for nuclear reactions in imploding bubbles', *Science*, vol. 295, pp. 1850–1868.

Curtin University (2002) 'Natural fossil fission reactors', Western Australian Isotope Science Research Centre, http://www.curtin.edu.au/curtin/centre/waisrc/OKLO/ [accessed 10 October 2002].

Frisch, O. R. (1979) *What Little I Remember*, Cambridge, Cambridge University Press.

Goodstein, D. (1994) 'Whatever happened to cold fusion?' http://www.its.caltech.edu/~dg/fusion.html [accessed 10 October 2002].

Gowing, M. (1964) *Britain and Atomic Energy, 1939–1945*, London, Macmillan, p. 389.

Hahn, O. (1950) *New Atoms*, New York, Elsevier.

Hahn, O. and Strassmann, F. (1939) 'Concerning the existence of alkaline earth metals resulting from neutron irradiation of uranium', *Naturwissenschaften*, vol. 27, no. 11.

Levy, B. C. (2002) 'Scepticism greets claim of bubble fusion' http://www.physicstoday.org/pt/vol-55/iss-4/p16.html [accessed 10 October 2002].

Macfarlane, A., von Hippel, F., Kang, J. and Nelson, R. (2001) 'Plutonium disposal the third way', *Bulletin of the Atomic Scientists*, vol. 57, no. 3, May/June.

Manley, J. H. (1962) 'Atomic energy', *Encyclopaedia Britannica*, London.

Meitner, L. and Frisch, O. R. (1939) 'Disintegration of uranium by neutrons: a new type of nuclear reaction', *Nature*, vol. 143, no. 239.

Merriman, B. (1996) 'An attempted replication of the CETI cold fusion experiment', http://www.math.ucla.edu/~barry/CF/reportcover.html [accessed 10 October 2002].

Pais, A. (1991) *Niels Bohr's Times*, Oxford, Oxford University Press.

Seaborg, G. T., McMillan, E. M., Kennedy, J. W. and Wahl A. C. (1946), 'Radioactive element 94 from deuterons on uranium', *Physical Review*, vol. 69, pp. 366–367.

UN (1946) *The Control of Atomic Energy*, United Nations, Department of Public Information, Lake Success (reproduced in Hahn, 1950).

Further reading

Hore-Lacey, I., *Nuclear Electricity*, 6th edn, Uranium Information Centre Ltd, http://world-nuclear.org/education/ne/ne.htm [accessed October 2002].

Useful web sites

Uranium Information Centre at http://www.uic.com.au

World Nuclear Association at http://world-nuclear.org/

Chapter 11

The Future of Nuclear Power

by David Elliott

Guided by electronics, powered by atomic energy, geared to the smooth effortless workings of automation, the magic carpet of our free economy heads for distant and undreamed of horizons. Just going along for the ride will be the biggest thrill on earth.

From an address to the American National Union of Manufacturers, quoted in J. Rose (1967) *Automation: its uses and consequences*, Oliver & Boyd, p.117.

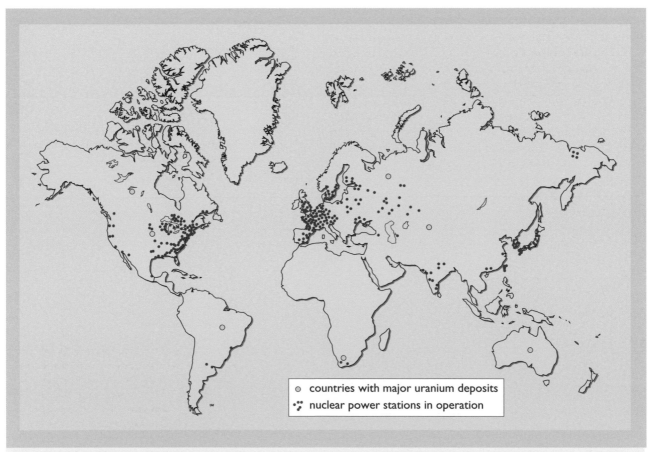

Figure 11.1 Map of reactor locations around the world. Locations and numbers of dots are approximate. There are currently 439 reactors operating in 31 countries

11.1 Introduction

As the previous chapter illustrated, nuclear technology emerged from the Second World War as a promising new source of energy. Nuclear power now provides around 6% of the world's primary energy, in the form of about 17% of the world's electricity, with, in 2002, some 439 plants installed around the world. Three countries obtain more than half their electricity from this source: France, 76%; Lithuania, 74%; and Belgium, 57% (see Figures 11.1 and 11.2). In addition to power plants, the expansion of the nuclear industry over the last 40 years or so has involved the creation of uranium mining, fuel fabrication and enrichment facilities in various locations around the world, backed up by interim waste storage facilities for the used fuel and wastes. Some countries (mainly the UK and France) also have reprocessing plants to extract plutonium from the used fuel, and some also have plans for long-term waste repositories. Nuclear power has become a major international industry.

However, the expansion of nuclear power has slowed in recent years, with fewer new plants being installed. Given that the first wave of plants are beginning to reach the end of their operational life, unless policies change, the nuclear contribution worldwide could reach a peak around the year 2010 and then decline, as is illustrated in Figure 11.3, which is from a joint study published in 1999 by the Royal Society and Royal Academy of Engineering. Based on current plans, the study concluded that the peak in the UK would be somewhat earlier, as Figure 11.4 illustrates.

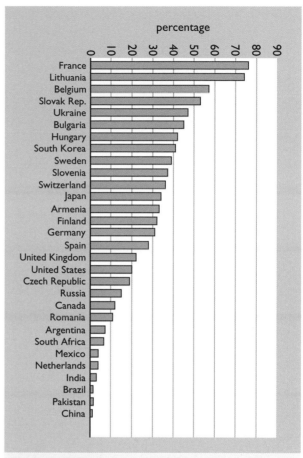

Figure 11.2 Percentage of electricity generated by nuclear plants in different countries in 2000

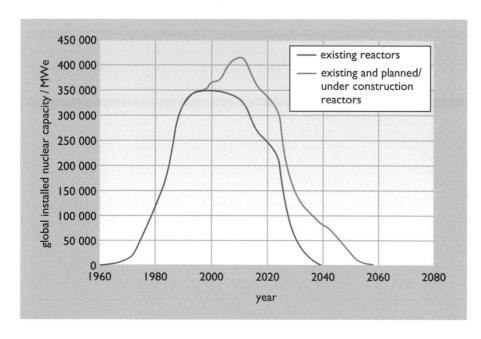

Figure 11.3 Rise and fall of nuclear power around the world – current and planned reactors. For a US government analysis which comes to a slightly more optimistic conclusion, but still talks of a 'levelling off' after 2010, see http://www.eia.doe.gov/oiaf/ieo/pdf/nuclear.pdf [accessed 18 June 2002]

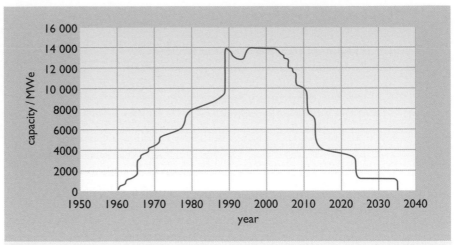

Figure 11.4 Rise and fall of nuclear power in the UK in the absence of new construction

The main aim of this chapter is to look at how this reversal in fortunes has come about, and provide a balanced overview of the facts about nuclear power.

Because the operation of nuclear plants does not generate substantial amounts of carbon dioxide gas, nuclear power could be seen as having the potential to play an increasing role in combating global warming and climate change. As we shall see, however, there are arguments to suggest that it may not be a safe or economically viable option. The debate on nuclear power is often a heated one, but in this chapter we aim to give as objective an account as possible.

11.2   Reasons for decline

Our first task is to explore why the prospects for nuclear power seem to have begun to dim in recent years. This will set the scene for looking at whether the problems that have led to this situation can be overcome.

The most familiar explanation for the evident decline in support for nuclear power is that fears about safety have grown in response to a series of increasingly worrying accidents. Fears about radiation hazards generated in relation to nuclear weapons have always meant that the general public has been nervous about the risks of nuclear technology, and the accidents have seemed to confirm that there could also be major problems with civil power plants. The result has been that, in general, nuclear power has become unpopular with the general public and, therefore, more difficult for governments to support.

As we shall see, though, this argument is not the whole story. Arguably, the reduced support for the nuclear option has mostly been the result of the poor economics of nuclear power, as compared with rival energy options like natural gas. That said, the increased concerns about safety have significantly increased the cost of nuclear power, as additional safety features had to be added. So, in addition to resulting in public disquiet, and associated political difficulties, accidents have undermined the economics of nuclear power.

Nuclear accidents

In the early years, nuclear power was widely seen as a 'wonder' technology. It was going to be cheap, safe and provide power more or less forever on a vast scale and in most sectors of the economy. There were fantasies of atomic-powered cars and aircraft (for a fascinating account of the latter see http://www.megazone.org/ANP/atomair.shtml [accessed 18 June 2002] for an article by Vincent Cortright, from *Aviation History*, March 1995). The reality has proved somewhat different. Some nuclear-powered ships have been built, mainly submarines and aircraft carriers, and some icebreakers, but otherwise, apart from some industrial and medical spin-offs from the use of radioisotopes, the only significant non-military application has been in the production of grid electricity.

Initially power production was seen by most people as an exciting idea, but then, following a spate of accidents, came the realization that, when and if things went wrong with nuclear technology, there could be very serious health implications. The result was that opposition to nuclear power began to grow.

It is not our intention to provide a full account of all the nuclear accidents that have occurred around the world. Instead we will focus just on some of the major ones, since these had most impact on public attitudes to nuclear power.

The three accidents we will look at are the fire at the Windscale military plutonium production plant in the UK in 1957, the Three Mile Island accident in the US in 1979, and the Chernobyl accident in the Ukraine in 1986.

The fire at the Windscale plutonium production pile in Cumbria (the site is now known as Sellafield) resulted in the release of radioactive material, including some 1000 TBq of radioactive iodine (see Table 10.1 in the previous chapter). Crops over an area of nearly 800 square kilometres were destroyed as a precaution, and two million litres of milk were poured away (Arnold, 1992).

The loss-of-coolant accident at Three Mile Island in Pennsylvania in 1979, which led to the partial melting of the fuel in the reactor core, was contained, although there were some releases of radioactive material and at one time the mass evacuation of at least 630 000 local people seemed imminent (Garrison, 1980). For an account of the accident from the industry perspective see http://www.world-nuclear.org/info/inf36.htm [accessed 18 June 2002].

In neither of these two cases were there any direct casualties, although there have been concerns expressed about the long-term health impacts. For example, it has been suggested that some early (i.e. premature) deaths, might have occurred subsequently in the population living down-wind from the Windscale plant. (See McSorley, 1990 and Chapter 13.)

The Chernobyl accident on 26 April 1986 was, in contrast, much more severe, and led to worldwide concern as the radioactive plume spread across north-west Europe and, although by then much weakened, to North America and around the northern hemisphere. The accident occurred at one of the four reactors on the site, when it was running at very low power during an

BOX 11.1 Chernobyl death estimates

Morbidity and mortality estimates for the Chernobyl accident were produced for an international conference *One Decade After Chernobyl*, organized by the International Atomic Energy Agency, the World Health Organization and the European Commission on the tenth anniversary of the accident (http://www.iaea.org/worldatom/Programmes/Safety/Chernobyl/ [accessed 18 June 2002]).

In summary, the analysis was as follows:

■ During the accident, three workers were killed. Among plant workers and firemen who responded to the event, 237 were suspected of acute radiation sickness (ARS) and admitted to hospitals for observation, where 134 cases of ARS were confirmed. Of those people treated for ARS, 28 have died within 10 years of the accident as a result of their injuries. The report noted that 'the others have survived but are in generally poor health and psychological condition'.

■ Subsequent to the initial event, a workforce of approximately 200 000 'liquidators' were hired or ordered to gather and bury radioactive material released by the blast. It was projected that this population of workers will suffer an excess burden of 2500 cancers as a result of their clean-up work.

■ Residents of communities off-site may be expected to suffer an excess burden of 2500 cancers as a result of exposure to fallout from the accident. Within the first 10 years, approximately 500 cases of thyroid cancer amongst people who were children at the time of the accident have been confirmed, and it was estimated that there could be 4000–8000 cases in total in the coming years. However, 90% or more of these cases were thought to be likely to be curable (surgical removal of the thyroid was widely adopted).

Some observers claim that the eventual death toll might be very much higher than these figures suggest. For example, a 1990 study by Medvedev suggested that there could be up to 40 000 early (i.e. premature) deaths (Medvedev, 1990). Given that it can take up to 20 years for cancers to emerge, it will still be some while before predictions like this can be assessed. But certainly some horrific reports have emerged. For example, in a retrospective review of the accident, the Guardian Weekly (19 December 2000) noted that 'Ukrainian government figures state that more than 4000 clean-up workers have died and a further 70 000 have been crippled by radiation poisoning. About 3 400 000 people, including 1 260 000 children, are

suffering from fallout-related illnesses. Unofficial statistics put the casualty rates much higher.'

However, it has, perhaps rather harshly, been suggested that some of these figures may have been inflated in order to attract aid, and certainly it can be hard to obtain reliable data, given that the break up of the USSR followed soon after the accident at Chernobyl.

A UN assessment, produced in 2000, suggested that, apart from the initial deaths and 1800 cases of (potentially treatable) thyroid cancer in children, there was 'no evidence of a major public health impact'. For the full report of the UN Scientific Committee on the Effects of Atomic Radiation, see http://www.unscear.org/pdffiles/gareport.pdf, and in particular, the detailed annex at http://www.unscear.org/pdffiles/annexj.pdf. For a summary of the report by the Nuclear Energy Agency see http://www.nea.fr/html/rp/chernobyl/chernobyl-update.pdf [all sites accessed 18 June 2002].

It has also been suggested in a recent UN report that some of the illnesses subsequently emerged as the result of the stress of over-zealous evacuation and forced relocation, and that some could even be put down to psychosomatic effects (UNDP and UNICEF, 2002).

Clearly, what initially might seem like a simple, if grim, exercise in body counting, turns out to be a far more complex and potentially conflict-laden activity, fraught with statistical and conceptual problems, and disagreements over the interpretation of data will no doubt continue. That was certainly the case with one early US government-backed attempt to assess the risk of nuclear plants, the methodology of which was criticized (Rasmussen, 1973, 1974, 1975; see also Chapter 13). Similar problems also face attempts to compare the risks of nuclear energy with those of other technologies, as was demonstrated by the debate over another early study for the Canadian Atomic Energy Control Board (Inhaber 1978, 1982; Holdren 1979). We will be looking at comparisons of the risks associated with various energy technologies in Chapter 13, but it is worth noting here the argument that, for example, major hydro dam failures can kill many more people than nuclear accidents, as can the routine emissions from coal-fired plants. On this basis supporters of nuclear power often claim that, compared with its alternatives, nuclear power is a *relatively* safe option: see for example the analysis by the Uranium Information Centre at http://www.uic.com.au/ne6.htm#6.3 [accessed 18 June 2002].

attempt to check the safety system, and only about 6% of the radioactive material it contained was released into the atmosphere (Read, 1993; NEA, 1995). Even so, quite apart from the 31 deaths on site and amongst fire-fighters, there have been reports of subsequent deaths and serious illness in the region around the plant. Although, as is discussed in Box 11.1, there have been some recent revisions and re-assessments, initial estimates of the eventual death toll varied from a few hundred to several thousand premature deaths, whilst some reports suggested that the health of many tens of thousands of people could be adversely effected (McSorley, 1990; Medvedev, 1990).

Clearly the Chernobyl accident raised major problems for the nuclear industry. However, as is discussed in Chapter 13, significant and catastrophic though major accidents may be for those involved, statistically the *total* health impact of occasional large nuclear accidents can be less than, for example, the cumulative and continuing impact of gaseous emissions from coal-fired plants.

Certainly the risks associated with the *routine* operation of nuclear plants are usually considered to be relatively small, especially for the general public. It should perhaps be remembered in this context that the nuclear industry is one of the most tightly regulated industries in the world, with the levels of emissions being very carefully controlled, although there are concerns that these controls may not be equally effective in all countries. However, there have been some concerns that, quite apart from the impacts of accidental leaks of radioactive materials, the scale and the health impact of the low-level routine emissions allowed from nuclear facilities have been underestimated – see, for example, the arguments presented by the Low Level Radiation Campaign at http://www.llrc.org [accessed 18 June 2002].

The nuclear industry argues that levels of exposure to man-made radiation sources, averaged out across the community, are very small compared with natural levels of exposure. The Low Level Waste Campaign, however, argues that average figures are of little meaning for people living near nuclear facilities, who are likely to receive more than average exposures, even assuming that emissions are kept within the agreed limits. Moreover, some people believe that all additional exposures over natural background levels should be avoided. These issues will be taken up in Chapter 13.

While issues like this have led to local campaigns, in general it has been the accidents and major leaks that have had most impact on the general public's views of nuclear power, in part because accidents naturally receive a lot of media attention. But this is not simply a media-led fear with nuclear power being picked out for special attention. The impacts of nuclear accidents and major leaks are widely perceived as different in kind from the hazards to the public associated with most other industrial activities, not least because they can lead to illness many years later, and possibly even across generations. Certainly concern over the potential long-term consequences of accidents has been one reason why nuclear power has been opposed by most environmentalists.

Public opposition to civil nuclear developments has been consistently high in most countries of the world, and at some sites there have been major confrontations involving tens of thousands of people. In the UK opposition has increased over the years, reaching around 75% after the Chernobyl accident in 1986. Interestingly, this opposition did not fade away subsequently – instead it increased. By 1991, 78% of respondents to a Gallup poll either wanted 'no more nuclear plants' or for the use of nuclear power to be halted (SHE, 1994). Since then opposition seems to have reduced slightly, possibly reflecting increasing concerns about climate change, but even so, resistance remains quite strong. For example, in a poll carried out by the British Market Research Bureau in 2001, 68% of those interviewed said that they 'did not think that nuclear power stations should be built in Britain in the next ten years' (BMRB, 2001). It is wise to be cautious about interpretations from opinion polls, but it does seem that, in general, nuclear power is not popular with most people.

Nuclear economics

Despite the risks and the public concerns, nuclear power might be seen as the ultimate 'technical fix' for energy supply. Large amounts of energy can be released from small amounts of relatively abundant and cheap material. Certainly, in the period after the Second World War, it looked very promising. There were even suggestions that nuclear electricity would be 'too cheap to meter'.

The reality has proved somewhat different. Fifty years on, after very large investment around the world, the cost of the electricity produced remains high. This might seem surprising, since uranium is a very concentrated energy source. In theory, one kilogram of natural uranium will yield about 20 000 times as much energy as the same amount of coal. Consequently, compared with coal, not much is needed to run a reactor and the costs of the fuel are therefore relatively low – typically the cost of raw uranium represents about 2% of the overall generation costs. The raw uranium does have to be processed and enriched to make usable fuel and this adds to the cost, but even so the cost of fully fabricated fuel is only around 20% of overall generation costs, equivalent to about a third of the fuel costs for coal-fired plants.

The main problem is not the cost of the fuel, but the cost of the plant. The capital (plant construction) costs are much higher for nuclear than for fossil fuel plants, typically by a factor of around three, with light water reactor costs in 2000 ranging between $1700 and $3100 per kW (WEA, 2000). This is mainly because of the need to use special materials, and to incorporate sophisticated safety features and back-up control equipment. These can contribute up to half the nuclear generation cost. This cost problem has sometimes been made worse by lengthy and expensive construction delays and overruns, the result, at least in part, of the complexity of the technology (see the World Nuclear Association's analysis at http://www.world-nuclear.com/info/inf02.htm [accessed 18 June 2002]).

The situation may improve with newer technology, but a recent US government study put the cost even for advanced nuclear plants at $2188 per kW installed, compared with $1092 per kW for coal plants and around $500 per kW for gas-fired plants (EIA, 2000). Operational costs are also large and have escalated as concerns about safety have increased. Typically, operation and maintenance costs for nuclear plants in the US are around three times those for fossil fuel plants (WEA, 2000).

As we shall see, there is much debate about the precise figures for nuclear costs. However, there is no disagreement that the cost of electricity from the rival energy technologies has fallen. For example, since the 1980s, the average price of electricity produced by coal plants has fallen by a factor of two in the US in real (inflation adjusted) terms. More recently, nuclear power has been seriously challenged by the so-called 'dash for gas' – the rapid adoption of low-cost electricity generation using highly efficient gas-fired turbines. So while some existing nuclear plants have, in effect, paid off their construction costs, and depending on market conditions, should be able to generate power reasonably economically, it is generally accepted that using current technology it would not be economically viable in most countries to build new nuclear plants, since they could not compete with combined-cycle gas turbine power.

The cost issue came to a head in the UK in 1990, following the privatization of the electricity industry and the government's introduction of the Fossil Fuel Levy. Electricity consumers had, in effect, to pay a surcharge of around 10% on their electricity bills to meet the extra cost of the 20% or so of the electricity that the UK's nuclear plants were supplying at that point.

Although this simple historical fact seems very clear, unfortunately the whole area of nuclear economics is fraught with complexity and contention. In part this is because, in the UK, nuclear power grew up in a subsidized public sector environment, protected from market considerations and overseen by the publicly owned Central Electricity Generating Board (CEGB), which, it is alleged, produced unrealistic, and possibly biased, costings for nuclear power. Thus, in its report on the cost of nuclear power in 1990, the House of Commons Select Committee on Energy commented that there had been 'a systematic bias in CEGB costings in favour of nuclear power, both in ignoring risk and failing to provide adequately for contingencies, and in respect of investment, in putting forward "best expectations" rather than more cautious estimates' (House of Commons Select Committee on Energy, 1990).

However, the initial relatively positive assessment of nuclear economics was not just a matter of bias. Operating in the public sector made the economics of nuclear power look reasonably favourable, with the government and the CEGB making use of a 'test discount rate' of 5% as a measure of the economic viability of investments (see Chapter 12). When an attempt was made to privatize the nuclear industry, as part of the wider privatization of the electricity industry in 1990, the economics started to look somewhat different. For example, Lord Marshall, then Chairman of Central Electricity Generation Board, warned that, whereas under the public sector the costs for pressurized water reactors were around 3.22 pence per kWh, under private sector conditions, with much higher rates of return being expected, they would rise to 6.25p per kWh. The economic assessments are actually much more complex than this implies, and include the issue of the cost of waste processing, storage and eventual plant decommissioning (Box 11.2). For a full, but rather tortuous, analysis of UK nuclear costs see the report *The Costs of Nuclear Power*, produced by the House of Commons Select Committee on Energy (1990).

BOX 11.2 Waste disposal and decommissioning costs

One of the reasons why estimates of nuclear costs have varied over time is that, initially, the industry paid less attention to the costs of the so called 'back-end' part of the nuclear fuel cycle, involving waste processing, interim storage and ultimate disposal, and eventual plant decommissioning. Subsequently however, awareness of the importance of these activities has increased, as have estimates of their cost.

Initially, the **waste management** costs were generally seen as a relatively small part of the overall cost, since the volumes of waste produced are small. For example, in the US utilities pay only $0.001 per kWh for the management and disposal of their spent (used) fuel, which amounts to around 2 or 3% of their generation costs (WEA, 2000). However, as we shall see later (Box 11.3), the waste management issue has become increasingly problematic and expensive, as local communities have resisted having waste disposal sites near them. The World Nuclear Association now puts the cost of managing and disposing of nuclear power plant wastes at about 5% of the total cost of the electricity generated (WNA, 2002a).

The situation in the UK is complicated by the fact that spent fuel from reactors is sent to Sellafield to be reprocessed, in order to separate out the plutonium and uranium it contains from the other radioactive materials before they are stored. This is an expensive activity. For example, in 1999 it cost British Energy (BE) around £300 million to have its spent fuel reprocessed by British Nuclear Fuels Ltd (BNFL) at the THORP plant at Sellafield. But, even without reprocessing, BE has indicated that just storing its output of spent fuel would have cost around £100 million. BNFL has also estimated that the total running and maintenance cost of radioactive waste storage facilities over the next 50 years at Sellafield will be £1100 million (Wilson, 2002).

Plant decommissioning, which involves a long process of dismantling reactors and their containment, and storing the large volumes of resultant scrap, is an even more complex, open-ended and costly activity. The World Nuclear Association estimates that decommissioning can add 9–15% to the initial capital cost of a nuclear power plant. However, these decommissioning costs only occur at the end of the plant's lifetime, and discounted over this long period, the WNA reports they 'contribute only a few per cent to the investment cost and even less to the generation cost. In the USA they account for 0.1–0.2 cents per kWh, which is no more than 5% of the cost of the electricity produced.' See http://www.world-nuclear.org/info/inf02.htm [accessed 18 June 2002] and Chapter 12.

In the UK, the nuclear industry has, over the years, put money aside to meet eventual decommissioning costs.

For example, British Energy, which runs the advanced gas-cooled reactors and the Sizewell pressurized water reactor, has been putting aside £16 million a year into a segregated fund to meet its long-term nuclear liabilities. However, with estimates of the cost of decommissioning rising (for example, BE's decommissioning liabilities for its nuclear plants have been put at around £8 billion) it has been argued that this level of forward investment might not prove to be sufficient. That rather depends on the route to decommissioning taken. The UK industry's preference is to remove the non-active parts of the plant, but leave the reactor core *in situ*, sealed with weatherproof cladding in a so-called 'safe store', for 135 years or so. It would then be less active and therefore safer and cheaper to demolish. (For more discussion of the possible technical options for decommissioning see http://www.world-nuclear.com/education/ne/ne5.htm#5.6 [accessed 18 June 2002].)

Postponing full decommissioning is not popular with environmentalists, who argue that this amounts to leaving it to future generations to deal with. The UK Nuclear Installation Inspectorate have suggested that, in fact, the safety case may only imply a deferral for around 40–50 years, at least for some reactors (NII, 2001). That is evidently the sort of timescale most other countries are considering. However, a longer delay has the economic attraction that, if relatively small amounts of money are put aside from current nuclear earnings, then, given that this money will be invested over a long time period, sufficient capital will be available to deal with final decommissioning. For example, if decommissioning the reactor cores can be delayed for say 135 years, it has been estimated that the discounted liability for decommissioning the Magnox and advanced gas-cooled reactors would reduce by between 45% and 50% respectively (NAO, 1993). Putting it another way, a more recent estimate was that if British Nuclear Fuels had to complete decommissioning of its Magnox reactors within 40 years, that would add £1.5 billion to the cost. With BNFL's total historic liabilities for plant decommissioning and site clean-up (including the Sellafield site) having been put at up to £34 billion, the attraction of discounting costs into the future becomes clear – it can produce significant changes in the financial burden, as we will be discussing further when we look at the idea of discounting in Chapter 12.

As can been seen from this brief survey, the overall 'back-end' costs of nuclear waste management and plant commissioning do not represent a very large proportion of total nuclear costs. That said, they are not trivial, and as concerns about the impacts of waste disposal and plant decommissioning have increased, so has the issue of how the costs are to be accounted for in the economic evaluation of nuclear power operations.

Subsequently, in 1995, after a major review, the UK Conservative Government concluded that 'providing public sector funds now for the construction of new nuclear power stations could not be justified on the grounds of wider economic benefits and would not therefore be in the best interest of either electricity consumers or tax payers' (HMSO, 1995).

By this time it was clear that, in the context of the newly privatized electricity industry, natural gas was the market's preferred option. The 'dash for gas' had a major impact on the fortunes of the coal industry – whereas in the 1980s around 80% of the UK's electricity had come from coal burning, by 1999 it only produced 28% – with gas supplying 38%. The nuclear share held up at 25%, but the longer term impact of gas looked likely to be dramatic. In effect it made further nuclear expansion very unlikely.

In 1998, the UK Labour Government confirmed the view that 'at present nuclear power is too expensive to be economic for new capacity and in current circumstances it is unlikely that new proposals for building nuclear plants will come forward from commercial promoters. It therefore seems unrealistic to expect an increasing contribution from new nuclear capacity in the medium term – indeed the contribution from nuclear is expected to decrease in the first decade of the next century as existing capacity is retired (and in all probability [be] replaced by gas-fired plants)' (Trade and Industry Committee, 1998).

In effect, the plan was to let nuclear power phase itself out as plants were retired but not replaced, with a policy of 'diminishing reliance' being established. The Royal Commission on Environmental Pollution, in a report on Energy Policy in 2000, noted that all the UK plants apart from the Sizewell PWR, would, based on current plans, be closed by 2025.

Figure 11.5 1.2 GW Sizewell 'B' pressurized water reactor on the Suffolk coast

Nuclear decline worldwide

Similar developments have occurred elsewhere in the world. Economic problems, coupled with public opposition, have meant that, in many countries around the world, nuclear power programmes, private or state-run, have been winding down or have been halted. Nuclear plants provide around 20% of the electricity in the US, but the accident at Three Mile Island, coupled with the worsening relative costs of nuclear power, effectively halted expansion. Greece had already decided not to go nuclear in 1975, on the grounds of seismic safety, and the late 1970s and early 1980s saw Denmark, Austria, Sweden, Spain and Norway variously elect not to go nuclear, stop building new reactors, or to phase out their existing plants. In 1986, the Chernobyl disaster led Russia to cancel or abandon several nuclear power plant projects (though it currently has plans for new, improved plants to replace the ones shut down), and subsequently most of the remaining western European countries have decided to halt further expansion or to develop phase-out plans. Following a referendum Italy decided to abandon nuclear power in 1987; Poland halted construction of a new plant in 1990; and in 1994 the Netherlands decided to phase out its Borselle plant by 2003. Germany has also decided to phase out its nineteen nuclear plants, with the current plan being for a complete phase-out by 2032. In 1999, Belgium likewise decided on a nuclear phase-out by 2025, and in 2000 Turkey decided not to invest in nuclear power.

In contrast, France still has a major nuclear programme: it obtains around 76% of its electricity from nuclear power. However, in 1997, the newly elected Socialist administration imposed a temporary moratorium on new nuclear developments while it considered the future of this technology.

Finland has decided to build a new nuclear plant, the first in western Europe for many years, and attempts are being made to keep the nuclear programme going in central and eastern Europe. In addition there are plans for expansion in some countries in the Far East. Japan already has a significant nuclear programme with 54 reactors, and China, with only three, would like to follow suit, as would India, which has fourteen power reactors at present.

Assuming that the currently planned projects go ahead, there could be moderate overall expansion. However, in its publication *International Energy Outlook*, the US Energy Information Administration suggested that, after 2010, there could be a levelling off (EIA, 2001). After that, unless policies change, the overall picture could be one of slow decline. (Refer back to Figure 11.3 at the start of this chapter.)

But policies could change. Nuclear reactors do provide a way of generating electricity without directly producing CO_2, and so around the world there are signs of interest in the nuclear option as one possible response to climate change. It is this idea, put forcefully by the nuclear industry, and also by, amongst others, the Royal Society and the Royal Academy of Engineering, to which we now turn.

11.3 **Nuclear power: a long-term answer to climate change?**

At a conference organized by the Uranium Institute (now the World Nuclear Association) in 1997, Dr Hans Blix, director general of the International Atomic Energy Agency, commented that 'it may seem puzzling that the nuclear option is largely ignored at a time when there is ever greater worry that the burning of fossil fuels might lead to global warming, and when it seems clearly unrealistic to expect that greater energy efficiency or greater use of renewables will go very far to help us restrain CO_2 emissions'.

The fission process that generates the energy in nuclear plants results in no conventional pollution or greenhouse gases such as carbon dioxide. On that basis, the industry argues, it should be supported as a way to help limit global warming. How valid is this argument?

Perhaps the first point to make is that it is not strictly true that nuclear power plants do not generate *any* carbon dioxide. *All* power plants require energy for their construction, and for the fabrication of materials used in construction. If this energy comes from fossil fuels it entails carbon emissions. In addition to this, and unlike the case for most other power plants, the raw material used to make fuel for nuclear plants has to be extensively processed (from ore), and once it has been used in a reactor it is (in the UK and France at least) reprocessed. These are energy-intensive activities. For example, as we shall see later in this section, depending on the grade of the ore and the technology used for enrichment, nuclear fuel processing can involve the consumption of the equivalent of more than 3% of the electricity eventually produced by the reactor over its lifetime. In most countries, the bulk of the power for this process will be from fossil fuel plants, which will generate carbon dioxide.

Even so, the complete nuclear power process *does* produce much less carbon dioxide than fossil plants. One estimate (Meridian, 1989) puts the total fuel cycle emissions for nuclear power at 8.6 tonnes of carbon dioxide per gigawatt hour, compared with 1058 tonnes for coal plants and 824 tonnes for combined-cycle gas fired plants. However, such estimates are very sensitive to the assumptions upon which they are based. We shall look again at such comparisons in Chapter 13, but for some recent overviews of the issues see: http://www.hydroquebec.com/environment/comparaison/pdf/ang1.pdf and http://www.world-nuclear.com/education/ueg.htm [both accessed 18 June 2002].

On the basis of calculations like this, it was suggested by the UK Atomic Energy Authority (UKAEA) that a programme aiming to increase the global nuclear contribution by nearly threefold, up to around 50% of world electricity requirements by 2020, would result in a 30% reduction in global carbon dioxide emissions (Donaldson and Betteridge, 1990).

However, as we have seen, there are also some major drawbacks. Basically, if it is to meet this challenge, the nuclear industry would have to deal with the problems that led to its decline. The problems we have looked at so far mainly relate to immediate issues such as the risk of accidents and the economics of nuclear generation. We will be looking at them again later. But there are also some more fundamental, longer-term issues that would have to be faced if nuclear power is to be relied on as a long-term solution to the problem of climate change.

The UN-supported World Energy Assessment, produced in conjunction with the World Energy Council, published in 2000, concluded that 'for nuclear energy to make a significant contribution to coping with climate change, nuclear capacity must be increased by at least an order of magnitude during the next 100 years' (WEA, 2000, p. 307).

If expansion on a large scale is to be considered, then there are two key long-term issues which have to be addressed:

- are long-term solutions to the problem of waste disposal available?
- will there be sufficient uranium reserves to sustain such a programme?

We will look at the waste issue first.

BOX 11.3 Nuclear waste disposal

It would be morally wrong to commit future generations to the consequences of fission power on a massive scale unless it has been demonstrated beyond reasonable doubt that at least one method exists for the safe isolation of these wastes for the indefinite future.

RCEP, 1976

More than 25 years on, the UK is still not much closer to finding a solution to the nuclear waste problem. The main difficulty has been in finding acceptable sites for waste repositories. An attempt by the UK Atomic Energy Authority to find suitable sites for disposing of vitrified high-level waste in Scotland, Cornwall, Wales and Northumberland was abandoned in 1981, following local opposition. An attempt to assess the suitability of sites at Billingham on Teeside, and Elstow, near Bedford, for low- and intermediate-level waste repositories, was abandoned in 1987, after extensive protests by people in the selected areas (Blowers *et al.*, 1991).

In 1997, the proposal by the UK nuclear waste company Nirex to build an underground repository for high-level wastes at Sellafield also failed to obtain government approval. The proposal was the subject of a 5-month long public inquiry which ended in February 1996. In March 1997, the then Secretary of State for the Environment, John Gummer, rejected Nirex's planning application. He justified his refusal saying that he remained 'concerned about the scientific uncertainties and technical deficiencies in the proposals presented by Nirex [and] about the process of site selection and the broader issue of the scope and adequacy of the environmental statement.'

The UK Labour Government, elected shortly after Gummer's decision, was therefore confronted with the need for a new technical approach and policy on nuclear waste management. It promised a wide-ranging consultation, which was launched in September 2001. However, the consultation and review exercise is not scheduled to be completed until 2006, so it will be some time before the issue can be resolved.

In the US, pursuant to his commitment to resuscitating nuclear power as an energy option, Vice-President Dick Cheney in May 2001 indicated support for using Yucca mountain, in a remote part of Nevada, as a site for the United States' wastes. This site has actually been under investigation for nearly twenty years, but there are still many political and regulatory hurdles to negotiate before it can be guaranteed as a viable resting place for the US's high-level nuclear wastes.

Sweden has a repository in operation for the less active low- and medium-level nuclear waste, and an interim store for spent fuel, having adopted what some call a 'Rolls Royce', no-expense-spared approach following widespread and sophisticated public consultation, linked to the plan to phase out the use of nuclear power. No long-term repository for the most dangerous high-level wastes exists as yet. Possible sites for such a deep repository for spent fuel are being investigated (see Figure 11.6), but a decision on location has yet to be made and such a repository could not be in operation until the mid-2010s. France has similar plans.

Finland used to have its spent fuel reprocessed in Russia, but now plans to encapsulate and store it all underground in bedrock. Like Sweden, it has carried out extensive public consultation on this issue. The construction of its final disposal facility is scheduled to start in the 2010s and the facility should be operational after 2020.

Even if there is no more expansion of nuclear power around the world, with temporary stores, usually on site at nuclear plants or at reprocessing plants like Sellafield, beginning to fill up, this is clearly an issue that the world's nuclear power users will have to face. Interestingly, Russia has offered its services as a repository for other countries' nuclear wastes, but, so far, no one has taken up this option.

Nuclear wastes

All nuclear plants generate dangerous **nuclear wastes**. The problem is not just a question of the economic costs of nuclear waste disposal, which, as was argued in Box 11.2, are not necessarily overwhelming – it also a matter of the longer-term viability of the nuclear option in safety and security terms. As has already been indicated in the previous chapter, some nuclear wastes remain dangerous for thousands of years (see also Section 10.6). It is claimed that it is technically possible to develop long-term engineered containment systems, which can be backed up by geological barriers to provide extra protection, via underground disposal in areas with suitably stable geological structures. 'Vitrification' techniques for converting the most dangerous part of the wastes, the so-called 'high-level wastes', into a

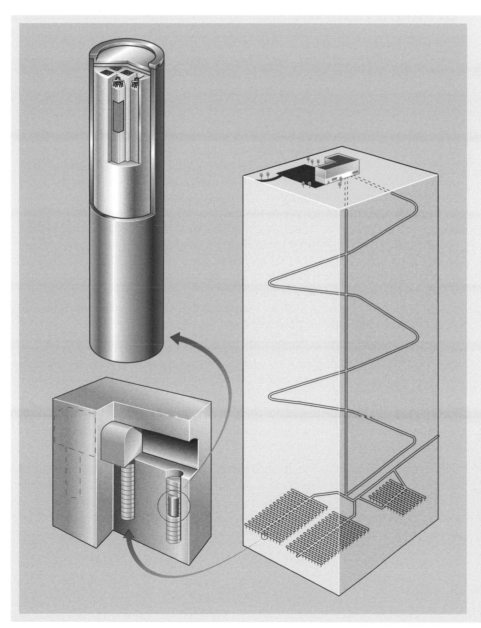

Figure 11.6 Schematic illustration of a deep repository, as proposed in Sweden. A system of tunnels with vertical deposition holes is built at a depth of about 500 metres, and sealed copper-walled canisters containing the spent fuel in a cast iron insert are placed in the holes and embedded in bentonite clay. Each canister is about 5 metres long and 1 metre in diameter (N.B. figure is not to scale)

glassified form have been developed in France and the UK. Some countries have developed stores for the less active low- and intermediate-level wastes. However, although some plans do exist, no one has yet developed a long-term repository for high-level wastes, with most projects and plans having run into significant local opposition. (See Box 11.3.)

The problem is that, although the engineers and geologists claim that wastes can be made safe, not everyone is convinced that they can be successfully contained over such long periods of time. Even in the short term, few communities are willing to accept waste repositories near them, and yet more wastes are being produced all the time. It is a difficult issue, which many environmentalists feel can only really be tackled if no more waste is created.

Uranium reserves

The next longer term issue is that, whereas the wastes will be with us for a very long period, nuclear power generation may be a relatively short-term option, because uranium reserves are not infinite – or to put it more precisely, because there are limits on how much economically extractable uranium is available.

This issue first came to the fore in the 1990s, when concerns about climate change first emerged. It was suggested that if the use of nuclear power was expanded in order to respond to climate change there might be shortages of fuel in the medium-term future. For example, in 1990, the UK Atomic Energy Authority suggested that *if* an attempt was made to *expand* nuclear power dramatically on a worldwide basis in response to the threat of global warming using conventional 'burner' (or 'thermal') reactors, 'the world's uranium supplies that are recoverable at a reasonable cost would be unlikely to last more than fifty years' (Donaldson *et al.*, 1990, p. 29). The UKAEA's journal *Atom* was even more specific, carrying an article in 1990 which suggested that 'for a nuclear contribution that expands continuously to about 50% of demand, uranium resources are only adequate for about 45 years' (Donaldson and Betteridge, 1990, p. 19).

It has to be said that these comments were made at a time when the UKAEA and others were trying to promote the development of the fast breeder reactor, which, they argued, could lift these resource limits. Subsequently, enthusiasm for fast breeder reactors declined around the world – in part because, given the slowdown of the nuclear programme worldwide, there was no immediate shortage of uranium. There have also been new finds of uranium.

Table 11.1 indicates the distribution of *known* reserves of uranium on the basis of estimates published in 1999. As Box 11.4 demonstrates, there is a complex relationship between the figures quoted for total available uranium resource at any one time, and the economics of nuclear power, with there being no absolute figure for 'reserves'.

Table 11.1 Known recoverable resources of uranium[1] (1999 data)

	Tonnes	% of world total
Australia	889 000	27
Kazakhstan	558 000	17
Canada	511 000	15
South Africa	354 000	11
Namibia	256 000	8
Brazil	232 000	7
Russian Federation	157 000	5
US	125 000	4
Uzbekistan	125 000	4
World total	**3 340 000**[2]	

1 'Known' reserves include two categories of uranium resources: reasonably assured resources (RAR) – uranium that occurs in known mineral deposits of such type that it could be recovered within specified cost ranges, using current technology; and estimated additional resources (EAR) – uranium that is expected to occur in extensions of well-explored deposits, little-explored deposits, and undiscovered deposits believed to exist along well-defined geological trends with known deposits, such that the uranium can subsequently be recovered within specified cost ranges. Resources generally occur as uranium oxides (U_3O_8) and pure U_3O_8 product contains about 85% uranium metal.

2 The 2001 joint estimate by the Office of Economic Co-operation and Development, Nuclear Energy Agency and International Atomic Energy Agency for world total was slightly lower, at 3 111 000 tonnes. 2001 usage of natural uranium was about 65 000 tonnes, which implies a known fuel reserve of around 48 years.

Source: OECD, NEA and IAEA (2000)

BOX 11.4 Uranium reserves

Although uranium is a constituent of many rocks, the concentration levels are low. For example, the uranium content in high-grade ore is around 2% or 20 000 parts per million; in low-grade ore it is around 0.1% or 1000 ppm; while in granite and sedimentary rock it is only 4 ppm and 2 ppm, respectively. Clearly it will be harder, and more expensive, to get uranium from the low-grade ores than from the high-grade ores.

The extent of economically available mineral resources depends on the economics of extraction relative to the economics of energy production and sale, and on the level of investment in exploration to find new reserves. The figures in Table 11.1 are only for the high-grade ore economically available at that time. However, the Australia-based Uranium Information Centre has argued that:

[C]hanges in costs or prices, or further exploration, may alter measured resource figures markedly. Thus, any predictions of the future availability of any mineral, including uranium, which are based on current cost and price data and current geological knowledge are likely to be extremely conservative.

[...]

Presently known resources of uranium are enough to last for half a century, considering only the lower cost category, and with it used only in conventional reactors. This represents a higher level of assured resources than is normal for most minerals. Further exploration and higher prices will certainly, on the basis of present geological knowledge, yield further resources as present ones are used up. A doubling of price from present contract levels could be expected to create about a tenfold increase in measured resources.

Uranium Information Centre, 2000

In addition to the 'known' reserves, there are also *speculative* reserves, anticipated on the basis of indirect evidence and geological extrapolations. The existence of these has yet to be confirmed, but if they do exist their costs will in many cases be higher than those for present reserves, because of the added difficulty of identifying and obtaining them. A 1991 estimate from the UKAEA suggested that while there were at that time only 81 years worth of *known* reserves *at then-current use rates*, there were 333 years of *speculative* reserves. Finds since then have pushed this figure still higher, with some reports suggesting that reserves of even low-cost ore are much larger than had been thought, so that supplies could be sufficient until 2100 and beyond, even with high nuclear growth rates: see the *World Energy Assessment* produced by United Nations and the World Energy Council (WEA, 2000).

Moreover, as reserves of high quality uranium ore become more scarce, reserves could be stretched by using lower grade uranium ores. That would also push the price up, although in their joint report the Royal Society and Royal Academy of Engineering (1999) argued that this might be acceptable since the cost of the raw uranium is currently only a small part (about 2%) of the overall costs of nuclear generation. The report suggested that, whereas at a price of US$130 per kg the total known reserves were around 4.4 million tonnes, if a price of 'several hundred dollars per kg' could be accepted, reserves would be 'of the order of 100 million tonnes,' implying a much longer period of fuel availability, possibly increased by a factor of twenty or more.

This assessment may be thought to be on the optimistic side, especially given the increasingly competitive nature of energy generation. Even the estimates quoted in Box 11.4, of a tenfold increase in the uranium resource if the price were doubled, might be thought to be a little speculative. However, even if the economics were to be acceptable, so that extra resources thereby became available, another problem could emerge. When and if lower grade ores had to be used, at some point the net energy balance might not be favourable – that is, with very low-grade reserves, we might have to use more energy to produce the fuel than could be generated from it.

As we have noted, uranium ore contains only a very small amount of the type of radioactive uranium needed for fission reactors (the uranium-235 isotope) and this has to be extracted and enriched to make usable fuel. This is an energy-intensive process, and using lower grade ores would make it even more energy intensive. At some stage the so-called 'point of futility' is reached, when, as lower and lower grades of uranium have to be used, more power is required in order to process the fuel than can be generated from it.

This type of energy analysis is the subject of some dispute, not least because it is difficult to know where to draw the boundaries for a full life-cycle analysis of energy inputs and outputs, and because the energy efficiency of the technologies for uranium concentration and processing continue to improve. For example, enrichment by centrifuge techniques is much less energy-intensive than the diffusion techniques originally used. An analysis presented by the World Nuclear Association suggests that, at worst (i.e. using diffusion), the total energy 'debt' for the complete fuel cycle, including fuel production, plant construction and operation and waste storage, is only about 4.8% of the energy the plant would produce in its lifetime (with fuel processing accounting for three quarters of the energy debt), while using centrifuge enrichment the total debt energy would be only 1.7% of the energy produced (with enrichment then only accounting for about a third of the energy debt). (See http://www.world-nuclear.com/info/inf11.htm [accessed 18 June 2002].)

Either way, on the basis of this analysis we are a long way from the 'point of futility'. However, this analysis assumes the use of high-grade ore. If low-grade ore had to be used then the energy needed for fuel fabrication could be more nearly comparable to the eventual power produced. As was noted in Box 11.4, uranium concentrations for low-grade ores are a factor of twenty less than for high-grade ores, and the concentrations for other materials are even less (e.g. 1000 times less for granite). The energy needed to process this material might not rise proportionately, but it would certainly increase.

Even before the 'point of futility' was reached, the carbon dioxide emissions produced by the fossil fuelled plants that would generate this energy could be significant. One study suggested that, as a result, if a major nuclear programme was launched in response to concerns about global warming, then, well before the point of futility was reached, the carbon dioxide produced by the uranium processing activity could undermine the advantage offered by nuclear power, i.e. the avoidance of carbon dioxide emissions. The situation could of course change. Nuclear reactors might eventually provide the energy for nuclear fuel processing. But in the initial phase, the bulk of the energy for processing the progressively lower grade uranium ores would have to come from fossil fuels (Mortimer, 1990).

Similar energy/carbon dioxide balance issues could arise from another idea sometimes offered by the nuclear industry – extracting uranium from sea water. The uranium resource in sea water is very large – sufficient, in theory, to supply the nuclear industry for several centuries, but the concentrations of uranium are very low, around three milligrams per tonne (0.003 parts per million). If the cost of uranium from uranium ore rose sufficiently, though, it might, at some point, become economically attractive to extract it from sea water. The UN/WEC *World Energy Assessment* quotes a recovery cost of US$100–300 per kilogram, the high figure being more than three times current uranium costs (WEA, 2000). However, even if this cost were acceptable, the energy required to extract the uranium, and the emissions therefore generated, could make the exercise futile, at least while the bulk of this energy was still provided by fossil fuels.

In general, the sea water option might be thought to be as a very speculative long shot. A less speculative option for providing new sources of fuel, at least for the short term, is to use plutonium extracted from spent reactor fuel, together with uranium, in mixed oxide fuel, MOX (see Section 10.6). We will be looking at MOX in a little more detail in Section 11.4. Certainly there is plutonium available from the reprocessing activity, and it has also been suggested that plutonium from redundant weapons could also be used, though the volumes available from the latter source are not large. For example, the World Nuclear Association has noted that that 'world stockpiles of surplus weapons-grade plutonium are reported to be some 260 tonnes, which if used in mixed oxide fuel in conventional reactors would be equivalent to a little over one year's world uranium production' (WNA, 2002b).

While the nuclear fuel reprocessing industry is keen to promote MOX, given the current glut of uranium and the higher cost of MOX, commercial interest in this new fuel is limited at the moment. Some reactors cannot use MOX without modifications, and, so far, there are no plans to use MOX in the UK's existing reactors. However, British Nuclear Fuels has contracts with Japan and elsewhere to supply MOX, although, embarrassingly, arrangements have had to be made for one shipment to be returned after faults in the quality assurance documentation were discovered. For the Nuclear Installations Inspectorate's report on this incident see http://www.hse.gov.uk/nsd/mox1.htm [accessed 18 June 2002].

Stretching reserves: the fast breeder reactor

More dramatically, uranium reserves could be used more efficiently in fast breeder reactors (FBRs), mentioned in Chapter 10. Fast breeders are reactors which can 'breed' plutonium from the otherwise wasted parts of the uranium. In principle they can extract 50 to 60 times more energy from the uranium, so that the availability of fissile material could, in theory, be extended by up to 50 or 60 times. Using figures like this, and assuming current reserves are put at 100 years, nuclear enthusiasts sometimes talk in terms of uranium reserves being stretched thereby to 'thousands' of years, although more guarded commentators limit it to around 1000 years (Eyre, 1991).

Claims that uranium reserves could be extended in practice by factors of 50 or 60 could be seen to be somewhat optimistic, especially since we are talking about as-yet hypothetical systems involving networks of fast breeder reactors and reprocessing plants.

In this context, it is important to realise that it would also take time for a breeder programme to make a significant contribution in energy terms. Despite its name, the breeding process is not 'fast' – 'fast' refers only to the speeds of the neutrons that cause the fission. In fact it can take years to breed as much plutonium as is initially input to the reactor. The so-called 'doubling time' can be in excess of 20 years, perhaps up to 30 years, especially when the fuel cooling, reprocessing and refabrication processes are taken into account (Mortimer, 1990).

In a breeder, the fast neutrons convert the non-fissile part of uranium (U-238) into plutonium, a process which also occurs in conventional reactors, but less efficiently. As was noted above, to get access to this plutonium the old, 'spent' fuel must be reprocessed, so that a fast breeder-based system would require significantly enlarged reprocessing facilities – that is, many more plutonium extraction plants like that at Sellafield. Not only would this significantly increase the amount of nuclear waste produced, it would also, presumably, increase the number of shipments of spent fuel and plutonium between the various reactors and reprocessing plants. That would open up a whole range of safety and security problems, not least the risk that, despite all the precautions that would no doubt be taken, plutonium could potentially be stolen for bomb-making activities.

So, although it has some attractions, and could extend the lifetime of the uranium resource, there are a range of problems with the fast breeder option. Around the world fast breeder projects have been shut down. In the US in the 1970s, President Jimmy Carter was evidently concerned about the potential safety and security problems of plutonium proliferation, and a moratorium was imposed on the fast breeder programme in 1977. In Germany the prototype fast breeder at Kalkar, the scene of major protests by objectors in the late 1970s, was finally abandoned in 1991. The UK government was alarmed at the costs of the FBR project at Dounreay, and at the long delay before a commercially viable technology might emerge, with the result that the FBR programme was halted in 1994. France abandoned its FBR programme in 1997, leaving only Japan with a major programme, although this has had technical problems, with an accident occurring at Japan's Monju plant in 1995 which led to a review of its nuclear programme.

Nuclear fusion: the ultimate answer?

The only other significant, albeit very long-term, nuclear option is nuclear **fusion**. As we have seen in Chapter 10, this process produces much more energy per kilogram of fuel than fission, and can in principle also be carried out under controlled conditions. Progress is being made with experimental devices like JET, the Joint European Torus at Culham in the UK (Figures 10.20 and 11.7). However, so far, the conditions for a fusion reaction have only been sustained for a few seconds. A more advanced machine, the so-called 'international thermonuclear experimental reactor', is planned, at a cost of around €4 billion, in an attempt to get nearer to the conditions for a sustained fusion reaction and for a net energy gain.

Figure 11.7 The Joint European Torus (JET) at Culham, UK

Working with the radioactive plasmas involved has implications which are discussed in Section 11.4. But for the moment, what we are concerned with here are the fuel resource aspects, and the likely scale of output from such a system. If a full-scale power-producing fusion reactor could be built, then in principle, fusion could be better in resource terms than fission since, instead of uranium, it uses materials which are relatively abundant. The basic fuels in the most likely configuration to be adopted would be deuterium, an isotope of hydrogen, which is found in water, and tritium, another isotope of hydrogen, which can be manufactured from lithium. Water is plentiful but lithium reserves are not extensive. Even so, it is claimed that they might provide sufficient tritium for perhaps 1000 years, depending on the rate of use (Keen and Maple, 1994).

However, despite funding on a very large scale over the years, (some £20 billion has been spent worldwide so far) there is some way to go before fusion can be seen as anything more than a long shot option. The physics has still to be fully resolved and a workable commercial device is at best decades away. There is a range of operational problems, not least the question of how to extract energy from the device.

At first sight this looks like an impossible task. On one hand there is a plasma at several million degrees Kelvin, and on the other hand, assuming a conventional power engineering approach, we need to boil water to generate steam for a turbine. The mis-match seems vast: you can hardly put a hot water pipe through the plasma. However, fortunately, it is not the heat of the plasma that would be tapped. Rather, it is the energy in the intense neutron emissions that emerge from the fusion reaction that needs to be absorbed in some way and converted into heat, this in turn presumably being extracted by a conventional heat exchanger.

At some point in the future it may be possible to convert the energy emerging from the fusion reaction directly into electricity, but for the foreseeable future, if a fusion reactor can be built, it will, perhaps rather strangely given its high-tech nature, still have to rely on a traditional steam-raising boiler to generate power.

Given that it will be a very long time before a workable fusion reactor is available, it is sometimes argued that fusion is irrelevant in strategic terms, since, by the time a workable system has been developed, if ever, the world's energy and climate change problems will have to have been solved anyway. In which case, while it might be reasonable to continue some research as a long-term insurance, it could be argued that, given the inevitable scarcity of funds and the urgency of our environmental problems, rather than trying to build a fusion reactor on earth, with unknown economic viability, it might be more effective to make use of the one humanity already has – the sun. This option is described in detail in the companion volume to this book, *Renewable Energy*.

11.4 New nuclear developments

Our analysis so far has sought to explore some of the longer term limitations that may face nuclear power as a way of responding to climate change. We have looked briefly at some fundamental operational issues – the level of carbon dioxide emissions, the need for long-term waste disposal and the availability of uranium resources. The latter issue led us to look at some technologies that could avoid resource scarcity – including the fast breeder reactor and, in the longer term, fusion.

However, there are other issues that would also need to be addressed if nuclear power, of whatever type, were to make a significant contribution to responding to climate change. These are, obviously, improved **economics** and improved **safety.**

Safer nuclear power

Looking first at safety, in recent years there have been attempts to develop new, safer fission technologies. For example, in an attempt to assuage public concerns over safety, the industry has talked of developing so-called 'safe integral' or 'passive' reactor technology, designed to be fail-safe (e.g. in terms of emergency cooling) under all conditions, rather than relying on complex 'active' safety systems, which can fail. As a result of this there has been some interest in smaller reactors, since passive cooling by natural convection is more viable in these; if the cooling system fails there is less chance of the reactor core overheating and the fuel melting-down.

A report produced in 1999 for the European Parliament by the Directorate General for Research (1999) reviewed the safety and proliferation implications of these new systems and suggested that the key, politically sensitive risk of major accidents, defined as unplanned releases requiring evacuation of local populations, might be reduced in some cases by such new reactor design concepts. However, it noted that most of the other problems remain, most notably the problem of waste disposal, although some of the new reactor technologies should produce less waste.

The European Parliament report also notes that, in some cases, as with the new advanced **European Pressurized Reactor** (EPR), all that was being attempted was the redesign of the standard pressurized water reactor (PWR). In the US, Westinghouse has also developed it own 'evolutionary' PWR-based designs, the AP600, and the AP1000, 'AP' standing for 'advanced-passive'. However, there are also some more original PWR-based designs. One example is PIUS, the **Process Inherent Ultimate Safety** system, with the reactor core immersed in a tank of borated water which floods the core and stops the chain reaction if there is any loss of coolant or pressure.

A more radical option is the **High Temperature Reactor** (HTR), which uses graphite as a moderator and helium gas as a coolant. Several HTR systems have been tested over the years, (including the Dragon reactor built at Winfrith in the UK), although without much success. However Eskom, the South African electric utility, is developing a small 100-MW HTR reactor, known as the **Pebble Bed Modular Reactor** (PBMR), so-called due to the fuel being encased in thousands of billiard ball-sized silica-covered spheres. These balls are contained in a hopper through which helium gas is passed (Figure 11.8). High thermal energy conversion efficiency (perhaps up to 40%) is expected since the hot helium gas, exiting the core at 900 °C, directly drives the gas turbines – there is no secondary heat exchanger system. This could actually be a weakness of the design as, if some of the balls were to rupture, the whole system would be contaminated with radioactive material, although it is claimed that the low power density of the device (250 MW thermal output) would ensure that a fuel meltdown would be

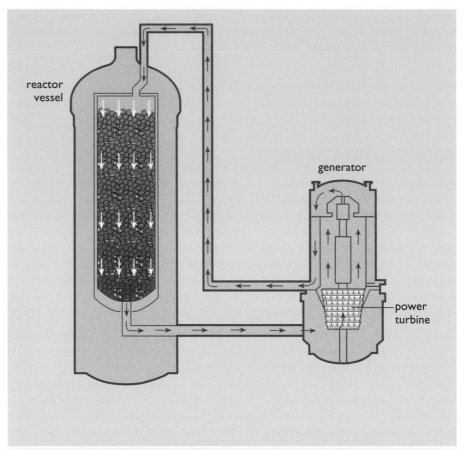

Figure 11.8 The pebble bed modular reactor. The PBMR consist of a vertical hopper (on the left) containing thousands of billiard ball-sized uranium spheres coated in silicon carbide. The heat from the nuclear reaction is extracted by passing high pressure helium gas through this stack of balls, from the top, to the turbine and generator (on the right), which generates electricity. The cooled gas is then returned to the reactor core. This whole system is contained in a pressure vessel

unlikely. All being well, the spent fuel would remain safely encapsulated, ready for eventual disposal, without reprocessing (see http:// www.pbmr.co.za [accessed 18 June 2002]).

In addition to the claimed safety aspects, it is argued that small modular reactors like the PBMR could be more easily mass-produced and, in theory, could be located nearer centres of energy demand, making them more suited to applications in the developing world.

One way to allow for the use of small compact cores is to use liquid metals such as sodium for cooling. For example, the Japanese Central Research Institute of Electrical Power is developing a very small reactor, of around 200 kW, known as the Rapid L, which would use molten lithium as a moderator and liquid sodium as a coolant, running at 530 °C. Liquid sodium has the attraction of high heat-transfer densities without the risk of boiling and it has been used for fast breeder reactors which have small cores. However, given that sodium is chemically a highly reactive material, its use has proved to be problematic, with accidental releases and fires halting some projects.

As we have seen, most FBR projects have now been shelved, since there is a glut of plutonium, and no need to breed more. Indeed, in recent years, the aim seems to have been to try to come up with designs that minimize or even eliminate plutonium production, since, in some forms, it can be used for nuclear weapons.

In light of this, given that a uranium fission reactor inevitably produces plutonium, there is still some interest in using another material, thorium, a weakly radioactive material which is actually more abundant than uranium. Thorium occurs in nature as the isotope Th-232, which is fertile (see Chapter 10). As with U-238, absorption of a fast neutron starts a process that leads to a fissile nucleus, in this case the uranium isotope U-233. However, in order to establish a chain reaction and generate power, the thorium has to be mixed with a fissile material such as plutonium. Even so, the result of the overall operating cycle is that more plutonium is used up than is created and so the use of thorium–plutonium fuelled reactors has been seen as one way to get rid of spare plutonium. Indeed, it is surprising that this idea has never really been taken up, unless, that is, plutonium production is the main attraction of reactors. However, it seems that it *is* still possible to extract other weapon-making materials from thorium-based systems, so perhaps this option too is flawed. Nevertheless, the European Parliament report still sees it as a useful alternative to the conventional fuel cycle, although it admits that the complete fuel cycle infrastructure would have to be reconstructed – a huge task.

One of the most exotic new ideas is the so-called **energy amplifier**. Developed as a concept by Carlo Rubbia at CERN, this uses a proton accelerator firing a beam of neutrons into a sub-critical assembly of thorium coupled with plutonium, which can be used to transmute some long-lived radioactive isotopes into less dangerous forms. The energy output is seen as a by-product – the main aim is to convert some nuclear wastes into less hazardous forms. The system is seen as fail-safe, since the chain reaction can be instantly halted by switching off the accelerator.

Waste transmutation has obvious attractions, but the European Parliament's report notes that it is perhaps 'just nice physics' rather than an economically and technically practical option. It would be a slow process, and only some waste would be suited to transmutation. In addition to the energy amplifier transmutation plants, a complete power system designed on these principles would also require a complex and very expensive network of fuel processing and waste reprocessing plants if significant amounts of waste throughput were to be treated in this way. That would, the EP report says, only be possible, even in theory, for 'countries with a huge nuclear industry'.

Finally, although it is clearly a long shot option, what about the safety, hazards and proliferation implications of nuclear **fusion**? Although no radioactive fission products are produced, the high neutron fluxes inside fusion reactors would induce radioactivity in some of the components. These activated materials would only have 'half lives' of around 10 years, so their radioactivity would decay much more rapidly than that of some of the wastes generated in fission reactors, though significant amounts of these materials would still remain dangerous for perhaps 100 years (Keen and Maple, 1994). At some stage, since this can affect reactor performance and component function, these active materials will need to be removed and stored. So there would still be some wastes to deal with.

And while fusion reactors would not produce fissionable material that can be used in weapons, they can provide a way to test out the *designs* for fusion weapons. The high neutron fluxes in fusion reactors could also, in theory, be used to breed plutonium from low-grade uranium. There could also be safety problems with the fusion reactor itself. Fusion reactions are difficult to sustain, so in any disturbance to normal operation the reaction would be likely to shut itself down very rapidly. It is conceivable, though, that some of the radioactive materials might escape, if for example the super-hot high-energy 'plasma' beam accidentally came into contact with and punctured the reactor containment before the fusion reaction died off. The main concern is the radioactive tritium that would be in the core of the reactor: any escapes could be very serious. Tritium, which is also used in nuclear weapons, is an isotope of hydrogen, and, if accidentally released, could be easily dispersed in the environment as tritiated water. To put it simply, this could then reach parts of the human body that other isotopes cannot (Fairlie, 1993).

For a useful review of the various new nuclear technologies see the relevant chapters in the Royal Society and Royal Academy of Engineering's 1999 report *Nuclear Energy: the Future Climate* at http://www.royalsoc.ac.uk/policy/reports.htm [accessed 18 June 2002].

In the short- to medium-term, interest has been shown (for example by British Energy) in some of the evolutionary options, like the upgraded Westinghouse PWR system and also in the well-established CANDU heavy water moderated reactor, as developed in Canada and exported overseas, for example to India. The CANDU has the advantage that it does not need enriched uranium, but it necessarily produces some plutonium, and this has led to concern about weapons proliferation issues. As for the more novel reactor concepts mentioned above, few are likely to be taken up on any significant scale in the immediate future. Most of these systems are still at an early stage of development, and some are just speculative. Moreover, even if they were adopted, there would still be the problems of waste disposal and proliferation risks, although these might be reduced if the pebble bed modular reactor were used, and it is conceivable that Rubbia's transmutation idea might also play a role in reducing these waste-related problems to some extent, albeit at significant cost.

Waste management and reprocessing

Finding a way to deal with nuclear waste is certainly a key issue that would have to be resolved if nuclear power were to expand significantly. But even without any new nuclear plants being planned, it remains a major problem. Over the next few decades, many of the first generation of reactors around the world will have to be decommissioned, and this will add to the problems of finding disposal routes for the radioactive materials. For example, as the UK Department for Food, Environment and Rural Affairs (DEFRA) has noted, in addition to the 10 000 tonnes of nuclear waste of various kinds already in existence in the UK, 'even if no new nuclear plants are built, and reprocessing of spent fuel ends when existing plants reach the end of their working lives, another 500 000 tonnes of waste will arise during their clean-up over the coming century' (DEFRA, 2001).

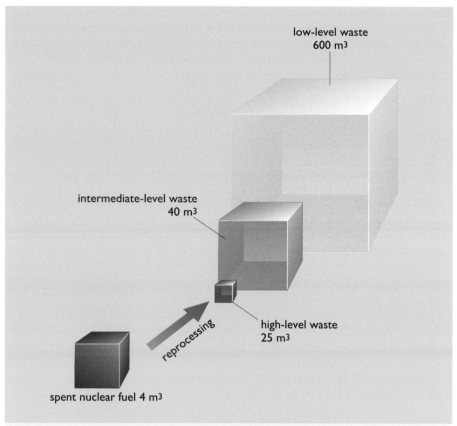

low-level waste
600 m³

intermediate-level waste
40 m³

high-level waste
25 m³

reprocessing

spent nuclear fuel 4 m³

Figure 11.9 Nuclear waste: comparison of volumes of high-, intermediate- and low-level wastes produced from reprocessing spent fuel. The volumes shown are for the spent fuel produced by a 1000-MW PWR over one year

At present, as we noted earlier (see Section 10.6 and Box 11.3), no permanent facility exists in the UK for the long-term disposal of high- and intermediate-level nuclear waste. BNFL claims that 'until one does, wastes can be stored safely and securely above ground for periods in excess of 50 years without the need for extensive store replacement and refurbishment'. However, this claim is not accepted by everyone, with, for example, the Nuclear Installations Inspectorate and the Health and Safety Executive expressing concerns about safety (NII, 2000; HSE, 2001). A more recent concern has been the risk and impact of a terrorist attack. For example, whereas nuclear facilities have usually been designed to withstand accidental aircraft crashes, few experts envisaged the risk of large passenger aircraft being deliberately used as weapons. The consequences of a successful attack could be very serious. A report to the European Parliament suggested that 'the long-term consequences of a release from the Sellafield high-level waste tanks could be much greater than the consequences of the Chernobyl accident, due to the large amounts of caesium-137 and other radioisotopes in the Sellafield tanks' (STOA, 2001).

An allied issue is the question of what to do with the plutonium that has been extracted from spent fuel by reprocessing. So far it has not been seen as a 'waste' but as a valuable material either for bombs, or for fuelling fast breeder reactors to generate yet more plutonium. However, as we have seen, there are no longer any major FBR programmes, and the end of the Cold War means that there is less need for weapons material. In any case, there are alternative techniques to reprocessing for obtaining military plutonium.

Some of the reactor systems mentioned earlier in this section could use plutonium, but currently, as noted earlier, the favoured route is to mix plutonium oxide with uranium oxide to produce mixed oxide fuel (MOX) for conventional reactors. As with conventional spent fuel, the resultant highly radioactive spent MOX fuel would make accessing the new plutonium difficult. Moreover, at present it seems that there are no plans to reprocess spent MOX fuel, since this is difficult and would require modifications to existing reprocessing plants.

However, not everyone is convinced that MOX is a good or viable option, not least since MOX is much more expensive than conventional uranium fuel, by perhaps a factor of five (Garvin, 1998), and using it in reactors inevitably generates more plutonium from the uranium-238 part. Critics also claim that, far from providing an easy way to deal with plutonium stocks, only relatively recently reprocessed plutonium can be used for MOX. Plutonium that has been stored for longer than around 4 or 5 years, including, presumably, plutonium from redundant nuclear weapons, cannot be used for MOX unless it is specially treated, due to the build up of contamination from radioactive decay products, thus adding further to the cost (OECD and NEA, 1989).

It is also argued that there are other ways to immobilize and secure plutonium so as to keep it safe from misuse, for example by mixing it with nuclear wastes (or of course leaving it in un-reprocessed spent fuel); see also Box 10.11. Certainly the European Parliament's (EP) report on safe reactor options mentioned earlier was not convinced that burning MOX was a very useful route to making plutonium secure. Although it would generate energy from what would otherwise be wasted material, not only is this expensive, moving it around to reactors is risky, since it is possible to use it for weapons production (it is easier to get at the plutonium in unused MOX fuel than in the fiercely radioactive spent fuel rods). Consequently, shipments have to be heavily guarded. But the crucial point is that the use of MOX fuel will only make a small reduction in the plutonium stockpile, since this is growing all the time due to the continuance of reprocessing operations.

For a useful briefing paper on MOX, see *Mixed Oxide Nuclear Fuel*, produced by the UK Parliamentary Office of Science and Technology (2000) at http://www.parliament.uk/post/pn137.pdf [accessed 18 June 2002].

As the long history of accidental leaks at Sellafield (and at its French equivalent in Brittany) has illustrated, reprocessing is arguably the main weak point in the 'closed' nuclear fuel system, whatever type of reactor is used. Only a minority of EU countries now favour reprocessing – dry storage of used fuel rods is becoming the favoured option. The main reason for reprocessing was to get access to the plutonium in spent fuel, but this process creates a lot of low- and intermediate-level wastes (see Figure 11.9). This process also has major implications for radiation exposure both to nuclear workers and to the general public – it involves a series of chemical

Figure 11.10 Cooling pond for spent fuel at Sellafield

separation stages which can increase the risk of minor accidents and leaks and involves more handling of radioactive material of various grades by workers. The EP report on reactor options mentioned earlier notes that, as a result, nearly 80% of the collective radiation dose associated with the complete nuclear fuel cycle comes from reprocessing activities. By contrast, electricity generation in nuclear plants accounts for only around 17% of the collective dose, uranium mining and milling only about 2% and waste disposal only just over 1% – all measured on the basis of the 'dose per kWh of power finally generated' (UNSCEAR, 1993).

On this basis, it is clear why some people see reprocessing as a problem, and why strenuous efforts have been made to impose tougher regulations on the levels of allowed emissions from reprocessing plants. For example, the current European policy is that liquid discharges should be reduced 'to near zero' by 2020. However, some opponents do not believe that even this target, which will cost the industry a lot to meet, will be sufficient, since there is still the risk of accidents. They would prefer reprocessing to be abandoned altogether.

The counter-argument is that, since reprocessing separates out the plutonium and uranium, it leaves a smaller amount of high-level nuclear wastes to deal with. While that may be true, the chemical separation process and associated handling activities lead to large amounts of low- and intermediate-level wastes, and, overall, it increases the potential level of radiation exposure to workers and the public. It would be far safer, opponents argue, to leave the spent fuel un-reprocessed and store it. That would have the added bonus that the plutonium would then be much more secure from theft. Moreover, this would also probably be the cheaper option. Certainly, reprocessing is expensive, and it would seem unwise to incur the significant costs and risks, unless the aim is to obtain access to the plutonium for weapons purposes or to produce MOX.

Cheaper nuclear power?

The crucial prerequisite for a revival of nuclear power would be improvement in the economics. The first line of attack would be to improve the operational efficiency of the existing plants. Certainly, since privatization in 1996, the newly established nuclear company British Energy has been successful at significantly improving the productivity of its plants. However, although the government sold BE the AGRs and the Sizewell PWR at, arguably, quite favourable rates, the economics are still marginal, especially given the fact that overall electricity prices have fallen dramatically since privatization. In 2002, the UK government announced plans to help deal with these economic problems, including a loan of £650 million to British Energy. However, with BE facing liquidation, at the time of writing, the prospects for the continued operation of the existing range of plants looks uncertain. This is a problem also faced by BNFL (Brown, 2001).

The main hope for the future is new technology. Some of the new nuclear plants currently being developed, as described earlier, are claimed to be likely to generate at much more competitive prices. For example, it has been estimated that the pebble bed modular reactor might have a capital cost of US$1000 per kW installed and generate at between 2.4–4.3 cents per kWh (WEA, 2000). However, so far these are just claims, not based on commercially operating systems, and it will be some time before a proper judgement can be made as to the technical and economic viability of this system. A more developed option is the 600 MW Westinghouse AP600, a redesign of the PWR concept. BNFL, which owns the license to the design, has claimed that it will offer 'significantly lower generation costs than any other nuclear station currently operating' (Hugh Collum, Chairman of British Nuclear Fuels plc, speaking at the 5th Nuclear Congress, 6 December 2000), and it is similarly enthusiastic about the larger, still to be fully developed, 1000 MW version, the AP1000, which it says might generate at between 2.2 and 3p per kWh (BNFL, 2001).

For the moment however, these figures involve a considerable amount of speculation. In order for new nuclear plants to achieve viability, most analysts of nuclear economics also have to assume some form of subsidy, or increases in the cost of rival technologies.

For example, at present, Finland is the only western European country planning nuclear expansion, having concluded that new nuclear plants, including building, running and decommissioning costs, would be competitive in terms of price with all other alternatives, including combined-cycle gas turbines. However, this analysis was based on the assumption of a 50% rise in gas prices. This may well happen, but if it does then many other technologies would also become competitive.

In its submission to a review of energy security carried out by the House of Commons Trade and Industry Select Committee in 2001, the British Nuclear Industry Forum noted that, even given the new technology that might be available within 10 years, there is still a gap between the wholesale electricity price of around 1.8p per kWh and the cost of new nuclear-generated electricity at about 2.5p per kWh. What is therefore required, it argued, is some sort of financial assistance to bridge that gap, preferably one that will provide a subsidy of about 1p per kWh (House of Commons Trade and Industry Select Committee, 2002).

In terms of subsidies, the nuclear industry around the world has done well over the years (see Figure 11.11). Indeed, it could be argued that it would not exist without subsidies and other forms of financial aid, some of which continue to protect it from competition. For example, in 2001, the UK government proposed that around £42 billion in historical liabilities (e.g. for waste clean-up and plant decommissioning) that had been built up by the state-owned company BNFL and the Atomic Energy Authority, should be transferred to a new public body, thus presumably paving the way for the subsequent partial privatization of BNFL, possibly via a public–private partnership arrangement.

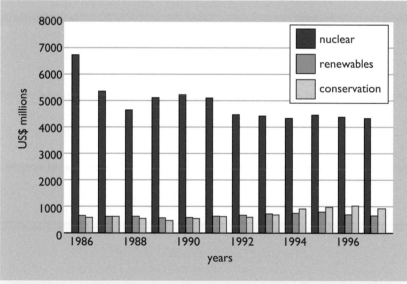

Figure 11.11 R & D budgets in industrialized countries for nuclear power, renewables and energy conservation. 'Nuclear' includes fast breeder and fusion. Prices are in US$ million, based on 1997 prices and exchange rates

The nuclear industry also benefits from legislation exempting nuclear companies from unlimited liability for accidents, a cover currently provided in the US by the Price-Anderson Act and by similar arrangements in the EU. For example, in the UK, under the Nuclear Installations Act 1965, any operator's liability is limited to a maximum £140 million per accident and, if damage caused exceeds that amount, public funds would be made available for the payment of compensation to a total amount of £300 million. These arrangements are currently under revision, with a view to substantially increasing the amount of aid available to victims of nuclear accidents.

UK government support for nuclear research and development (R & D) has continued despite the decline in interest in nuclear power. For example, nuclear R & D (including fusion) was allocated £24 million out of the total UK budget for energy research and development of £50.7 million in 2000/1, and the projections for 2003/4 are that the nuclear share will rise to £52.3 million out of a total energy budget of £106.1 million. Some of this money is for work on decommissioning and clean-up operations (including, £32.2 million in 2003/4 in support of the Russian nuclear programme), but, even so, nuclear energy is still attracting around half the total energy R & D budget, with renewables trailing behind at £13.6 million in 2000/01 and £23 million in 2003/4 – less than half the nuclear allocation. See the UK government's Expenditure Plans 2001/02 to 2003/04 and Main Estimates 2001/02, DTI website, http://www.dti.gov.uk/expenditure/chapter9/page1.htm [accessed 18 June 2002].

Despite the problems discussed in earlier sections, some subsidies could perhaps be justified on the basis of the contribution that nuclear power might make to limiting climate change, by avoiding carbon emissions. However, independent studies have suggested that even with a significant economic boost from some form of carbon tax or carbon credit, it would still be hard for nuclear to be competitive with gas in the UK's liberalized electricity market (Pena-Torres and Pearson, 2000). For example, the Climate Change Levy, which was introduced in the UK in 2001, imposes a 0.43p per kWh surcharge on the electricity used by most companies. However, if electricity from nuclear power plants were to be exempted from this levy, as some have proposed, the resultant price reduction would still not be sufficient to make nuclear power competitive with gas plants. As we noted earlier, the nuclear industry has suggested that it would need a subsidy of around 1p per kWh to bridge the gap.

The situation elsewhere might be different and the nuclear industry was very keen to be allowed to obtain support for new projects overseas from the Clean Development Mechanism (CDM), the emission credit scheme first proposed at the UN Climate Change conference held in Kyoto in Japan in 1997. The CDM is meant to support projects in developing countries which avoid greenhouse gas emissions, and award the developers 'credits' for so doing. Given that these credits can be traded, in effect the cost of the project is reduced. It has been estimated that this might reduce the cost of nuclear plants by up to 20 or 30%. However at the Conference of Parties to the Kyoto agreement held in Bonn in 2001, it was decided, after pressure from the EU, that nuclear projects should not be eligible for CDM credits, with opponents to nuclear inclusion arguing that it was not a clean, safe or sustainable option, nor a useful tool for economic development (NIRS, 2000).

Development issues and proliferation

Given the poor prospects for new nuclear plants in their domestic markets, it is perhaps not surprising that the main emphasis for western nuclear companies currently seems to be in export orders, possibly using new nuclear technologies, with developing countries being the obvious target.

This raises the issue of the role of nuclear power in economic development. Some developing countries evidently see nuclear power as part of the industrialization process. However, it can also be argued that importing capital-intensive nuclear power technology does not seem to be the best option for those 'Third World' countries which are struggling with already large foreign debts, or those whose widely distributed populations in the main need cheap, simple, locally accessible sources of power. There are also worries that part of the attraction of 'going nuclear' is that it can provide a means of developing nuclear weapons. There are other routes to obtaining the necessary nuclear materials, including non-reactor options, but civil nuclear programmes provide one potential source, and, as far as is known, most of the weapons material that has been produced so far has come from reactors of various types.

Although most, but not all, countries around the world have signed non-proliferation agreements for nuclear weapons, any expansion of nuclear power raises concerns about the problem of illegal production or diversion of weapon-making materials like plutonium. These materials are carefully controlled, but a black market has grown up, particularly since the collapse of the former Soviet Union. Adding to the security problem by building more reactors around the world could be unwise. For a discussion of the way in which nuclear weapons proliferation might be guarded against, see the UN and WEC *World Energy Assessment* (WEA, 2000). For a more critical view, see the Nuclear Control Institute website: http://www.nci.org/ [accessed 18 June 2002].

The ex-Soviet nuclear programme, which provides around 12% of the electricity used in the region, presents special problems. The safety of some of the reactors worries many observers, but few of the new central and eastern European states can afford to clean them up, and equally they cannot shut them down, as they need the energy. Fortunately, some of the older reactors are now being closed, including the undamaged reactors on the Chernobyl site, with some funding being provided from the EU and other external sources. In parallel, Russia has plans for new nuclear plants to replace them, based on improved designs.

Concerns about safety, security and proliferation have clearly grown since the terrorist attacks on the US in September 2001. Now, it is not just the theft of nuclear materials that we need to worry about, but also possible terrorist attacks on nuclear plants and facilities. France responded by putting ground-to-air missiles around its reprocessing plant in Brittany, while, perhaps less effectively, the US removed its nuclear plants from maps, e.g. on the World Wide Web. There have been suggestions that any new nuclear plants should be built underground.

However, it might also be argued that, instead of focusing on new plants in developing countries or anywhere else, the nuclear industry should be focusing on plant decommissioning and securing the dangerous waste products, including plutonium. Putting this argument more positively, whatever happens in the future, there will be tonnes of nuclear materials to store securely and all the current nuclear facilities will have to be decommissioned when they have reached the end of their operational life. Around the world, many of the first generation of plants will reach this point in the next decade or so. So regardless of what happens in terms of the nuclear expansion or contraction, the nuclear industry will have plenty to do in the years ahead.

11.5 The future: conflicting views

As we have seen, the prospects for nuclear power are somewhat mixed, and forming a balanced view on what is likely to happen is difficult given that there are strong and polarized views about nuclear technology. Clearly the nuclear industry is keen for the use of nuclear technology to increase, and sees climate change concerns as its possible saviour. At the very least, the industry does not want the nuclear contribution to diminish. Thus, speaking at a World Nuclear Association Conference on 6 September 2001, BNFL's Chief Executive Norman Askew argued that

> Nuclear energy must continue to play a significant role in the UK's base-load electricity generation. Without nuclear's contribution this country cannot have a continued secure, diverse and environmentally-friendly energy supply.

But equally, the environmental opponents of nuclear power are keen to see it contract – and argue that it is foolish to try to resolve one problem (climate change) by introducing another (pollution by nuclear radiation). They believe the money would be better spent on renewables and energy conservation.

Thus the US-based Nuclear Information and Resource Center has argued that nuclear expansion would be a 'lose–lose' option for the environment:

> Not only will there be an expanded nuclear industry, with increased production of radioactive waste and the constant risk of catastrophic accidents, but every dollar spent on nuclear power will be diverted from the development of sustainable energy systems and effective measures to combat climate change.

NIRS, 2000

The view from the other side of the argument has become somewhat less assertive in recent years. For example, in a paper to an international conference on energy organized by the World Energy Council in May 2000, Steven Kidd from the Uranium Institute (now known as the World Nuclear Association) commented that, although nuclear plants produced around 17% of the world's electricity at present 'given that there are relatively few reactors currently under construction and that some of the early plants are scheduled for closure, further increases in world electricity output will mean that this share is set to fall'. He was not sanguine about opportunities for renewal of nuclear growth – except possibly in China: 'Although the economic turbulence in this region has not been helpful, this region appears likely to continue to be the main growth area. In particular, China could feasibly embark on a construction programme similar to that of France in the 1970s and 1980s.'

Kidd continued 'Elsewhere the economics of new nuclear plants do not look good, unless one takes a pessimistic view of fossil fuel prices and utilizes a low discount rate. Combined-cycle gas turbine plants of similar generating capacity can be built for one third of the capital cost of a nuclear plant, and will also be open several years earlier. This is a powerful set of arguments for private financiers.' He concluded 'hopes for renewed nuclear growth are now largely based upon the connection with the global warming debate on emissions targets'. But he felt this was not a reliable route since it is 'unclear how far various countries will go down the road of carbon taxes, emissions permit trading and the like'.

About the only ray of hope for nuclear power that Kidd seemed to offer, apart from redoubling efforts to generate clean, cheap and safe nuclear technology, was that renewables might not work: 'High hopes are placed on renewable energy resources and it will take time to demonstrate that their possible contribution is insecure, to say the least.'

In another paper to the WEC conference, Hans Holgar Rogner and Lucille Langlois from the International Atomic Energy Agency were also somewhat muted on the prospects of a nuclear renewal based on concerns about climate change.

> It is true that nuclear power offers governments the opportunity to achieve a number of national policy goals, including energy supply security and environmental protection, particularly by reducing air pollution and greenhouse gas emissions. But these policy-related 'benefits' are vulnerable to policy change, and are insufficient by themselves to assure a nuclear future. Similarly, the further internalization of externalities [i.e. the inclusion of environmental costs in the generation costs] – to a large extent already imposed on nuclear power – is a policy decision and it is unclear when and to what extent such policies will be implemented. Those who pin their hopes for nuclear growth on externalities or on the Kyoto Protocol – and ignore reform and the need to innovate – will be doomed to disappointment. The nuclear industry has to bootstrap itself to economic competitiveness by way of accelerated technological development and innovation.

Clearly these representatives from the nuclear industry were not too confident about the immediate future of nuclear power. There could of course be political breakthroughs, particularly in the US, where, in 2001, the Bush administration indicated a desire to rethink its policies on nuclear power. In particular, US Vice-President Dick Cheney has backed nuclear expansion, seeing nuclear power as a 'safe, clean and very plentiful energy source' (for the report of the National Energy Policy Development Group, see http://www.whitehouse.gov/energy/ [accessed 18 June 2002]).

However, in the UK, the comprehensive review of energy policy over the next 50 years, carried out in 2001 by the Cabinet Office's Performance and Innovation Unit (PIU), saw nuclear power as likely to remain relatively expensive. Even with the new reactors that might be ready for commercial use within 15–20 years, generation costs for new plants were put at between 3–4p per kWh. Consequently, the PIU, in effect, relegated nuclear power to a longer-term 'insurance' role, to be called on in case its preferred options, renewables, combined heat and power and energy efficiency, failed to deliver. The PIU concluded that 'there is no current case for public support for the existing generation of nuclear technology' although it added 'there are however good grounds for taking a positive stance to keeping the nuclear option open'. The full PIU report can be accessed at http://www.piu.gov.uk/2002/energy/report/TheEnergyReview.pdf [accessed 18 June 2002].

This 'leave the nuclear option open' policy was clearly also supported by the major *World Energy Assessment* carried out by the UN Development Programme, the UN Department of Economic and Social Affairs and the

World Energy Council. It concluded that 'if the energy innovation effort in the near-term emphasizes improved energy efficiency, renewables, and the decarbonised fossil energy strategies, the world community should know by 2020 or before much better than now if nuclear power will be needed on a large scale to meet sustainable energy goals' (WEA, 2000, p. 318).

11.6　Conclusion

The nuclear issue is a complex one. This overview has tried to illustrate the following basic points:

- Nuclear power is a major energy option, which supplies reliable power on a large scale without directly producing any of the emissions (such as carbon dioxide gas) associated with fossil-fuelled power plants. However, recent years have seen a decline in the industry as a result of safety and economic concerns. It could possibly still be revived at some stage, as one response to the problem of climate change.

- Currently, nuclear power is not economically competitive with gas-fired generation, and few governments are willing to subsidize it.

- Accidents can happen even with the best designed technologies: although the chances of major nuclear accidents are low, the impacts can be very serious.

- Nuclear operations, particularly those associated with spent fuel reprocessing, involve the permitted routine release of low-level radioactive materials, and many people remain concerned about the impacts.

- The problem of nuclear waste disposal remains unresolved, whilst a fraught political climate means that fears of weapons proliferation and terrorist attack have increased.

- Concerns about safety, security and economics have led to a decline in enthusiasm for the nuclear option in some, but not all, countries. Public acceptance of nuclear power is generally low in most countries.

- Although there is currently a glut of uranium, reserves might become scarce and expensive if an attempt were made to respond to climate change by building large numbers of conventional nuclear reactors.

- Fast breeder reactors could extend uranium reserves, but at uncertain cost, and the waste and plutonium proliferation problem would be increased.

- Nuclear fusion could, if successfully demonstrated, offer a new energy source with a reasonably long life, but it is a technological 'long shot' with its own safety and economic problems.

It remains unclear therefore whether nuclear power can make a major contribution to responding to climate change.

However, the nuclear industry is developing new reactor technology designed to be fail-safe and more economic, and there are still areas of potential growth for nuclear power, e.g. in Asia, and possibly, given the changed political climate, in the US.

In addition, if concerns about the impacts of climate change grow, and other non-fossil options cannot be developed sufficiently, it could be that backing for nuclear power will revive and grow.

References

Arnold, L. (1992) *Windscale 1957: Anatomy of a Nuclear Accident*, London, Macmillan.

Blowers, A., Lowry, D. and Solomon, B. (1991) *The International Politics of Nuclear Waste*, London, Macmillan.

BMRB (2001) Public opinion survey carried out for the Royal Society for the Protection of Birds, British Market Research Bureau.

BNFL (2001) Submission to the PIU Energy Review by British Nuclear Fuels Ltd, Cabinet Office.

Brown, P. (2001) 'Nuclear fallout', *The Guardian, Analysis*, 14 December 2001.

DEFRA (2001) *Government Looks for Public Consensus on Managing Radioactive Waste*, Press Release, Department for Food, Environment, and Rural Affairs, London, 12 September 2001.

Donaldson, D. and Betteridge, G. (1990) 'Carbon dioxide emissions from nuclear power – a critical analysis of FOE 9', *Atom*, vol. 400, February, pp. 18–22.

Donaldson, D., Tolland H. and Grimston, M. (1990) *Nuclear Power and the Greenhouse Effect*, UK Atomic Energy Authority.

EIA (2000) *Assumptions to the Annual Energy Outlook 2001*, Energy Information Administration, US Department of Energy, DOE/EIA 0554, Table 43, p. 69.

EIA (2001) 'Nuclear power' in *International Energy Outlook 2001*, Energy Information Administration, US Department of Energy.

European Parliament, Directorate General for Research (1999) *Emerging Nuclear Energy Systems, Their Possible Safety and Proliferation Risks*, Working Paper, Energy and Research Series, ENER 111EN.

Eyre, B. (1991) 'The longer term direction for the nuclear industry', *Atom*, vol. 411, March, pp. 8–14.

Fairlie, I. (1993) 'Tritium: the overlooked nuclear hazard', *The Ecologist*, vol. 22, no. 5, pp. 228–232.

Garrison, J. (1980) *From Hiroshima to Harrisburg*, London, SCM Press.

Garvin, R. L. (1998) 'The nuclear fuel cycle', *Pugwash Conference*, Paris, December 1998.

HMSO (1995) *The Prospects for Nuclear Power in the UK*, Cmnd 2860, London.

Holdren, J. P. (1979) *Risk of Renewable Energy Sources: A Critique of the Inhaber Report*, University of California, Report ERG 79-3.

House of Commons Select Committee on Energy (1990) *The Costs of Nuclear Power*, Fourth Report, Session 1989–90, HC 205-1, London.

House of Commons Trade and Industry Select Committee (2002) *Security of Energy Supply*, Second Report, Session 2001–02, HC364-I, London.

HSE (2001) *HSE Enforces Waste Reduction at Sellafield*, Health and Safety Executive Press Release, 31 January 2001.

Inhaber, H. (1978) *Risks of Energy Production*, Canadian Atomic Energy Control Board, AECB-1119.

Inhaber, H. (1982) *Energy Risk Assessment*, New York, Gordon & Breach.

Keen, B. E. and Maple, J. H. C. (1994) *JET and Nuclear Fusion*, JET Publications Group, Culham, UK. See also the leaflet *Energy; The Importance of Fusion Research*, JET Publications, Culham, UK.

McSorley, J. (1990) *Living in the Shadow*, Pan Books.

Medvedev, Z. (1990) *The Legacy of Chernobyl*, Oxford, Blackwell. For related articles see also 'Chernobyl ten years on', *Safe Energy*, vol. 108, March–May 1996; and Tucker, A. (1996) 'Chernobyl ten years on', *The Guardian*, 17 February 1996.

Meridian (1989) *Energy System Emissions and Material Requirements*, prepared for the Deputy Assistant Secretary for Renewable Energy, US Department of Energy, Washington DC, by the Meridian Corporation.

Mortimer, N. (1990) *The Controversial Impact of Nuclear Power on Global Warming*, NATTA Discussion Paper 9, Network for Alternative Technology and Technology Assessment, Milton Keynes.

NAO (1993) *The Cost of Decommissioning Nuclear Facilities*, London, National Audit Office.

NEA (1995) *Chernobyl: Ten Years On – Radiological and Health Impacts*, Paris, Nuclear Energy Agency, OECD.

NII (2000) *The Storage of Liquid High Level Waste at BNFL Sellafield*, London, HM Nuclear Installations Inspectorate.

NII (2001) *A Review by HM Nuclear Installations Inspectorate of the British Energy plc's Strategy for Decommissioning of its Nuclear Licensed Sites*, London, NII.

NIRS (2000) 'CDM – a new nuclear subsidy?' *Climate Change and the CDM Briefing Note*, Washington DC, Nuclear Information and Resource Service.

OECD and NEA (1989) *Plutonium Fuel: An Assessment. Report by an Expert Group*, Paris, Office of Economic Co-operation and Development and Nuclear Energy Agency.

OECD, NEA and IAEA (2000) *Unranium: Resources, Production and Demand, 1999*.

Pena-Torres, J. and Pearson, J.G. (2000) 'Carbon abatement and new investment in liberalised electricity markets : a nuclear revival in the UK', *Energy Policy*, vol. 28, pp. 115–135.

Rasmussen, N. (1973) *The Safety of Nuclear Power Reactors (Light Water Cooled) and Related Facilities*, WASH-1250.

Rasmussen, N. (1974) *An Assessment of Risks in US Commercial Nuclear Power Plants*, WASH-1400.

Rasmussen, N. (1975) *Reactor Safety Study: An Assessment of Accident Risks in US Commercial Nuclear Power Plants*, WASH 400/NUREG 75/014, Washington DC.

Read, P. R. (1993) *Ablaze: The Story of Chernobyl*, Secker and Warburg.

Royal Commission on Environmental Pollution (1976) 'Nuclear power and the environment', *Sixth Report of the Royal Commission on Environmental Pollution* (*The Flowers' Report*).

Royal Commission on Environmental Pollution (2000) *Energy – The Changing Climate*, 22nd Report, Cmnd 4749, HMSO.

Royal Society and Royal Academy of Engineering (1999) *Nuclear Energy: the Future Climate,* http://www.royalsoc.ac.uk/policy/reports.htm [accessed 18 June 2002].

SHE (1994) Gallup Poll data quoted by Stop Hinkley Expansion in its evidence to the UK government's nuclear review.

STOA (2001) *Possible Toxic Effects from the Nuclear Reprocessing Plants at Sellafield (UK) and Cap de La Hague (France)*, WISE-Paris report for the European Parliament Scientific and Technological Options Assessment programme, Strasbourgh.

Trade and Industry Committee (1998) *Conclusions of the Review of Energy Sources for Power Generation and the Government Response to the Fourth and Fifth Reports of the Trade and Industry Committee*, Cmnd 4071.

UNDP and UNICEF (2002) *The Human Consequences of the Chernobyl Nuclear Accident*, United Nations Development Programme and UN Children's Fund

UNSCEAR (1993) Report to the General Assembly by the UN Scientific Committee on the Effects of Atomic Radiation, New York.

Uranium Information Centre (2000) *Nuclear Power*, 6th edn, WNA website, http://www.world-nuclear.org/education/ne/ne3.htm [accessed 18 June 2002].

WEA (2000) *World Energy Assessment: Energy and the Challenge of Sustainability*, UN Development Programme, UN Department of Economic and Social Affairs and the World Energy Council.

WNA (2002a) 'Waste management in the nuclear fuel cycle', *Information and Issue Briefs,* World Nuclear Association web site, http://www.world nuclear.com [accessed 18 June 2002].

WNA (2002b) 'Military warheads as a source of nuclear fuel', *Information and Issue Briefs,* World Nuclear Association web site http://www.world-nuclear.org/info/inf13.htm [accessed 18 June 2002].

Wilson, B. (2002) *Hansard*, column 511W, 18 January 2002.

Chapter 12

Costing Energy

by Bob Everett

12.1 Introduction

As you progress through this book and its companion volume, *Renewable Energy*, you will realize that there are a wide range of possible ways of providing society's requirements for energy services, and also of conserving energy. Which methods are actually used in any particular place is a matter of local circumstances, the perceived cost of the energy supply, and attitudes to investment in energy projects. These topics are discussed in this chapter. It must be said that some of these concepts are not easy to grasp and the mathematical equations may appear daunting at first sight. Try not to be put off by this, as these are complex issues and you will not be alone if you find some of the ideas difficult to follow at first.

First, this chapter looks at recent energy prices in the UK and compares them with prices in the countries we have discussed in earlier chapters. Next, in Section 12.3 we look at how prices in the UK have changed over the years in real terms and in relation to earnings. This requires some explanation of the retail price index and how inflation affects the perceived value of money when used as a unit of account. This leads us to the question of the affordability of energy and the topic of fuel poverty.

Section 12.4 then looks at the problems of investing money in projects both to *supply* energy and to *save* it. If the 'project' is something simple like a low-energy light bulb with a lifetime of a few years, then it may only be necessary to consider the time taken to get our invested money back – the payback time. If, however, something more substantial is being developed, like a power station, then money will have to be borrowed from investors and financial institutions, which will want to charge interest. The money borrowed will have to be repaid over the working life of the station, which may be 30 years or more.

This kind of analysis involves understanding 'real' interest rates and discounted cash flow calculations. As examples, we compare the economics of a proposed nuclear power station and a rival combined cycle gas turbine scheme. This raises additional questions of how financial risk can be included in the calculations, and how high interest rates and short investment lifetimes may predispose investors to one technology rather than another.

The section then goes on to explore how different people and organizations may have different financial outlooks, and looks at the difficult question of what constitutes an equitable discount rate to make sure that the present generation is not getting energy benefits at the expense of future generations.

Finally, Section 12.5 looks at some of the real-world complications with the basic free market outlook on energy, including: the perceived need for countries to subsidize their own fuel industries in order to maintain security and diversity of supply; external costs due to pollution (dealt with in more detail in the following chapter); and the problems of using subsidies to encourage new technologies.

12.2 Energy prices today

We have seen in the previous chapters that different countries have different magnitudes and patterns of energy use. They also have different energy prices and taxation policies, and varying levels of subsidy between the different sectors of the economy. Let us start with the prices that usually provoke the most complaints from consumers – those of petrol and diesel.

Petrol and diesel fuel

Crude oil is a globally traded commodity, so there is a **world price**, currently in the region of about $25 per barrel. In 2000 the cost of transporting oil in a supertanker from the Persian Gulf to the UK added a further $2 a barrel, giving a **beach price** of about $27 a barrel.

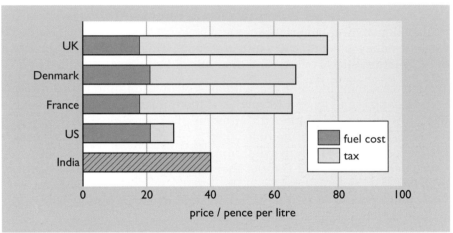

Figure 12.1 Comparative petrol prices, 2000. Note: tax breakdown not available for India (sources: IEA, 2000a and TERI, 2002)

BOX 12.1 Converting energy prices

Different forms of energy are traditionally measured in different units.

Electricity has always been sold by the kilowatt-hour. Gas has usually been *sold* by volume in cubic metres or cubic feet, but always *priced* by its energy content, now normally expressed in kWh on gas bills in the UK. Typically one cubic metre of natural gas as normally supplied in the UK contains about 38 MJ of energy. However, in many other places, such as the Channel Islands, town gas can have a lower calorific value. (Remember that 3.6 MJ = 1 kWh.)

Petroleum is still sold by volume. Crude oil, which varies in energy content depending on its source, is sold by the barrel (159 litres), containing roughly 5.7 GJ. Petrol and diesel fuels are sold by the litre. Diesel fuel is denser than petrol. A litre of petrol contains about 32 MJ and a litre of DERV (DiEsel for

Road Vehicles) contains about 38 MJ. Coal is sold by the tonne, typically containing about 28 GJ per tonne, though different grades, especially power station coal with high ash content, may have a lower energy content.

Although most of the prices in this book are quoted in pounds and pence, oil is traditionally priced in dollars. In mid-2000 approximate exchange rates were £1.00 = $1.50 = €1.60, and these have been fairly stable over the period from 1998 to the time of writing in mid-2002.

So, when reading this chapter remember that:

£1 per GJ equals precisely 0.36p per kWh

$25 per barrel roughly equals $4.50 per GJ, 10p per litre, £3 per GJ or 1p per kWh

Crude oil is refined into a whole range of petroleum products. As pointed out in Chapter 7, petrol requires careful refining and blending, so it is not surprising that its final price in 2000 on leaving the refinery was about 20p per litre or £6 per GJ. This final price is largely the same across different countries (see Figure 12.1 opposite). However, what *is* different is the level of tax charged by the different governments.

In the UK in 2000, petrol cost about 76p per litre, of which 77% was tax. This petrol price was almost the highest in Europe, second only to Norway. Denmark and France also had very high tax rates on petrol, with only slightly lower prices than the UK. In contrast, in the US petrol sold for 28p per litre, of which only 24% was tax. In India, where the indigenous oil industry absorbs a large subsidy, petrol prices were low at around 40p per litre in 2000.

The UK is unusual in having little difference between the price of petrol and road diesel fuel. In most other countries, diesel is cheaper than petrol. In India, for example, in 2000 diesel was sold for only about 22p per litre.

Domestic energy prices

In the domestic sector, UK prices for gas and heating oil are similar to those in France and the United States, in the range £3.50–5.50 per GJ or 1.3–2p per kWh (see prices for 1998 in Figure 12.2). Denmark, however, has had a policy of taxing these fuels as part of their programme of energy conservation, giving much higher prices of £10.60 per GJ or 3.8p per kWh. The high energy taxes in Denmark caused considerable friction with the European Commission in Brussels over their policy of 'harmonization of energy prices'. The Danish government argued that the taxes had been applied to protect the environment, something which was also a key policy objective of the European Union. Even Denmark's prices are low, however, compared to the prices in Japan, which are about £20 per GJ or 7p per kWh.

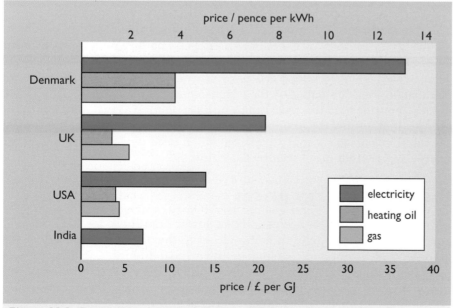

Figure 12.2 Comparative domestic fuel prices, 1998. Note that oil prices were generally low in that year and have risen considerably since then (sources: IEA, 2000b and TERI, 1999)

Average UK domestic electricity prices are about £20 per GJ or 7p per kWh, again similar to those in France, but far lower than the highly taxed rates in Denmark and Japan (13–14p per kWh). The United States has lower domestic electricity rates, of about 5p per kWh.

In India, domestic electricity is actually subsidized down to about half the price paid by industrial and commercial consumers. Kerosene and liquefied petroleum gas (LPG) are also widely used for cooking and lighting, and these too are subsidized and priced to compete with wood. In 1998 kerosene sold at about 5p per litre, equivalent to about £1.40 per GJ, and LPG in cylinders sold at about £3 per GJ.

It is worth remembering that the two heating fuels, oil and gas, have to be burned in devices such as boilers with a limited efficiency, usually between 60% and 90%. This means that the *useful* energy price will be higher than the *delivered* energy price. For example, if gas at 2p per kWh is burned in a boiler with an efficiency of 67%, then the useful heat energy output will cost 2/0.67 = 3p per kWh. Even worse, burning coal at 2p per kWh in an open fire with an efficiency of 25% gives a useful energy price of 2/0.25 = 8p per kWh. It would be cheaper to use electricity at 7p per kWh and an electric fire with an end-use efficiency of virtually 100%.

It should also be noted that figures quoted for domestic electricity are only *average* prices. At night, off-peak electricity for heating purposes is sold at a considerably lower rate, currently about 2.5p per kWh in the UK (as compared to the average cost of 7p).

Therefore, it is perhaps not surprising that there is an increasing use of electricity for heating, particularly in France and the US, where in the recent past heating fuels have usually been coal and oil. In France, these more traditional fuels are being displaced by nuclear-generated electricity. However, in the US it is increasingly coal-fired electricity that is being used. The resulting CO_2 emissions pose considerable environmental problems.

The figures above are for mains electricity. Yet we also regularly purchase electricity in another form – batteries. A D-cell battery for a torch or radio costs about £1, yet it only holds about 6 watt-hours of electrical energy. This is equivalent to over £160 per kWh or £46 000 per GJ. The energy from small long-life batteries for watches and cameras can be fifty times more expensive than this, reaching a million pounds a gigajoule! For these applications it is not the *quantity* of energy that is important, but the *quality* – a reliable, portable supply available when needed. This, it seems, is an energy service that has an enormous value.

Industrial energy prices

Industrial energy prices are usually far lower than those for the domestic consumer. They also tend to be far closer to the world average prices for globally traded commodities such as oil and coal. Figure 12.3 below shows recent variations in UK industrial energy prices. These exclude value added tax (VAT) at 17.5% and the Climate Change Levy, introduced in April 2001, which adds 0.15p per kWh to gas and coal prices and 0.43p per kWh to electricity prices.

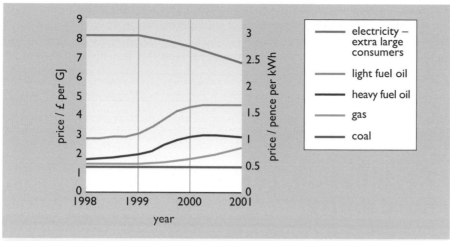

Figure 12.3 Recent trends in UK industrial energy prices (source: DTI, 2002)

UK coal prices, at about £1.20 per GJ (0.43p per kWh) are just slightly above the world market price of coal, which is roughly £1 per GJ. These prices are largely determined by the supplies from Australia, South Africa and South America. Only a few UK pits can compete with these prices, and even they require a small subsidy from the government to do so. This is a far cry from the position in 1913, when the UK was the world's leading coal exporter.

During 1998 and 1999 gas and oil prices were low. North Sea gas was selling at about £1.40 per GJ (0.5p per kWh). Light fuel oil for office heating sold at just under £3 per GJ and heavy fuel oil (used in power stations) sold for £1.60 per GJ. Then in late 1999 oil prices rose and gas followed suit. By 2001 these fuels were 60–70% above their 1998 prices.

Strangely, over the same period UK electricity prices to large consumers actually fell, from £8.10 per GJ (2.9p per kWh) to £6.70 per GJ (2.4p per kWh), and this trend has continued into 2002. The fierce competition in the electricity market has cut profits to the point where many suppliers, and notably the nuclear industry, are experiencing financial difficulties. The United States has similarly low industrial energy prices. However, in Denmark and Japan even industrial energy has been taxed and prices are much higher. In 1998 Japanese average industrial electricity prices were over twice those in the UK and four times those in the US!

In many countries, there is a basic assumption that all electricity should, within reason, have the same price to all consumers. This is not so in India, where there is a general policy of subsidizing electricity differently for different consumers. In 2000 the average electricity price was about £8.60 per GJ (3.1p per kWh). Domestic consumers only paid about 80% of this, but electricity for agriculture and irrigation was almost given away. Over 100 TWh was sold at an average price of 0.4p per kWh. Indeed in the state of the Punjab, it was actually supplied free of charge! This, of course, required a massive subsidy cash flow of over £5 billion per year. Much of this money was recovered by charging far higher prices to the industrial and commercial sectors, and particularly the railways.

There also appear to be many other people in India who get their electricity for nothing. Electricity transmission and distribution losses in India are over 20% of the total generated, compared to only about 8% in the UK. Indeed in Delhi, in 2000, the figure was 45%, and it is admitted that most of the discrepancy is simply due to theft (Mathews, 2001). Unlike the UK, where electricity cables are mainly underground, in India most of the distribution to residential areas is via 240-volt overhead cables on wooden poles. Obtaining an unauthorized supply is a simple (but hazardous) matter of climbing a pole and clipping on a couple of leads.

12.3 Inflation, real prices and affordability

The section above has expressed energy prices in terms of pounds and dollars. These different currencies are freely exchangeable for each other (and euros) and can be used to purchase energy, or other goods or services. Here we are talking about money being used for its principal purpose, as a **medium of exchange**, without which we would have to resort to barter.

Another use of money is as a **unit of account** enabling us to talk about values at different times. Currently (2002), in London in the UK, domestic electricity costs 6.6p per kWh. Fifteen years ago it cost 5.7p per kWh. Obviously the price of electricity has risen, yet government statistics say that domestic electricity in the UK is now cheaper than it has ever been in 'real' terms. This needs some reflection, and we will look at this idea now.

Later in this chapter, in Section 12.4, we will look at a third use of money, as a **store of value over time**.

The value of money

Since around 1920, the pound sterling, like almost all world currencies, has suffered from **inflation**. This is best thought of as a 'disease of money' which progressively decreases its purchasing power in real terms from year to year. We normally think of money as the benchmark against which we assess the value of goods and services. But in reality it is the other way round: money is the medium through which we exchange things and talk about their value.

In order to use money as a unit of account in different years, we should really ask how much of it is required in order to buy the same goods and services in different years.

In the UK, as in most other industrial economies, government statisticians regularly assess prices and compile statistical analyses of their year-by-year changes. The **retail price index** (RPI) is an economic indicator which specifically reflects the prices of a representative mixture of typical household goods at a point in time. Indices such as these are usually expressed as being equal to 100 in a specific year, the base year (the year 2000 in Figure 12.4). From the figure we can see that goods that would cost £100 to purchase in 2000 could have been purchased for £40 in 1980 and only £7.30 (7 pounds and 6 shillings in 'old' money, pre-decimalization) in 1960.

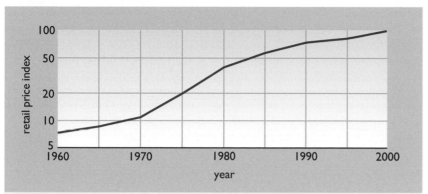

Figure 12.4 Changes in UK retail price index 1960–2000 (year 2000 = 100) (source: National Statistics, 2002)

Figure 12.5 A sample batch of everyday household items. It is these (and a few others) that enable us to measure the 'real' value of money

Inflation is usually described as being *the rate of change of the retail price index per year*. Mathematically speaking, this is the slope of the RPI–time chart. Note that Figure 12.4 has been plotted with a logarithmic vertical axis, so that an equal percentage increase each year would appear as a straight line. The line here is not straight and increases more rapidly during the period of high inflation of the 1970s.

Since the 1920s inflation in the UK has chiefly been positive and often quite high (see Figure 12.6 below), reaching over ten per cent per annum for much of the late 1970s and early 1980s. For most people in the UK, the fourteen-fold price increase since 1960 has not involved social hardship, because average UK wages have risen even faster – in money terms by a factor of twenty-five over the same period. This increase in earning power has been matched in virtually all industrialized nations. Looking back further, there have been times, such as in the economic depression following the First World War, when inflation levels have fluctuated widely, and have even been negative, with both prices and wages falling. Although price cuts are always popular, wage cuts are certainly not, and can provoke social unrest.

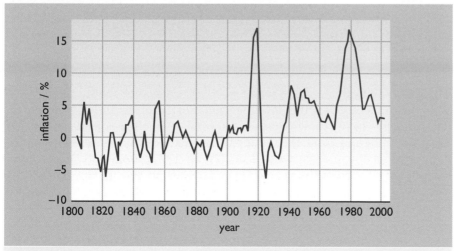

Figure 12.6 UK annual inflation rate 1800–2000 (source: *Guardian*, 5 May 2001; data from Financial Services Authority)

If inflation gets out of control, as it did in Germany in the early 1920s (see Box 12.2 below) it can become extremely difficult to know how to carry out economic analysis at all. Low inflation is always a prime economic political objective for governments.

BOX 12.2 Hyperinflation

In 1922 and 1923 inflation in Germany reached staggering proportions, peaking at 2500% per month. At the beginning of 1922 the mark was worth $2.38. Eventually, in November 1923, the exchange rate was stabilized at 4 200 000 000 000 marks to the dollar. The money was so worthless that it required thick bundles of banknotes to buy even basic groceries. Yet life went on and workers were paid in similar thick bundles of notes. The real value of the goods did not change, just the value of the money used as a medium of exchange. Traders and banks used the US dollar ('real money') as a unit of account to keep track of what was going on. The mark became useless as a 'store of value' and many people's life savings were wiped out.

Figure 12.7 This 1922 German shopkeeper had to keep his money in a tea-chest because there wasn't room for it in the cash register

---

Thus, to be strictly accurate when quoting prices, we should also always state the monetary year that is being used. For example, in 1971 on average a dozen eggs cost 22 (new) pence – that is £0.22 in 1971 pounds or £(1971). Between 1971 and 2000, the retail price index rose by a factor of 8.4, so we can say that those eggs, valued at 22p in 1971, are worth 8.4 × £0.22 = £1.85 in pounds of the year 2000 or £(2000). In fact, in the year 2000 a dozen eggs cost on average £1.72, so in real terms eggs had become slightly cheaper in relation to the other items in the statisticians' reference batch of commodities (bread, potatoes, etc.).

An egg is an egg, whether the year is 1971 or 2000, but we are measuring its value in different money.

If you look back to Figure 2.6 in Chapter 2, you will see a chart of world oil production and prices, where the prices are expressed in US$(2000). The actual dollar prices of the day have been converted into year 2000 values by deflating them using a dollar retail price index. This figure shows that between 1905 and 1972 the price of oil rarely rose above $20 per barrel in year 2000 dollars, yet in the money of the day at the beginning of the century it was sometimes being sold at 10 cents per barrel. Converting the value of the oil into money of a particular standard year shows historically how serious was the sevenfold increase in real price between 1973 and 1979, and how, by 1998, prices had almost fallen back to their real pre-1972 levels.

To take another example, we can look at how real UK petrol and diesel prices have varied in recent years. Between 1989 and 2001 the UK government had a policy of progressively increasing these by applying a varying rate of tax. This is shown in Figure 12.8 below. Leaded petrol was also given a higher rate of tax than unleaded, in order to encourage its phasing out.

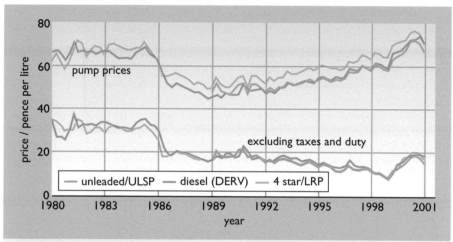

Figure 12.8 UK real petrol and diesel prices 1980–2001, expressed in £(1995) (source: DTI, 2001b)

This tax policy was, of course, unpopular with many consumers while the *world* oil price was falling between 1986 and 1999; they felt that these benefits should be passed on to them. However, Figure 12.8 demonstrates that by 2000 pump prices were almost exactly the same in real terms as they had been in the early 1980s.

In a similar fashion, let us now look back further in time at how energy prices in Great Britain have changed over the whole course of the twentieth century. The values, expressed in £(2000) in Figure 12.9 overleaf, are for Great Britain, rather than for the whole UK, because natural gas has not been available in Northern Ireland until very recently.

The figure shows quite clearly why solid fuel was the preferred option in Great Britain at the beginning of the twentieth century – seen in today's money, it was cheap! Town gas, the convenience fuel made from coal by a quite labour-intensive process, sold for about five times the price of solid

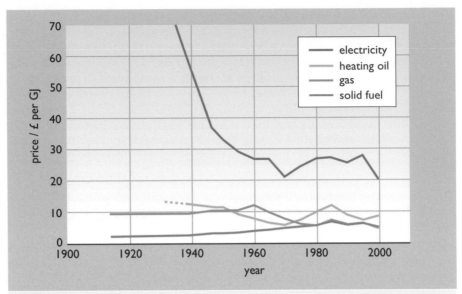

Figure 12.9 Real GB domestic fuel prices, 1914–2000, expressed in £(2000) (source: updated from Evans and Herring, 1989)

fuel, and electricity was a very expensive luxury fuel. In 1885 the Brighton Electric Light Company was selling electricity for 1 shilling (5p) per kWh. This is equivalent to over £3 per kWh or almost £900 per GJ in year 2000 money! Fortunately for their customers, the company had cut its prices by a factor of twelve by 1898. Even so, however, electricity prices only creep onto the range at the top of Figure 12.9 in the 1930s. Since then, continuous improvements in technology have brought the price of electricity down, with it falling to below £30 per GJ in 2000 money terms in the late 1950s. The fall between 1995 and 2000 was largely a consequence of the extension of competitive electricity supply to the domestic sector and a continuous price war between suppliers.

Oil for heating was not a major fuel in Britain before the Second World War, so its price only enters the chart in 1939. Its price fell during the 1950s and by the 1960s it had become cheaper than town gas. Sales of oil-fired central heating expanded and gas sales contracted. The electricity industry also started marketing electric central heating using off-peak storage heaters and cheap night-time electricity. Had this era of cheap oil continued, it is possible that the market for gas as a fuel in Britain could have collapsed completely, as it actually did in Northern Ireland. However, in the 1970s the position was turned around when UK oil prices rose, although the availability of North Sea oil to some extent cushioned the UK consumer against the enormous jump in world prices shown in Figure 2.6. In addition, new supplies of North Sea gas were being brought into Britain, selling at half the price of town gas and competing heavily with coal. The scene was set for a massive boom in sales of gas central heating and a continuing decline in the use of oil and coal.

Affordability and fuel poverty

Although the retail price index allows us to compare the value of goods from one year to another, it says little about their affordability. As we saw above, the monetary price of eggs rose by a factor of eight between 1971 and 2000. However, average wages increased by a factor of almost fifteen over the same period.

In 1971 the average UK gross weekly wage was £28.70. This sum would have bought 28.70 ÷ 0.22 × 12 = 1565 eggs. By 2000, the average weekly wage had risen to £420, which would have bought 420 ÷ 1.72 × 12 = 2930 eggs, almost twice as many. Put another way, eggs were twice as affordable in 2000 as they were in 1971.

Energy use, like everything else, is likely to be determined by the ability of consumers to buy it, which means it has to be judged in terms of earning power. Just as government statisticians compile a retail price index, they also compile a similar one of average year-by-year earnings. This can be used to show the change in relative affordability of energy. Figure 12.10 below shows energy prices for Great Britain deflated by the average earning power in each year and expressed in £(2000).

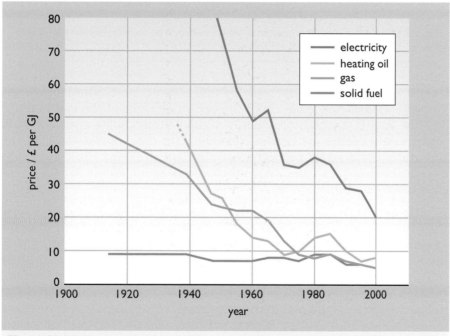

Figure 12.10 Energy prices in Great Britain relative to earnings, 1914–2000, expressed in £(2000) (source: updated from Evans and Herring, 1989)

These declining curves clearly show how heating oil, gas and electricity have become more affordable over the last century, though the post-1973 bump in the oil price curve suggests that these trends are not totally irreversible. The fact that solid fuel prices on this basis did not change much over the course of the twentieth century probably reflects the labour costs involved. The coal price largely followed the earnings of the miners, which in turn kept pace with general earnings. Both town gas (until the

1960s) and electricity (until the 1990s) were largely produced from coal. The declining price of these fuels reflects the improvements in technology. An average week's wages in 2000 would have bought almost four times as many kilowatt-hours of electricity as a week's wages in 1950.

The percentage of household expenditure on fuel and power peaked in the 1960s at about 6% (see Figure 12.11). By 2000 it had fallen to only 3.3%. However, it should be noted that this figure *does not* include expenditure on fuel for the motor car!

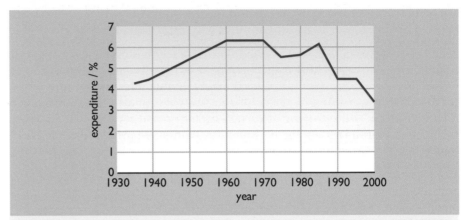

Figure 12.11 UK percentage of household expenditure on fuel and power 1930–2000 (sources: Stone, 1996 and national Statistics, 2002)

It is all very well to say that energy has become more affordable to the 'average worker', but what about the poorer sections of the community? The UK Family Expenditure Survey divides the population according to income into tenths or 'deciles'. In 2000, for the poorest tenth of the population, fuel and power accounted for 6.5% of the household expenditure; 3.6% of the total income was spent on electricity. Food accounted for a further 21%.

A household is defined as being in **fuel poverty** if it needs to spend more than 10% of its income on energy in order to achieve a satisfactory degree of comfort. This is defined as having a minimum temperature of 21 °C in the living room and 18 °C elsewhere in the house. It has been estimated (DTI, 2001a) that in 1998 approximately one in six households in England suffered from fuel poverty. The statistics show that the poorest section of the community, which includes pensioners and single parents, could not afford to spend 10% of its income on fuel. Instead, this section of the community only used as much energy as it could afford and the comfort standards were not achieved.

Fuel poverty creates a difficult energy pricing problem. It is easy to say that energy, and especially electricity, is too cheap and that higher prices would encourage conservation, but raising prices with environmental taxes is likely to hit the poor hardest. The most satisfactory solution is to improve the insulation standards and heating systems of the housing stock, so that the satisfactory degree of comfort can be achieved with an affordable amount of energy. How exactly this is to be done in the UK is currently a matter of considerable debate (see, for example, DTI and DEFRA, 2001).

If fuel poverty is a problem in the UK, it is an even worse one in India. The country has a rapidly rising population and an acknowledged serious energy shortage. Wood-fuel for cooking is undoubtedly being gathered on an unsustainable basis in many areas and electricity shortages cause frequent blackouts. According to a 1993/4 survey (NSSO, 1996) the annual average expenditure in India on fuel and lighting amounted to 7.4% of household income in rural areas and 6.6% in urban areas. At first sight these figures appear similar to those for the UK, but it must be remembered that these are proportions of a far lower income. In Chapter 3 it was pointed out how expensive and inefficient fuel-based lighting is compared to modern electric lighting with fluorescent lamps – yet according to another 1993/4 survey, 62% of the Indian rural population and 16% of the urban population still used kerosene for lighting at that time. The Indian government has an overall policy of extending the electricity grid to as many villages as possible, and statistics suggest that 85% of villages (though not individual homes) had some sort of connection by 1997. There is also a special programme, the *Kutir Jyoti*, run by the Gujarat government, for providing single-point light connections for the rural poor. By March 1997 this had provided 2.4 million households with lighting supplies.

The question is sometimes asked 'how can the poor afford electric light?', and the answer is 'with great difficulty' – but electric light is more cost-effective than that from kerosene lamps.

12.4 **Investing in energy**

This next section describes the textbook theory of investment in energy, where energy is seen as a simple, tradable commodity in a perfect investment market place. However, there are many complications with this view, some of which we will go on to look at in Section 12.5.

Price and cost

The previous sections have described the price of energy today and over the past century. The next question is, how much would it cost individuals or organizations to produce their own energy, or to invest in energy saving?

At first sight the two words 'price' and 'cost' would seem to mean the same thing. A **price**, though, is something *determined in the market place*. Conventional economic theory says that if there is a shortage of a particular freely traded commodity then the market price will rise so that supply matches demand.

A **cost**, at least as used in this chapter, assumes that an energy producer (a private individual, a company, or perhaps a government department) can make some **profit** out of a particular investment. Normally a cost figure would include such things as materials consumed and labour, and also repayments on invested capital at rates that would be competitive with other projects. The cost is therefore *a measure of the minimum amount for which the producer must sell energy in order for its production to make a profit*. Put simply, for the purposes of this chapter:

price = cost + profit

If something is being sold at a price equal to the cost of producing it, then there is no profit. However, 'profit' is a word which should be used carefully, not least because profits tend to be taxable. Readers who wish to pursue the finer points of the nature of profit are referred to standard textbooks on economics such as Wonnacott and Wonnacott (1986).

The distinction between cost and price is demonstrated clearly in the case of crude oil from the wells of Persian Gulf states. Here the cost consists of:

- a large historic capital investment in exploration and drilling until an oil field has actually been discovered, followed by other investments in setting up the wells, pipelines and port facilities to ship it out;
- a continuing marginal operational cost of labour and energy to actually pump out the oil and load it onto tankers.

At the time of writing (2002) the world crude oil *price* is about $25 per barrel. This is the amount of money that consuming countries are prepared to pay for the limited supply of oil available. Yet the marginal *cost* of pumping another barrel in the Gulf States is estimated to be less than $1.50 (EU, 2000). The remaining $23.50 is available to pay off the capital cost of the original drilling and exploration (much of which took place decades ago, and has probably been paid off completely by now), and for profit.

The extraordinary instability in world oil prices since 1972 has been largely a matter of how much profit the Middle Eastern countries wish to make in selling their oil, which, after all, they can only ever do once. The ability and willingness of these countries to produce oil has been coloured by the continuing political conflicts in the region, which show no sign of abating.

Balancing investment against cash flow

The Middle East is fortunate to have such large energy reserves. Elsewhere, practical energy economics is largely a matter of carefully balancing a capital investment in a project against the flow of money for that project over a period of time. The return on investment can be expressed in a number of ways, of which the simplest is the payback time. Exactly the same financial analysis can be applied whether it is a matter of investing money to *produce* energy or to *save* energy.

A simple example is the investment in using a compact fluorescent lamp (CFL) like the one shown opposite.

Currently (2002) in the UK a good-quality 20-watt electronically ballasted CFL costs about £10 in the shops (though they are often available for considerably less under various marketing promotions), whereas an ordinary 100-watt incandescent bulb of equivalent light output costs only 50p. Clearly the incandescent bulb is cheaper, so why waste £9.50 on buying the CFL?

The answer is that CFLs use much less electricity and last much longer. There is thus a balance to be struck between an investment in capital equipment (the CFL) and the energy savings in kilowatt-hours spread over the life of the lamp. Depending on the level of use, the extra capital expenditure can be recovered in year or less. Given this short time-frame, we can ignore inflation and interest on money. The calculation is given in Box 12.3.

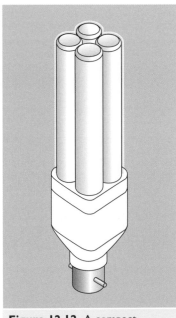

Figure 12.12 A compact fluorescent lamp

BOX 12.3 **Calculating the payback time on using a compact fluorescent lamp**

An electronically ballasted CFL typically uses only 20% of the electricity of a 100-watt incandescent bulb of equivalent light output. In addition, the CFL will last 8000 hours against only 1000 hours for an incandescent bulb.

In order to calculate the payback time we need to look at the expenditure over the entire life of a CFL. Currently UK electricity costs, on average, about 7p per kilowatt-hour.

20-watt compact fluorescent lamp
In its 8000-hour life a 20-watt CFL will consume:
8000 h × 20 W = 160 000 watt-hours = 160 kilowatt-hours of electricity.

At 7p per kilowatt-hour, this will cost: 160 × 7p or £11.20.

So, including the £10 cost of the lamp, the total cost over 8000 hours = £21.20.

100-watt incandescent bulb
In 8000 hours a 100-watt incandescent bulb will consume: 8000 h × 100 W = 800 000 watt-hours = 800 kilowatt-hours of electricity.

At 7p per kilowatt hour, this will cost: 800 × 7p or £56.00.

Each incandescent bulb lasts 1000 hours, so over 8000 hours we will need to buy 8 bulbs at 50p each. This adds another £4.00, bringing the total to £60.00.

So the overall profit in using a compact fluorescent lamp is £60.00 − £21.20 = £38.80.

This is nearly four times the initial cost of the CFL.

This sounds good, but how long will it take to get the investment back?

A simple way to work this out is to plot a comparative expenditure–time chart for the two options (see Figure 12.13 below).

The CFL initially costs £10. With an energy consumption of 20 watts, it will consume 20 kWh in every 1000 hours of use. This amount of electricity will cost 20 × 7p or £1.40. So after 1000 hours of use, our total expenditure will have been £10 + £1.40 = £11.40 and after 2000 hours it will have been £10 + £1.40 + £1.40 = £12.80. This is plotted in Figure 12.13 as the purple line.

An incandescent bulb costs 50p and consumes 100 watts. In each 1000 hours it will consume 100 kWh of electricity, worth £7.00. After the first 1000 hours the total expenditure will have been £7.50. We will then have to buy a new bulb. In the next 1000 hours we will spend another £7.50, so after 2000 hours we will have spent £15.00.

When the two options are plotted, we can see that after about 1600 hours of continuous use, the costs of using the CFL have become less than those of the incandescent bulbs. Use for 1600 hours is about equivalent to 4.5 hours of use every day for a year. So at this level of use, we would have a payback time of one year. If the lamp is used more often, say for 9 hours per day, then the payback time would be only six months. This would normally be regarded as exceptionally good value for money.

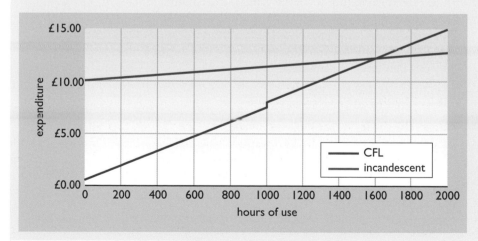

Figure 12.13 Expenditure–time chart for CFL and incandescent bulbs

A CFL is a low-cost investment and its practical life is relatively short. Other energy investments require a longer time-perspective, but the same considerations will apply whether we are looking at investment in a power station, a wind turbine, or cavity wall insulation in a house.

Let us now take a combined cycle gas turbine power station as an example. It requires a large capital investment at the beginning of the project, a continuing supply of gas fuel, a certain expenditure on operational staff and occasional expenditure on maintenance. It can be built within two years and once constructed it should be able to produce electricity for 20 years or more.

How do we calculate a cost for the electricity produced in pence per kilowatt-hour? As a first attempt, the capital cost of the plant can be annuitized (i.e. spread out over a number of years) by dividing it by the lifetime of the machine. The annual fuel, operation and maintenance costs can then be added to give the total annual cost. Finally, this value can be divided by the annual electricity output to give the cost in pence per kilowatt-hour. Wear and tear on this kind of equipment is often proportional to the number of hours run, so operation and maintenance (O & M) costs are often expressed as a certain sum per kilowatt-hour of electricity produced. Box 12.4 below illustrates this calculation. The figures come from a 1994 comparative study with a nuclear power station, which we will revisit later in this chapter.

BOX 12.4 A combined cycle gas turbine power station

The capital cost of a proposed gas-fuelled 400-MW combined cycle gas turbine (CCGT) is £180 million. It is expected to run for 7500 hours per year at full power and has a design lifetime of 20 years. Its electrical generation efficiency is 54%. Operating and maintenance (O & M) costs have been estimated at 0.33p per kWh of electricity generated. Gas is available at 0.75p per kWh.

Calculate the cost of the electricity in pence per kilowatt-hour ignoring any interest charges.

Solution

If the electrical output is 400 MW and the generation efficiency is 54%, the rate of gas consumption = 400 MW ÷ 0.54 = 741 MW.

If the station unit runs for 7500 hours per year, in each year it will consume 7500 h × 741 MW = 5 560 000 MWh of gas, and produce 7500 h × 400 MW = 3 000 000 MWh of electricity.

Annual costs and benefits:

  capital cost spread over 20 years = £180 000 000 ÷ 20 = £9 million

  gas costs at 0.75p per kWh = 5 560 000 000 × 0.75p or £41.7 million

  O & M costs at 0.33p per kWh of electricity generated
  = 3 000 000 000 × 0.33p or £9.9 million.

The total cost per year = £9 million + £41.7 million + £9.9 million
= £60.6 million

The cost per kWh = £60 600 000 × 100p ÷ 3 000 000 000 kWh = **2.02p per kWh**.

This approach seems reasonable enough if we just happen to have £180 million to spare. However, the capital expenditure has to be made now, whereas the benefits in electricity only arise in the future (and some of it will be in 20 years' time, which is a long time to wait). What happens when we have to borrow the capital from a bank or raise it from shareholders? They might be prepared to supply £180 million but will want to receive a steady stream of interest payments for the privilege. This makes the calculation more complex, and we will revisit this example and expand upon it in the next subsection.

Discounted cash flow analysis

Interest rates and discount rates

In practice, a pound tomorrow is not worth the same as a pound today. We must look at the important concept of the **time value of money**. There are several separate processes involved and it is important to understand them. They are:

1   The effect of **inflation**, already discussed, which progressively erodes the capacity of money for purchasing real goods.

2   The normal **time preference for money**. Given a choice, most people would generally rather have a pound today than a pound in the future. Put another way, we would need to be offered a pound plus some additional sum, say $x\%$, next year to forgo the use of one pound today. This preference will exist even when the inflation rate is zero.

3   The **opportunity cost** of a potential investment. We have the ability to lend out money and charge interest. This means that we can forgo the use of a pound today in order to have a pound plus an additional sum in the future. There is always likely to be a number of different opportunities for investing money. If we spend money now rather than invest it, or use it in a poor investment rather than a better one, then an economist would say that we have incurred a cost. This is known as the opportunity cost.

Different individuals' time preferences for money will vary according to their financial circumstances, but in general it is these that determine interest rates as quoted by banks. Someone opening a savings account at a bank is giving up the use of a pound today, but expects to be able to withdraw somewhat more than a pound at some time in the future. The theory assumes that they will only open the account if the rate of interest offered is sufficiently good to overcome the attractions of spending the money now.

Let us say that £100 is invested at a 10% rate of interest per annum. One would expect to be able to withdraw £110 in one year's time. Let us also suppose that this £110 has to be paid to settle some bill that is expected to arise at that time. By investing only £100 now we can pay off a larger bill in the future. If this bill is due quite a long time in the future then the effect of the investment can be very marked. After ten years' investment at 10% interest, £100 would have a **future value** of £260. Put another way, the **present value** of £260 in ten years' time at an interest rate of 10% is only £100.

Here, we are discounting future payments, effectively saying that sums of money in the future can be expressed in terms of smaller sums today. This concept leads to a technique of economic appraisal known as **discounted cash flow** (DCF) analysis. This allows us to express a series of bills at various times in the future as a single lump sum in the present. For example, we may have three bills: £100 today, £110 in one year's time, and £260 in ten years' time. As we have seen above, the separate present values of each of these is £100. We can add them together to have a **net present value** (NPV) of £300. This is the total amount of money that we must have available today to settle all these bills, given that we can invest the money at an interest rate of 10%.

If the interest rate were lower, say only 5%, then the net present value would be different. A net present value calculation is a convenient tool for dealing with payments over long periods of time.

In practice the terms 'discount rate' and 'interest rate' tend to be used interchangeably. They are not necessarily quite the same thing, however. The discount rate is literally the rate at which we 'dis-count' the value of future income or expenditure. It may be similar to a bank's quoted interest rate, but with an extra allowance to cover project risks. This will be discussed later in this section.

We can now go back and look at the power station example from Box 12.4 through the eyes of a potential investor – would it be better to invest in this plant or in something else? The value of investing in this project must be competitive with the opportunity cost, the value of the next best opportunity forgone in financing a given investment. If money can be safely invested to produce an 8% return in say, chemicals or electronics, or even just a bank savings account, then why bother with gas-fired power stations unless these can produce better returns?

Historically, bank interest rates in the UK (and other countries) have varied widely from year to year since the 1940s (see Figure 12.14 below). Looking back further, though, the rates were remarkably low (between 3% and 6%) and stable through most of the eighteenth and nineteenth centuries, the key period of the Industrial Revolution. Also, as we have seen in Figure 12.6 above, inflation rates were also very low, approximately zero over the whole of the nineteenth century.

In practice, for a new investment, the interest rate is likely to be something set by a bank, and this value in turn is likely to have been politically and economically determined by the government treasury.

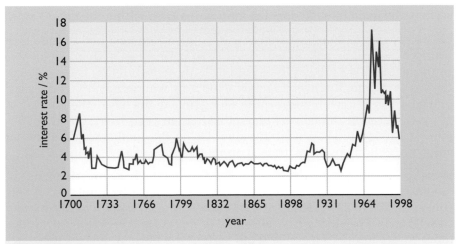

Figure 12.14 UK bank interest rates 1700–1998 (source: *Guardian*, 5 May 2001; data from Merrill Lynch)

We also have to take into account the effects of inflation. It is very important to realize that discounting and inflation are completely separate things. Inflation describes how prices rise with time, eroding the real value of the very money for which we are attempting to perform the calculation. Discounting, in contrast, describes the fact that because of uncertainty about the future, people usually prefer to have money today rather than the promise of money tomorrow; and that if we do have money now it can always be invested in order to make even more money later.

In order to adjust for the effects of inflation, we need to calculate a real interest rate, rather than the purely monetary one. This is simple enough:

real interest rate = monetary interest rate − rate of inflation

It is worth noting that since both monetary interest rates and inflation rates vary with time, so can the real interest rate, which is the difference between the two. Although during the 1970s real interest rates were actually negative, since about 1982 the rate has been positive and has varied between 3% and 8% since 1985 (see Figure 12.15 below).

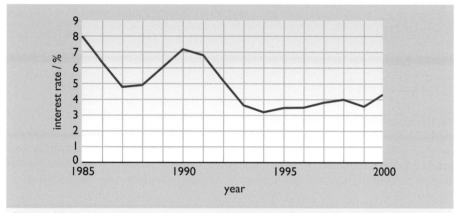

Figure 12.15 UK real interest rates net of inflation, 1985–2000 (Bank of England base rate – change in RPI) (source: National Statistics, 2002)

Project lifetime

Generally if we borrow money, we agree to pay it back over a certain number of years. For example, mortgage payments for a house might be spread over 15 or 25 years. Financial institutions are unwilling to consider loans spread over longer periods than this because of the essential uncertainty about the future. Yet the working life of a nuclear power station might be 40 years and that of a hydroelectric station in excess of a hundred. Although this does not often create problems of analysis for most energy projects, it is worth remembering that the capital repayment time (to pay off a loan) may be shorter than the physical working lifetime of the scheme.

Basic discounting formulae

Although the actual choice of interest rates and repayment times for discounting is fraught with complications, its mathematical application is relatively straightforward. Although the formulae below may look daunting, they are standard functions in computer spreadsheet packages.

Present value

The present value of a future sum of money is an important quantity. Its general formula is fairly easy to explain.

Given a discount rate of 10%, then a sum of £100 today (i.e. its present value) is equivalent to a future value of

£100 × (1 + 0.1) in one year's time

£100 × (1 + 0.1) × (1 + 0.1) in two years' time

£100 × (1 + 0.1) × (1 + 0.1) × (1 + 0.1) in three years' time, or

£100 × (1 + 0.1)^n in n years' time

In general for an interest rate r, the value of a sum V_p in the present is equivalent to a sum V_n in n years' time, where:

$$V_n = V_\mathrm{p} \times \left(1 + r\right)^n$$

or

$$V_\mathrm{p} = \frac{V_n}{\left(1 + r\right)^n}$$

which alternatively can be written as:

$$V_\mathrm{p} = V_n \times \left(1 + r\right)^{-n}$$

Net present value

The above is just the present value of *one* sum of money at some time n years in the future. If we have a whole sequence of sums of money spread over a long time, then the total present value of all of these is the net present value (NPV).

If we denote these sums of money spread over n years as V_1, V_2, V_3, ..., V_n, then

$$\mathrm{NPV} = \frac{V_1}{\left(1 + r\right)} + \frac{V_2}{\left(1 + r\right)^2} + \frac{V_3}{\left(1 + r\right)^3} + \cdots + \frac{V_n}{\left(1 + r\right)^n}$$

or

$$\mathrm{NPV} = \sum_{i=1}^{n} \frac{V_i}{\left(1 + r\right)^i}$$

This has become a standard function in computer spreadsheets.

Annuitization

If all of the repayments on a loan are equal, of value A, then we have an annuity, continuing for a fixed number of years to repay an initial capital sum and all its interest. The present value of this annuity is:

$$V_\mathrm{p} = A \times \left(\frac{1}{\left(1 + r\right)} + \frac{1}{\left(1 + r\right)^2} + \frac{1}{\left(1 + r\right)^3} + \cdots + \frac{1}{\left(1 + r\right)^{n-1}} + \frac{1}{\left(1 + r\right)^n} \right) \tag{1}$$

Simplifying this involves a little mathematical ingenuity. Multiplying both sides by $(1 + r)$:

$$\left(1 + r\right) \times V_p = A \times \left(1 + \frac{1}{\left(1 + r\right)} + \frac{1}{\left(1 + r\right)^2} + \frac{1}{\left(1 + r\right)^3} + \cdots + \frac{1}{\left(1 + r\right)^{n-1}}\right) \quad (2)$$

Subtracting (1) from (2):

$$r \times V_p = A \times \left(1 - \frac{1}{\left(1 + r\right)^n}\right)$$

or

$$V_p = A \times \frac{\left(1 - \left(1 + r\right)^{-n}\right)}{r}$$

Inverting this, we can say that A is the annuitized value of a present capital payment V_p:

$$A = V_p \times \frac{r}{1 - \left(1 + r\right)^{-n}}$$

Putting it more simply, if an amount V_p is borrowed from the bank now, a sum A must be paid back every year for n years to pay off the loan.

We can tabulate values of A as a function of r and n, as shown in Table 12.1 overleaf. Although nowadays project costings are worked out on computer spreadsheets, in the past banks and finance houses used simple books of pre-calculated tables such as these, and they are still useful to give quick answers.

BOX 12.5 Annuitization: a simple method for costing

The concept of capital expenditure having an annuitized value depending on the discount rate and the project's financial lifetime is extremely useful. Table 12.1 shows the annuitized values for every £1000 of capital for various different real discount rates and project lifetimes. Using the table, it is easy to calculate the annuitized cost of a given capital sum over a given period. To this must be added the annual running cost (fuel, maintenance, etc.) When this is done, the average cost of energy is given by the simple formula:

$$\text{cost of energy} = \frac{\text{annuitized capital cost} + \text{average running costs}}{\text{annual average energy produced}}$$

This is sufficiently accurate for many purposes, though not in the case where annual costs are highly variable, when it is necessary to use a full NPV calculation.

Table 12.1 Annuitized value of capital costs (annual cost in £ per £1000 of capital) for various discount rates and capital repayment periods

Capital repayment period /years[1]	Real discount rate /%						
	0	2	5	8	10	12	15
5	200	212	231	250	264	277	298
10	100	111	130	149	163	177	199
15	67	78	96	117	131	147	171
20	50	61	80	102	117	134	160
25	40	51	71	94	110	127	155
30	33	45	65	89	106	124	152
35	29	40	61	86	104	122	151
40	25	37	58	84	102	121	151
45	22	34	56	83	101	121	150
50	20	32	55	82	101	120	150
55	18	30	54	81	101	120	150
60	17	29	53	81	100	120	150

1   This is not necessarily equal to the total physical lifetime of the project.

We can now revisit the example of the power station from Box 12.4, this time taking into account interest charges to show that the cost of 2.02p per kWh calculated previously is now an underestimate – see Box 12.6 below.

BOX 12.6

Recalculate the cost per kWh of electricity from the power station considered in Box 12.4, assuming a discount rate of 10% per year and using the figures in Table 12.1.

Solution

First we find the annuitized value of the capital cost. From Table 12.1 the annuitized value of £1000 over 20 years at 10% per year = £117 per year.

Therefore for a borrowed sum of £180 million the annual repayments will be 180 000 × £117 = £21.06 million. This is far higher than the annual repayment figure of £9 million used in Box 12.4, which can be derived using the annuitized value of £1000 over 20 years at 0% per year = £50 per year. 180 000 × £50 = £9 million.

Our total annual costs now read:

    capital repayments over 20 years = £21.06 million

    gas costs = £41.7 million

    operation and maintenance costs = £9.9 million

The total cost per year = £21.06 million + £41.7 million + £9.9 million = £72.66 million.

The cost per kWh = £72 660 000 × 100p ÷ 3 000 000 000 kWh = **2.42p per kWh**.

This is somewhat higher than the earlier figure of 2.02p per kWh. In a highly competitive market environment, this could make the difference between a project being accepted as profitable, or rejected.

A discounted cash flow calculation in detail

A full calculation for a major project will require considerable detail. In Boxes 12.4 and 12.6, we used two major cost categories: the capital costs (those occurring at the start of a project), and running costs (those recurring throughout the project lifetime) consisting of O & M and fuel costs.

In practice, a scheme is more accurately described by a list of costs or a 'cost breakdown'. The two major cost categories still apply – capital costs and running costs – with the addition of an important third category, decommissioning and disposal costs, which occur at the end of the project. Each of these cost categories occurs at different times throughout the project.

Let us look at the example of a nuclear power station. In 1994, the UK government conducted a review into the future of nuclear power (DTI, 1995). As part of this, the privatized UK company Nuclear Electric proposed the construction of a giant 2.6-GW twin pressurized water reactor, based upon the existing 1.2-GW Sizewell B reactor and to be known as **Sizewell C**. The costs below are based on Nuclear Electric's submission to the review (1994) and a detailed critique (Sadnicki, 1994) prepared for a client in the coal industry. It was estimated that the power station would take seven years to build, followed by a 40-year operational life, after which the station would have to be decommissioned. Decommissioning would be done in a number of stages: Stage 1 would involve removing the fuel; Stage 2 would then involve the dismantling of all the non-radioactive parts. The plant would then be left for 40 years to allow the high-level radioactive material inside the reactor to decay away, before Stage 3, the final dismantling stage, a century after the work started on construction. The costs involved are summarized in Table 12.2 below.

Table 12.2 Sizewell C nuclear power station cost breakdown

Item		Period/years	Amount
Construction costs		7	£3450 million total
Running costs	fuel supply	40	0.3p per kWh
	fuel disposal	40	0.1p per kWh
	operation and maintenance (O & M)	40	0.7p per kWh
	total running costs	40	1.1p per kWh
Decommissioning and disposal costs	Stage 1 – de-fuelling	5	£50 million
	Stage 2 – removal of outside shield	5	£160 million
	surveillance during 'cooling down' period	40	£1 million per year
	Stage 3 – back to green-field site	10	£190 million
	total decommissioning and disposal costs		£440 million

There is obviously an enormous expenditure involved. If we add up the total construction, running and decommissioning costs they amount to over £12 billion spread out over a period of more than a century. Balancing this is a large output of electricity. Nuclear Electric's proposal assumed that the station would operate at a load factor of 85% (i.e. it would run at full power for 85% of each year), producing approximately 19 TWh of electricity per year. These flows of cash and electricity are shown in Figure 12.16 below.

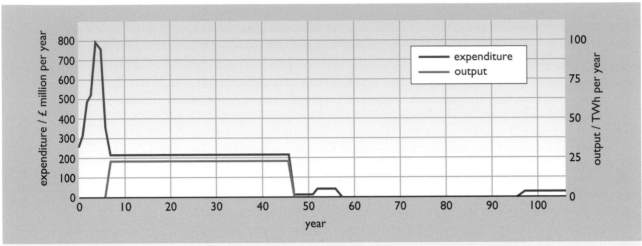

Figure 12.16 Sizewell C – assumed expenditure £(1993) and electrical output (source: Sadnicki (1994))

The aim is to produce a single figure for the cost of the electricity per kilowatt hour. The first step is to calculate the net present value of every year's financial expenditure. This requires making an assumption about the discount rate, or rather discount rates, involved.

Addressing the costs involved in decommissioning in particular has been a difficult problem. These costs amount to a total sum of £440 million to be spent over a period of 60 years, starting over 40 years into the future. Nuclear Electric argued that this should be dealt with by setting up a pot of money of only £270 million on the last day of operation of the station. This could be put into a 'safe investment' which would then earn sufficient interest to pay off the total £440 million. Such a long-term investment could only be expected to earn 2% interest.

The concept of decommissioning costs over such a timescale raises difficult questions about (a) the reliability of the use of money as a store of value and (b) intergenerational equity – i.e. issues of whether or not economic benefits enjoyed by this generation are being taken at the expense of future generations. These topics are discussed later in this section.

For the moment we will assume that decommissioning can be paid for with £270 million. In the calculation this and all the remaining expenditure of construction and operation are discounted back to year zero, the time when work first started, using the commercial interest rate. The value of this was a subject of considerable debate at the time, with suggestions in the range 8–12%. Taking a figure of 10%, we can calculate a single figure for the net present value of all the costs as £3.36 billion. Note that this is far less than the undiscounted figure of £12 billion.

The next step is to calculate the net present value (in kilowatt-hours) of every year's electricity generation, again discounted back to year zero. This amounts to 96.3 TWh (1 TWh = 10^9 kWh). It may seem odd to discount electricity generation in kilowatt-hours, but this has an equivalent real monetary value which can only be assigned when the calculation is complete.

Then we can work out the cost of electricity in pence per kWh as:

$$\frac{\text{NPV of capital costs, fuel, O \& M and decommissioning (£)} \times 100}{\text{NPV of electricity (kWh)}}$$

$$= \frac{\text{£3 360 000 000} \times 100}{\text{96 300 000 000}} = 3.49 \, \text{p per kWh}$$

We can also break the answer down into its different components to show how the initial capital cost dominates the final electricity cost, as in Table 12.3.

Table 12.3 Breakdown of electricity cost[1]

Component	Cost/p per kWh
Capital charges	2.38
Fuel costs	0.40
Operation and maintenance	0.70
Decommissioning	0.01
Total	3.49

1 In 1993 prices.

Having arrived at this answer it is worth noting that it can be expressed in various other ways.

- NPV of the whole project: Using the discount rate of 10% and this electricity price, then the net present value of all the costs is equal to the net present value of all the benefits (i.e. the electricity). We can say the net present value of the project as a whole = NPV (costs) − NPV (benefits) = 0. If the electricity could be sold at more than 3.49p per kWh, or alternatively the costs could be reduced, then this project NPV would become greater than zero and there would be some room for profit. Generally where this kind of analysis is carried out, projects with overall NPVs greater than 0 are worth doing and the larger the NPV the better.

- Percentage Internal Rate of Return (IRR%): This is the value that the discount rate must take in order to make the overall project NPV equal to zero, i.e. in this case 10%. It is worth borrowing money at this rate to carry out the project. However, if the cost of borrowing were higher than this, then the project would become uneconomic.

- Benefit–Cost Ratio (B/C): This is the ratio of the net present value of all the benefits to the net present value of all the costs. At a discount rate of 10% and an electricity price of 3.49p per kWh this equals 1. Generally projects with a B/C ratio greater than 1 are worthwhile considering, and the higher the B/C, the better.

How discounted cash flow affects the perceived costs of different forms of energy

The debate surrounding the 1994 government review on nuclear power concerned the relative commercial merits of nuclear electricity and those of its chief rival, the combined cycle gas turbine. The key financial factors (as given by Nuclear Electric in 1994) can be put simply, as in Table 12.4.

Table 12.4 Financial characteristics of different electricity generation technologies

	Construction time	Capital costs £(1993)	Operation and fuel costs	Operational lifetime
Sizewell C	7 years	£1340 per kW	1.1p per kWh	40 years
CCGT	2 years	£450 per kW	1.7p per kWh	20 years

The nuclear power station detailed in Table 12.2 has high capital costs and a long construction time, but relatively low running costs over a long operating lifetime. The competing CCGT could be built quickly and cheaply, had minimal decommissioning and dismantling costs, but had higher running costs over a shorter life. Which is best?

Setting aside the many objections as to whether Nuclear Electric's cost and performance estimates were correct, the decision rests largely on attitudes to finance.

If we ignore discounted cash flow (or effectively say we have a discount rate of 0%), then we are saying that we are prepared to wait patiently for the returns on a particular energy investment. As the discount rate is increased we are saying that we are progressively less prepared to do so. We want quick returns and may be unwilling to invest in projects that require a long construction time.

We can repeat the electricity cost calculation above for Sizewell C with different discount rates and compare the results with similar calculations for a CCGT station. The results are plotted against discount rate in Figure 12.17 below.

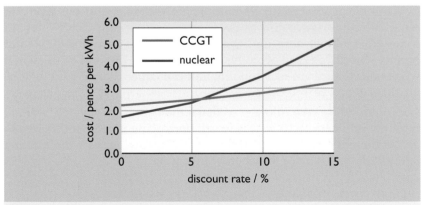

Figure 12.17 Comparison of electricity costs with discount rate for CCGT and nuclear plants

Because the cost of nuclear electricity is dominated by the large capital costs of the initial construction, it is very dependent on the discount rate, whereas the cost of electricity from CCGT generation is dominated by the price of the gas. In this example, if a discount rate of zero per cent (i.e. a long return time) is acceptable, then nuclear power is cheaper than gas. If a discount rate of 10% is chosen then the position is reversed. So the choice as to which technology is most economic is largely dependent on the cost of borrowing money.

The 1994 government review on nuclear power also prompted debate on whether or not the financial markets could actually put up almost £3.5 billion for a nuclear power station at all, as this would require a considerable degree of faith in the success of one project. A further problem was whether or not private investors would be prepared to lend money over the full 40-year life of the station. If they insisted on the loans being paid off in 15 or 20 years, then the cost of the electricity would have to go up to 5p per kWh to meet the costs.

A CCGT project, on the other hand, was considered to be a far more attractive prospect for investors. An entire 400-MW power station could be bought for a modest £180 million and its operating life of 20 years is more in line with normal expectations of loan repayments. The conclusion of the UK government's nuclear review (DTI, 1995) was that the Sizewell C station was the best option that the nuclear industry could offer in the short term, but that it could not compete with new gas-fired stations. Sizewell C itself was never built, but since then, many new CCGT stations have been constructed.

As mentioned in the previous chapter, in 2001 British Nuclear Fuels Ltd was proposing an AP600 system with a much lower capital cost of $1200 per kW, or around £800 per kW (BNFL, 2001). If this price could be achieved with the same performance, fuel and operation costs as Sizewell C, then it would give electricity costs of about 2.5p per kWh rather than 3.5p per kWh. That said, the capital costs of CCGT stations have now actually fallen from the figure of £450 per kW quoted for 1993, to £270 per kW in 2001 (PIU, 2002), which in turn reduces their generation costs by about 0.3p per kWh.

Dealing with normal financial risk

The calculations above have assumed that there is a single commercial discount rate for the project, plus, of course a very low discount rate for the 'safe storage' of the decommissioning money. In practice, financial institutions have to safeguard their investors, so they only like to lend money on low-risk investments. If there is any possibility that a project may fail or not produce the promised returns, then there has to be an extra allowance to cover this risk. Often the borrowing may be split into two parts. A bank may supply a part of the money at a normal interest rate. This is called **debt**. The remainder is raised by selling shares to shareholders, with a prospect of likely good returns, but no guarantees. This is called **equity**. The debt always gets paid out first and the shareholders carry the project risks. Depending on the level of risk, shareholders may want potential returns of 20% or more.

Some types of energy project always carry some risk, but others may be seen as sufficiently risk-free to be paid for by debt alone. The financial risk on such projects can be estimated by sensitivity analyses. If the cost per kilowatt-hour of energy generation can be reduced to a simple discounted cash flow equation, the most likely values can be put in to give a best estimate of the energy costs. It can then be analysed to establish which factors have most influence. For example, the cost of electricity from a CCGT project could be expressed in terms of its capital cost, discount rate, gas fuel costs, operating and maintenance costs and the load factor. Each particular value can be varied around a central estimate (marked 100% on the x-axis) to give a spider diagram such as shown in Figure 12.18 below.

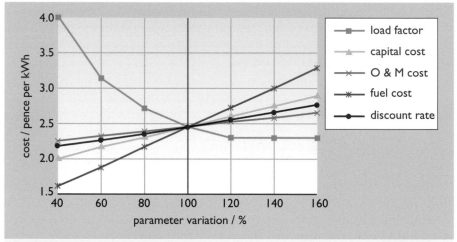

Figure 12.18 A spider diagram showing the sensitivity of electricity cost to different parameters

This diagram shows that two parameters, the fuel costs and the load factor, are more important than the others. Variations in these costs make a bigger difference to the cost than variations in other parameters. In a competitive electricity market a margin of 0.5p per kWh or less may make the difference between commercial success and failure. A rise in gas prices of about 35% would force up the electricity cost by this amount, as would a reduction of about 30% in the running hours. The worst possible case scenario might be if a rise in gas prices were to make the station unable to compete with others in the marketplace, automatically reducing the load factor. A wise management would thus make sure that it had a firm contract for long-term fuel supply and a ready market for all its electricity.

In the extreme case, a rise in fuel prices could lead to the mothballing of the plant, a fate which has befallen the giant Grain oil-fired power station in the Thames Estuary (Figure 12.19).

Figure 12.19 An unwanted investment – the 1.4 GW Grain oil-fired power station on the Thames estuary is virtually unused for want of cheap oil to burn

What are acceptable discount rates and investment lifetimes?

Conventional textbook economics assumes that all investors have free access to information and free access to credit. Anything less than this is a 'distortion of the market'. We have discussed how discount rates are based on opportunity cost and bank interest rates. However, in practice different individuals and organizations have different discount rates and time perspectives. It is worth looking at the time preference for money more carefully. Often, discount rates are high and time perspectives short. This can lead to an economic mismatch between the judgements made by consumers who might conserve energy, and large utilities and corporations interested in supplying it.

The private individual

It must be said that for domestic consumers energy investments are not a very exciting prospect. Often the attractions of spending money *now* will outweigh the attractions of savings in the future. Attractions for expenditure like food or holidays compete with investments such as compact fluorescent lamps and cavity wall insulation, and these two technologies in particular have required subsidies in the UK in recent years, despite very short payback times. Conversely, sales of double-glazing, with a much poorer economic justification have been surprisingly high. This would appear to be because double glazing is extremely visible and adds to the perceived value of a house in a way that 'invisible' cavity wall insulation does not.

Many private individuals, especially those on low incomes and in rented accommodation, have high perceived discount rates and short investment time horizons. At the extreme, those living on state benefits or low pensions may have no access to credit. Given the choice of a pound today or a pound tomorrow, they will always say they need it now. Their discount rate is infinitely large. For those living in rented accommodation, investment is also limited by the length of their lease or the probable time when they will move on. Yet it is these very people who are likely to suffer most from fuel poverty and need energy savings.

That said, even the poorest individuals may save money to provide an income in old age (some people would continue to do this even if their savings earned zero interest), or to provide for their children's education. The UK state pension system does not actually have any accumulated cash reserve; today's pensioners are paid for by today's workers. However, private pensions are different. An individual pays into a fund over their working life, building up a 'store of value' to be repaid in their old age. The contents of private pension funds make up an enormous body of invested capital, potentially available for energy projects.

Commercial companies

Many commercial companies may find investment in energy financially difficult. Many still treat energy costs as 'running costs' to be paid out of a particular annual budget, or the accounting system may simply not have an 'energy investment' budget. For this reason, at worst, companies may only consider energy projects with a payback time of less than one year. They may also be in rented offices, with no incentive to invest in capital equipment that lasts longer than their lease.

If they are in a very competitive market, companies may also feel that they need to plough as much money as possible into their latest product. Anything less than this could lead to commercial failure. This can lead to very high perceived discount rates and short financial time horizons.

Public sector organizations

In its *Green Book*, the UK Treasury suggests that public sector organizations should make a real return of 6% on investments, though this point is qualified with a long discussion on the proper evaluation of risk and 'optimism bias' (HM Treasury, 2003). This figure has changed slightly over the years, with the 6% figure being lower than the 8% quoted at the time of the 1994 government nuclear review, but higher than an earlier figure of 5% used in the 1970s. Public sector bodies are also encouraged to enter into public–private partnerships (PPP), however, in which case higher commercial interest rates will apply.

Intergenerational equity

Very-long term investments, such as those involved in paying for the decommissioning of nuclear power stations, may span one or more generations. This raises the question of whether or not today's generation is paying its full share of the costs. This concept of fairness across generations is known as **intergenerational equity**.

It may seem strange that in the Sizewell C calculation above the £440 million of decommissioning costs amount to nearly 4% of the total £12 billion undiscounted lifetime costs of the station, but only feature as 0.2% of the electricity cost. This is because the 10% discount rate used 'dis-counts' both benefits and disbenefits a long way into the future. In this case, the costs of the final decommissioning only arise over a century after the start of construction. There are, of course, other long-term elements that have not been entered into the equation: for example, the benefits of avoided CO_2 emissions, which may last for a hundred years or more; and the disbenefits of the nuclear waste, which could last for thousands of years.

In Section 12.3 of this chapter we looked at how the value of money is measured against real goods, such as a dozen eggs. Money is used as a unit of account for the real value of the eggs. A dozen 1971 eggs have the same food value as a dozen eggs in the year 2000, yet they are assigned different values in the money of their respective years, 22p in 1971 and £1.72 in 2000. Likewise, a derelict nuclear power station requiring £440 million of dismantling work must be considered as a 'real', unpleasant object, with real impacts on its surroundings. And yet, at a discount rate of 10%, on paper the strange magic of discounted cash flow can dis-count a 'future' derelict power station down to a negative asset of a mere £2 million in present day values. This gives the distinct impression that there must be something wrong with such an accounting system.

Mathematically speaking, discounted cash flow sets up a value function, expressed in money, and the economic analyst aims to maximize it. However, this may not always give very acceptable answers. For instance, in 1973 the mathematician Colin Clarke took investment in the whaling industry as an example. He demonstrated that using DCF it could be considered economically preferable for a fishing fleet to hunt blue whales to extinction rather than let the population of the species produce a sustainable annual catch. All that was required was for the discount rate used to be more than twice the breeding rate of the whale population. If the natural rate of population increase was 5% per year and the whaling company used a discount rate of over 10%, then the ruthless pursuit of profit maximization would lead to the extinction of the species (Clarke, 1973).

The notion of a discount rate can be justified on the basis of our time preference for money – whether or not we want 'jam today' or 'jam tomorrow'. This seems to be a morally acceptable question. The question of whether we would prefer 'waste today' or 'waste tomorrow' is an altogether more difficult one, highlighted by the fact that large parts of the UK are still littered with coal mining and other waste from previous generations.

Firstly, there is the question of whether or not future generations will tolerate today's standards of waste management. Given the progressive tightening of legislation on particulate smogs, acid rain, petrol containing lead, and general aspects of health and safety, etc., it seems unlikely that they will. It is possible that in the future society will insist on a far more thorough and expensive clean-up of today's technologies than is budgeted for at present.

Secondly, there is whole philosophical question of the ethics of discounting the future. A conventional short-term economic analysis may only consider the maximization of benefits to a small group of investors in the present generation. However, many economists and philosophers believe that the analysis should really deal with society as a whole and include the welfare of future generations. This could imply that the real discount rate applied should be very low. In his mathematical theorem about blue whale fishing, Clarke showed that the use of a zero discount rate would lead to the maximum sustainable catch. Another study of the philosophical, rather than mathematical, basis of discounting also came down in favour of a rate of zero (Broome, 1992).

Others have argued that, hopefully, economic growth and technological innovation should make future generations wealthier in real terms. A societal discount rate of around 2% (the actual rate of real growth in incomes in the UK over the past 40 years) might be more appropriate for economic analysis (Pearce *et al.*, 1992).

However, Broome (1992) does issue a warning that if future climate change were to cause the world economy to start contracting, then an appropriate discount rate might become negative. This would imply that it would pay to invest now to stave off future disaster. Broome adds, though, that 'it is not plausible that any of this will be reflected in people's present expectations of interest rates'.

In the extreme case there is a small but finite risk that an economic catastrophe akin to the inflation in Germany in the 1920s or a Wall Street Crash could wipe out the store of value in any decommissioning fund, leaving future generations to foot the bill out of their own resources.

Discounted cash flow – summary

Discounted cash flow analysis is a simple technique for taking account of the time value of money in cost calculations. It was originally developed for the comparison of the relative economics of different projects where some form of borrowing has to be made. The difficulties in its use are mainly those of choosing an appropriate discount rate (particularly one taking account of inflation) and an investment lifetime, which may often be shorter than the actual operational lifetime of the project. So when a particular figure of x pence per kWh has been produced, it is vital that all the key assumptions are clearly stated. It is all too easy to produce low figures using unrealistically low discount rates, long project lifetimes and inflated performance estimates.

It must be said that the basic concept of the time value of money is difficult to grasp. Even when a discounted cash flow calculation has been explained in detail it often provokes the exasperated question 'Can't you just tell me in simple terms what the payback time is?'.

Those readers who have that feeling can take comfort from this quote from Lord Peston during the debate on electricity privatization in the UK parliament in 1989:

> When I was a junior economist in the Treasury ... I remember preparing a paper which was to go to Ministers. It contained a formula. The people who looked after me asked me whether my formula had to go in ... I said 'Yes, my formula must go in'. They were horrified that a document going to Ministers would contain a formula ... It was a discounted cash flow formula, as I was the first one to think that discounted cash flow might apply to government matters ... the old Ministry of Power had not the slightest idea of what discounted cash flow or compound interest were.
>
> *Peston*, 1989

This is not to say that at the time Peston is referring to (probably the late 1940s) Treasury officials were particularly ignorant. The notion of maximizing return on investment was only introduced in the Dupont and General Motors Corporations in the United States in the 1920s. Before then, the accounting systems in use did not distinguish clearly enough between interest charges and general operating costs to allow this sort of analysis to be carried out at all.

12.5 **Real world complications**

In the UK at present, energy is being treated as a freely traded free-market commodity. This is in line with the European Union's programme on the harmonization of energy markets. There is a world price for oil and a similar one for coal. There has also been progressive legislation towards the 'liberalization' of electricity markets, with the aim that eventually all consumers, industrial or domestic, should have the choice of being able to buy their electricity from anywhere, including from another country.

This view is considerably different to that prevailing thirty years ago, and even now is only grudgingly accepted by many European countries for a whole range of reasons, some of which are described below.

Security and diversity of supply

It is easy to argue that energy is just a tradable commodity if you have plenty of it or reliable access to someone else's supplies. It is less easy to do so if you haven't. During the twentieth century, two World Wars gave Europeans ample experience of severe food and fuel shortages. Since the Second World War, politicians, particularly in the European Union, have been very careful to ensure that the Common Agricultural Policy keeps the larders full, even if it does require an extensive subsidy system. In India, a country with a long history of droughts and famines, subsidized electricity is almost given away to the agricultural sector.

Fuel shortages are extremely unpopular and are likely to provoke rapid political action. Current International Energy Agency policies insist that each member state has 90 days' worth of reserves of fuel supplies, but even so, quite minor petrol shortages can provoke panic buying. Any wise government will make sure that it has adequate energy supplies to hand.

Since the end of the nineteenth century, industrial countries such as the UK and Germany have manufactured goods for export. Until recently, the factories have been run on coal (either directly, or on coal-generated electricity). It has thus been in the interests of their governments to promote indigenous coal production to provide reliable energy supplies for home industry and ensure full employment. In wartime, the need to run the energy system flat out became even more important.

In other countries, such as France, the Soviet Union and India, throughout the twentieth century the rapid expansion of heavy industry and the electricity industries has been seen as essential to modernize and compete. The overall result was that these countries developed extensive subsidy systems to their coal-based industrial infrastructure. This in turn created a political power base of large numbers of miners and industrial workers interested in maintaining the status quo. In Germany, for example, there was a law saying that only German coal could be burnt in power stations, even though this required a large subsidy, the *kohlepfennig*.

However, this was increasingly difficult to justify against a background of falling world coal prices from large pits in Australia, South Africa and South America. Coal from these has been available since 1985 at under £30 per tonne. In Germany by 1995 the coal subsidy was running at nearly £70 per tonne, or over £5 billion per year. The German government strongly defended this on the grounds of security of supply, though the subsidy was reduced to £2.8 billion per year by 2000. France has also had a similar coal subsidy scheme, running at £600 million per year in 2000.

In the UK, between 1980 and 1990 government grants to the loss-making British Coal company totalled almost £15 billion. The choice between subsidized home-produced coal or cheaper imported coal resulted in a show-down in 1984 between the then-Conservative government and the miners. There was a long and bitter miners' strike. The government won the day, by running oil-fired power stations flat out to keep the lights on. Eventually the government paid out large sums in redundancy and retraining allowances to miners, equivalent to many years' subsidy payments. Since then, UK coal production has fallen dramatically and the subsidy payments drastically cut.

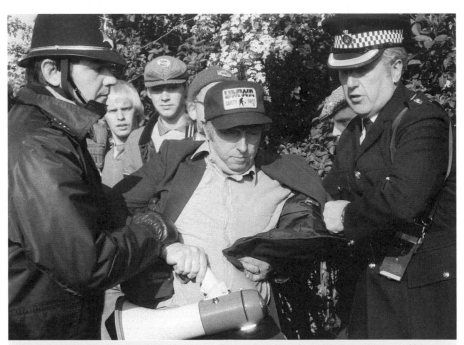

Figure 12.20 The political price of coal. The president of the National Union of Miners, Arthur Scargill, is arrested at the Orgreave coking works picket, Sheffield, during the miners' strike, 30th May 1984

More surprisingly, subsidies can also be set up as part of a programme to implement free markets, as has been described in previous chapters. In 1989, the UK government started the break-up and privatization of the state-owned Central Electricity Generating Board (CEGB). The aim was to reduce electricity costs by promoting competition within the industry. However, this exposed the true costs of the existing nuclear power plants, in particular the older Magnox plants, which had to be hastily withdrawn from the privatization and remain in state hands. A tax system, the Non Fossil Fuel Obligation (NFFO), levied on fossil-fuel generated electricity, had to be set up to subsidize these nuclear plants. Once in place, though, a small proportion of the funds were used to support new renewable energy projects. However, the sums paid out to renewable energy were in terms of millions of pounds, rather than the billions paid out to the nuclear industry or spent on coal subsidies.

Also, since 1989 there has been a massive switch to burning North Sea gas to generate electricity. However, these gas reserves are very limited and the projection for the immediate future is that by 2020, the UK could be importing up to 90% of its gas. The current UK policy is that energy should come from a wide range of sources, in order to minimize the problems of disruption to any one of them. This policy includes expanding the supply from renewable energy sources. The NFFO mechanism for the support of renewable energy was withdrawn in 2001 and replaced by a Renewables Obligation on electrical utilities to buy a certain proportion of their electricity from renewable sources. If they do not meet this obligation then suppliers have to pay what is effectively a fine, currently 3p for every kWh of renewable energy that they fail to purchase.

The energy situation in the United States is not much better, since it has been a net oil importer since 1960 and oil imports have been steadily growing since then (see Chapter 2, Figure 2.15). The 1991 Gulf War, fought largely over the ownership of the oil wells of Kuwait, is widely seen as being about maintaining US oil supplies. It has been suggested that the costs of this war constituted an external cost of $23.50 per barrel (almost 1p per kWh) to the oil imported into the US (House of Commons Select Committee on Energy, 1992).

The fact that governments are willing to make very large sums of money available to the 'military–industrial complex' to keep the supplies of fossil fuels flowing suggests that it might not be unreasonable to deploy similar sums to put the energy economies onto a more sustainable footing.

Externality costs of pollution and disaster

A further question is whether or not all of the costs of a particular energy technology are genuinely included in the economic equations. These include costs resulting from air pollution, noise pollution and those resulting from the risk of major accidents in mines or power stations. Although the effects of these are described in the next chapter, assigning a monetary value to such things can be difficult and the subject of much debate. However, over the past sixty years, detailed analysis has resulted in a whole range of clean-up programmes being implemented; these are discussed in Chapter 13 and 14. For example, the extra costs of fitting flue-gas desulphurization equipment to coal- and oil-fired power stations, which added about 0.5p per kWh to their generation cost, were far less than the estimated acid rain damage to trees, buildings and human health, estimated at 4–5p per kWh. The costs associated with climate change are still a matter of fierce debate but an attempt to internalize the resulting costs has been made with the new UK Climate Change Levy. This tax, introduced in April 2001 adds 0.43p per kWh to business electricity bills and 0.15p per kWh to business bills for gas or coal.

New technologies, economies of scale and 'market washing'

Another difficulty is that a particular technology may have high costs simply because of its low volume of production. If only mass production and consumption could be encouraged, then the technology would be considered economic.

Although this provokes criticism from dedicated free-marketeers, there is a good argument in favour of government subsidies. In the UK the relatively high price of compact fluorescent lamps (CFLs) has been the subject of a number of campaigns, notably by the Energy Savings Trust, an organization set up to promote energy efficiency. Subsidies have boosted CFL sales, encouraging production and driving down prices to the point where consumers will choose to buy them normally. The argument in favour of subsidies has been that despite the obvious short payback times, domestic consumers find it very difficult to make investments to cut energy consumption. Nor, following the privatization of electricity in the UK in 1989, has it been in the commercial interests of the electricity industry to encourage cuts in electricity sales.

A similar good example of this kind of market intervention has been in Thailand, a country heavily dependent on imported oil for electricity generation. There, the government has made a concerted effort to phase out the use of old-style R40 tubular fluorescent lamps in favour of the more efficient R8 types. Local manufacture of the new type of lamp was promoted and imports of the old type banned. The programme is delightfully described by a phrase which translates from the Thai as market washing – removing the 'dirty' products (Birner, 2000).

Again, to take an example from the field of renewable energy, solar electricity generation from photovoltaic panels is also a good candidate for this kind of market intervention. A recent study for Dutch Greenpeace (KPMG, 1999) concluded that electricity from photovoltaic panels could be competitive with that generated from fossil fuels if the demand for panels could be raised to a level great enough to support a very large production plant. In order to have the required economies of scale, it would need to produce 500 MWpk (or roughly five square kilometres) of panels per year, approximately three times the 1999 world demand for PV panels. Yet, if PV power could be made competitive with fossil fuels, the world's insatiable demand for electricity could easily absorb the output of this plant and many, many more. The problem, therefore, is to get the first plant built.

12.6 Conclusions

Energy is obviously a vital commodity for the modern industrial society. As such its prices are subject to a host of economic and political pressures. The most obvious of these in recent years has been the political control exerted by OPEC over oil prices. Within individual countries energy prices are subject to extensive taxes and subsidies. Some, like petrol taxes, are clearly stated, while others, such as industrial subsidies to coal or nuclear industries, are less clearly defined.

Despite the continual grumbles about the high costs of energy, it must be said that the real price of energy is almost at an all-time low in relation to the earning power of workers in the UK and other industrial countries.

When working on the costings for any particular energy project, especially one that requires borrowing money, it is vital to appreciate the processes of inflation and of discounted cash flow calculations. It is also necessary to understand that there may be considerable disagreements about the results because different organizations might have different perspectives on the values of discount rate and project lifetime to be used. These should always be clearly stated in any particular calculation.

There are still unanswered questions as to whether or not security of supply and self-sufficiency in energy should be ends in themselves and what kinds of costs that might impose on energy prices. External costs due to pollution or potential disaster are a thorny issue, since they may be difficult to estimate and the perception of their importance may vary from country to country. These perceptions also affect programmes to promote clean technology, energy efficiency or renewable energy, all of which are bound to require taxes and subsidies.

Attempts to levy environmental taxes to 'internalize' these 'external' costs may incur the wrath of those dedicated to free markets. They may see these taxes as 'interference in the free market'. The converse view is that they are 'corrections to an imperfect market'.

Although we have talked about 'real' prices in this chapter, it is important to remember that money is only used as a medium of exchange, a unit of account and, somewhat imperfectly, a store of value. It is used to facilitate the provision of energy services such as warm homes, cooked dinners, illumination and transport. These are the 'real' end products.

References

Birner, S. (2000) 'How Thailand washed away wasteful lighting', *International Association for Energy-Efficient Lighting Newsletter*, 1-2/00 vol. 9, issue 24.

BNFL (2001) *BNFL Submission to the Performance and Innovation Unit's Review of UK Energy Policy*, British Nuclear Fuels Ltd, http://www.theenergyreview.com/docs/BNFL submission to the Energy Review.pdf [accessed 12 August 2002].

Broome, J. (1992) *Counting the Cost of Global Warming*, Report to the Economic and Social Research Council, White Horse Press.

Clarke, C. (1973) 'The economics of over-exploitation', *Science*, vol. 181, pp. 630–4.

DTI (1995) *The Prospects for Nuclear Power in the UK – Conclusions of the Government's Nuclear Review*, HMSO.

DTI (2001a) *Fuel Poverty in England, 1998*, Department of Trade and Industry, http://www.dti.gov.uk/energy/consumers/fuel poverty/1998estimates england.pdf [accessed 7 August 2002].

DTI and DEFRA (2001) *The UK Fuel Poverty Strategy*, Department of Trade and Industry and Department of Environment, Food and the Regions, http://www.dti.gov.uk/energy/consumers/fuel poverty/strategy.shtml [accessed 7 August 2002].

House of Commons Select Committee on Energy (1992) *Renewable Energy, Fourth Report*, HMSO.

KPMG Bureau voor Economische Argumentatie (1999) *Solar Energy: From Perennial Promise to Competitive Alternative*, Hoofddorp, Netherlands.

NSSO (1996) *Sarvekshana*, National Statistical Survey Office, vol. XIX-3, issue 66.

Nuclear Electric plc (1994) 'Further nuclear construction in the UK', volume 1 of a submission from Nuclear Electric to the Government's *Review of Nuclear Energy* (DTI, 1995).

Pearce, D., Bann, C. and Georgiou, S. (1992) *The Social Costs of Fuel Cycles*, Report to Department of Trade and Industry.

Peston, M. (1989) *Hansard*, 16 May 1989, col. 1108.

PIU (2002) *The Energy Review*, Performance and Innovation Unit, UK Cabinet Office, http://www.cabinet-office.gov.uk/innovation/2002/energy/report/1.html [accessed 7 August 2002].

Sadnicki, M.J. (1994) *Nuclear Review Background Paper, Sizewell B and Sizewell C*, Hoskins Group plc.

Statistical data sources

Average Earnings Index, 1963–2000, National Statistics web site, HM Government, http://www.statistics.gov.uk/themes/labour_market/pay_and_earnings/default.asp [accessed 7 August 2002].

BRE (1998) *BRE Domestic Fact File*, UK Building Research Establishment.

DTI (2001b) *UK Energy in Brief*, UK Department of Trade and Industry.

DTI (2002) *Digest of UK Energy Statistics (DUKES)*, UK Department of Trade and Industry.

Economic and Financial Trends (2002) National Statistics web site, HM Government http://www.statistics.gov.uk/nsbase/ukinfigs/economy.asp [accessed 7 August 2002].

EIA (2002) *International Energy Outlook 2002 – Coal*, US Energy Information Administration, http://www.eia.doe.gov/oiaf/archive/ieo00/coal.html [accessed 7 August 2002].

EU (2000) *Annual Energy Review 2000*, Office for Official Publications of the European Communities.

Evans, R.D. and Herring, H. (1989) *Energy Use and Energy Efficiency in the UK Domestic Sector up to the Year 2010,* Energy Efficiency Office, HMSO.

IEA (1999) *Natural Gas Information, 1999 edition*, OECD, Paris.

IEA (2000a) *Energy Policies of IEA Countries, 2000 Review,* OECD, Paris.

IEA (2000b) *Oil Information 2000,* IEA Statistics, OECD, Paris.

Indian Ministry of Information and Broadcasting (2002) *India: A Reference Annual – 2002.*

Mathews, M. M. (2001) *India, Facts and Figures*, Sterling Publishers Private Ltd, New Delhi.

Mersey, R. (2001) *Pole Power*, Talleyrand Books.

National Statistics (2002) http://www.statistics.gov.uk [accessed 7 August 2002].

Stone, R. (ed.) (1966) *The Measurement of Consumers' Expenditure and Behaviour in the UK, 1920–1938*, National Institute of Economic and Social Research.

TERI (1999) *Tata Energy Directory and Data Year Book (TEDDY), 1998/9,* Tata Energy Research Institute.

TERI (2002) *Tata Energy Directory and Data Year Book (TEDDY), 2001/2,* Tata Energy Research Institute.

Twigger, R. (1999) *Inflation, the Value of a Pound, 1750–1998*, Research Paper 99/20, Economic Policy and Statistics Section, House of Commons Library, http://www.parliament.uk/commons/ [accessed 7 August 2002].

Further reading

HM Treasury (2003) *Appraisal and Evaluation in Central Government (The Green Book)*, HMSO – a very readable investment guide.

Available at http://greenbook.treasury.gov.uk [accessed 04 June 2003]. This 2003 issue 'unbundles' the 6% figure into one of 3.5% based on social time preference plus an element for systematic optimism ('optimism bias') in projects. See especially Annex 4 on risk and uncertainty and Annex 6 on discount rate.

OANDA Corp. (2003) *The Currency Site*, http://www.oanda.com – has an online currency converter between all major currencies including past exchange rates [accessed 10 March 2003].

Wonnacott, P. and Wonnacott, R. (1986) *Economics*, McGraw-Hill – an excellent teaching textbook on economic theory with interesting comments on the 1973–1979 oil price rises.

Chapter 13

Penalties: Assessing the Environmental and Health Impacts of Energy Use

by Godfrey Boyle, Janet Ramage and David Elliott

13.1 Introduction

The preceding chapters on fossil and nuclear fuels will have left readers in no doubt that our use of these fuels, whilst conferring numerous benefits, also incurs very substantial penalties. These include adverse impacts on the Earth's ecosystems and climate, together with deleterious effects on the health of humans and many of the other species with which we share the planet. These effects are not only detrimental in themselves, in ways that can often be quantified, but they in turn create substantial additional *monetary* costs to society over and above the simple market prices of the fuels themselves.

In this chapter we shall concern ourselves mainly with the environmental and health impacts of fossil and nuclear fuels: those of hydroelectricity, wind power and the other renewable energy sources are also mentioned here briefly, but are described in more detail in the companion volume, *Renewable Energy*.

We start by classifying and identifying the impacts of energy use and then go on to examine a number of ways of assessing and comparing them quantitatively. However it should be stressed at the outset that there is no comprehensive assessment method enabling *all* of the widely-varying and qualitatively-different impacts involved in our use of energy to be compared together and 'objectively'. Chalk cannot be compared with cheese, no matter how hard we try.

We then examine the issue of accidents and risk, concentrating on nuclear energy. Finally, we compare the environmental impacts of various electricity generating systems and look at various attempts to calculate in monetary terms the costs to society of their environmental and health effects.

13.2 Classifying the impacts of energy use

What, then, are the principal impacts of human energy use? There are various ways in which these impacts can be analysed and classified. Here we look at three approaches to classification:

- by source,
- by pollutant,
- by scale.

Classification by source

Table 13.1 (overleaf) lists the main environmental and social concerns associated with the principal energy *sources*. It illustrates the extremely wide variety of quantitatively and qualitatively different impacts that humanity's use of energy has on the environment and on society.

Table 13.1 Environmental and social concerns associated with the principal energy sources

Source	Potential causes for concern
Oil	global climate change, air pollution by vehicles, acid rain, oil spills, oil rig accidents
Natural gas	global climate change, methane leakage from pipes, methane explosions, gas rig accidents
Coal	global climate change, acid rain, environmental spoliation by open-cast mining, land subsidence due to deep mining, spoil heaps, ground water pollution, mining accidents, health effects on miners
Nuclear power	radioactivity (routine release, risk of accident, waste disposal), misuse of fissile and other radioactive material by terrorists, proliferation of nuclear weapons, land pollution by mine tailings, health effects on uranium miners
Biomass	effect on landscape and biodiversity, ground water pollution due to fertilizers, use of scarce water, competition with food production
Hydroelectricity	displacement of populations, effect on rivers and ground water, dams (visual intrusion and risk of accident), seismic effects, downstream effects on agriculture, methane emissions from submerged biomass
Wind power	visual intrusion in sensitive landscapes, noise, bird strikes, interference with telecommunications
Tidal power	visual intrusion and destruction of wildlife habitat, reduced dispersal of effluents (these concerns apply mainly to tidal barrages, not tidal current turbines)
Geothermal energy	release of polluting gases (SO_2, H_2S, etc.), ground water pollution by chemicals including heavy metals, seismic effects
Solar energy	sequestration of large land areas (in the case of centralized plant), use of toxic materials in manufacture of some PV cells, visual intrusion in rural and urban environments

Source: based on Ramage, J. R., 1997

Classification by pollutant

Instead of listing the effects of each energy *source* in turn, as in Table 13.1, another approach to categorizing the environmental impacts of human energy use is that taken by Holdren and Smith (2000), who list various 'environmental insults' caused by human activities, as shown in Table 13.2. **Insults** are the physical *stressors*, such as air pollution, produced by an energy system, in contrast to the **impacts** of the system, which are the potential *outcomes* (negative or positive) affecting humanity, such as respiratory disease or forest destruction. Holdren and Smith identify those insults that are due to 'commercial' and 'traditional' energy supplies and distinguish them from those created by agriculture, manufacturing and other activities. They then create a **Human Disruption Index**, which is the ratio of the human-generated flow to the natural ('baseline') flow of a particular substance. Note, however, that 'insults' due to nuclear energy are not included in Table 13.2.

Table 13.2 Environmental insults due to human activities by sector, mid–1990s

Insult	Natural baseline (tonnes per year)	Human disruption index[a]	Share of human disruption caused by: commercial energy supply	traditional energy supply	agriculture	manufacturing, other
Lead emissions to atmosphere[b]	12 000	18	41% (fossil fuel burning, including additives)	Negligible	Negligible	59% (metal processing, manufacturing, refuse burning)
Oil added to oceans	200 000	10	44% (petroleum harvesting, processing, and transport)	Negligible	Negligible	56% (disposal of oil wastes, including motor oil changes)
Cadmium emissions to atmosphere	1400	5.4	13% (fossil fuel burning)	5% (traditional fuel burning)	12% (agricultural manufacturing, refuse burning)	70% (metals processing
Sulphur emissions to atmosphere	31 million (sulphur)	2.7	85% (fossil fuel burning)	0.5% (traditional fuel burning)	1% (agricultural burning)	13% (smelting refuse burning)
Methane flow to atmosphere	160 million	2.3	18% (fossil fuel harvesting and processing)	5% (traditional fuel burning)	65% (rice paddies, domestic animals, land clearing)	12% (landfills)
Nitrogen fixation (as nitrogen oxide and ammonium)[c]	140 million (nitrogen)	1.5	30% (fossil fuel burning)	2% (traditional fuel burning)	67% (fertilizer, agricultural burning)	1% (refuse burning)
Mercury emissions to atmosphere	2500	1.4	20% (fossil fuel burning)	1% (traditional fuel burning)	2% (agricultural burning)	77% (metals processing, manufacturing, refuse burning)
Dinitrogen oxide flows to atmosphere	33 million	0.5	12% (fossil fuel burning)	8% (traditional fuel burning)	80% (fertilizer, land clearing, aquifer disruption)	Negligible
Particulate emissions to atmosphere	3100 million[d]	0.12	35% (fossil fuel burning)	10% (traditional fuel burning)	40% (agricultural burning)	15% (smelting, non-agricultural land clearing, refuse)
Non-methane hydrocarbon emissions to atmosphere	1000 million	0.12	35% (fossil fuel processing and burning)	5% (traditional fuel burning)	40% (agricultural burning)	30% (non-agricultural land clearing, refuse burning)
Carbon dioxide flows to atmosphere	150 billion (carbon)	0.05[e]	75% (fossil fuel burning)	3% (net deforestation for fuel wood)	15% (net deforestation for land clearing)	7% (net deforestation for lumber, cement manufacturing)

Notes: The magnitude of the insult is only one factor determining the size of the actual environmental impact.

a The human disruption index is the ratio of human-generated flow to the natural (baseline) flow.

b The automotive portion of anthropogenic lead emissions in the mid-1990s is assumed to be 50 percent of global automotive emissions in the early 1990s.

c Calculated from total nitrogen fixation minus that from dinitrogen oxide (nitrous oxide).

d Dry mass.

e Although seemingly small, because of the long atmospheric lifetime and other characteristics of carbon dioxide, this slight imbalance in natural flows is causing a 0.4 percent annual increase in the global atmospheric concentration of carbon dioxide.

Source: Holdren and Smith, 2000

For example, as can be seen from the first row of Table 13.2, the natural 'baseline' level of lead emissions to the atmosphere is estimated as 12 000 tonnes per year, and the amount annually released to the atmosphere by human activities is estimated to be 18 times the natural level. Similarly, the last row of Table 13.2 shows that the natural baseline level of carbon emissions to the atmosphere is around 150 billion tonnes of carbon per year, and that human activities annually release only one-twentieth (0.05 times) as much carbon to the atmosphere. The Human Disruption Index is therefore 18 in the case of lead emissions to the atmosphere, and 0.05 in the case of atmospheric carbon dioxide flows. This does not, of course, mean that lead emissions from energy use are more deleterious to humanity than CO_2 emissions. Each 'insult' has an important effect, but in a different way, in different amounts and on a different scale. For example, even relatively small quantities of carbon dioxide added to the atmosphere can have a relatively large impact on the earth's climate – as we shall see in Section 13.7 below.

Classification by scale

Holdren and Smith also categorize the impacts of human energy use according to the different *scales* at which these impacts principally manifest themselves. These are:

- **Household scale** – for example, emissions of smoke and other pollutants from domestic wood burning.
- **Workplace scale** – for example, the health hazards entailed in coal mining or oil and gas extraction.
- **Community scale** – for example, emissions of air pollutants from vehicle engines in urban areas.
- **Regional scale** – for example, acid rain caused by emission of sulphur dioxide and oxides of nitrogen from power stations, which can have adverse effects many thousands of miles away from the emission sources.
- **Global scale** – for example, emissions of carbon dioxide and other greenhouse gases produced by fossil fuel combustion, which can cause climate changes at a global level.

Some insults, of course, produce impacts at several scales: for example, emissions of some of the pollutants from domestic wood burning not only affect the health of humans in households, but also make a contribution to community, regional and global pollution levels. Another example is the possible impact on human health of electric and magnetic fields from power lines (see Box 13.2), which manifests itself mainly at household, workplace and community scales. We shall follow Holdren and Smith's general approach here (in Sections 13.3 to 13.7) drawing on some of their analysis and on other sources where relevant. Space limitations make the discussion inevitably a somewhat abbreviated and selective one; but further references to some of the extensive literature on the environmental and social impacts of energy use can be found at the end of this chapter.

However, there is one issue that we cannot leave aside. In reviewing the various impacts, it is important to be aware of the limitations of the data and of interpretations of it. Box 13.1 attempts to put some of the methodological and statistical issues in perspective.

BOX 13.1 Assessing the evidence

One thing becomes obvious as soon as we start to look at the health and environmental effects of energy systems. The available information, the factual background, is often very different from the facts about, say, the energy content of a fuel or even the national energy 'statistics' of a country. The significant difference is that the available data here are often *essentially statistical*, and such data can be interpreted in different ways.

Causes, effects and correlations

First, there is the question of the *effects* of our energy systems. Quite often the only way to link an effect with its cause is to look carefully for correlations in the variations of the two. It is believed, for instance, that one particularly bad smog in London in 1952 caused 4000 deaths in five days (see Box 13.4). This was not because 4000 otherwise healthy people suddenly fell dead, but because there were 4000 more deaths from respiratory and related diseases than normal for that population in that place during that period. It could of course have been a random fluctuation, but because there was other statistical evidence relating air pollution and respiratory illness, and at least the beginnings of an understanding of the link, the smog was held responsible.

The above example is uncharacteristic, however. The detrimental effects commonly appear at much lower levels, or are diffuse or distant in space or time from their original causes. There are of course cases where a particular health or environmental effect is attributable to a known unique cause, but many of the figures that appear in this chapter are the result of statistical analysis, correlating data on a low-level effect and its potential cause or causes. Space does not permit the inclusion of details of these analyses, but it is wise to be cautious about the results, particularly when the effect is only a marginal change against an existing background.

Types of data

It is also important to be aware of the types of data that form the basis for the statistics. It is useful to distinguish three aspects of any energy system that determine its effects on human health or the environment.

1 **Routine operation**. The construction and normal operation of any energy system usually has some deleterious effects. Ideally these should be predictable and measurable, and they are often subject to regulation.

2 **Accidents**. These are a feature of all human activities, and most of the energy industries are sufficiently well-established to have accumulated data on 'normal' accident rates and their consequences.

3 **Catastrophes**. These are large-scale, individually unpredictable, and rare.

The first two categories properly supply the data needed to estimate the effects of an existing or proposed system, or to compare the effects of different systems, such as the various ways in which we generate electric power.

Whether the consequences of catastrophes should be similarly included in the statistics is a subject of debate. The data they provide are certainly different: instead of averages over periods of time and many similar systems, the only available information might come from a single event, which may or may not be repeated in the future.

One way to deal with this situation is to assess the risk from *potential* catastrophes – those that have not yet occurred as well as those that have. The results of such assessments can then take their place with the other data in assessing the environmental or health effects of a system. Risk assessment is a two-stage process, requiring estimates of the probable consequences of each event and of the probability that it will occur. As we shall see in Section 13.8 below it relies heavily on statistical reasoning.

13.3 Household-scale impacts

Wood burning in developing countries

It is estimated (Reddy, Williams and Johansson, 1997) that perhaps as much as half of humanity still cooks its food using fires fuelled by biomass, usually wood. Often, these fires are located indoors and emit smoke and other pollutants that are particularly harmful to health, especially at the high concentrations found in confined spaces.

For example, an estimated 80% of Indian households in the early 1990s used biomass as their main cooking fuel. The adverse health effects of this included increased incidence of acute respiratory infection, chronic respiratory disease, tuberculosis, lung cancer, heart disease and adverse pregnancy outcomes. Overall, it is calculated that some 5 to 6 % of premature deaths of Indian women and children (the groups most exposed to indoor pollution) at that time were attributable to biomass fuel use in households (Smith *et al.*, 2000, cited in Holdren and Smith, 2000). And even before it is burned, the gathering and cutting of wood is usually a laborious and sometimes a hazardous business, which often results in additional adverse effects on human health.

Furthermore, as mentioned in Chapter 1, domestic cooking fires seldom achieve complete combustion of the wood fuel and so emit significant quantities of greenhouse gases that contribute to global warming. Figure 13.1 shows total greenhouse gas emissions (including methane, non-methane hydrocarbons and dinitrogen oxide, as well as carbon dioxide) arising from the burning of renewable sources such as wood, dung and crop residues, expressed in terms of the equivalent number of grams of carbon per megajoule of cooking energy delivered. It is clear from the Figure that the greenhouse gas emissions from these renewables, under typical (non-optimum) conditions, can be higher than those for non-renewable sources such as kerosene or liquefied petroleum gas – although greenhouse

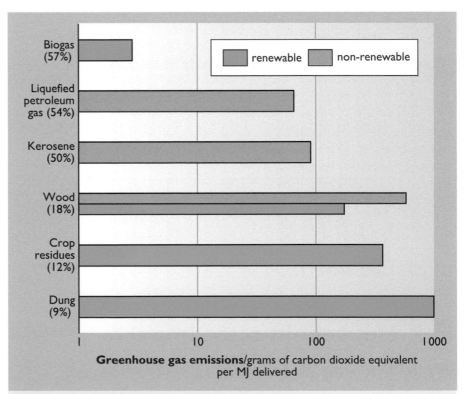

Figure 13.1 Greenhouse gas emissions from household fuels in India. Includes warming from all greenhouse gases emitted: carbon dioxide, methane, non-methane hydrocarbons, and dinitrogen oxide (dinitrogen oxide is commonly known as nitrous oxide.), weighted by stove distribution in India. Percentages shown in parentheses are average energy efficiencies of stoves used for fuel combustion (source: Smith *et al.*, 2000)

gas emissions from the use of one renewable source, biogas, are much lower. We shall examine the subject of greenhouse gas emissions and climate change in more detail in Section 13.7.

The subject of wood fuel use in developing countries and its environmental and health impacts is dealt with in more detail in the companion volume *Renewable Energy.*

13.4 Workplace-scale impacts

Biomass harvesting and forestry

As already noted, the manual gathering and harvesting of traditional biomass sources by villagers in developing countries can be a hazardous occupation. Even when wood fuel is harvested on a commercial basis, taking advantage of mechanization in such forms as the chain saw, working conditions among forestry workers in many countries are still often very dangerous and unhealthy. Fully mechanized forestry as practised in some developed countries is considerably less hazardous, but still involves some manual handling of timber, with attendant safety risks. Globally, the annual mortality rate for forestry workers is estimated to lie between 32 000 and 160 000, though by no means all of these are related to energy production (Holdren and Smith, 2000).

Hydro and wind power

Hydro power is a very substantial contributor to world electricity supplies. In 2000 it provided some 17% of the estimated 15 379 TWh of electricity generated world-wide (IEA, 2002). Wind power generation, though growing fast, still makes a much smaller contribution. Total world wind power generating capacity at the end of 2002 was just under 30 GW, contributing less than 1% of world electricity supplies.

Workplace hazards in the hydroelectricity industry mainly occur when large dams are under construction, and are likely to involve the same levels of fatalities and injuries to workers as those on other large construction projects. There are other substantial hazards due to the occasional catastrophic effects of dam failures, but these are mainly to the public rather than workers and are discussed in Section 13.5: *Community-scale impacts.*

In comparison with hydroelectricity, wind power is a relatively new technology. By the end of 2001 there had been approximately 20 operator deaths, most of them due to falls and injury by blades. There were no known injuries to the public (Gipe, 2001).

As renewable energy sources, wind and hydro power are treated in detail in the companion volume, *Renewable Energy.* The accounts of their effects in this chapter should therefore be regarded as brief summaries only.

Coal, oil and gas

The extraction of coal from the ground, whether by surface (open cast) or underground mining, and the extraction of oil and gas from beneath the earth's surface, entail substantial risks of death or injury to the workers in the industries involved – even in countries like the UK where safety considerations are today given a very high priority.

Table 13.3 gives data on injuries for some of the relevant industries in Britain for the period 1997 to 1999. The inclusion of the water supply industry in the statistics will obviously have somewhat inflated the figures. But it seems clear that there is a continuing annual incidence of around a dozen fatal injuries and several hundred serious but non-fatal injuries in Britain's energy-related industries.

Table 13.3 Injuries to workers in energy-related industries, UK, 1997–9

Type of industry	Fatal injuries 1997–8	Fatal injuries 1998–9	Fatal injuries 1999–2000	Major injuries 1997–8	Major injuries 1998–9	Major injuries 1999–2000
Mining and quarrying of energy-producing materials*	14	7	5	440	346	313
Electricity, gas and water supply	4	4	2	244	209	193
TOTAL: Energy and water supply industries*	18	11	7	684	555	506

* Includes the number of injuries in the oil and gas industry recorded under offshore installations safety legislation

Source: Office for National Statistics, 2002. (Data from the UK Health and Safety Executive.)

For the world as a whole, estimates of deaths and injuries due to coal mining seem to vary very widely between different sources. World-wide death rates from coal mining are very roughly estimated by Holdren and Smith to lie somewhere between 30 000 and 150 000 per year. They also derive another estimate, based on average annual US coal mining deaths between 1890 and 1939, which puts world wide deaths in coal mining at around 16 000 per year.

It is generally assumed that death and injury rates in the coal, oil, gas and nuclear industries are roughly proportional to the quantities of energy produced, but the proportions vary widely between the different industries. Table 13.4 gives one estimate (Nordhaus, 1997) of the number of occupational hazards, both fatal and non-fatal, due to the use of coal, oil and gas in electricity production, expressed in numbers of deaths and diseases per gigawatt-year of generation. (One gigawatt-year is equivalent to 8.76 terawatt-hours). Also shown are the occupational hazards of nuclear power generation, which are discussed below. Hazards of nuclear power to the *public*, also shown in Table 13.4, are discussed in Section 3.6.

Note, however, that the data in the table exclude 'severe accidents'. The statistical problems involved in comparing data on rare, catastrophic accidents have been mentioned in Box 13.1 above and will be discussed again in Section 13.8 below.

Table 13.4 Occupational hazards of electricity production by fuel cycle: number of deaths and diseases per gigawatt-year of output – including entire fuel cycle, excluding severe accidents

Fuel cycle	Occupational hazards per GWyr[**]		Public (off-site) hazards per GWyr[**]	
	Fatal	Non-fatal	Fatal	Non-fatal
Coal	0.2–4.3	63	2.1–7.0	2018
Oil	0.2–1.4	30	2.0–6.1	2000
Gas	0.1–1.0	15	0.2–0.4	15
Nuclear (LWR[*])	0.1–0.9	15	0.006–0.2	16

[*]LWR= Light Water Reactor, the generic term for reactors using ordinary water for cooling, like PWRs; [**]GWyr = gigwatt year; 1 GWyr = 8.76 TWh

Source: Nordhaus, 1997.

Nordhaus's figures suggest that the total number of occupational deaths attributable to fossil-fuelled electricity generation ranges from as little as 0.1 per gigawatt-year in the case of the best gas-fired generation to as much as 4.3 per gigawatt year in the case of the worst coal-fired generation.

Other estimates of the numbers of deaths from various methods of electricity generation are shown in Table 13.5. This gives the estimates of two authors, Hamilton and Morris, both from the Brookhaven National Laboratory in the USA. For coal-fired electricity generation, they calculate that the total number of deaths per gigawatt-year, among *both* workers and the public, ranges from as few as one to as high as 330 (the lower and upper ends of Morris's range). Their estimates of deaths among workers range from 0.21 per gigawatt-year for gas-fired generation to more than 93 per gigawatt-year for coal-fired generation.

Table 13.5 Estimated deaths from power generation per gigawatt-year of output

	Estimates by Hamilton				Estimates by Morris			
	Occupational accidents	Occupational Disease	Public	Total	Occupational accidents	Occupational Disease	Public	Total
Coal	0.46	93	4–150	100–243	0.53–0.93	0.13–8.7	0–320	1–330
Oil	1.63		1.3–130	3–130				
Gas	0.21							
Nuclear	0.35	0.18	0.067	0.6	0.14–0.6	0–0.90	0.2	0.4–1.7

Source: Gipe, 1995

Nuclear power

Risks to workers in the nuclear industry can occur throughout the nuclear fuel cycle (see Chapter 10, Figure 10.18). The hazards of uranium mining, which in addition to normal mining hazards also entails exposure to radioactive radon gas, have been long known. The high incidence of lung cancer in the uranium miners of the Joachimstal in Bohemia was one of the earliest recognized cases of radiation-related industrial disease. Other occasions for exposure of workers arise throughout the fuel cycle: in enrichment and fuel fabrication, in the routine operation of nuclear power stations, and in the processes of waste storage and disposal or reprocessing. Estimates of fatal hazards to workers in the nuclear industry are shown in Tables 13.4 and 13.5. They suggest that occupational deaths in the nuclear industry range from 0.1 to 0.9 per GWyr (Nordhaus) to around 0.53 per GWyr (Hamilton); or 0.14 to 1.5 per GWyr (Morris). (Hamilton's and Morris's estimates include both deaths in occupational accidents and deaths due to disease.)

The presence of highly radioactive materials means that minor accidents, whose effect might be negligible in other industries, can have serious consequences. An example is the case of a worker dealing with liquid containing plutonium whose protective gloves were punctured by an unseen wire, which then penetrated his finger. Measurement showed that the plutonium activity in the wound was some 200 times the 'safe' level, and within a few hours the decision had been taken to remove a slice of flesh 20 mm long down to the tendon (Norwood, 1963).

However, awareness of the potential dangers has led to careful monitoring, and studies have shown that the occupational health risk to workers in the nuclear power industry is lower than the average for similar workers elsewhere (Morison, 1998). The figures in Tables 13.4 and 13.5 imply that the occupational hazards of the nuclear industry are broadly comparable to those involved in gas extraction, and substantially less than those of oil or coal extraction.

Until the Chernobyl catastrophe, the nuclear industry claimed that there was no known case of a worker fatality resulting from a reactor accident in a civil nuclear power station, anywhere. This claim may be a little disingenuous, however. It excludes mining accidents or those involving workers in fuel enrichment and fabrication facilities or reprocessing plants, as distinct from power stations. Some early deaths may have resulted as a consequence of these exposures, but the existence of confidential 'out of court settlement' schemes for claims makes it hard to be certain, since formal admission that any death or injury definitely had occupational causes related to nuclear hazards is usually avoided. (Risks to the general public caused during the various stages of the nuclear fuel cycle are discussed below in Box 13.6, *The health effects of radioactivity*.)

BOX 13.2 **The health effects of electromagnetic fields**

One impact of human energy use that manifests itself mainly at the household, workplace and community scales is the impact of electromagnetic fields. Ever since the use of mains electricity became widespread, concerns have been expressed by some that the electric and magnetic fields emanating from power cables might be detrimental to human health.

Essentially, there is an *electric* field around a cable whenever a voltage is present, even if little or no current is flowing. Electric field strengths are measured in volts per metre (V m^{-1}). A *magnetic* field is produced only if a current is flowing through the cable. Magnetic fields are measured in teslas (T), and for example the field strength ten metres from a conductor carrying a current of 2000 A is 40 μT (microteslas). For comparison, the strength of the earth's magnetic field in which we all live is usually in the range 40–60 μT.

The *alternating* voltages and currents normally used in mains electricity distribution oscillate at a rate of fifty cycles per second, 50 Hz (or 60 Hz in the USA), so the electric and magnetic fields associated with them will be oscillating at this rate. But as we saw in Chapter 4, any alternating current also generates *electromagnetic waves*, so the electric and magnetic fields of these must be included in the total. However, their wavelength at a frequency of 50 Hz will be about 6000 km, and the contribution at a distance of, say, 60 metres from a cable is negligible – only 0.001% of the total. (It is proportional to the distance divided by the wavelength.) In practice, therefore, the radiation fields can be ignored, and any effects on people at normal distances from the cables would be due to the 'ordinary' electric or magnetic fields.

Electric fields are easily screened (reduced in intensity) by cable sheaths, the ground, vegetation and buildings. But magnetic fields are relatively unaffected by such obstacles and can only be screened by the use of substantial amounts of metal.

The intensities of the electric and magnetic fields decrease with distance from the cables. But in a typical electric transmission system, the current is flowing from the source though one conductor and the return current flows through another nearby conductor in the opposite direction. In this case the fields are in opposing directions and largely cancel each other out; and the overall field also decreases even more rapidly with distance, reducing the field strength in the 40 µT example above, for instance, to as little as 10 nT (nanoteslas).

Effects of electric and magnetic fields

The slowly varying electric and magnetic fields around transmission cables cannot produce the type of direct tissue damage that results from exposure to radioactivity (see Box 13.6). However, exposure to very high intensity electric fields can induce 'microshocks' in the body similar to the familiar effects of 'static' electricity. Very high magnetic fields can also induce currents in the body that, if strong enough, could be harmful by causing localised heating. To minimise such risks, the UK National Radiological Protection Board (NRPB) has specified a maximum limit for exposure to members of the public of 12 kV m^{-1} for electric fields and 1600 µT for magnetic fields associated with power lines (see IEE, 2001).

As Figure 13.2 shows, measured magnetic field levels in a typical home lie well below the 1600 µT limit. They normally range from a background level of around 10 nT rising to peaks of 100 nT or more in the vicinity of appliances like radios, TVs, transformers and fluorescent lights.

Electric field strengths only approach 12 kV m^{-1} directly below high-voltage transmission lines. Here, magnetic fields can be up to 100 nT, but are more typically below 10 nT, both well below the NRPB limits. However, the NRPB limits are somewhat higher than those set by the International Commission on

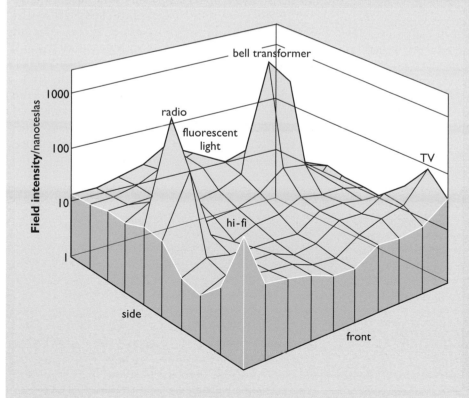

Figure 13.2 Variations in magnetic field intensity over the ground floor of a typical house, measured on a 1m grid at 1m above floor level (source: Swanson and Renew, 1994)

Non-Ionising Radiation Protection, which specifies a limit of 100 µT (see IEE, 2001 and Henshaw, 2002).

Given the rapidity with which field strengths decrease with distance, and given that only about 0.1% of the UK population lives within 50 metres of a high-voltage overhead power line (Figure 13.3), is seems clear that the vast majority of public exposure to low-frequency electric and magnetic fields results from low-voltage domestic wiring or local distribution cables (Swanson and Renew, 1994).

Figure 13.3 Only one person in a thousand in the UK lives in a house within 50 metres of a high-voltage overhead power line

Despite the apparent unlikelihood of direct bodily effects from electric and magnetic fields, there is some evidence that such fields could be harmful to health in more subtle ways. Several recent epidemiological studies have suggested a link between childhood leukaemia and exposure to magnetic fields. A recent review of the evidence in nine countries in Europe, North America and New Zealand (Ahlbom *et al.*, 2000) concluded that prolonged exposure to fields above 0.4 µT appears to double (from one in 1400 to one in 700) the risk of leukaemia occurring before age of 15.

The mechanism by which such cancers might be induced is unclear. Prof Denis Henshaw of Bristol University has suggested, citing work in the US and elsewhere, that although magnetic fields probably do not have a role in *initiating* cancers, they may have a role in *promoting* them (Henshaw, 2002). It appears that prolonged exposure to magnetic fields may suppress the body's nocturnal production, in the pineal gland, of the hormone melatonin. Melatonin is a powerful anti-oxidant and has been shown in some studies to suppress breast cancer, so its absence could weaken one of the body's natural protective mechanisms.

Prof Henshaw has also put forward a theory that could account for adverse health effects near high voltage power lines. He suggests that because the high-voltage lines cause ionisation of the surrounding air, the ionised air molecules could attach themselves to the minute droplets (aerosols) of pollutants such as sulphur dioxide (see Section 13.6 on acid rain) that are often present in the atmosphere. Such pollutants when ionised (i.e. given an electrical charge) would be more likely to lodge within the lungs of people living nearby, thus increasing their chances of ill-health.

Such theories are, however, still controversial and research into the effects of electric and magnetic fields continues. In effect, the 'jury is still out' on whether such fields are significantly detrimental to health, and if so, on what the causal mechanisms might be.

In cases where such uncertainty prevails, some countries and authorities have chosen to adopt the 'precautionary principle' – that is, in the absence of conclusive evidence, whilst debate and research continue, judgements regarding risk should err on the side of caution when hazards to human health or the environment are concerned. It was on this basis that Switzerland in 1999 adopted a precautionary maximum permitted level of exposure to magnetic fields of one microtesla (IEE, 2002).

If, as some evidence now appears to suggest, exposure to electric and magnetic fields can in some circumstances be significantly detrimental to human health, the effects appear at present to be small and restricted to those whose exposure to such fields is abnormally high.

However, it would be unwise to be complacent about such risks. It is possible that further epidemiological investigations will uncover further associations between illness and exposure to electric or magnetic fields, and that further biological research will suggest hitherto-unsuspected mechanisms by which such effects might be caused.

New advice to the UK Government on limiting exposure to electric and magnetic fields is due to be published for public consultation by the NRPB in 2003.

Further information on the subject can be found in the references cited and at the following Web sites:

Institution of Electrical Engineers: www.iee.org/Policy/Areas/BioEffects/emfhealth.pdf [accessed 1st March 2003]

Bristol University Physics Laboratory: Human Radiation Effects Group: www.electric-fields.bris.ac.uk [accessed 1st March 2003]

UK National Radiological Protection Board (NRPB): www.nrpb.org [accessed 1st March 2003]

International Commission on Non-Ionising Radiation Protection: www.icnirp.de [accessed 1st March 2003]

13.5 Community-scale impacts

The term 'community-scale' in Holdren and Smith's classification refers principally to towns and cities, in both developed and developing countries. In many such urban areas, **atmospheric pollution** has been a major environmental concern since at least the Middle Ages.

What is atmospheric pollution? Like any form of pollution, it is the presence of substances that are potentially harmful to the health and well-being of the human population or of other living things, or even to inanimate objects such as buildings. As described in Box 13.3, atmospheric pollutants can take a number of forms. Several gaseous pollutants have already been introduced in the discussions of fuels and their uses in earlier chapters. The main contributors to atmospheric pollution are:

Sulphur dioxide (SO_2). This results from combustion of almost any fuel, because almost all – not only fossil fuels but biofuels too – contain some sulphur. But the amounts vary greatly, as does the ease with which the sulphur can be removed before using the fuel. For these reasons coal-fired power stations are the main producers of SO_2. The main polluting effects result from the conversion of SO_2 into sulphuric acid (H_2SO_4) in the atmosphere. If it remains in the air, this can produce an aerosol of sulphuric acid droplets which we experience as an unpleasant haze; but it is also the main contributor to acid rain (see Section 13.6 below).

NO_x. This is the general term for nitrogen-oxygen compounds that are an inevitable result of burning any fuel in air. In the atmosphere, the NO_x become nitric acid (HNO_3) and provide a further contribution to acid rain. Their effects are therefore similar to those of SO_2, but as their main sources are internal combustion engines, which are often in close proximity to people, the immediate effects of NO_x on the respiratory system are more obvious.

Carbon monoxide (CO), which is a result of incomplete combustion of a fuel containing carbon. Although usually produced in small quantities, it is a very undesirable pollutant because it is a poison. A badly adjusted

---

BOX 13.3 Forms of pollutant

Pollutants in the atmosphere appear in different forms, and it may be worth distinguishing between these.

They can be **gases**, consisting of individual atoms or molecules – like the oxygen and nitrogen of the air itself. Gaseous atmospheric pollutants include SO_2, NO_x, CO and others. (See Section 13.7 below for the role of CO_2.) The term **vapour** is often used in this context. Strictly, a vapour is a gas at a temperature at which it can be liquefied under pressure. This is the case for water at normal temperatures (see Chapter 6, Box 6.3), so we are quite correct to talk of 'water vapour' in the air. The common use of 'vapour' for other air-borne substances that we can see, such as a coloured gas or a 'mist', is not strictly correct. An **aerosol** is quite different from a gas or vapour.

It consists of tiny droplets or particles as little as one micron (a thousandth of a millimetre) in diameter, but nevertheless each containing billions of atoms; much larger than the individual free atoms or molecules of a gas or vapour. A 'mist' is an aerosol of water droplets, i.e. water in the liquid rather than the vapour phase.

A class of pollutants that is usually distinguished from aerosols is the **particulates**. These are solid particles that are small enough to float in air. Only a few microns in size, they are individually visible as the 'motes' we see in a beam of sunlight. At higher concentrations they appear as smoke from chimneys, or as contributors to the brown layer seen from an aircraft above any large city.

combustion system such as a household gas boiler can produce enough CO to kill a person in a matter of hours in an ill-ventilated room. By far the major producer of carbon monoxide is the internal combustion engine, although the toxic effect is of course less dramatic out of doors. The toxicity of carbon monoxide is related to the fact that it is chemically very active – an important difference from a gas such as carbon dioxide which produces immediate physiological effects only at concentrations so high that the person starts to suffocate through lack of oxygen. The chemical activity of CO also means that it easily forms compounds with other substances, producing further atmospheric pollutants.

Other gaseous pollutants include the **VOCs** (**volatile organic compounds**), a range of chemical compounds of different kinds and from different sources. They include hydrocarbons, mainly resulting from incomplete combustion in internal combustion engines, and vapours from solvents and similar materials used in industry. Many of these are carcinogenic, and they contribute to the 'chemical' smog of urban areas such as Los Angeles or Mexico City.

Other forms of pollutant which give rise to concern are chlorinated compounds called **dioxins**, and the related **furans**. They are produced in small quantities in most combustion processes and as by-products in some industrial processes. As a recent report by the UK Department for Environment, Food and Rural Affairs explains:

> ...all forms of combustion, both industrial and domestic, may release dioxins if chlorine is present, even in trace quantities. This includes the incineration of wastes (including municipal, medical and hazardous wastes), the combustion of solid and liquid fuels, including coal, oil and wood, both on a large scale, such as in electrical power generation, and on a small scale in domestic stoves and fires, and in a range of other combustion processes such as garden-refuse burning, bonfires and accidental fires.
>
> DEFRA, 2002

Dioxins are highly toxic, even in very small quantities. They can be present in both the flue gases and the ash produced by incinerators. Though the volumes involved are normally small, they can lead to cancers and other adverse effects on the human immune and reproductive systems even at very low levels of exposure. In many Western nations there has been an extensive debate on the scale of heath risks associated with dioxins emitted from the combustion of municipal solid waste (MSW) (see Box 13.5).

Two remaining components of the atmosphere whose concentration gives rise to concern are ozone and carbon dioxide. Both are discussed in Section 13.6, *Global-scale Impacts*.

Urban air pollution in developed countries

In the urban areas of many developed countries, emissions of the main gaseous pollutants have generally fallen in recent decades. For example, in the UK between 1970 and 1998, as shown in Figures 13.4, 13.5 and 13.6, annual emissions of SO_2 fell from over 6 million tonnes to less than 2 million tonnes; emissions of NO_x reduced from around 2.5 million tonnes to about 1.7 million tonnes per annum; and annual CO emissions fell from

BOX 13.4 **A tale of two cities**

The effectiveness of some of the measures designed to reduce pollution can be seen in the following brief accounts of two cities in the mid-twentieth century.

For Londoners in the 1950s, coal was still the main household fuel and 'smog' was a common winter feature – the consequence of cold foggy days and high levels of atmospheric particulates and SO_2. During the winter of 1952, the concentrations of these pollutants reached several thousand micrograms per normal cubic metre of air (μg Nm^{-3}), ten times the already high average. The result, the 'great smog', was a yellow-grey blanket so impenetrable that pedestrians negotiated the streets by feeling their way along the walls of buildings. Its catastrophic consequences led to action, and the Clean Air Act of 1956 totally prohibited the burning of untreated coal in London. Within twenty years the annual average particulate level had fallen below 50 μg Nm^{-3}, and whilst London still has damp foggy days, the last real 'pea-souper' is now a distant memory.

The Angelinos of Southern California also suffered from air pollution in the mid-1950s. The LA smog, a thin yellow-brown, throat-catching, eye-watering haze in which the presence of hydrocarbons was only too evident, was quite unlike the London version. The causes were different too; not coal burning but automobiles with an *average* fuel consumption of about 13 miles per US gallon, together with heavy industries and an absence of controls exemplified by the burning of most domestic rubbish in backyard incinerators. These were made illegal in the late 1950s, and over the next two decades California also became a world leader in the control of vehicle emissions. Catalytic converters were mandatory for all new cars from 1975; 'clean gasoline' reduced the VOCs, and all cars were required to meet emissions standards on any change of ownership. Overall, the late twentieth century car was emitting less than a tenth of the pollutants of its predecessor twenty or thirty years earlier. Air quality in the Los Angeles area still fails to meet the US federal standards for much of the time, but a halving of the maximum ozone level over a period when the population rose by 40% and the number of cars by 60% should perhaps be commended.

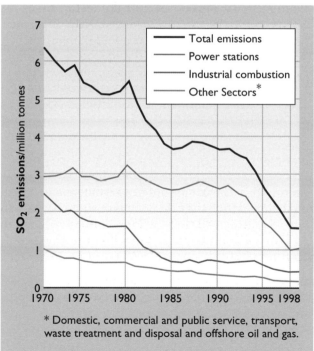

* Domestic, commercial and public service, transport, waste treatment and disposal and offshore oil and gas.

Figure 13.4 UK sulphur dioxide (SO_2) emissions, 1970–98. Data compiled on the basis of United Nations Economic Commission for Europe (UNECE) definitions. Source: Department of Trade and Industry (2000) *Digest of UK Energy Statistics (DUKES)*, Annex B: Energy and the Environment, Chart B7. p. 255

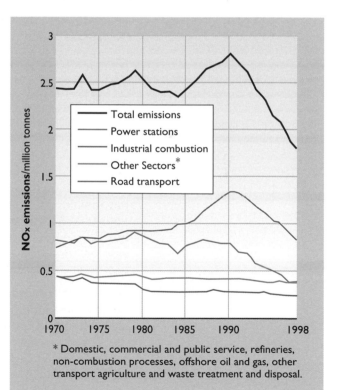

* Domestic, commercial and public service, refineries, non-combustion processes, offshore oil and gas, other transport agriculture and waste treatment and disposal.

Figure 13.5 UK nitrogen oxides (NO_x) emissions, 1970–98. Data compiled on the basis of United Nations Economic Commission for Europe (UNECE) definitions. Source: Department of Trade and Industry (2000) *Digest of UK Energy Statistics (DUKES)*, Annex B: Energy and the Environment, Chart B9, p. 257

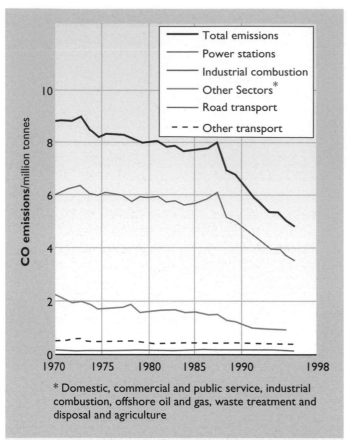

Figure 13.6 UK carbon monoxide (CO) emissions, 1970–98. Data compiled on the basis of United Nations Economic Commission for Europe (UNECE) definitions. Source: Department of Trade and Industry (2000) *Digest of UK Energy Statistics (DUKES)*, Annex B: Energy and the Environment, Chart B11, p. 258

* Domestic, commercial and public service, industrial combustion, offshore oil and gas, waste treatment and disposal and agriculture

about 9 million to around 5 million tonnes. The measures that have led to such reductions vary between communities and countries (see for instance Box 13.4). They may be due to the increasing use of nuclear power, replacing fossil-fuelled plants; or result from regulations requiring pollution control equipment to be fitted to power plants to reduce SO_2 and NO_x emissions; or be due to a switch to cleaner and more efficient natural gas for heating and electricity generation; or to the imposition of more stringent controls on vehicle exhaust emissions, leading to the use of cleaner fuels and engine technologies. (These aspects of vehicle emissions were discussed in Chapter 8.)

Emissions of particulates (Box 13.3) have also fallen in recent years in most developed countries. In urban areas these are mainly from vehicle exhausts – especially those of diesel engines. The particulates that cause concern are often very small, less than 10 microns in diameter, and consist mainly of carbon. Measurements of these **PM10** particles in most cities of the developed world show substantial reductions in recent decades. However, it is now considered that the even smaller **PM2.5** particles (i.e. those with a diameter of less than 2.5 microns) are more damaging to health than PM10s, but PM2.5 levels were not measured in many cities until relatively recently.

Particulates can lodge deep within the lungs and are carcinogenic. There appears to be no threshold below which they do not cause some damage to health (see Box 13.5 below), so ultimately these will need to be reduced to zero or 'near zero' levels. This will probably require the introduction of new fuels and technologies, as will be discussed in Chapter 14: *Remedies*.

Another significant pollutant for which levels have reduced in recent decades is **lead** from vehicle exhausts, which accumulates in the brain and is believed to impair the cognitive development of children. In most developed countries, leaded petrol (gasoline) has been or is being phased out and emissions have fallen sharply, but in many developing countries it is, unfortunately, still in use.

However, despite these reductions in pollutant levels, they remain a cause for serious concern in developed countries, and there is still a need for further cuts.

BOX 13.5 Health effects of atmospheric pollution

What is known about the health effects of atmospheric pollutants? At what levels are they harmful? As mentioned in the main text, two pollutants that continue to cause concern are the particulates and the dioxins, so we shall take these as our examples.

Particulates

We might start by considering the great London smog of 1952 (Box 13.4). It lasted for five days and is believed to have caused about 4000 deaths: 800 additional deaths per day in a population of about 10 million, or a daily rate of 80 per million people. The normal annual death rate from all causes at that time would have been roughly 10 000 per million per year: a daily rate of about 27 deaths per million people. So the *increase* in the death rate was an astonishing 300%.

But this case is well-known precisely because it is not typical. Can we apply the results to today's much lower pollution levels? In 1952, the levels of SO_2 and particulates reached perhaps 3000 μg Nm^{-3}, and the death rate increased by 300%. Does it follow that a rise of 10 μg Nm^{-3} in the concentrations of these pollutants would increase the death rate by 1%? This question, whether the results obtained from the effects at high concentrations still apply at levels hundreds of times smaller, has been the subject of much debate. The current view, based on more recent measurements correlating health effects and particulate concentration, is that the effect is proportional to the exposure down to the lowest levels: there is no 'safe threshold' below which the pollutant has no effect at all (WHO, 1999).

If this is indeed the case, what are the implications? Consider a city of 10 million people. The normal annual death rate from all causes will be about 100 000 per year. Suppose now that the concentration of PM_{10} particles is about 40 μg Nm^{-3}, which is typical for cities in Western Europe in the early twenty-first century. If this increases the death rate by 4%, we have an additional 4000 deaths a year, mainly as a result of vehicle emissions, in a city the size of London. In countries such as China and India, where urban particulate concentrations can be as high as 200 μg Nm^{-3}, the consequences are proportionally more severe.

Dioxins

The levels at which the dioxins are considered a health hazard can be seen in the current regulations governing their release. UK Environment Agency regulations require concentrations of less than 1.0 *nanogram* TEQ (toxicological equivalent) per normal cubic metre (<1.0 ng TEQ Nm^{-3}). These regulations have resulted in the closure of older Municipal Solid Waste (MSW) incineration plants, where emissions could be as high as 45 ng TEQ Nm^{-3}. New EU requirements are even lower: <0.1 ng TEQ Nm^{-3}, which implies a 98% reduction in dioxin emissions from MSW combustion. It is claimed that modern, state-of-the-art energy-from-waste (EfW) plants easily meet these requirements, with emissions in the range 0.02–0.4 ng TEQ Nm^{-3} (Porteous, 1999).

Following the imposition of increasingly stringent controls, overall dioxin emissions in the UK fell by 70% during the 1990s, from 1142 g TEQ in 1990 to 345 g TEQ in 1999. Emissions from MSW incineration were cut from 51% of all emissions in 1990 to 1% in 1999. Other sources, such as accidental fires, bonfires and the iron and steel industry, are now believed to be responsible for a much greater share of total dioxin emissions (DEFRA, 2002).

The UK Royal Commission on Environmental Pollution has concluded that even with very large increases in the amount of MSW going to energy from waste plants, the total pollution load from modern plants under these new standards would not be a cause for concern (RCEP, 1994).

However environmental organizations such as Greenpeace and Friends of the Earth are not convinced by these arguments (see Allsop *et al.*, 2001 and FoE, 2002). They contend that emissions of dioxins and similar substances are harmful to health even at the extremely low levels permitted by current regulations. They also suspect that incinerator emission control and monitoring systems are not always effective, allowing emission levels occasionally to exceed the permitted maxima. They believe that waste reduction, re-use and recycling are more appropriate approaches to waste management than incineration.

The debate between proponents and opponents of incineration currently remains unresolved.

Urban air pollution in developing countries

In the urban areas of developing countries, emissions of most of the pollutants discussed above far exceed those found in similar areas of the developed world. For example, particulate (PM10) levels in many cities of the developing world are extremely high. The principal causes include poorly-tuned and -maintained engines, particularly diesels and the polluting two-stroke engines often used in motorcycles and motorized 'rickshaws'; and the uncontrolled burning of solid fuels (wood and coal) and refuse. The effects of such emissions on the environment and human health are profound. In China, it is estimated that there are some 170 000 to 230 000 premature deaths per year from urban air pollution (Holdren and Smith, 2000).

Some economists point to the fact that in those countries we now call 'developed', levels of pollution also initially rose during the early stages of development, but later fell as higher levels of affluence gave rise to higher environmental standards, resulting in the introduction of cleaner technologies and practices. The implication is that increasing living standards will in themselves, given time, lead to reductions in environmental impact. However it remains to be seen whether rising affluence alone will be sufficient to enable the polluted cities of the developing world eventually to reduce their emissions. Other, more active approaches may prove necessary.

Community impacts of hydroelectricity

The construction of large dams has flooded about half a million square kilometres of land (roughly the total area of Spain), and is estimated to have led to the displacement of some 30 to 60 million people during the 20th century. In China alone, according to World Bank estimates, 10 million people were displaced by reservoirs in the period from 1950 to 1989, and more than 1 million people are expected to be displaced by the Three Gorges Dam, the word's largest hydro-electric project with a capacity of over 18 GW (Holdren and Smith, 2000) Not all the world's dams were built for hydroelectricity, of course; many are for irrigation or other purposes. Nevertheless, most of the very large systems have been primarily for electric power.

The adverse effects of large hydro schemes are not confined to displacement of population. The anaerobic decomposition of submerged vegetation produces methane, a potent greenhouse gas (see Section 13.7 below), and it has been claimed that in some circumstances hydroelectric plants can produce greenhouse gas emissions with a greater global warming effect than would be produced by a coal-fired plant of the same generating capacity (World Commission on Dams, 2000).

The collapse of a dam can be a major catastrophe for the local communities. During the twentieth century, some 200 dam failures outside China are thought to have resulted in the deaths of more than ten thousand people. And within China, in one year alone, 1975, it is estimated that almost a quarter of a million people perished in a series of hydroelectric dam failures (Sullivan, 1995).

As mentioned above, the ecological and social effects of hydroelectricity are discussed further in the companion volume, *Renewable Energy*.

13.6 Regional-scale impacts

Acid deposition

It is well established that SO_2 and NO_2 in the atmosphere can lead to the production of sulphuric and nitric acids. Measurements in clouds have detected water drops with the acidity of vinegar, and when these are incorporated into rain or snow the result is **acid precipitation**. More commonly called **acid rain**, this has damaging effects on vegetation, lakes and fish, buildings and structures. It can cause respiratory diseases in humans, especially the vulnerable.

As early as the 1960s, it was found that some lakes in Scandinavia and the eastern USA were so acidic that they could no longer support life, and the late 1970s saw growing signs of an even more wide-spread potential disaster: *waldsterben*, or the death of the forests. By the mid-1980s, along a swathe from Scandinavia through the Low Countries, Germany and Poland and into the Czech Republic, up to half of all forest trees were reported to be damaged in varying degrees; about one in twenty seriously so. Similar effects were being observed in the eastern regions of North America.

As techniques for tracing pollutants through the atmosphere improved, it became generally recognized that the change from gases to acidic aerosols takes place in long plumes which can stretch over a thousand miles downwind from the source. For North America and Western Europe, this means eastwards.

By the 1980s, however, acid rain levels were already falling in most of the developed countries, the effect of increasingly stringent limits to permitted levels of its main precursors, SO_2 and NO_x emissions. The fact that the deleterious effects appeared when SO_2 emissions were already falling led to two rather different interpretations: that these were delayed effects of the acid rain of earlier years, or that acid rain was irrelevant. However, no other single cause of the *waldsterben* has been found, and the current view of many experts is that acid rain entering the soil initiates a process that ultimately weakens the trees, making them more vulnerable to other stresses such as those caused by the periods of low rainfall during the 1980s.

In Asia the problem of acid precipitation is increasing. China, one of the world's major coal consumers, has experienced many of the problems that Europe encountered twenty years ago. In the industrial city of Chongqing, metal and concrete structures corrode so fast that rust removal and re-painting are needed every one or two years, and in coniferous forests of the region up to half the trees are dying.

Acid rain may also soon become a problem in South America and Africa if fossil fuel use and vehicle ownership levels in those regions continue to increase.

Public impacts of nuclear power

As in the case of the *workers* in the nuclear industry (see Section 13.4 above), exposure of the *public* to radioactivity can arise within each stage of the fuel cycle. In order to assess the significance of this exposure, it is useful to compare it with the natural background radiation to which we are all subjected continuously in varying degrees.

BOX 13.6 The health effects of radioactivity

As the high-speed particles or penetrating radiation from a source of radioactivity pass through any material, living or otherwise, they tear apart the atoms or molecules, and the energy dissipated in doing this is a measure of the damage they cause. However, the biological damage is about ten times greater for energy delivered by heavy particles – alphas or neutrons – than by betas, gammas or X-rays. It also depends on other factors such as the type of tissue and the rate at which the energy is delivered. The unit that takes these factors into account is the **sievert** (Sv). In simplified form, the relationship between the sievert and the energy deposited in tissue is as follows:

Beta particles, gamma rays and X-rays deliver a radiation dose of one sievert in depositing 1 J (one joule) of energy per kilogram of tissue.

Alpha particles and neutrons deliver a radiation dose of one sievert in depositing 0.1 J of energy per kilogram of tissue.

How large a dose is 1 sievert? What effect does it have? Data come from three main sources: laboratory experiments on tissue samples or on animals, studies of those unfortunate groups of people who have been exposed to high radiation levels, and epidemiological studies of larger populations at low doses. To summarize the often-controversial results of many studies over many years may mean dangerous over-simplification. Nevertheless, it is perhaps useful to give some idea of the orders of magnitude.

- **Single large doses in a short period:** A dose of 10 Sv or more almost certainly means death within hours or days. Doses of 1–10 Sv lead to radiation sickness and disability for weeks or months, and can be fatal. Below 1 Sv the symptoms decrease, until at about 0.1 Sv there may be no immediately obvious effects.

- **Long-term effects of lower doses:** Over a wide range of doses and population sizes, the long-term rate of induced cancers appears to lie in the range *between one and two cancers per 100 person-sieverts*. The total number of genetic effects down to about the tenth generation is thought to be of the same order.

Discharges and exposures

Figures such as those above provide the basis for international standards which are used by many countries in legislation governing permitted exposures and permitted discharges from nuclear plants into the air or water. The latter must of course take into account the nature of different isotopes. Are they air-borne? What is their half-life? By what routes, if any, can they enter the human food chain?

To see in outline how permitted exposures and permitted discharges are related, we'll consider just one fission product: iodine-131 (I-131). Although its half-life is only 8 days, this radioisotope is particularly hazardous. Emitted as a gas, it condenses on vegetation and is easily taken up by grazing cattle, entering their milk. This provides a short route into the human body. The iodine then concentrates in the thyroid gland, and concentration of any radioisotope means an increased dose to the surrounding tissue. Estimates of routine emissions of I-131 from a nuclear power station tend to fall in the range 1–10 MBq a day. To convert these figures into health effects requires data on weather, population distribution, etc., and a more detailed analysis than is possible here; but the general conclusion is that the annual population dose from this source is unlikely to reach even 1 person-sievert. This implies that, on average, there would not even be one induced cancer over the normal lifetime of the plant.

See the main text for discussion of the significance of the data on low doses.

The background arises from cosmic radiation reaching the Earth from outer space, from radioactive rocks and from naturally occurring radioisotopes in our food. The data in Box 13.6 allow us to assess its effect. The average individual dose from these sources is slightly more than one millisievert (1 mSv) a year. If the dose from medical and industrial uses of radioactive materials and X-rays is added, the total becomes about 1.6 mSv a year. (The figure can be appreciably greater for people living in areas with uranium-rich rocks, or those who spend time at high altitudes.)

If the average individual dose is 1.6 mSv, the total annual dose to the UK population will be just under 100 000 person-sieverts. The data in Box 13.6 then imply that natural background radiation is responsible for 1000–2000 cancers per year. The annual death rate from all cancers in this population is of the order of 100 000, so natural radiation accounts for at most 1–2% of the total. We should note, however, that this calculation assumes that the effect is proportional to the total dose, no matter how small, and as mentioned in Box 13.5, not everyone accepts such extrapolations as valid.

Estimates of the average individual dose due to emissions of radioactive substances during the routine operation of nuclear power plants tend to be less than 0.01 mSv. This implies a UK population dose of about 600 person-sieverts, and an additional 6 cancers per year – an undetectable increase, statistically speaking. It has however been pointed out that national averages may be inappropriate when the emissions arise from a relatively small number of plants. For example, objectors argue that exposure levels are likely to be higher near nuclear facilities. There are however conflicting opinions on the evidence for these 'cluster' effects.

The effects of mine tailings and of spent fuel are also the subjects of controversy. As explained in Chapter 10, the mine tailings are initially highly radioactive and require suitable storage until a safe radiation level is reached. The issue lies in the word 'safe'. Some of the radioisotopes have very long half-lives, and although the activity may be low, the cumulative dose to the public continues to rise, and can reach a very high level after a long period. The question is whether it is legitimate to 'count' a dose that accumulates over a period much greater than any one human lifetime.

Similar reasoning could be applied to the long-lived isotopes in spent fuel, but there is also particular concern with the immediate or shorter-term effects. It is generally agreed that the public exposure due to a reprocessing plant is greater than from the routine operation of a power station. The 'cluster' argument may also have more force in this case, but the relatively small number of civil reprocessing plants means that reliable data are still rather sparse. The final issue, the ultimate disposal of spent fuel, has been discussed in the earlier chapters, where we have seen that it remains unresolved.

13.7 Global-scale impacts

Global climate change

Probably the most important global-scale impact of humanity's fossil fuel use is that resulting from the associated emissions of 'greenhouse' gases such as carbon dioxide (CO_2), methane (CH_4) and dinitrogen oxide (N_2O, also known as nitrous oxide). As described in Chapter 1, Box 1.2, these human-induced or 'anthropogenic' greenhouse gas emissions are now considered to be the principal cause of a process of global climate change that has already led to a rise in the earth's mean surface temperature of around 0.6 °C during the twentieth century.

In this section we expand on the summary given in Chapter 1 and consider some of the effects of fossil-fuel induced climate change in more detail.

Carbon dioxide is the most important anthropogenic greenhouse gas implicated in global warming. The majority of anthropogenic carbon dioxide emissions are due to fossil fuel combustion, as Table 13.6 (overleaf) shows, but not all of the emitted carbon remains in the atmosphere. Some of it is absorbed on land, for example by photosynthesis in growing trees and vegetation, and some is absorbed by the oceans. The proportion that remains in the atmosphere has been rising, however. As a result, atmospheric CO_2 concentrations have increased from about 285 parts per million at the end of the pre-industrial era, around 1850, to just under 370 parts per million in 2000 – an increase of some 30% in 150 years (see Chapter 1, Figure 1.18). The mean residence time of CO_2 in the atmosphere is of the order of 100 years, so even if humanity is successful in reducing anthropogenic emissions dramatically in coming decades, it will take a long time for atmospheric concentrations to decrease.

Table 13.6 Net flows of carbon to and from the atmosphere, in billion tonnes of carbon, 1980–99, based on measurements of CO_2 and O_2. Positive values are flows to the atmosphere; negative values (in brackets) represent uptake from the atmosphere.

Flows of carbon	1980 to 1989	1990 to 1999
Atmospheric increase	3.3 ± 0.1	3.2 ± 0.1
Emissions (fossil fuel, cement manufacture)	5.4 ± 0.3	6.3 ± 0.4
Flows from atmosphere to oceans	(1.9 ± 0.6)	(1.7 ± 0.5)
Flows from atmosphere to land	(0.2 ± 0.7)	(1.4 ± 0.7)

Source: Intergovernmental Panel on Climate Change (2001b) Climate Change 2001, the Scientific Basis: Summary for Policy Makers, Table 2, p. 39

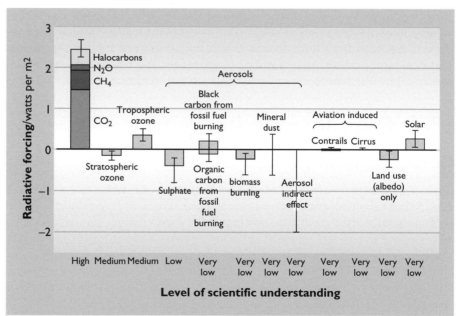

Figure 13.7 The many factors that force climate change: global mean radiative forcing of the climate system by various factors for the year 2000, relative to 1750.
Source: IPCC, 2001b, Figure 3, p. 8

Notes: Radiative forcing is the change in the balance between radiation coming into the atmosphere and radiation going out.

The albedo of a surface is the fraction of solar radiation that is reflected by that surface. The term 'land use (albedo only)' refers to changes in the fraction of solar radiation reflected from the earth's surface due to changes in land use.

Solar refers to changes in solar radiation intensity

The other anthropogenic greenhouse gases, though individually less important, are together estimated to increase by about 75% the global warming effect created by anthropogenic CO_2 emissions (see Figure 13.7). Among the most important of such gases are methane (CH_4) and dinitrogen oxide (N_2O), which are estimated to have increased by some 145% and 14% respectively since 1850. A significant proportion of these increased emissions is estimated to be due to energy-related activities.

The **halocarbons**, which include chlorofluorocarbons (CFCs) widely used until recently in refrigeration systems, are entirely anthropogenic and not found in nature. They cause depletion of the earth's protective ozone layer in the upper atmosphere – the stratosphere. However, the halocarbons are also potent greenhouse gases, although this strong global warming potential is fortunately offset by the fact they are emitted in relatively small quantities. It is also fortunate that the phasing-out of their use in order to protect the ozone layer will also reduce their contribution to global warming.

Ozone is another important greenhouse gas. It is found in the lower atmosphere – the troposphere – as well as in the stratosphere. It is formed in the troposphere through the interaction of oxygen, nitrogen oxides and certain hydrocarbons, as described in Box 13.7. It is estimated that the warming effect due to formation of ozone in the troposphere is greater than the cooling effect of ozone depletion (caused by halocarbons) in the stratosphere (see Figure 13.7).

In contrast to the global warming effects of the above gases, the aerosols and suspended fine particles that are generated in the atmosphere by burning fossil and biomass fuels (and periodically by volcanic eruptions) have a negative global warming effect – i.e. they lead to global cooling. This is because they tend to reflect incoming solar radiation but absorb little outgoing infra-red radiation. At present, their cooling effect offsets almost half of the warming effect of the other greenhouse gases, as shown in Figure 13.7. However, this arguably-beneficial influence is likely to diminish in coming decades as emissions of the precursor substances,

---

BOX 13.7 **Ozone in the troposphere and stratosphere**

Ozone is a gas whose molecules consist of three oxygen atoms: its chemical symbol is O_3, rather than the O_2 of normal air, which has two oxygen atoms. It is an important indicator of atmospheric pollution and features in environmental concerns in two entirely different ways. In the lower atmosphere, the **troposphere**, ozone is a by-product rather than a direct product of our use of motor vehicles. It is generated at or near ground level in a series of complex interactions, energized by sunlight, between oxygen and two other precursor pollutants: NO_x and VOCs. It is itself a pollutant, chemically active, and damaging to the health of crops, trees and human beings. The ozone concentration is often used as a general indicator of the level of atmospheric pollution.

As in the case of the acid precipitation discussed above, most developed countries are making efforts to reduce ozone concentrations by limiting emissions of its precursors, but in many developing countries levels are rising.

At a much greater height, twenty to thirty thousand metres above the surface of the Earth, in the **stratosphere**, is the natural phenomenon known as the ozone layer. Ozone is produced here by the effects of sunlight, but unlike the ozone near the ground, it is of great positive value because it absorbs much of the ultra-violet part of the solar radiation reaching the outer atmosphere. Too much ultra-violet radiation is harmful to our eyes and may lead to skin cancer, so the discovery of 'holes' in the ozone layer has given rise to concern. In this case, however, it is the release of certain organic substances (principally the halocarbons) that causes the problem, and not the world's energy systems. The **ozone hole** thus remains outside the scope of this book.

principally SO_2 and NO_x, are reduced by regulations designed to diminish the other environmental and health effects of these pollutants.

In addition to the anthropogenic greenhouse gas emissions mentioned above, another greenhouse gas, water vapour, also plays a very important role in the 'natural' greenhouse effect that maintains the earth's temperature at a level suitable for life, some 33 °C higher than it would be if all atmospheric greenhouse gases were absent. Human-induced emissions of water vapour are too small to affect directly the enormous quantities involved in the global water cycle, but there is a significant indirect effect. Increased global temperatures due to emissions of *other* greenhouse gases like CO_2 and CH_4 cause increased evaporation of water from the oceans, leading to increased water vapour in the atmosphere, which then leads to further warming and further evaporation – a 'positive feedback' effect.

The successive reports of the Intergovernmental Panel on Climate Change have communicated the increasingly confident conclusions of the world scientific community that the increased concentrations of anthropogenic greenhouse gases in our atmosphere are having a significant effect on the earth's climate. The 'signals' of climate change are becoming increasingly discernible amid the 'noise' of natural climate variability.

The changes that have been observed in a wide range of key atmospheric, climatic and biophysical indicators during the twentieth century, as summarized by the Intergovernmental Panel on Climate Change in its Third Assessment Report (IPCC, 2001a), are shown in Table 13.7.

Table 13.7 Twentieth century changes in the earth's atmosphere, climate and biophysical system

Indicator	Observed changes
Concentration indicators	
Atmospheric concentration of CO_2	280 parts per million (ppm) for the period 1000–1750 to 368 ppm in year 2000 (31±4% increase).
Terrestrial biospheric CO_2 exchange [carbon stored in the biosphere]	Cumulative source of about 30 gigatonnes of carbon (GtC) between the years 1800 and 2000; but during the 1990s, a net sink of about 14±7 GtC.
Atmospheric concentration of methane (CH_4)	700 parts per billion (ppb) for the period 1000–1750 to 1750 ppb in year 2000 (151±25% increase).
Atmospheric concentration of dinitrogen oxide (N_2O)	270 ppb for the period 1000–1750 to 316 ppb in year 2000 (17±5% increase).
Tropospheric concentration of ozone (O_3)	Increased by 35±15% from the years 1750 to 2000, varies with region.
Stratospheric concentration of O_3	Decreased over the years 1970 to 2000, varies with altitude and latitude.
Atmospheric concentration of HFCs, PFCs, and SF_6	Increased globally over the last 50 years.
Weather indicators	
Global mean surface temperature	Increased by 0.6±0.2°C over the twentieth century; land areas warmed more than the oceans (*very likely*).
Northern Hemisphere surface temperature	Increase over the twentieth century greater than during any other century in the last 1000 years; 1990s warmest decade of the millennium (*likely*).
Diurnal surface temperature range	Decreased over the years 1950 to 2000 over land: night-time minimum temperatures increased at twice the rate of daytime maximum temperatures (*likely*).

Number of hot days/heat index	Increased (*likely*).
Number of cold/frost days	Decreased for nearly all land areas during the twentieth century (*very likely*).
Continental precipitation	Increased by 5–10% over the twentieth century in the Northern Hemisphere (*very likely*), although decreased in some regions (e.g., north and west Africa and parts of the Mediterranean).
Number of heavy precipitation events	Increased at mid- and high-northern latitudes (*likely*).
Frequency and severity of drought	Increased summer drying and associated incidence of drought in a few areas (*likely*). In some regions, such as parts of Asia and Africa, the frequency and intensity of droughts have been observed to increase in recent decades.

Biological and physical indicators

Global mean sea level	Increased at an average annual rate of 1 to 2 mm during the twentieth century.
Duration of ice cover of rivers and lakes	Decreased by about 2 weeks over the twentieth century in mid and high latitudes of the Northern Hemisphere (*very likely*).
Arctic sea-ice extent and thickness	Thinned by 40% in recent decades in late summer to early autumn (*likely*) and decreased in extent by 10–15% since the 1950s in spring and summer.
Non-polar glaciers	Widespread retreat during the twentieth century.
Snow cover	Decreased in area by 10% since global observations became available from satellites in the 1960s (*very likely*).
Permafrost	Thawed, warmed, and degraded in parts of the polar, sub-polar, and mountainous regions.
El Niño events	Became more frequent, persistent, and intense during the last 20 to 30 years compared to the previous 100 years.
Growing season	Lengthened by about 1 to 4 days per decade during the last 40 years in the Northern Hemisphere, especially at higher latitudes.
Plant and animal ranges	Shifted poleward and up in elevation for plants, insects, birds, and fish.
Breeding, flowering, and migration	Earlier plant flowering, earlier bird arrival, earlier dates of breeding season, and earlier emergence of insects in the Northern Hemisphere.
Coral reef bleaching	Increased frequency, especially during El Niño events.

Economic indicators

Weather-related economic losses	Global inflation-adjusted losses rose by an order of magnitude over the last 40 years. Part of the observed upward trend is linked to socio-economic factors and part is linked to climate factors.

Notes:

1 This table provides examples of key observed changes and is not an exhaustive list. It includes both changes attributable to anthropogenic climate change and those that may be caused by natural variations or anthropogenic climate change.

2 Confidence levels ('likely', 'very likely' etc.) are reported where they are explicitly assessed by the relevant Working Group. For definitions of confidence levels, see IPCC, 2001a.

3 Terrestrial biospheric CO_2 exchange is the exchange of CO_2 between the atmosphere and the terrestrial biosphere (i.e. the land-based biosphere, excluding the oceans). When the net flow of CO_2 is *from the atmosphere to the land* the terms 'sink' is used; when the flow is *from land to atmosphere* the term used is 'source'. See also Table 13.6 above.

4 HFCs are hydrofluorocarbons; PFCs are perfluorocarbons; SF_6 is sulphur hexafluoride.

5 'El Niño' is a warm water current that periodically flows along the coast of Ecuador and Peru, disrupting the local fishery. It is associated with fluctuations in the circulation of the Indian and Pacific oceans, called the 'Southern Oscillation'. During El Niño events the prevailing trade winds weaken and the equatorial counter-current strengthens, causing warm surface waters in the Indonesian area to overlie the cold waters of the Peru current. Such events have climatic effects throughout the Pacific region and in many other parts of the world. (IPCC, 2001b).

Source: IPCC (2001a), pp. 45–6, Table 2-1

The changes in global weather that are being observed are consistent with the predictions made by sophisticated computer models of the world's climate. These models calculate the effects of increasing greenhouse gas concentrations, taking into account the effects of suspended fine particles, variations in solar activity, positive feedback effects and numerous other factors. Such models predict further increases in global mean temperature during the 21st century of between 1.4–5.8 °C (depending on on the assumptions regarding future emission levels and the climate's sensitivity to them) together with a rise in mean sea levels of between 9 and 88 cm (see Table 13.8) due to thermal expansion of the oceans (see IPCC, 2001a).

The consequences of climate change

A rise in global mean temperature of a few degrees C, coupled with a rise in sea levels of a few tens of centimetres by the end of the 21st century, might seem relatively unproblematic. But these rising average levels do not reflect the increases in *extremes* of temperature, sea levels, rainfall and other weather-related phenomena that are likely to accompany them. Increasing mean temperature means increasing heat energy within the planet's weather system and this is likely to manifest itself in an increasing *frequency* of more extreme weather events. Model projections also suggest that future temperature increases will be greater on land than in the oceans; greater inland than in coastal regions; and greater at high latitudes than at lower latitudes.

Increased temperatures will lead to increased evaporation of water, which will in turn lead to increased precipitation (rain and snow). It is estimated that every degree C of temperature increase will lead to about a 2.5% increase in precipitation. Again, the increases in precipitation are likely to manifest themselves in more extreme events, such as flooding. But paradoxically, despite such increases in rainfall, increased evaporation in summer is likely to lead to droughts in some regions.

Increased evaporation means that many regions are likely to become more humid and this, coupled with increases in temperature, will lead to significant increases in the so-called 'heat index' (an indicator of discomfort) and thus to additional heat-related illnesses and deaths – though in some regions these could be at least partially offset by decreases in winter deaths due to cold.

Melting ice is another major consequence of global warming. Mountain glaciers world-wide appear to be retreating, and there is considerable evidence of ice melting in the Arctic, but this has little impact on global sea levels since Arctic ice is mostly floating on water. However, much of the ice in the Antarctic is land-based. Very substantial global sea level rises would result if the supports that anchor the enormous West Antarctic Ice sheet to land were to melt, causing the sheet gradually to slide into the sea. This, it is estimated, could cause global sea levels to rise by several metres over the next 1000 years (IPCC, 2001).

However, the various phenomena are not fully understood and there is considerable uncertainty as to many (though not all) of the consequences of climate change. This is due in great measure to the **non-linear** nature of the response of the climate systems to changes in temperature and other inputs. To say that a system is *non-linear* means that there is no simple relationship between its inputs and its output: in some circumstances, a small change in inputs might give rise to a large change in output; in other circumstances a large change in input might cause only a small change in output. And this sensitivity of outputs to inputs may also change with time. In a **linear** system, by contrast, there is a an explicit ('analytical') mathematical relationship between any output and the relevant inputs, and if you know this, it is in principle always possible to predict the behaviour of the system. Predicting the behaviour of non-linear systems, by contrast, is notoriously difficult.

Some of the ecological consequences of global warming can be predicted with reasonable confidence; others are highly uncertain. Some species are unlikely to be able to adapt quickly enough to increasing temperatures and may decline or become extinct in some regions. Increasing temperatures are likely to lead to pests and diseases (such as mosquitoes and malaria) that were once confined to the tropics becoming prevalent in previously-temperate regions. Warmer winters in many regions are likely to mean that more pests survive winter than in the past.

Table 13.8 (overleaf) gives a summary by the Intergovernmental Panel on Climate Change (IPCC, 2001) of the likely consequences of climate change during the 21st century, for human health, ecosystems, agriculture and water resources, if current climate policies remain unchanged.

Global warming and climate change are perhaps the most challenging and profound of the many impacts of human energy use on the planet. In the next chapter we shall discuss some of the measures that humanity could take during the coming decades to reduce and sequester emissions of climate-changing greenhouse gases from fossil fuel combustion. However, even if the world is successful in stabilising atmospheric carbon concentrations during the 21st century, global mean temperatures and sea levels will still continue to rise gradually for hundreds of years thereafter, due to the enormous inertia inherent in the global climate system (IPCC, 2001a).

13.8 Accidents and risk

The energy industries are no exception to the rule that fatal accidents are a feature of life. Mining accidents lead to tens of thousands deaths a year world-wide; oil rigs have experienced serious fires; liquefied natural gas tanks have exploded and dams burst, sometimes killing hundreds in a single incident. Air disasters bring similar numbers of fatalities, and world-wide we kill ourselves in our hundreds of thousands on the roads every year. The nuclear power industry has experienced one serious accident, which resulted in 31 immediate or near-immediate fatalities. Nevertheless, it is the thought of a catastrophic accident that is at the centre of much of the public concern about nuclear power. Is there a rational justification for this?

In this section we look at the third of the categories of data identified in Box 13.1: the catastrophes. The difference between these and the 'normal' background of industrial accidents is not that the latter are less serious – any fatality is of course serious – but that the catastrophes are *rare*. Being rare, they don't provide sufficient 'historic' data to calculate annual averages

Table 13.8 Consequences of climate change for human health, ecosystems, agriculture and water resources by the years 2025, 2050 and 2100 if no climate policy interventions are made

	2025	2050	2100
CO_2 concentration	405–460 ppm	445–640 ppm	540–970 ppm
Global mean temperature change from the year 1990	0.4–1.1 °C	0.8–2.6 °C	1.4–5.8 °C
Global mean sea-level rise from the year 1990	3–14 cm	5–32 cm	9–88 cm
Human health effects			
Heat stress and winter mortality	Increase in heat-related deaths and illness (*high confidence*). Decrease in winter deaths in some temperature regions (*high confidence*).	Thermal stress effects amplified (*high confidence*).	Thermal stress effects amplified (*high confidence*).
Vector- and water-borne diseases		Expansion of areas of potential transmission of malaria and dengue (*medium to high confidence*).	Further expansion of areas of potential transmission (*medium to high confidence*).
Floods and storms	Increase in deaths, injuries, and infections associated with extreme weather (*medium confidence*).	Greater increases in deaths, injuries, and infections (*medium confidence*).	Greater increases in deaths, injuries, and infections (*medium confidence*).
Nutrition	Poor are vulnerable to increased risk of hunger, but state of science very incomplete	Poor remain vulnerable to increased risk of hunger.	Poor remain vulnerable to increased risk of hunger.
Ecosystems effects			
Corals	Increase in frequency of coral bleaching and death of corals (*high confidence*).	More extensive coral bleaching and death (*high confidence*).	More extensive coral bleaching and death (*high confidence*). Reduced species biodiversity and fish yields from reefs (*medium confidence*).
Coastal wetlands and shorelines	Loss of some coastal wetlands to sea-level rise (*medium confidence*). Increased erosion of shorelines (*medium confidence*).	More extensive loss of coastal wetlands (*medium confidence*). Further erosion of shorelines (*medium confidence*).	Further loss of coastal wetlands (*medium confidence*). Further erosion of shorelines (*medium confidence*).
Terrestrial ecosystems	Lengthening of growing season in mid and high latitudes; shifts in ranges of plant and animal species (*high confidence*). Increase in net primary productivity of many mid- and high-latitude forests (*medium confidence*). Increase in frequency of ecosystem disturbance by fire and insect pests (*high confidence*).	Extinction of some endangered species; many others pushed closer to extinction (*high confidence*). Increase in net primary productivity may or may not continue. Increase in frequency of ecosystem disturbance by fire and insect pests (*high confidence*).	Loss of unique habitats and their endemic species (e.g., Cape region of South Africa and some cloud forests) (*medium confidence*). Increase in frequency of ecosystem disturbance by fire and insect pests (*high confidence*).
Ice environments	Retreat of glaciers, decreased sea-ice extent, thawing of some permafrost, longer ice-free seasons on rivers and lakes (*high confidence*).	Extensive Arctic sea-ice reduction, benefiting shipping but harming wildlife (e.g., seals, polar bears, walrus) (*medium confidence*). Ground subsidence leading to infrastructure damage (*high confidence*).	Substantial loss of ice volume from glaciers, particularly tropical glaciers (*high confidence*).

Agricultural effects

Average crop yields	Cereal crop yields increase in many mid- and high-latitude regions (*low to medium confidence*). Cereal crop yields decrease in most tropical and subtropical regions (*low to medium confidence*).	Mixed effects on cereal yields in mid-latitude regions. More pronounced cereal yield decreases in tropical and subtropical regions (*low to medium confidence*).	General reduction in cereal yields in most mid-latitude regions for warming of more than a few °C (*low to medium confidence*).
Extreme low and high temperatures	Reduced frost damage to some crops (*high confidence*). Increased heat stress damage to some crops (*high confidence*). Increased heat stress in livestock (*high confidence*).	Effects of changes in extreme temperatures amplified (*high confidence*).	Effects of changes in extreme temperatures amplified (*high confidence*).
Incomes and prices		Incomes of poor farmers in developing countries decrease (*low to medium confidence*).	Food prices increase relative to projections that exclude climate change (*low to medium confidence*).

Water resource effects

Water supply	Peak river flow shifts from spring toward winter in basins where snowfall is an important source of water (*high confidence*).	Water supply decreased in many water-stressed countries, increased in some other water-stressed countries (*high confidence*).	Water supply effects amplified (*high confidence*).
Water quality	Water quality degraded by higher temperatures. Water quality changes modified by changes in water flow volume. Increase in saltwater intrusion into coastal aquifers due to sea-level rise (*medium confidence*).	Water quality degraded by higher temperatures (*high confidence*[*]). Water quality changes modified by changes in water flow volume (*high confidence*).	Water quality effects amplified (*high confidence*).
Water demand	Water demand for irrigation will respond to changes in climate; higher temperatures will tend to increase demand (*high confidence*).	Water demand effects amplified (*high confidence*).	Water demand effects amplified (*high confidence*).
Extreme events	Increased flood damage due to more intense precipitation events (*high confidence*). Increased drought frequency (*high confidence*).	Further increase in flood damage (*high confidence*). Further increase in drought events and their impacts.	Flood damage several-fold higher than 'no climate change scenarios.'

Other market sector effects

Energy	Decreased energy demand for heating buildings (*high confidence*). Increased energy demand for cooling buildings (*high confidence*).	Energy demand effects amplified (*high confidence*).	Energy demand effects amplified (*high confidence*).
Financial sector		Increased insurance prices and reduced insurance availability (*high confidence*).	Effects on financial sector amplified.
Aggregate market effects	Net market sector losses in many developing countries (*low confidence*). Mixture of market gains and losses in developed countries (*low confidence*).	Losses in developing countries amplified (*medium confidence*). Gains diminished and losses amplified in developed countries (*medium confidence*).	Losses in developing countries amplified (*medium confidence*). Net market sector losses in developed countries from warming of more than a few °C (*medium confidence*).

Judgments of confidence use the following scale: *very high* (95% or greater), *high* (67–95%), *medium* (33–67%), *low* (5–33%), and *very low* (5% or less).
Source: IPCC (2001a), Tables 3-1, 3-2, 3-3 and 3-4, pp. 69–72. (See original Table footnotes for further details.)

of the type that have appeared in the graphs and tables in earlier sections of this chapter. If they are to be taken into account, other methods must be used to assess their likelihood and their likely consequences. Questions about the past health or environmental effects per year or per kilowatt-hour must be replaced by questions about the assessment of probabilities of these effects in the future.

We concentrate on accidents in the nuclear industry, in part because this field attracts the most public attention but also because it has been subject to the most detailed analysis. It is not the intention here to assess the probabilities of specific accidents, or their consequences if they do occur: space does not permit the necessary detail. But what we can do is to ask in general terms how probabilities such as that quoted in Box 13.8 are obtained. In other words, we play the role of person **D**. We then look at some results of these methods, and compare them with the few available data.

BOX 13.8 **Not in my back yard**

It is proposed to build a nuclear power station near your home. When you express your concern about the possibility of a serious accident, you are informed that the chances are one in ten million per year of an accident serious enough to cause perhaps ten immediate deaths and 10 000 ultimate deaths from cancer. How do you react?

A　That would be terrible. I'd rather live without electricity.

B　So what? In ten million years I'll be dead anyway.

C　£5000 to a penny that it'll survive ten years.

D　How do they know?

Probabilities and consequences

We start with the obvious point that probabilities are not certainties. The chances are 1 in 6 that your dice will show a 2; but this doesn't mean that every sixth throw will be a 2, nor even that a 2 will turn up once in every six throws. It is perfectly possible that you get no 2s at all in a dozen throws, or that you throw half a dozen in succession. In this sense, respondent **B** of Box 13.8 shows a poor appreciation of probabilities, whilst the gambler **C** understands what they mean (and evidently hopes to make a profit).

How is a figure obtained for the probability of a certain type of accident? One method is to use past history where it is available: 'How frequently has this type of switch failed?' The probability of different accidents can then be assessed — or reduced at the design stage — by accumulation of such information on all the component parts of a system. But history may not repeat itself. One minor reactor accident in the USA was complicated by switches which stuck, a fault traced to a previously reliable supplier who had taken on new, inexperienced workers. On the whole, however, the approach has proved sufficiently reliable to be used routinely for many years by designers of complex systems throughout industry.

Risk assessment like this has moved from the design office to the public domain largely with the rising concern about reactor accidents; indeed it could be seen as a counter to the response of person **A** in Box 13.8. As people became aware of the appalling consequences of the worst conceivable accident, it became necessary to point out (or to find out, according to your view) just how small the probability was that such an accident would occur. A rational person, it is argued, will reach decisions on a basis of *probability multiplied by consequences*. If you accept a situation with a one in a thousand chance of 5 fatalities a year, then you should accept a one in a million chance of 5000 fatalities, and so on. And similar reasoning should determine your choice between alternatives.

Accidents in complex systems are caused by combinations of circumstances, so we need combinations of probabilities to assess their likelihood. If one in every 500 switches is faulty, and one in every 200 indicator lights, then the probability that a switch and its indicator light both fail should be one in 500 times 200, or 100 000. If we know the individual probabilities we can calculate overall probabilities for complex sets of events – like a switch failing to open *and* a valve failing to close *and* ....

The reasoning can run either way. You can say that, in order for the reactor core to melt, the following *and* the following *and* the following must all happen, and then calculate the probability of this conjunction. Or you can ask what happens if X fails, and in addition Y fails, and so on, and in this way find in principle the probabilities and consequences of all possible accidents.

There are complications of course. If an electrical surge caused both the switch and the indicator light to fail, the result would be the unfortunate consequence of what is known as a **common mode failure**. Allowing for common mode failures is essential in assessing risk, and eliminating them is an important aspect of design. The best-known example of what *not* to do is probably the 1975 boiling water reactor accident at Brown's Ferry in Alabama, USA, where a relatively small fire caused by technicians using a candle to test for leaks in one air duct affected hundreds of cables and put out all the carefully planned emergency cooling systems.

Two other difficulties are 'unknowns' and people. There can be parts of the system for which probabilities based on past history are not available. Risks must then be assessed by calculation, by analogy with similar cases, and where possible by experiment. The behaviour of the massive steel pressure vessel of a pressurised water reactor under the stresses which might result from an accident, and the behaviour of emergency cooling water as it hits the hot core, are examples where there is little history to go on, and this has led to controversy over both.

People present similar problems. Brown's Ferry is only one of many cases where human ingenuity eventually *prevented* serious consequences, but the incident at Three Mile Island seems initially to have been the result of a determined effort by all present to counter the actions of the safety systems, and the Chernobyl catastrophe was also in part the result of human error (see Chapters 10 and 11 and *The Evidence*, below). It should perhaps be added that the TMI engineers were in a control room with at one moment *over a hundred* alarms sounding and lights flashing. Like those at Chernobyl who lost their lives or subsequently put themselves at severe risk, they were heroes to be there at all. Any risk assessment that ignores the human factor will obviously be of little value, yet to give figures for the chances that people will behave in certain ways clearly introduces another spectrum of uncertainties.

Despite all these difficulties, estimates are made. The Rasmussen Report (Rasmussen, 1975) remains one of the most comprehensive risk assessment exercises for any major part of the nuclear industry. It comprises three years' work, at a cost of three million dollars, producing a multi-volume analysis of the probabilities and consequences of accidents in light-water reactors (PWRs and BWRs). Its conclusions have been much criticized for the degree of certainty they claim, and the Report and its successors have been attacked on the more general ground that most major industrial accidents have been caused by totally unforeseen factors rather than combinations of known possibilities. Nevertheless it may be worth looking at a few of the Rasmussen figures, if only to obtain an idea of its assessments of the orders of magnitude of the risks and to compare these with actual events.

Table 13.9 Estimates of accident probabilities and consequences

Event	Activity released (Tbq)	Total dose to population (person-sieverts)	Consequent number of deaths		Probable number of events per million reactor-years	Predicted number of deaths per thousand reactor-years
			immediate	delayed		
Meltdown without major breach of containment	0–500	0–1000	0	0–10	10–100	0–1
Meltdown with breach of containment, under average conditions	500 000–5 million	0.1–1 million	1–10	1000–10 000	0.1–10	0.1–100
Meltdown with breach of containment under worst conditions	500 000–5 million	1–10 million	1000–10 000	10 000–100 000	0.001–0.01	0.001–1

See the main text for further explanation and comment on these figures

Source: Ramage, 1997, adapted from Rasmussen, 1975.

Table 13.9 shows simplified versions of the probability-times-consequences conclusions for just three potential accidents: a core meltdown with little or no release of radioactivity, a meltdown with an explosion violent enough to breach the containment, and the second of these again but with the reactor in a densely populated area and weather conditions that maximize the effect, exposing a population of 10 million or so. The ranges in the third column allow for considerable uncertainties about the worker and population doses that would result from each event. The next two columns show the effects expected from these doses. (Column 5 follows from Column 3 and the data in Box 13.6, with the assumption that all induced cancers lead to premature death.) The figures in the sixth column are determined by combining the probabilities of many events to give a range of probabilities for each accident, and the final column shows the result of multiplying probabilities by consequences. It should be noted that this is just part of a very much larger set of conclusions – and that this analysis is for essentially one type of reactor.

The evidence

Fortunately the number of serious reactor accidents has been far too small for any statistically valid test of these estimates, but it may be worth looking briefly at three accidents already mentioned in earlier chapters – noting however that only one was in fact in a light-water reactor. The first is the 1957 fire in the graphite moderator of the Windscale reactor. The containment was not breached in the usual sense but, as described in Chapter 11, there was a release of radioactive gases including some 1000 TBq of iodine-131. The accident was initially claimed not to have caused a single death. If the relationship between activity released and consequences shown in Table 13.9 is at all correct, this seems statistically unlikely. In the late 1980s the official view changed, with the statement that the event probably led to a few tens of deaths from cancer – far too small a number, of course, to be detected in the mortality statistics, but now in rough agreement with the consequences of such a release as estimated by Rasmussen (see also Arnold, 1992 and McSorley, 1990).

What about the probability of the event? An event with similar release of radioactivity is given a probability of 10–100 per million reactor-years, and Windscale had been in operation for about one year. Does this tell us anything? Yes, it tells us that the chances are greater than zero. Otherwise extremely little; not because this reactor wasn't a PWR but because a single event says very little else about the statistical chances, one way or the other.

Iodine-131 features largely in concerns about reactors. As we have seen in Box 13.6, it is an isotope that easily enters the food chain and that presents particular health hazards. But it was shown there that routine I-131 emissions from a nuclear plant, of the order of a million becquerels a day, are considered unlikely to cause detectable harm. However, the total I-131 content of a reactor can be exabecquerels: several million, million times these routine amounts; and it is the possibility of an appreciable fraction of this escaping that causes concern.

Iodine certainly featured in the accident at Three Mile Island, mentioned in the earlier chapters. Here a combination of improbable events – a switch failing to open, a valve failing to close, misinterpretation of meter readings – resulted in a chemical explosion within the reactor. To the credit of the designers, the containment was not breached, but about 1 TBq of I-131 was released, less than a millionth of the total content. Although this was a million times the daily norm, the estimates in Table 13.9 suggest that it should not have caused even a single additional cancer or death. TMI is however an interesting case when we look at probabilities. One accident like this per 300–30 000 reactor-years was the prediction, and by 1979 these PWRs had accumulated a total of about 400 reactor-years. We cannot however agree with the statement (in a major energy study, surprisingly) that 'the range has proved to be correct'. The fact that your sixth throw of the dice is a 2 does *not* confirm the view that 1 throw in 6 will be a 2.

The most serious accident known to date in a nuclear power plant was at Chernobyl in 1986 (see Chapters 10 and 11). The 'lid' of the reactor was blown out and fell back tilted, leaving the interior exposed, and about 6% of the core content was released. (The floor, blown downwards, was eventually found four metres lower, in the sub-structure.) The 31 people who received lethal radiation doses in the first few days included pilots who flew helicopters low over the open reactor in an attempt to drop neutron absorbers into the core; a useless action, as almost no fuel remained there – but this wasn't known until later.

The radioactivity of 6% of the core of an operating reactor is of the order of tens of millions of TBq, perhaps ten times the largest release shown in Table 13.9. Extrapolating the data there would suggest a number of immediate deaths in the range 10–100, which was indeed the case. As mentioned above, the Chernobyl reactor was not a PWR, and there are doubts about the Rasmussen estimates – and further doubts about extrapolating them. Nevertheless, this agreement suggests that we might look at the 'delayed deaths' column. Similar extrapolation would predict a total of between 10 000 and 100 000 premature deaths of workers or the public. It is interesting to compare these figures with the estimates of delayed deaths in Box 11.1 of Chapter 11, which range from about 5000 to 40 000. It is also worth comparing them with the estimates of fatalities due to fossil-fuelled electricity generation given in Tables 13.4 and 13.5 above.

The responses

We have now looked at the methodology behind the statistics, playing the role of person **D** in Box 13.8, who asked: 'how do they know?' Are we in a better position to assess the other three responses offered there?

We saw in an earlier section that routine operation of nuclear plants is estimated to account for about 6 of the 100 000 deaths a year from cancer in the UK. If we take the 'worst case' from Table 13.9, with the present number of reactors, accidents might add another 1 or 2 – on average.

And there's the rub. When all the sums are done, how should we respond to probabilities? Is one of the first three responses in Box 13.8 the 'correct' one?

- If the rational response is to compare probability-times-consequences for different alternatives, then as far as routine emissions and reactor accidents are concerned, the figures – if even remotely reliable – place nuclear power amongst the safest energy sources, and the gambler **C** is on good ground.

- But perhaps this response is not appropriate for low-probability but potentially large-scale catastrophes? It has been argued that, as individuals or societies, we normally don't bother at all about risks whose chances are less than perhaps one in a million a year; that we do in practice adopt the light-hearted approach of person **B**.

- On the other hand, there are those who argue for response **A**: that if the consequences are sufficiently awful we shouldn't accept the risk at all, no matter how low the probability.

The answer may not be a simple one. It is arguable that the degree of risk we accept depends on the degree of control that we have as individuals. (Despite the accident statistics, we may prefer driving to flying.) And if the risk is imposed, our criteria may be even more severe: we may reject an involuntary and imposed risk but accept an appreciably higher one that is voluntary.

To dismiss any of these views as 'not rational' seems, to us at least, neither justifiable nor particularly useful.

13.9 Comparing the impacts of electricity generating systems

We have reviewed some of the many and varied impacts of our energy use on our environment and on human health.

How might these various impacts be compared quantitatively? Although, as stated at the beginning of this chapter, there is no *overall* methodology that allows all aspects to be measured and compared objectively, there are a number of partial measures of the impacts of energy systems that can be compared. These give some insight into the relative merits and demerits of different energy sources. In this section, we concentrate on electricity, and compare various generating systems in terms of (a) their carbon emissions, (b) their SO_2 and NO_x emissions (c) their land use implications, and (d) their energy payback time.

Comparing carbon emissions

In Table 13.10, Holdren and Smith summarise the results of a variety of estimates of greenhouse gas emissions from electric power generation.

Table 13.10 Greenhouse gas emission intensities for selected fuels (grams of carbon dioxide equivalent per kilowatt-hour)

Conventional coal	Advanced coal	Oil	Gas	Nuclear	Biomass	Photovoltaic	Hydroelectric	Wind
960–1300	800–850	690–870	460–1230[a]	9–100	37–166[a]	30–150	2–410	11–75

Note. These estimates encompass a range of technologies and countries as described in Pearce and Bann, 1992; Fritsche, 1992; Yasukawa and other, 1992; ORNL and RFF, 1992–9; Gagnon and de Vate, 1997; and Rogner and Kahn, 1998.

a Natural gas and biomass fuel cycles were also analysed in cogeneration configurations, with the heat produced credited for displacing greenhouse gas emissions from gas heating systems. That approach reduced greenhouse gas emissions to 220 grams of carbon dioxide equivalent per kilowatt-hour for natural gas, and to minus 400 grams of carbon dioxide equivalent per kilowatt-hour for biomass (Fritsche, 1992) Other cycles could incorporate cogeneration and be analysed in this manner.

Source: Holdren and Smith, 2000, Box 3.10, p. 103.

As Table 13.10 demonstrates, the range of estimates of emissions is very wide. Holdren and Smith explain the variations as follows:

> Some of the variability arises from the different conversion efficiencies of the technologies assessed – for example, biomass configurations include a wood steam boiler, an atmospheric fluidized bed combustor and an integrated gasifier combined cycle turbine. But methodological issues and assumptions associated with activities outside the generation stage account for a large portion of the variability. For example one study credits product heat from cogeneration cycles for displacing greenhouse gases from gas heating systems (Fritsche, 1992). In this framework, the greenhouse gas intensity of biomass can become negative, and that of natural gas cycles can be reduced 50% below the next lowest estimate.

Among fossil fuels, the greenhouse gas intensity of natural gas is most variable, primarily due to different assumptions about methane emissions during drilling, processing and transport.

For non-fossil fuels, estimates generally span at least an order of magnitude, primarily because of the sensitivity of these cycles to assumptions on the operational life of the facility and the greenhouse gas intensity of the electric and manufacturing sectors on which equipment production depends. In addition, the hydroelectric cycle is sensitive to the area of land flooded and, for projects with multiple generating units per reservoir, the boundary of the system considered.

Holdren, J.P. and Smith, K.R. (2000)

Comparing SO_2 and NO_x emissions

The **acid emissions** issue was discussed in Section 13.5. Box 13.9 gives estimates, from a study for the International Energy Agency by the Canadian utility Hydro-Québec, of the levels of sulphur dioxide and NO_x, the main precursors of acid rain, emitted from a variety of electricity generating systems (Gagnon *et al.*, 2002). As can be seen from Figures 13.8 and 13.9, the Hydro-Québec study distinguishes between three types of generating system: (a) sources that are suitable for both base-load and peak-load operation, (b) sources with limited flexibility that are suited to base-load operation, and (c) sources that generate intermittently and need a backup supply.

Land use

Another way to assess the merits of energy systems on a quantitative basis is to compare their land usage. Given that there are competing uses for land, including obviously food production, but also increasingly housing, industry and leisure, the land use requirements of energy systems could become an important issue. The results of Hydro-Québec's study of the land use requirements of various electricity generating systems are shown in Box 13.10.

As can be seen from Figure 13.10, biomass plantations emerge as the most land-intensive energy option. Energy crop growing may also have serious implications for biodiversity, depending of the approach adopted. These issues are examined in more detail in the companion volume *Renewable Energy*.

BOX 13.9 Comparing the precursors of acid precipitation: SO_2 and NO_x emissions of electricity generating systems

Figures 13.8 and 13.9 (overleaf) present the results of studies concerning the two major precursors of acid precipitation:

- The main precursor is SO_2, which leads to the formation of sulphuric acid.

- The other precursor is NO_x, which leads to the formation of nitric acid. Before contributing to the formation of acid precipitation, NO_x can also be involved in other chemical reactions with volatile organic compounds (VOCs), causing smog.

Notes

In the case of forest productivity, impacts of pollutants are numerous and sometimes indirect (Godish, pp. 108–12):

- Acids will tend to remove some essential nutrients from soils (potassium, calcium, magnesium).

- Acids may mobilize toxic metals such as aluminum, which can damage roots.

- Adding nitrogen, the main nutrient of plants, may create an imbalance in resources and make trees more vulnerable to diseases and frost.

Impacts of other atmospheric pollution must be also considered:

- Photochemical smog can damage the leaves.

- Climate change may increase heat stress or intensity of droughts.

When looking at Figure 13.8, the reader should keep in mind that SO_2 emissions can vary significantly, according to the following factors for each fossil fuel:

- For coal, the sulphur content can vary from 0.5% to 5% and even more in exceptional cases.

- For oil, average sulphur content in light oil/diesel is about 0.2% and 2% for heavy oil, but these percentages can vary significantly from one region to another.

- Commercial natural gas has virtually no sulphur, because it is removed in processing plants after extraction. Depending upon sulphur concentrations and regulations, this process can create high or low SO_2 emissions.

- There is a wide variety of technologies to reduce emissions at plant, with differing performance. Some commercial scrubbing technologies that are currently available are capable of removing about 90% of SO_2 emissions. But these technologies have not been widely implemented.

It is therefore normal that studies arrive at a wide range of results.

Understanding the results of studies on NO_x

Studies on NO_x emissions can also arrive at a wide range of results, but these variations are more dependent upon combustion technology than on fuel:

- Most NO_x emissions are caused by the fact that oxygen is required for any form of combustion and that the main source of oxygen is ambient air, which is composed of 79% nitrogen (N). Therefore, the conditions of combustion are the main determinant in the level of NO_x emissions.

- Technologies that involve compression of air, such as diesel engines, will normally produce high levels of NO_x emissions.

- In the case of coal, significant amounts of nitrogen are also part of the fuel, thereby increasing NO_x emission factors.

Main findings concerning acid precipitation

Emission factors for hydropower and nuclear energy are hundreds of times less than those of coal based power generation systems without scrubbing. Considering both SO_2 and NO_x, coal, oil and diesel based generation systems are important contributors to acid precipitation.

Biomass has a low emissions factor for SO_2 but a very high factor for NO_x. It is therefore a significant source of acid precipitation. Natural gas, when considering the processing of fuel and NO_x emissions, can also be a significant source of acid precipitation.

The benefits of wind power are dependent upon network conditions and more difficult to assess. If wind power reduces the use of oil fired plants (which themselves can compensate for wind fluctuations), this would result in a reduction in net emissions; however, in some cases, wind power may increase the use of oil-fired plants (as backup)[*].

Source: Gagnon *et al.*, 2002

[*] Note: the issue of the backup requirements of wind power and other intermittent electricity generating systems is a complex one, addressed more fully in the companion volume *Renewable Energy*.

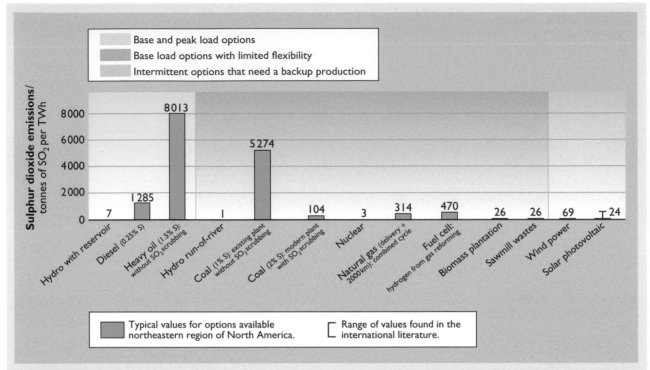

Figure 13.8 SO$_2$ emissions of energy options

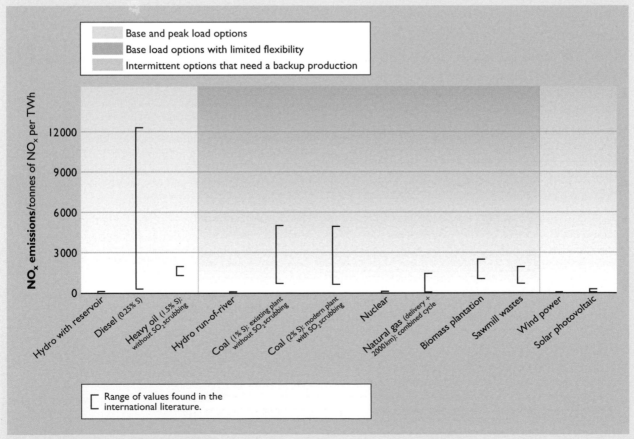

Figure 13.9 NO$_x$ emissions of energy options

BOX 13.10 Comparing the land requirements of electricity generation systems

All electricity generation systems affect land use, either directly or indirectly.

Examples include:

- For hydropower, the transformation of forests/land into aquatic ecosystems.
- For coal, the large areas affected by mining activities and fallout of acid rain.
- For biomass, the area of forests that are exploited.

This issue will gain in importance in the future for many reasons:

- With population growth, more land is required for farms, cities and industries and land for other uses is becoming scarce.
- Emerging renewable sources of energy, such as wind or solar power, require large land areas.
- Alternative sources of biofuels, such as ethanol from crops, require large areas of farmland.

Understanding the results of studies

This type of assessment must be considered with caution, because it does not consider the intensity of the impact, nor the degree of compatibility of generation options with other land use. Moreover, the data in Figure 13.10 considers only the direct use of land. It does not consider indirect impacts, such as losses related to climate change (e.g. losses of land due to increase in sea levels).

The results for hydropower vary significantly because of site-specific conditions. The figures shown are for projects designed mainly for power generation. In many countries, such as the United States, most reservoirs were created for purposes of irrigation and water supply. Many of these reservoirs involve very little power generation and would have even higher land use factors per TWh.

For fossil fuels, very little data exists and some upstream activities are not considered. For example, surface mining of coal would require much more land than underground mining, but the data does not allow for such distinctions.

Main findings concerning land requirements

Nuclear energy has the lowest land requirements, if we do not consider the land required for long-term waste

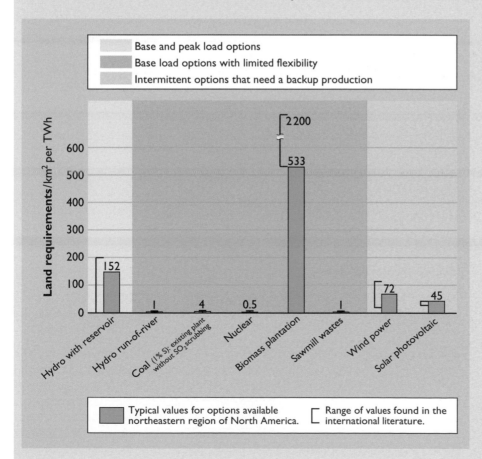

Figure 13.10 Direct land requirements of energy options

disposal. The inclusion of this use of land would seriously increase the land requirements, because a small area of land is needed, but for many thousands of years. For example, if 0.1 km² per TWh is required for waste disposal, multiplied by 30 000 years, applied to 30 years of generation, the factor would increase from 0.5 km² per TWh per year to 100 km² per TWh per year.

Biomass plantations require the most land per unit of energy produced. Other renewable sources (hydropower, wind power and solar power) have similar high land requirements, which can vary significantly according to site-specific conditions. Data on hydropower is based on the total area of reservoirs, not the flooded areas, which would necessarily be smaller (before impoundment, rivers and lakes already occupy a significant area).

Available data concerning fossil fuels show that they require much less land than any renewable source of energy. But this is an assessment based on direct land requirements only (of power plants and mining activities). Indirect 'use' of land, related to fallout of atmospheric emissions or related to the impacts of climate change, is not included in the data. These areas are huge and can multiply the land 'use' factors of fossil fuels.

To show the order of magnitude of these indirect land 'uses', it is possible to calculate a factor of land modified by acid rain. In North America, even after Canadian and American SO_2 reduction plans are implemented in 2010, 800 000 square kilometres in Canada will still receive more than the critical load of 8 to 20 kg of sulphates per hectare per year (Environment Canada, 1998). Above this level, aquatic and forest productivity are seriously reduced.

If we consider that all coal-fired plants in the region (producing 500 TWh per yr) are responsible for this excess, it means that each TWh of coal generation reduces ecosystems productivity over 1600 km². This is ten times the highest estimate of the area used by hydro projects for the same power generation.

Future performance of energy systems

For all options, it is unlikely that technological development will lead to serious reductions in land use. For coal and oil, it would be possible to reduce areas affected by acid precipitation, but no new technology is required to achieve that.

At first sight, future development of renewable sources could be constrained by land requirements. But this issue is specific to each project, because compatibility with other land uses will differ widely. The constraints will depend on many factors:

- population density;
- compatibility of a project with other land use such as recreation, forestry or agriculture;
- and for hydropower, the need for other uses of water, such as irrigation, water supply or flood control.

One key issue will be competition with food production, but most renewable projects have little negative impacts on agriculture:

- For windpower, the land around the windmills can still be used for agriculture.
- Solar energy can be developed on rooftops or arid areas without agriculture.
- Hydropower can be developed in mountainous or cold climate areas, and hydro reservoirs can improve food production through irrigation.

One future energy option could however be severely limited by competing land uses: biomass plantations for energy production (either by direct burning of biomass or by conversion to fuels such as ethanol). In this case, there is direct competition with farmland.

Source: Gagnon *et al.*, 2002

Energy payback ratios

Another important way of comparing power plants is in terms of their **energy payback ratio**, that is, the ratio of the energy generated over the lifetime of the plant to the energy required to construct and maintain it.

The Hydro-Québec study described in Box 13.11 below suggests that the overall energy payback ratio for nuclear power, for example, is 16, compared to 80 for wind. This means that the energy produced by a nuclear power plant over its lifetime is about 16 times the energy used in its construction and in the associated production of materials. Wind turbines produce about 80 times as much energy over their lifetime as is required to construct them. The energy payback ratio for coal plants is 5 to 7, depending on whether SO_2 scrubbing equipment is included, and for combined cycle gas turbine plant the ratio is 5, though this includes gas transmission over 2000 km. For hydro dams the ratio is 205 – presumably because of the long lifetimes and large outputs of hydro plant. PV solar and biomass plantations have much lower payback ratios, 9 and 5 respectively, reflecting the relatively large energy debt incurred in PV cell manufacture and the mechanical energy used in the harvesting and transportation of energy crops. However, in the case of PV, more advanced manufacturing methods, which are less materials- and energy-intensive, are likely to result in more favourable energy ratios for photovoltaics in future.

BOX 13.11 Comparing the energy payback ratios of electricity generating systems

For each power generation system, the 'energy payback' ratio is the ratio of energy produced during its normal life span, divided by the energy required to build, maintain and fuel the generation equipment.

If a system has a low payback ratio, it means that much energy is required to maintain it and this energy is likely to produce many environmental impacts. For fossil fuels, it means environmental impacts from the extraction, transportation and processing of fuels. For renewable sources, it means environmental impacts from building the facility. If an option has a ratio of close to one, it means it consumes almost as much energy as it produces, so it should not be developed.

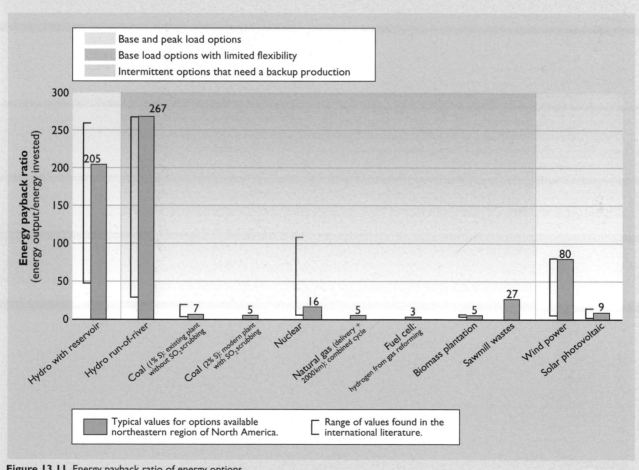

Figure 13.11 Energy payback ratio of energy options

Understanding the results of studies

The data in Figure 13.11 shows that payback ratios do not vary much for fossil fuels but vary significantly for renewable energies. This is due to variable site conditions (topography for hydro, quality of the wind, intensity of solar radiation for solar energy).

Life-cycle Assessment

Some studies on energy options consider emissions at the power plant only, while the values cited here are based on a life-cycle assessment (LCA).

For each energy option, the whole energy system is studied.

For fossil fuel or nuclear systems this includes: exploration and resource extraction, preparation, transportation, storage of fuel, construction of plant, combustion, maintenance and dismantling.

For renewable systems, this includes: construction of plant, maintenance and dismantling.

Main findings concerning 'energy payback ratios'

Hydropower clearly has the highest performance, with energy payback ratios of 205 and 267. The chosen typical figures, 205 for hydro with reservoir and 267 for run-of-river, are for projects in Québec, assessed over a period of 100 years.

At the other extreme, coal has a ratio of 5 to 7 and the ratio for combined cycle gas turbines (including long-distance gas transmission) is 5. The figure for nuclear is 16. In between these extremes, as Figure 13.11 shows, wind power, for the best wind sites, has a good performance ratio, 80). However, this ratio is overestimated because the calculations did not consider the need for backup capacity to compensate wind fluctuations.

Biomass also has a good performance (ratio of 27) when electricity is produced from forestry wastes. But when trees are planted for the purpose of producing electricity, the ratio is much lower (about 5), because the exploitation of these biomass plantations requires many energy inputs. For these two biomass options, the distance between the source of biomass and the power plant must be short, otherwise the energy payback ratio drops to very low values.

Future performance of energy systems

Fossil fuels already have low energy payback ratios and these will be declining over the next decades. This is due to multiple factors:

- As the best fossil reserves are depleted, they tend to be replaced by wells that require a higher energy investment (located in far away regions or under the sea).

- Other considerations may involve selecting resources that are located at greater distances. For example, transportation of coal by train in the US has increased in the last decade because users tend to select Western low-sulphur coal.

- In the future, there will be more energy wasted in fossil-fuelled power plants to reduce emissions. Scrubbing to remove sulphur reduces the efficiency of a plant. If capture and sequestration of CO_2 becomes commercially available (see Chapter 14), this will involve spending substantial amounts of energy in the operation of CO_2 scrubbing and disposal equipment.

- The only fossil fuel that should maintain its performance is natural gas, because the above negative factors are likely to be offset by more efficient combustion turbines.

Source: adapted from Gagnon *et al.*, 2002

13.10 Comparing the external costs of electricity generating systems

In Chapter 12 we looked at the monetary costs of energy, measured in conventional economic terms. We mentioned that attempts have been made to calculate the so-called 'external' social and environmental costs of energy use – that is, those additional costs to the environment and society as a whole that are not fully reflected in the conventional market prices of the energy from various sources. These extra costs are sometimes called 'environmental adders', and Table 13.11 summarizes the results of one attempt to calculate them (Pearce *et al.*, 1992). Some of the difficulties entailed in such calculations are discussed later in this section.

Table 13.11 Illustrative estimates of the environmental external costs (in pence per kilowatt-hour) for electricity production from selected energy sources

Cost category	Old coal	New coal	Oil	Gas	Nuclear	Solar	Wind	Hydro
Health								
– Mortality	0.32	0.32	0.29	0.02	0.01	0.07	0.04	0.03
– Morbidity	0.12	0.12	0.12	0.04	0.01	0	0	0
– Disaster	NE	NE	NE	NE	0.45	0	0	0
Crop damage	0.10	0.05	0.05	0.02	0	0	0	0
Damage to forests	0.84	0.07	0.98	0.03	0	0	0	0
Reduction of biological diversity	NE	NE	NE	NE	NE	NE	NE	NE
Damage to buildings	3.22	0.28	3.77	0.11	0	0	0	0
Noise	NE	NE	NE	NE	NE	NE	NE	NE
Global warming damage	0.40	0.34	0.35	0.16	0.01	0	0	0.01
Visibility impact	NE	NE	NE	NE	NE	NE	NE	NE
Water pollution	0.40	0.04	0.049	0.01	0	0	0	0
Land contamination	NE	NE	NE	NE	NE	NE	NE	NE
TOTAL	5.40	1.22	6.05	0.39	0.48	0.07	0.04	0.04

NE: not estimated but probably positive.

(Source: Adapted from Pearce *et al.*, 1992)

In Table 13.12 (overleaf), the results of number of such studies are summarized. As can be seen, there is a wide variation in the estimated external costs, due to the wide range of possible assumptions upon which they are based.

Another indication of the widely-varying estimates of the external costs of electricity production can be seen from Table 13.13, complied by the US Environmental Protection Agency in 1992.

A more recent report, from the European Union's 'EXTERNE' (Externalities of Energy) study, published in 2001, has attempted to put a price on the environmental and social impacts of electricity production in the EU. The

Table 13.12 Assessment of external costs by different studies

Author	Fossil fuels	Nuclear
Hohmeyer	0.8–3p kWh^{-1}[a]	1.3–6.7p kWh^{-1}
Friedrich and Voss	0.17p kWh^{-1} [b]	0.02–0.03p kWh^{-1}
Ottinger	Coal 1.5–3.4p kWh^{-1} [c]	
	Oil 1.5–3.9p kWh^{-1}	
	Gas 0.5–0.6p kWh^{-1}	
Stocker et al.	Coal 2p kWh^{-1} [d]	
	Gas 1p kWh^{-1}	
	Combined cycle gas 0.75p kWh^{-1}	
Hagen and Kaneff	Combined cycle coal 0.6p kWh^{-1} [e]	2.9p kWh^{-1}
Ferguson	(Acid pollution) £0.02–£2 kWh^{-1}	0.2–5p kWh^{-1} [f]
	(Global warming) £0.1–£10 kWh^{-1} [g]	
Koomey	0.9–2.4p kWh^{-1} [h]	

Sources:

Hohmeyer, O. (1988) *The Social Costs of Energy Consumption*, Springer.

Hohmeyer, O. (1990) 'Latest results of the international discussion on the social costs of energy – how does wind compare today?', *Proceedings of the EC Wind Energy Conference*, Madrid, Spain, September, pp. 718–24.

Friedrich, R. and Voss, A. (1989) 'Die Sozialen Kosten der Elektrizitatserzeugung', *Energie Wirtschaftliche Tagestragen*, vol. 39, no. 10, pp. 640–9.

Ottinger, R. L. (1990) *Environmental Costs of Electricity*, Oceana Publications.

Stocker, L., Harman, F. and Topham, F. (1991) *Comprehensive Costs of Electricity in Western Australia*, Ecologically Sustainable Development Working Group on Energy Production, Canberra, Australia.

Hagen, D. and Kaneff, S. (1991) *Application of Solar Thermal Technology in Reducing Greenhouse Gas Emissions – Opportunities and Benefits for Australian Industry*, ANUTECH Pty.

Ferguson, R. (1990) Newcastle University, UK.

Koomey, as cited in Hill, R. and Baumann, A. (1991) *External Cost Benefits of Energy Technologies*, Newcastle Polytechnic, UK.

Notes:

(a) Includes estimates of costs of clean-up of acid rain and the greenhouse effect.

(b) Does not include estimates of climate damage.

(c) Costs of clean-up of acid rain and the greenhouse effect.

(d) Pollution costs only.

(e) Pollution costs only.

(f) Acidic deposition and global warming damage costs.

(g) Health/catastrophes.

(h) Pollution costs.

Source: House of Commons Select Committee on Energy, 1992

results, summarized in Table 13.14, suggest that the cost of producing electricity from coal would double and the cost of electricity production from gas would increase by 30% if external costs such as damage to the environment and to health were taken into account. The methodology used in the EXTERNE study to calculate external costs is called the impact pathway approach, and is illustrated in Figure 13.12. It starts by measuring

Table 13.13 Estimates of external costs (1990 US cents per kWh)

	Range	Hohmeyer	Pace	BPA	Tellus	JBS
Gas turbine	0.1–6	0.6–2.9	0.7–1	0.1	6	1.6–4.1
Oil turbine	0.3–10.3	0.6–2.9	2.6–6.9	0.3	10.3	
Coal	0.6–10	0.6–2.9	2.6–5.9	0.7–1.1	4.5–10	2.8–8.2
Nuclear	0–5.7	0–5.7	3			
Photovoltaics	0–0.4	0–0.2	0–0.4			
Wind	0–0.1	0	0–0.1			
Biomass	0–0.7		0–0.7			
Geothermal	0					
Solid waste	3.7–48.2		2.9	3.7–48.2		

Source: Gipe, 1995, Table 12.13, p. 433.

emissions (including applying uniform measuring methods to allow comparison), then estimating the dispersion of pollutants in the environment and the subsequent increase in ambient concentrations. Then, impacts on, for example, crop yields or health are evaluated. The methodology finishes with an assessment of the resulting cost. This phase is perhaps the most difficult. Some direct damage costs will be clear, but in other cases various techniques need to be used to try to attach prices to the loss of amenities – for example, the so-called 'willingness to pay' approach involves asking people likely to be affected how much they would be willing to pay to avoid some particular amenity loss.

The external costs of electricity generation vary widely between different countries in the EU, depending on the mix of fuels used for generation and the efficiencies and ages of plant. Table 13.14 shows the external costs for various systems averaged over the EU as a whole. As can be seen, on the basis of this analysis, fossil fuels have much larger environmental costs than any of the other options, with coal clearly being the worst. By contrast, wind seems relatively benign, having four times lower environmental costs than nuclear.

Figure 13.12 shows the results of applying the ExternE approach to estimating the external costs of the various energy sources used in the UK's electricity generating system, based on data for 1998 (Watkiss, 2002). Once

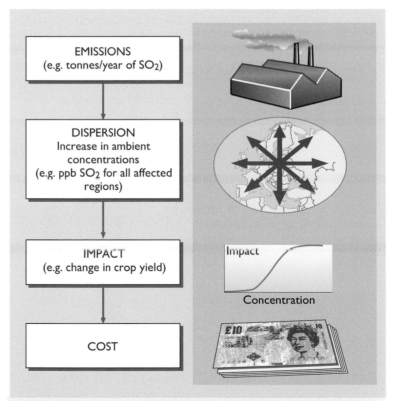

Figure 13.12 The Impact Pathway approach that forms the basis of the EU's EXTERNE methodology, as applied to the assessment of crop damage from SO_2 exposure. SO_2 concentration is expressed in parts per billion (ppb) (source: Watkiss, 2002, Figure 1, p. 28)

Table 13.14 Estimated external costs of electricity generation from various primary energy sources: European Union average

Primary energy source	External costs (Euros per kWh)
Coal	0.057
Gas	0.016
Biomass	0.016
PV solar	0.006
Hydro	0.004
Nuclear	0.004
Wind	0.001

External costs are additional to conventional electricity costs, assumed to be 0.04 euro per kWh (EU average)

Source: EU EXTERNE study / European Commission, 'Externalities of Energy' reports on the EXTERNE programme, DG12, L-2920 Luxembourg, 2001

again coal and oil can be seen to have the largest costs, followed by gas, nuclear, hydro and wind. This analysis also suggests that for coal and oil, the external costs of air pollution are greater than the estimated costs of global warming, but that for gas the costs of global warming outweigh those of pollution.

The results of assessments like these are interesting, but are fraught with methodological problems, for example concerning how to calculate the cost of specific types of damage. Simple economic assessment, based on insurance replacement costs for example, may not provide a realistic measure of, or proxies for, the human value of amenity loss or health damage, much less the ecological value of any disruption. Even more contentious is the value put on human life, which – inequitably but seemingly inexorably – differs around the world. However, even though the absolute magnitudes of the external costs of energy sources have proved to be highly-sensitive to changing assumptions and research results, the overall conclusions on the *relative* ranking of energy sources in terms of their external costs have not altered. As Krewitt (2002) has commented in a 10-year review of the various ExternE studies: 'even under different background assumptions, electricity generation from solid fossil fuels is consistently associated with the highest external costs, while the renewable energy sources cause the lowest externalities. The robustness of this ranking is an important finding of the ExternE study, which implies a clear message to the decision-maker.'

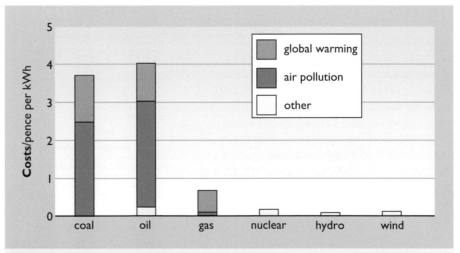

Figure 13.13 Breakdown of the global warming, air pollution and other external environmental costs of UK electricity generating systems (source: Watkiss, 2002, Figure, p. 29)

13.11 Summary and conclusions

In this chapter, we have reviewed the environmental and health impacts of humanity's use of fossil and nuclear fuels. We have seen that these impacts can be classified in a variety of ways. One approach is simply to list the concerns associated with each energy source. A second approach involves estimating the various substances emitted in the course of energy use and comparing these with the natural flows of such substances. A third approach involves looking at the different scales, from local to global, on which the various impacts principally manifest themselves.

We then pursued the last of these approaches in more detail, examining the impacts of various key forms of energy use on five different scales: domestic, workplace, community, regional and global.

At the **domestic scale**, we found that the emissions produced by wood fuel combustion under typical conditions in developing countries are a cause for serious concern, on both health and climate change grounds.

At **the workplace scale** we found that, although estimates vary widely, occupational fatalities in the coal and oil industries appear to be significantly worse than in the natural gas industry. Fatal hazards to workers in the nuclear industry seem comparable to those of natural gas – although deaths due to radiation are difficult to prove and may have been understated in the statistics.

At the **community scale**, we focused on atmospheric pollution in cities, a cause of many premature deaths among urban populations. We examined the health effects of the various pollutants and found that pollutant levels in the cities of developed countries have been falling in recent decades, but rising in the cities of developing countries.

At **regional scale**, we examined firstly the problem of acid precipitation and secondly the impacts of nuclear radiation on the health of the general population.

On a **global scale**, we returned to look in more detail at the problem of greenhouse gas emissions and global climate change that was examined briefly in Chapter 1. There appears to be little remaining doubt that energy-related emissions of greenhouse gases have caused the earth's mean temperature to increase significantly during the 20th century and that these are likely to cause further substantial increases during the 21st century. These increases will probably be accompanied by rising sea levels, increased frequency of extreme weather events and other changes in the world's climate and ecosystems, most of them adverse in their overall effects.

We then turned to the controversial issue of **risk assessment** in the nuclear power industry. We discussed the difficult problem of how to compare the probabilities and consequences of extremely rare, but catastrophic, accidents with those of a more normal and regular kind, and concluded that rational individuals can take a variety of views on the subject, none of which should be dismissed as illegitimate.

We then examined four different ways of comparing the impacts of different **electricity generating systems**: in terms of (i) their greenhouse gas emissions, (ii) their SO_2 and NO_x emissions, (iii) their land requirements, and

(iv) their energy payback ratios. Broadly, the fossil fuels, and especially coal, emit the largest quantities of CO_2, SO_2 and NO_x per unit of energy produced, with nuclear power and renewables (apart from biomass) generating virtually no such emissions.

The land use requirements of the renewables, and especially of biomass, appear to be substantially greater than those of fossil and nuclear fuels, although under certain circumstances (such as the use of solar on roof-tops, or the growing of crops around wind turbines) the land requirements of renewables might be greatly reduced.

The energy payback ratios of most of the renewable sources were substantially better than those of the fossil and nuclear fuels.

Finally, we looked at some attempts to estimate the 'external' costs of energy use: the social and environmental costs that are not reflected fully in the monetary prices we pay for our energy. Overall, it appears that coal combustion gives rise to the largest external costs, followed by oil, with natural gas and nuclear power being significantly less expensive and renewables the least expensive, in external cost terms. In the companion volume, *Renewable Energy*, we examine in more detail the potential of the renewable energy technologies to contribute to a more sustainable energy future.

In this chapter we have reviewed the many ways in which our society currently incurs substantial environmental and social penalties through its use of fossil and nuclear-fuelled based energy systems, ultimately rendering them unsustainable in the sense defined in Chapter 1. We now turn, in the next and concluding chapter, *Remedies*, to examine various ways in which our use of fossil fuels might be made more sustainable in the 21st century, through the introduction of new, cleaner technologies.

References and further reading

Ahlbom, A., Day, N., Feychting, M. *et al.* (2000) 'A Pooled Analysis of Magnetic Fields and Childhood Leukemia', *Br J Cancer* 83, pp. 692–8.

Allsopp, M., Costner, P. and Johnston, P. (2001) *Incineration and human health: State of knowledge of the impacts of waste incinerators on human health,* Greenpeace Research Laboratories, University of Exeter, UK, March, pp. 84.

Arnold, L. (1992) *Windscale 1957: Anatomy of a Nuclear Accident*, London, Macmillan.

Barker, A. T., Coulton, L. A. *et al.* (2002) 'EMF Health Effects – Are They Real?' *IEE Engineering Science and Education Journal*, August, pp. 122–3.

DEFRA (2002) *Dioxins and dioxin-like PCBs in the UK environment: a consultation document*, Department for Environment, Food and Rural Affairs, HMSO, 98pp.

EC (1995) *ExternE: Externalities of Energy, vol. 2 Methodology*, Bruxelles, European Commission, Directorate-General XII, Science Research and Development.

Environment Canada (1998) *1997 Canadian acid rain assessment, volume 1: Summary of results*, Ottawa, Ontario, Minister of Supply and Services, 16pp.

FoE (2002) *Incineration and Health Issues,* Briefing paper, London, Friends of the Earth, May, 11pp.

de Fre, R. and Wevers, M. (1998) Underestimation in dioxin emission inventories, Organohalogen Compounds, 36, pp. 17–20.

Fritsche, U. (1992) 'TEMIS: a Computerised Tool for Energy and Environmental Life Cycle Analysis'. Paper presented at International Energy Agency expert workshop on Life-Cycle Analysis of Energy Systems: Methods and Experience, Paris.

Gagnon, L. and van de Vate, J. F. (1997) 'Greenhouse gas emissions from hydropower, the state of research in 1996', *Energy Policy*, 25, 1, pp. 7–13.

Gagnon, L. *et al.* (2002) 'Life-cycle assessment of electricity generation: The status of research in the year 2001, *Energy Policy*, 30, 10, pp. 1267–78.

Gipe, P. (1995) *Wind energy comes of age*, New York, John Wiley & Sons, Inc.

Gipe, P. (2001) 'Wind related deaths data base', *WindStats Newsletter*, vol. 14, no. 4, Autumn.

Godish, T. (1997) *Air Quality*, Chelsea MI, Lewis Publishers Inc.

Henshaw, D. L. (2002) 'Adverse Health Effects from Power-frequency EMFs – They Are Starting to Look Real', *IEE Engineering Science and Education Journal*, June, pp. 82–3.

Holdren, J. P. and Smith, K. R. (2000) 'Energy, the environment and health' in Goldemberg, J. (ed.) *World Energy Assessment: Energy and the Challenge of Sustainability*, New York, United Nations Development Programme, United Nations Department of Economic & Social Affairs and World Energy Council, pp. 61–110.

House of Commons Select Committee on Energy (1992) *Fourth Report: Renewable Energy*, HMSO.

IEA (2002) *Key World Energy Statistics from the IEA*, Paris, International Energy Agency, www.iea.org/statist/keyworld2002/key2002/keystats.htm [accessed February 2003].

IEA/OECD (1996) 'Comparing Energy Technologies', *IEA Report*, Paris, OECD.

IEE (2001) *IEE Fact File: Electromagnetic Fields and Health,* Institution of Electrical Engineers, London, March. Available at: www.iee.org/Policy/Areas/BioEffects/emfhealth.pdf). [accessed January 2003].

Intergovernmental Panel on Climate Change (IPCC) (2001a) *Climate Change 2001: Synthesis Report*, Cambridge, United Kingdom and New York, USA, Cambridge University Press.

Intergovernmental Panel on Climate Change (IPCC) (2001b) *Climate Change 2001: The Scientific Basis*, Cambridge, United Kingdom and New York, USA, Cambridge University Press, p. 87.

International Commission on Non-Ionising Radiation Protection (ICNIRP) (1998) 'Guidelines for Limiting Exposure to Time-Varying Electric, Magnetic and Electromagnetic Fields (up to 300 GHz)', *Health Physics*, April, vol. 74, no. 4, pp. 494–522.

Krewitt, W. (2002) 'External costs of energy – do the answers match the questions? Looking back at 10 years of ExternE', *Energy Policy*, 30, 10, pp. 839–48.

McSorley, J. (1990) *Living in the Shadow*, Pan Books.

Morison W. G. (1998) 'Nuclear Power Generation' in Stellman, J. M. (ed.) *Encyclopaedia of Occupational Health and Safety*, 4th edn, Geneva, International Labour Organization.

National Statistics (2002) *Annual abstract of statistics*, London, Stationery Office, Table 9.8, p. 136.

Nordhaus, W. (1997) *The Swedish Nuclear Dilemma: energy and the environment*, Washington DC, Resources for the Future Press.

Norwood W. D. (1963) 'Removal of Plutonium and other Transuranic Elements from Man' in *Proceedings of the Scientific Meeting on the Diagnosis and Treatment of Radioactive Poisoning*, Vienna, International Atomic Energy Agency.

ORNL and RFF (Oak Ridge National Laboratory and Resources for the Future) (1992–8) 'US-EC Fuel Cycle Study: Background Document to the Approach and Issues', and other reports on Estimating Externalities of Fuel Cycles, 1994–8. Oak Ridge, Tennessee, USA.

Pearce, D., Bann, C. and Georgiou, S. (1992) *The Social Costs of Fuel Cycles* (report to UK Department of Trade and Industry), London, HMSO.

Porteous, A. (2000) *Dictionary of environmental science and technology*, Chichester, Wiley, 3rd Edition, pp. 157–64.

Ramage, J. R. (1997) *Energy: a Guidebook*, Oxford, Oxford University Press.

Rasmussen, N. (1975) *Reactor Safety Study: An Assessment of Accident Risks in US Commercial Nuclear Power Plants* (WASH-400) Washington DC, US Nuclear Regulatory Commission.

RCEP (1994) 'Incineration of waste: Dioxin emissions', Letter from the Secretary to the Royal Commission on Environmental Pollution to the Director, Pollution Control and Wastes, Department of the Environment, December 1994. available at: www.rcep.org.uk/news/95-1.html). [accessed 1st March 2003]

Reddy, A., Williams, R., Johansson, T. (1997) *Energy after Rio: prospects and challenges*, New York, United Nations Development Program.

Rogner, H. H. and Kahn, A. (1998) 'Comparing Energy Options: Progress Report on the Inter-Agency DECADES Project', *IAEA Bulletin*, vol. 40, no. 1.

Smith, K. R. *et al.* (2000) 'Greenhouse implications for household stoves: an analysis for India', *Annual Review of Energy and Environment*, 25, pp. 741–63.

Swanson, J. and Renew, D. C. (1994) 'Power-frequency Fields and People' *IEE Engineering Science and Education Journal*, April, pp. 71–9.

Sullivan, L. (1995) 'The Three Gorges Project: damn if they do?', *Current History*, September.

Watkiss, P. (2002) 'Electricity's Hidden Costs', *IEE Review*, November, pp. 27–31.

Watson, R. T. *et al.* (eds) (2002) *Climate Change 2001: Synthesis Report,* Cambridge, Cambridge University Press.

WHO (1999) *Provisional Global Air Quality Guidelines*, Geneva, World Health Organization.

Yasukawa, S., *ct al.* (1992) 'Life Cycle CO_2 Emissions from Nuclear Power Reactor and Fuel Cycle System'. Paper presented at in International Energy Agency expert workshop on Life-Cycle Analysis of Energy Systems: Methods and Experience, Paris.

Chapter 14

Remedies: Making Fossil Fuel Use More Sustainable

by Godfrey Boyle

14.1 Introduction

In the preceding chapters we examined the main energy sources currently employed by humanity, principally the fossil and nuclear fuels. We looked at their origins and history, the scientific and technological principles underlying their use, their economics – and in particular, their sustainability problems. In Chapters 10 and 11, on nuclear energy, the problems of sustainability of *nuclear* fuel use were discussed, together with some potential solutions to those problems. These topics will not therefore be re-examined in this chapter.

In the companion volume to this book, *Renewable Energy*, we examine the *renewable* energy sources, including their sustainability problems and various approaches to resolving them. For this reason, possible solutions to some of the problems of renewable energy described in Chapter 13 *Penalties*, such as the impacts of emissions from wood fuel use in developing countries or the impacts of hydro and wind power in the community or the workplace, will not be covered here.

In this final chapter, therefore, we focus on existing and emerging solutions to the sustainability problems of *fossil* fuel use, concentrating in particular on ways of reducing or 'sequestering' carbon emissions from fossil fuel combustion – or, better still, as we shall see, using carbon-free fuels and associated technologies that enable the combustion process and its attendant problems to be avoided entirely.

In reducing the impacts of *fossil* fuel use, three distinct approaches can be identified:

The first is to improve the energy conversion efficiency of fuel-based energy *supply* systems, so that less fuel is required to achieve a given level of energy output. This entails such technologies as the high-efficiency combined cycle gas turbine (CCGT) for electricity generation; the use of combined heat and power (CHP) to enable the 'waste' heat from electricity generation to be usefully employed; and the use of high-efficiency condensing boilers in buildings, or high-efficiency engines in vehicles, to allow less fuel to be burned for a given level of useful output. These approaches have been discussed in previous chapters and will not be pursued further here. In this chapter, and indeed this book, we also do not intend to address in detail the topic of energy *demand* management, which of course is a further way of improving the sustainability of human energy use. This very important but separate issue is covered in, for example, Beggs (2002) and Eastop and Croft (1990).

The second approach involves various ways of 'cleaning-up' fossil fuel combustion, to reduce the levels and impacts of pollutant emissions. Earlier chapters have discussed various ways of reducing emissions of SO_2, NO_x and particulates from fossil fuel combustion, for example in power stations or vehicle engines. Here we concentrate, in Section 14.2, on reducing CO_2 emissions by switching to lower-carbon fuels, and in Section 14.3 on various ways of 'sequestering' the CO_2 emissions from fossil fuel combustion so that they do not enter the atmosphere and do not contribute to climate change problems.

The third approach, to which we turn in Section 14.4, is to use an energy conversion device that can extract useful energy from fossil fuels directly, avoiding combustion and its associated emissions – namely the fuel cell.

Fuel cells are normally fuelled by hydrogen, a carbon-free fuel, and in Section 14.5 we examine the prospects for a 'hydrogen economy' based on hydrogen derived from fossil fuels, with capture and sequestration of the associated carbon emissions.

Finally, in Section 14.6, we conclude with an assessment of the extent to which fossil fuel use can in future be made more sustainable, with respect to the sustainability criteria described in Chapter 1.

14.2 Reducing pollutant emissions from fossil fuel combustion

In Chapter 5 we saw how levels of SO_2, NO_x and particulates from coal combustion could be substantially reduced, using such techniques as fluidized bed combustion, flue gas de-sulphurization, low-NO_x burners and electrostatic precipitators. In Chapter 8 we also saw how polluting emissions from oil (gasoline and diesel) combustion in engines can also be cut substantially, for example through the use of low sulphur fuels, catalytic converters and 'lean burn' engines.

Progress in reducing such emissions still further seems likely in coming decades, for example through the use of cleaner fossil fuels, improved combustion techniques, particulate filters, better engines and more sophisticated engine control systems. But there remains the persistent problem of carbon dioxide emissions, which are unavoidable when carbon-based fuels undergo combustion.

Fuel switching

Substantial reductions in carbon dioxide emissions can be achieved by switching to fuels with a lower carbon content, notably natural gas (see Table 14.1). Natural gas can now be burned in condensing boilers (mentioned in Chapter 7) which have extremely high efficiency (around

Table 14.1 Emission factors of fossil fuels in 1998 (grams per GJ of heat)

	Natural gas	Oil	Coal
Carbon dioxide[1]	14 000	19 000	24 000
Methane	3.7	3	20
Sulphur dioxide	1	400	840
Black smoke	0	15	55
Nitrogen oxides	55	100	230
Volatile organic compounds	5	2	18
Carbon monoxide	7	8	210

1 In terms of mass of carbon produced (see Table 14.2, Note 2)

Source: Department of Trade and Industry, 2001

90%) and this, combined with the gas's relatively low carbon content, leads to significant reductions in CO_2 emissions. Moreover, there are virtually no emissions of sulphur or particulates, though of course there will still normally be NO_x emissions as a result of the high nitrogen content of air.

Similarly, as mentioned in earlier chapters, switching from coal to natural gas for electricity generation using combined cycle gas turbine (CCGT) plant enables much higher generation efficiencies to be achieved and this, combined with the relatively low carbon content of natural gas compared to coal, allows substantial CO_2 emission reductions to be achieved. (Table 14.2) Here again, there are still some NO_x emissions but no emissions of sulphur or particulates.

Table 14.2 Greenhouse gas emissions for different types of electricity generating plant

Type of generating plant	kg of greenhouse gases (CO$_2$ equivalent) emitted per kWh of electricity
Modern coal plant (including FGD and low NO$_x$ burners)	1.1
Oil plant without FGD or low NO$_x$ burners	1.1
Combined cycle gas turbine	0.5
Nuclear	0.05

Notes:

1 FGD = Flue Gas Desulphurization

2 The emissions data in the above table are expressed in terms of the equivalent number of grams of *carbon dioxide* (CO_2) per kWh of electricity generated. To convert these figures to grams of *carbon* per kWh, they should be multiplied by 12/44, which is the ratio of the relative masses of carbon (12) and the carbon dioxide molecule (44).

Source: Eyre, 1990

Figure 14.1 in Box 14.1 shows the effect upon carbon emissions in the UK of the shift from coal to natural gas in electricity generation between 1990 and 2000. The figure also shows that a small contribution to these CO_2 emission reductions was made by increased use of nuclear power over much of the period.

But even though emissions of SO_2, NO_x and particulates from fossil fuel burning can all be reduced significantly by various technological means, and by switching to lower-carbon fuels, nevertheless all fossil fuels – even natural gas – still produce some CO_2 during their combustion. In the following section we look at ways in which these carbon emissions might be captured and 'sequestered', so that they do not contribute to global climate change.

BOX 14.1 Reducing carbon emissions by fuel switching: the UK's 'dash for gas'

As described in Chapter 9, the privatization of the UK electricity industry in 1989 precipitated the 'dash for gas'. Not only could CCGT stations produce electricity more cheaply than coal stations, they could also help cut national acid emissions without incurring extra expense. They had the further benefit of producing less CO_2.

The Kyoto protocol on climate change was signed in 1997. Under this the UK agreed to cut its emissions of greenhouse gases by 12.5%, relative to emission levels in 1990, by 2010. The UK government also proposed for itself a more ambitious, voluntary, target of a 20% cut in emissions by 2010. The government could afford

to be generous in its promises because national CO_2 emissions had already fallen by 7% between 1990 and 1997, and emissions from power stations had fallen by 27%, even though electricity demand had risen. A further contribution to the trend of declining emissions was provided by the improved performance of many UK nuclear stations which had not previously been achieving full output.

As we can see, however, the declining trend of CO_2 emissions stopped in 2000, following rising gas prices and concerns about the long-term availability of sufficient quantities of natural gas.

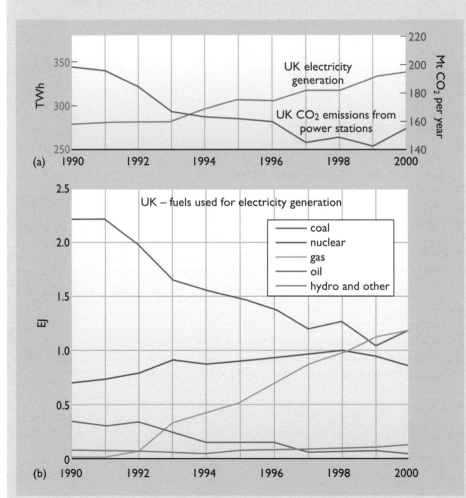

Figure 14.1 (a) Trends in UK electricity generation and carbon emissions, 1990–2000; (b) fuels used for UK electricity generation, 1990–2000 (source: DTI: Energy Trends, various issues)

14.3 Capturing and sequestering carbon emissions from fossil fuel combustion

The basic chemistry of combustion, as explained in previous chapters, inevitably means that burning fossil fuels produces carbon dioxide (CO_2), the principal anthropogenic greenhouse gas. And as we saw in Chapter 13, *Penalties*, worldwide emissions of CO_2 will almost certainly need to be greatly reduced during the twenty-first century if the consequences of global climate change are to be limited to an extent manageable by humanity and the other species of animals and plants with which we share the planet. However, although fossil fuel combustion always *produces* carbon dioxide, the CO_2 does not necessarily have to be allowed to *escape* into the atmosphere.

To try to reduce the concentrations of atmospheric CO_2, a number of approaches are being investigated. These involve the capture and **sequestration** of atmospheric carbon, either in forests or below the earth's surface or in its oceans.

Carbon sequestration in forests

One approach involves growing additional forests on a large scale, to take advantage of the fact that growing trees absorb carbon dioxide from the atmosphere. When trees die they cease to absorb CO_2, and eventually decay, re-emitting CO_2 to the atmosphere in the process. If new trees grow naturally, or are planted by humans, to replace those that have died, then little or no net CO_2 is emitted to the atmosphere. However if new trees are not planted then CO_2 will be emitted to the atmosphere, not only from the portion of the decaying tree that is above ground but also from the roots below ground.

So in order for afforestation to be a secure way of sequestering carbon, it would need to involve a permanent programme of tree planting and re-planting over a very long time scale. And in order to sequester significant quantities of carbon, it would have to be carried out on a very large scale. As mentioned in Chapter 1, the afforestation of a land area as large as Europe would be required in order to sequester the amount of carbon likely to be emitted from the burning of fossil fuels during the first half of the twenty-first century. The Intergovernmental Panel on Climate Change has calculated that a global programme to 2050 involving reduced de-forestation, enhanced natural regeneration of tropical forests and worldwide re-afforestation could sequester 60–87 billion tonnes of atmospheric carbon, equivalent to some 12–15% of projected CO_2 emissions from fossil fuel burning during the period (RCEP, 2000). Clearly, although carbon sequestration using forests can play a useful role, it is unlikely to provide a complete solution.

However, instead of simply growing trees to act as passive absorbers of carbon emissions from fossil fuel combustion, such trees could be harvested and burned as biofuels to substitute for fossil fuels. Provided such combustion was performed as completely and efficiently as possible, to minimize the production of non-CO_2 greenhouse gases (as mentioned in Chapter 1, Section 1.3), and provided new trees were planted to replace those that had been harvested, such a scheme could enable substantial reductions in carbon emissions to be achieved. The topic of biofuel use is covered in more detail in the companion volume, *Renewable Energy*.

Carbon capture and sequestration beneath the earth's surface

Another approach to sequestration that is receiving increasing attention involves capturing CO_2 from fossil fuel combustion and depositing it beneath the earth's surface, for example in depleted oil or gas wells, in deep coal seams, or in aquifers, which are subterranean zones of water-bearing rock or sand.

In order to be able to sequester CO_2 following combustion, it has to be captured from flue gases. There are three main options (see RCEP, 2000):

- **Absorption:** Absorption is the take-up of a gas into the interior of a solid or liquid. In **chemical absorption** the gas reacts chemically with the absorbing substance, changing its molecular structure. The ethanolamines, organic liquids related to ammonia, have long been used to absorb CO_2 in this way. The resulting compound can be heated later in order to recover the CO_2 and regenerate the original liquid for re-use. Sprayed into flue gases, these liquids can capture from 82% to as much as 99% of the CO_2, depending on the energy used for regeneration. An alternative method involves the **physical absorption** of CO_2 by organic solvents. There is no chemical reaction in this case, and the CO_2 molecules remain unchanged in the solvent. This method is more suitable for integrated gasification combined cycle (IGCC) plants (see Box 14.3).

- **Adsorption:** In adsorption, the gas is taken up in the form of a layer on the *surface* of a solid. It can be either **chemical adsorption** (or chemisorption), in which the gas is bonded to the surface molecules; or **physical adsorption**, in which the gas is held by much weaker electrical forces called van der Waals forces. Finely divided solids such as activated carbon or alumina, or porous substances such as zeolite, are the best adsorbers, as they offer large surface areas for a given mass of material. The adsorbed CO_2 is recovered, and the solid regenerated, by heating. Using this method, 95% of the CO_2 can be captured.

- **Gas separation membranes:** These are of various different types, including porous inorganic membranes and membranes incorporating palladium, polymers or zeolites. They all allow one component of a gas stream to pass through faster than the others, but do not usually achieve high degrees of separation, so multiple membranes or recycling of gases is required, which entails additional complexity, cost and energy consumption.

Having captured the CO_2 by one means or another, an obvious place to sequester it is in depleted or depleting oil or gas wells. Indeed, CO_2 injection is already used by some operators to increase the amount of oil that can be recovered from oil wells, a process known as **Enhanced Oil Recovery** (EOR). (In Chapter 7 this process was referred to as tertiary recovery.) In this case, the value of the additional oil generated through enhanced recovery is greater than the additional cost of pumping-in CO_2.

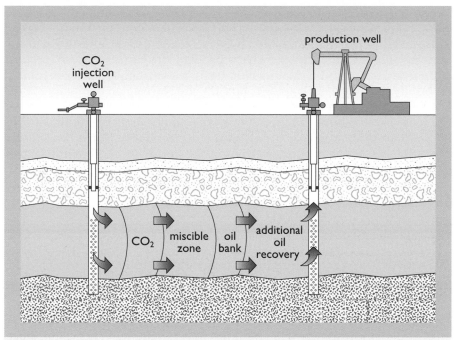

Figure 14.2 CO_2 injection for enhanced oil recovery (source: IEA, 2001)

CO_2 can also be sequestered in deep beds of coal that are too inaccessible for conventional mining. Such coal beds usually contain significant quantities of methane, which in suitable circumstances can be piped to the surface and used as a fuel. Injecting CO_2 into the coal bed can increase the quantity of methane produced, as well as sequestering the CO_2. Again, the value of the energy produced can offset the additional costs of CO_2 capture and sequestration.

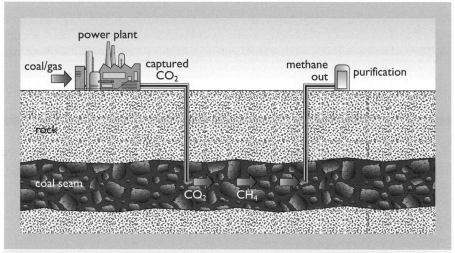

Figure 14.3 CO_2 injection for enhanced recovery of coal-bed methane (source: IEA, 2001)

Yet another option is to sequester the CO_2 in the many large, deep, saline aquifers that lie beneath the earth's surface. These aquifers are not unlike those used to provide water for human consumption, and are equally widespread, but lie in deeper layers beneath the earth's surface, around 800 metres or so, and contain saline (salt) water unsuitable for drinking. As briefly described in Chapter 1, Box 1.3, the Norwegian company Statoil has since 1996 been successfully sequestering some 1 million tonnes of CO_2 per year in a saline aquifer of this type, the Utsira formation, beneath its Sleipner West (Vest) gas production platform in the North Sea. In this case, the CO_2 being sequestered is not the result of combustion but of excessive CO_2 levels in the methane produced by this particular gas field, which have to be reduced in order to make the gas saleable. However, the project illustrates the feasibility of this method of storage.

Storage potential

The above options sound promising, but could a sizeable proportion of the CO_2 produced from the world's fossil fuel combustion be sequestered in such ways?

The Intergovernmental Panel on Climate Change has estimated that some 40 gigatonnes of carbon (GtC) could be stored in depleted oil wells, some 90 GtC in depleted gas wells and some 20 GtC via enhanced oil recovery (Goldemberg, 2000). Global carbon emissions in 2000 were some 6 GtC, which means that, at present levels, only about 25 years' worth of global emissions could be stored in this way.

The outlook for storage in saline aquifers seems better, however. The storage capacity available depends on the type of aquifer, but Statoil have calculated that the Utsira formation alone could store some 1000 million tonnes of CO_2 per year – roughly equivalent to the current total of CO_2 emissions from all of the EU's electric power plants – for the next 600 years. Power plants, however, are by no means the only sources of CO_2 emissions

Another concern is whether we can be sure that the CO_2, once stored, would not sooner or later find its way back into the atmosphere.

The most secure storage would be provided by those aquifers that have a cap of rock above them to prevent CO_2 escaping, but it is estimated that the capacity available in such 'sealed' aquifers is only about 50 GtC – enough to absorb only 8 years' worth of global CO_2 emissions at current rates. 'Open' saline aquifers, without cap rocks, have a far greater capacity – estimated at between 2900 and 13 000 GtC (Goldemberg, 2000) – but there is a greater risk that the CO_2 will escape back into the atmosphere. However, it has been shown that the CO_2 stored under pressure in such aquifers can react with minerals to form carbonates, making the CO_2 storage secure.

Ocean sequestration of carbon

A further method of sequestration under active investigation is to increase the amount of carbon dioxide absorbed by the earth's oceans. The oceans are by far the largest stores of carbon within the earth's natural carbon cycle. In this cycle, carbon is exchanged between the earth's soil, vegetation, atmosphere and oceans. As shown in Figure 14.4, anthropogenic emissions of CO_2 in 2000 were approximately 6 GtC. Of this, under normal conditions approximately 2 GtC is transferred naturally from the atmosphere to the oceans through the operation of the carbon cycle. Some 0.7 GtC is also sequestered on land (in forests etc.), but the remaining 3.3 GtC has to be absorbed by the atmosphere itself. This, of course, is the cause of increasing atmospheric CO_2 concentrations, as shown in Figure 1.18(a) of Chapter 1.

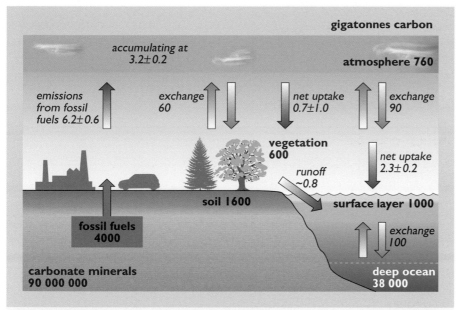

Figure 14.4 Carbon sources and sinks in the global carbon cycle (source: RCEP, 2000). Note: there are small discrepancies between the figures shown here and those given in Chapter 13, Table 13.6. This is because, as Table 13.6 illustrates, the flows change slightly from year to year

The *surfaces* of the world's oceans are already saturated with CO_2, which is gradually transferred to the deeper layers in a very slow process lasting hundreds of years. The ocean *depths*, however, have an enormous capacity to absorb CO_2. The dissolved inorganic carbon (DIC) content of the world's oceans is estimated to be some 38 000 GtC. In comparison, the world's total fossil carbon reserves, including conventional and unconventional deposits of oil, natural gas and coal, are estimated at around 6500 GtC (Goldemberg, 2000), so even if all of the earth's fossil fuels were burned and the CO_2 eventually sequestered in the deep oceans, the DIC content would only increase by about 17%, to 44 500 GtC.

There are two main ways in which the oceans could be used to store additional CO_2 from fossil fuel burning. One method is to capture the CO_2 and inject it deep into the ocean. The other method is to enhance the oceans' natural absorption of CO_2 from the atmosphere, and there are several ways in which this might be done. Both approaches can be viewed simply as ways of accelerating the existing processes of the natural carbon cycle.

Injecting CO_2 into the oceans

Carbon dioxide, having been captured in power stations or other substantial, stationary CO_2 emitters, could be transported by pipeline or tanker to a suitable ocean injection site. It would then be injected at depths greater

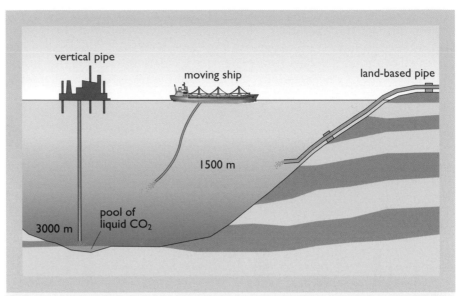

Figure 14.5 Methods for ocean CO_2 injection (source: International Energy Agency, 1999)

than about 800 metres, where the increased pressure would turn the CO_2 from a gas to a liquid. Studies suggest that it would be necessary to inject CO_2 at depths of 1000–1500 m in order to ensure that the majority of it remained in the ocean for a very long time.

A more challenging and costly approach, but one that should ensure the maximum effectiveness of sequestration, would be to inject liquid CO_2 into a very deep sea floor depression, at a depth of around 3–4000 metres. At this depth, it would form a 'lake' of CO_2 that should remain at the bottom of the ocean indefinitely (see Figure 14.5).

Enhancing natural ocean uptake of CO_2

There are two natural processes that, together, gradually remove CO_2 from the surface of the oceans and deposit it at greater depths.

The first process operates because CO_2 is highly soluble in the cold, dense waters found at high latitudes, and sinks to the bottom of the ocean. This gives rise to the so-called **thermohaline circulation**, also known as the '**Great Ocean Conveyor Belt**', in which cold, CO_2-rich water from the depths of the North Atlantic is 'conveyed' southwards to Antarctic regions before eventually surfacing in the Indian Ocean and the Equatorial Pacific. (Figure 14.6). There the CO_2 eventually escapes, but the time scale is of the order of 1000 years (IEA, 1999b).

The second process depends on the fact that, at the surface of the world's oceans, microscopic plants called phytoplankton grow by photosynthesis, harnessing the energy of sunlight and absorbing the CO_2 dissolved near the surface. Phytoplankton are then grazed by zooplankton, which in turn are consumed by marine animals, mainly fish. Eventually, when these die, a proportion (about 30%) of the carbon in their remains descends to the deep ocean, where it remains for as much as 1000 years before being eventually turned back into CO_2 by bacteria and returning to the surface waters. This process is sometimes termed the 'biological pump'.

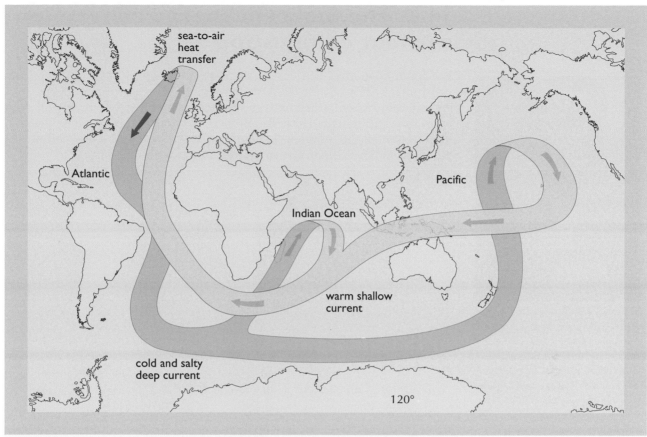

Figure 14.6 The 'Great Ocean Conveyor Belt' (source: International Energy Agency, 1999)

Several ways of accelerating these natural processes have been proposed. One involves the addition of nutrients, such as nitrates and phosphates, to large areas of ocean in order to increase the production of phytoplankton and the other organisms that depend on them, so increasing the amounts of CO_2 eventually deposited from their remains in the ocean depths. A second approach involves adding iron to specific areas of the ocean that are known to be lacking in this essential micro-nutrient. Again, this would have the effect of increasing the biological productivity of the ocean surface (and probably increasing fish catches), eventually resulting in increased CO_2 deposition in deep waters.

However, there are concerns that these approaches could upset the ecological balance of the oceans, and a great deal of additional research would be required before any large-scale projects could be undertaken – though small-scale pilot projects to investigate the potential of these techniques are likely.

Thus it seems likely that carbon capture and disposal beneath the earth's surface, in oil or gas reservoirs, coal seams or saline aquifers, together with sequestration in forests, will offer the only acceptable carbon sequestration options in the near future.

14.4 The fuel cell: energy conversion without combustion

We have seen that the combustion process inevitably involves the production of pollutants. Although these can be 'cleaned-up' to a considerable extent, this involves additional costs and low-level pollutant emissions still remain. However, there is an energy conversion device that does not entail the problems of combustion. Known as the **fuel cell**, it was invented by Sir William Grove (Figure 14.7) in 1850.

Figure 14.7 Sir William Grove (1811–1896) was both a lawyer and a physicist. He was called to the bar in 1835, but ill-health interrupted his legal career and he turned to science. In addition to inventing the fuel cell, he was the first to enunciate the principle of conservation of energy in 1846, a year before the German physicist Helmholtz. He returned to the legal profession after 1853, specializing in patent law, was knighted in 1872 and became a High Court judge in 1880. On retirement from the bench in 1887, he resumed his scientific studies

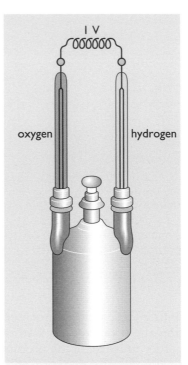

Figure 14.8 Diagram illustrating Grove's original apparatus

In Grove's apparatus (Figure 14.8), supplies of hydrogen and oxygen were combined in a container where they reacted continuously, with the help of a platinum catalyst, to produce a small but steady flow of electric current. Grove's work was regarded by most as merely a scientific curiosity until the 1950s, when fuel cell technology was further developed by the chemist Roger Bacon in Cambridge. In the 1960s fuel cells were successfully used to generate electricity, heat and drinking water for the US Gemini and Apollo spacecraft, but they were very expensive. Though still costly, in recent years there have been substantial falls in fuel cell prices and major improvements in performance, trends that lead many to believe they could soon become competitive with more conventional energy conversion systems. The basic principles of fuel cells are described in Box 14.2, opposite.

BOX 14.2 Fuel cells: the basic principles

How do fuel cells work? Many people will be familiar with school experiments in which water is split into its constituents, hydrogen and oxygen, by the process of 'electrolysis' – passing an electric current between two electrodes immersed in water. Fuel cells operate in a manner that is essentially the reverse of electrolysis – by combining, rather than splitting, hydrogen and oxygen. This process generates an electric current, water – and some 'waste' heat.

Fuel cells can use any two reactants that are respectively oxidising (i.e. a source of oxygen) and reducing (i.e. readily combine with oxygen), but the most common reactants are hydrogen and oxygen (or air, which is approximately 20% oxygen).

The principal characteristics of fuel cells and their mode of operation are shown in Figure 14.9. As can be seen, most fuel cells consist of two electrodes, an 'anode' and a 'cathode', which can be made from a variety of electrically-conducting materials. The electrodes are usually coated with platinum, or a platinum group metal such as palladium or ruthenium, which acts as a catalyst. Catalysts are substances that increase the rate of a chemical reaction without themselves undergoing any permanent chemical change. Between these is placed an 'electrolyte', of which again there are a variety of types. Normally hydrogen is the fuel fed to the anode, while oxygen (from air) is supplied to the cathode. Both the anode

and the cathode are porous, allowing the gases to flow through them. With the aid of the catalysts present on the surface of the electrodes, the hydrogen splits into hydrogen ions (i.e. protons) and electrons. The electrons flow away from the anode into an external electrical circuit where they can be made to deliver useful energy. Meanwhile, the hydrogen ions flow through the electrolyte to the cathode where (again with the aid of a catalyst) they combine with the oxygen supplied to the cathode and the incoming electrons from the external electrical circuit to form water vapour. Depending on the type of cell, some 30% to 60 % of the energy content of the input fuel is converted to electricity: the rest appears as heat, but this can often be used, either for space or water heating or to provide energy for the 'reformers' that may be required to convert, say, natural gas into the pure hydrogen required by the fuel cell.

The key advantage of the fuel cell as an energy converter is that electricity is produced *directly*. This means that its efficiency can be higher than the limits set by Carnot for heat engines (see Chapter 6). It also means that there are no emissions of the gaseous pollutants that are associated with combustion processes, such as SO_2, NO_x or particulates. If pure hydrogen is the fuel, there are no CO_2 emissions and the only other emission, apart from some 'waste' heat, is water vapour.

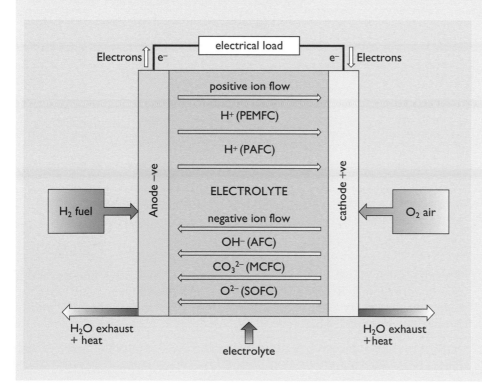

Figure 14.9 Principles of fuel cell operation. For key to abbreviations see Table 14.3 Notes (source: Laughton, 2002)

Fuel cell types

The science and technology of the fuel cell have advanced enormously since Grove's day. There is now a wide variety of fuel cell types, each using different arrangements of fuel, electrolyte, electrodes, etc. One way of classifying fuel cells is by the type of electrolyte used. A summary of the main fuel types and their characteristics is shown in Table 14.3.

Table 14.3 Classification of fuel cell types and characteristics

Characteristics	Fuel cell type					
	AFC	PEMFC	DMFC	PAFC	MCFC	SOFC
Electrolyte	aqueous potassium hydroxide (30–40%)	sulphonated organic polymer (hydrated during operation)	sulphonated organic polymer (hydrated during operation)	phosphoric acid	molten lithium/ sodium/ potassium carbonate	yttria-stabilized zirconia
Operating temperature	60–90	70–100	90	150–220	600–700	650–1000
Charge carrier	OH^-	H^+	H^+	H^+	CO_3^{2-}	O^{2-}
Anode	nickel (Ni) or platinum group metal	platinum (Pt)	platinum-ruthenium (Pt, Ru)	Platinum (Pt)	nickel/ chromium oxide	nickel/yttria-stabilized zirconia
Cathode	platinum (Pt) or lithiated NiO	platinum (Pt)	platinum-ruthenium (Pt, Ru)	platinum (Pt)	nickel oxide NiO)	strontium (Sr) doped lanthanum manganite
Co-generation heat	Low temperature	Low temperature	Low temperature	Temperature acceptable for many applications	High temperature	High temperature
Electrical efficiency, %	60	40–45	30–35	40–45	50–60	50–60
Fuel sources	H_2 Removal of CO_2 from both gas streams necessary	H_2 Reformate with less than 10 ppm CO	water/ methanol solution	H_2 reformate	H_2, CO, natural gas	H_2, CO, natural gas

Notes: AFC is Alkaline Fuel Cell. PEMFC is Proton Exchange Membrane Fuel Cell. DMFC is Direct Methanol Fuel Cell. PAFC is Phosphoric Acid Fuel Cell. MCFC is Molten Carbonate Fuel Cell. SOFC is Solid Oxide Fuel Cell. Reformate is pure hydrogen, or a hydrogen-rich gas, produced by reforming fossil fuels

Source: based on Laughton, 2002

The alkaline fuel cell (AFC) was the type used in the US Apollo spacecraft and later the Space Shuttle, where it provides both electricity and drinking water. It uses as its electrolyte a solution of potassium hydroxide and is fuelled by pure hydrogen.

Solid polymer fuel cells, as their name implies, use as their electrolyte specially treated polymers that only allow the passage of positive ions.

They were initially developed in the 1960s by General Electric for the US Gemini spacecraft. There are two types of Solid Polymer Fuel Cell (SPFC), both of which operate at relatively low temperatures: the Proton Exchange Membrane Fuel Cell (PEMFC), which uses hydrogen as its input fuel, and the Direct Methanol Fuel Cell (DMFC), which is fuelled by a water/methanol solution. The PEMFC technology has been refined and developed from the early spacecraft designs by a Canadian company, Ballard Power Systems, and has now been adopted by many of the world's leading motor manufacturers, such as Daimler-Chrysler, Ford and General Motors, as the basis of a new generation of 'zero-emission' cars (see Chapter 1 Figure 1.53). The Direct Methanol Fuel Cell is at a less advanced stage of development.

The Phosphoric Acid Fuel Cell (PAFC), fuelled by hydrogen, uses phosphoric acid as its electrolyte and operates at moderate temperatures. PAFC technology has become established as a reliable source of electricity and heat in buildings and other stationary applications.

The Molten Carbonate Fuel Cell (MCFC) uses molten lithium, sodium or potassium carbonate as its electrolyte. In the Solid Oxide Fuel Cell (SOFC) the electrolyte is a solid, yttria-stablilized zirconia, which also constitutes the anode. Both MCFCs and SOFCs operate at high temperatures and can use as their fuel either hydrogen, carbon monoxide or natural gas. Though still under development, they are regarded as promising technologies for future large-scale, low-emission electrical power production.

More than a century and a half after Sir William Grove's invention, the fuel cell now seems on the verge of widespread deployment in the world's energy systems, supplanting cruder, combustion-based technologies for converting the energy content of fossil fuels into useful electricity and heat.

Fuel cells are a key technology underpinning what some see as an emerging 'hydrogen economy' of the future – a topic to which we now turn.

14.5 **A fossil-fuel based hydrogen economy**

One of the first to envisage the possibility of a future world powered by hydrogen fuel was the science fiction writer Jules Verne. His 1874 novel *The Mysterious Island* includes the following prescient passage in which the engineer Cyrus Harding discusses with his companions what will happen when supplies of coal run out.

'And what will they burn instead of coal?'

'Water,' replied Harding.

'Water!' cried Pencroft, 'water as fuel for steamers and engines! Water to heat water!'

'Yes, but water decomposed into its primitive elements,' replied Cyrus Harding, 'and decomposed, doubtless, by electricity, which will then have become a powerful and manageable force…. Yes, my friends, I believe that water will some day be employed as a fuel, that hydrogen and oxygen, which constitute it, used singly or together, will furnish an inexhaustible source of heat and light of an intensity of which coal is not capable. Some day the coalrooms of steamers and the tenders of locomotives will, instead of coal, be stored with these two condensed gases, which will burn in the furnaces with enormous calorific power.'

Verne, understandably, left open the question of where the electricity required to 'decompose' water would come from.

The history of human fuel use over the twentieth century can be viewed as a process of gradual de-carbonization: a movement from coal to oil and more recently to natural gas – though of course, all three fuels are still in use.

By the middle of the twenty-first century it looks increasingly likely that further de-carbonization will have taken place and that hydrogen, a zero-carbon fuel, will be playing an important role in the world's energy systems. Some of this hydrogen will probably be generated by electrolysis from renewable electricity sources such as hydro, wind or solar power. But it is likely that much of it will be produced from fossil fuels such as natural gas or coal, with capture and sequestration of the CO_2 produced as part of the conversion process see Figure 14.10.

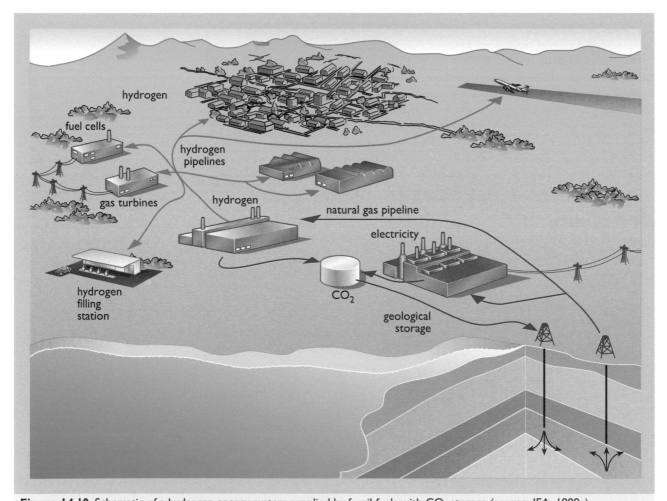

Figure 14.10 Schematic of a hydrogen energy system supplied by fossil fuels with CO_2 storage (source: IEA, 1999c)

There are two main reasons for converting fossil fuels to hydrogen. Firstly, it is easier to separate the CO_2 from fossil fuels during the process of hydrogen production than it is to capture CO_2 after the combustion of fossil fuels in conventional systems such as boilers. Secondly, as we have seen, the hydrogen can be converted in a fuel cell into electricity and heat with high efficiency and no pollutant emissions. Hydrogen, like conventional fuels, can also be simply burned in air to produce heat. Some NO_x is produced, but otherwise there are far fewer pollutants than with fossil fuel combustion.

Hydrogen, however, is not like fossil fuels that can be simply be extracted from the ground. Before it can be used, hydrogen needs to be extracted from the other compounds within which it is normally bonded in nature – principally water, or the hydrocarbons – and this separation requires energy. Hence hydrogen is not a primary but a secondary energy source. It is perhaps best viewed as an energy *carrier* – a convenient way of transporting or storing the energy originating from primary sources.

Over long distances, the energy losses involved in hydrogen distribution can be lower than those incurred when electricity, the most popular energy carrier, is used for distribution.

Hydrogen is already produced in substantial quantities worldwide, mainly by 'reforming' natural gas using steam, as described below and referred to in Chapter 7 in the context of converting one fossil fuel to another. World production in the late 1990s was approximately 500 billion cubic metres per year, with an energy content of around 6.5 EJ, equivalent to some 1.5% of world primary energy production (IEA, 1999c). At present, hydrogen is seldom used for energy purposes – with the exotic exception of its use as a fuel for space rockets. Its main use is in producing ammonia for fertilizer production, in oil refining and in various chemical processes. Hydrogen has for many years been transported reliably and safely over long distances. There are some 1500 km of pipelines carrying hydrogen in Europe, and some 700 km in the US. Pipelines carrying pure hydrogen need to be made of special steel alloys to avoid the hydrogen degrading the steel, but such materials are well established and understood.

Hydrogen is also currently produced by electrolysis. For example, Norwegian fertilizer production is based on hydrogen produced in this way from the country's low-cost hydro-electric plants. But at present the cost of electrolytically-produced hydrogen is usually greater than that of hydrogen derived from fossil fuels.

The use of renewable energy to produce hydrogen, in what has been called the 'solar hydrogen economy', will be discussed in more detail in the companion volume *Renewable Energy*. A solar hydrogen economy may well be feasible in the longer term, when technologies and economics are more favourable than at present; but in the meantime a hydrogen economy based on fossil fuels with carbon capture and sequestration could be a valuable transitional stage, facilitating the eventual emergence of a renewables-based hydrogen economy.

Producing hydrogen from fossil fuels, as we have seen, entails the production of carbon dioxide. But if the CO_2 from hydrogen production were captured and sequestered, using the techniques described in Section 14.2

above, the overall process would be a 'climate-friendly' one. It is considerably easier to capture CO_2 during the process of reforming natural gas, for example, than it is to capture it from the much more dilute stream of gases that are the result of combustion.

Reforming is currently the cheapest way of producing hydrogen. In this process, natural gas and superheated steam are passed over a catalyst at 900 °C to produce the mixture of hydrogen and carbon monoxide called synthesis gas, or syngas (see Section 7.12 of Chapter 7). The carbon monoxide is then reacted with more steam, in a so-called 'shift reaction', to produce more hydrogen and carbon dioxide. The hydrogen is then purified. The CO_2 can be captured using an ethanolamine solvent as described in Section 14.2 above. Heat generated by the process would be used to release the CO_2 and regenerate the solvent. This point highlights the fact that some energy is required for CO_2 capture and sequestration. The CO_2 would then be compressed, liquefied and transported by pipeline or tanker to the sequestration site. It is estimated that in a purpose-built plant the costs of capturing and sequestering the CO_2 in this way would add some 20-30% to the price of energy.

Oil and coal can also be used to produce syngas, as described in Section 7.12. The syngas can then be burned directly in a combined-cycle gas turbine plant similar to that used for natural gas. The whole process, including gasification, is described as an 'Integrated Gasification Combined Cycle' (IGCC) power plant (Box 14.3). If pure hydrogen is required, the mixture can be processed in a further 'shift reaction' as described for natural gas above. Capture and storage of CO_2 from such a process is, once again, estimated to add about 25% to the costs of hydrogen production, if the plant is purpose-built.

Hydrogen storage and use in transport

Hydrogen gas has a relatively high energy content per unit *mass* (120 MJ kg^{-1}, compared to 42 MJ kg^{-1} for gasoline) but has a very low energy content per unit *volume*. In other words, at normal pressures and temperatures energy storage using hydrogen involves relatively large volumes.

The principal methods of storage of hydrogen at present are (a) as a compressed gas, and (b) as a liquid. Compressed hydrogen gas can be stored in tanks made of steel or composite materials at pressures as high as 700 bar (70 MPa). In liquid form, hydrogen requires relatively less storage volume, but has to be stored at temperatures below −253 °C, which involves highly insulated tanks and significant cooling energy use.

Hydrogen can also be stored by allowing it to react with certain metal alloys to form metal hydrides. The hydrogen, when required, is then released by heating the hydride. But this method at present involves substantial expense and weight.

Research continues into other, more cost-effective methods of storing hydrogen, for example by adsorbing it to finely-powdered carbon particles.

A way of introducing hydrogen into the existing energy infrastructure would be to blend hydrogen, in modest proportions, into the distribution networks for natural gas (methane) that exist in many countries. The resulting blend

BOX 14.3 'Clean' coal-fired electricity: the integrated gasification combined cycle (IGCC) power plant

The Integrated Gasification Combined Cycle (IGCC) power plant (Figure 14.11) is one of the most promising approaches to reducing emissions from coal-fired electricity generation. Syngas produced from coal is burned in a combined cycle gas turbine/steam turbine power plant similar to that of a conventional CCGT (see Box 9.11 of Chapter 9). As with the CCGT, the efficiency of electricity production is high. But because energy is required to convert the coal to syngas, the overall efficiency is lower than for a CCGT – around 45%.

A further advantage of the IGCC plant is that sulphur and nitrogen compounds can be removed from the gas before it enters the turbine. Moreover, although carbon dioxide is produced during the re-formation process, because it is produced at high pressure it is relatively easy to capture, for subsequent 'sequestration' as described in Section 14.3 above. However, only a few

IGCC plants have been built and their costs are still comparatively high.

In an integrated gasification combined cycle (IGCC) power-generating plant, pulverized coal is fed into a gasifier at about 30 bar pressure, together with oxygen from an air separation unit (ASU). The raw fuel-gas (syngas) is produced in the gasifier at about 1300 °C and is cooled to about 200 °C before being scrubbed with water to remove compounds such as ammonia (NH_3) and hydrogen chloride (HCl). It is then further cooled and scrubbed with a solvent to remove sulphur compounds such as hydrogen sulphide (H_2S). The cleaned gas is then fired in a gas turbine. Ash in the coal is recovered as a mineral slag from the gasifier and the sulphur compounds removed from the gas are recovered as sulphur. Nitrogen (N_2) from the ASU is typically added to the fuel-gas in the gas turbine to control NO_x emissions (source: DTI, 1999).

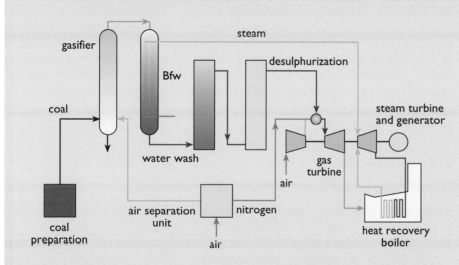

Figure 14.11 Diagram illustrating operation of IGCC Power plant. Note: Bfw is boiler feed water

Source: DTI, 1999

is sometimes known as 'hythane'. It is estimated that around 10–15% of hydrogen could be added to natural gas without noticeable effects on appliances. If this hydrogen were produced from fossil fuels in a process that included CO_2 capture and sequestration, this would enable a moderate reduction in overall greenhouse gas emissions to be achieved.

For supplying energy to buildings, hydrogen can be used in internal combustion engines or fuel cells to provide combined heat and power.

When used as a transport fuel, hydrogen can also be used either in a modified internal combustion engine or in a fuel cell. It has been estimated that hydrogen can be delivered to users at a cost, excluding taxes, roughly competitive with petroleum including taxation, in countries such as the

UK where fuel taxes are high. If a fuel cell is used, the overall energy efficiency is likely to be about double that of an internal combustion engine, mainly because of the fuel cell's higher part-load efficiency. This means that the hydrogen fuel could be up to twice the price of petroleum and still remain competitive.

The main problem is finding enough space for hydrogen storage in vehicles. For this reason, some of the first applications of hydrogen-powered vehicles are buses, where room for storage can be found – usually on the roof. A further advantage of buses is that they travel on pre-arranged routes and are refuelled at depots, so hydrogen only has to be supplied at certain fixed locations.

A few hydrogen-fuelled buses use internal combustion engines, but most are powered by fuel cells. Some use pure hydrogen from on-board storage tanks; in others, the hydrogen is produced by on-board reforming of fossil fuels. In order for hydrogen to become attractive to millions of car users, there will need to be an extensive network of hydrogen filling stations. This entails a familiar 'chicken and egg' problem: until there are enough filling stations, motorists will be unwilling to use hydrogen in their cars; and until enough motorists are using hydrogen in their cars, it will be uneconomic to provide filling stations and a hydrogen distribution network. As an interim option, conventional filling stations could install reformers to manufacture hydrogen from, for example, natural gas, which is already widely distributed in many countries. However, manufacturing hydrogen locally in this way would involve some CO_2 emissions, since capture and sequestration would probably be impractical on such a small scale.

Hydrogen could also be used as a zero-carbon fuel for aircraft. Jet engines can be run on hydrogen, as Russia demonstrated in 1988 when a modified Tupolev 155 airliner made a brief flight with one of its engines fuelled by hydrogen (see Hoffman, 2001). Burning hydrogen in jet engines does, however, lead to greenhouse gas emissions in the form of water vapour and NO_x. Of course, conventional jet engines fuelled by kerosene also emit water vapour and NO_x, but with hydrogen-fuelled jet engines the amount of water vapour emitted would be considerably greater. Water vapour emitted in the upper troposphere, where most commercial aircraft fly, is removed from the atmosphere by precipitation in a few weeks. However, for aircraft flying in the stratosphere, water vapor emissions persist.

Aircraft powered by hydrogen could be bulky in appearance, depending on the method of hydrogen storage used, but should have a lower take-off weight, due to hydrogen's low weight in relation to its energy content, and this should lead to improved fuel economy on long-distance flights. The European aircraft manufacturer Airbus Industrie has stated that it aims to demonstrate a hydrogen-powered aircraft early in the twenty-first century.

Hydrogen safety

Ever since the disastrous fire on the airship Hindenburg at Lakehurst, New Jersey, in 1937, hydrogen has been considered highly unsafe by many. But this reputation is probably unjustified.

In 1997 Addison Bain, former manager of the hydrogen program at NASA's Kennedy Space Center, presented to the annual meeting of the US National

Hydrogen Association the results of his extensive investigation of the Hindenburg disaster.

> Bain uncovered impressive evidence that persuasively exonerated hydrogen as the primary cause of the disaster and showed that static electricity and the presence of highly-inflammable materials on the airship's skin were to blame.
>
> Hoffman, 2001

Static electrical discharges, it seems, ignited the skin of the Hindenburg, which was made of a highly-flammable cellulosic material, and this then ignited the hydrogen. If the skin had been made of a non-combustible material the disaster would not have happened.

Figure 14.12 The German airship Hindenburg burst into flames at Lakehurst, New Jersey in May 1937, killing 36 people. Recent research has shown that hydrogen was not the primary cause of the disaster

Hydrogen is the lightest element in nature; it is highly reactive and difficult to handle. Fire and explosion are the main dangers, but on the other hand hydrogen dissipates very rapidly in confined spaces. In cars and buildings where hydrogen is already used, sensors are employed to detect concentrations of over 4% and open windows automatically.

Hydrogen flames are invisible, which increases the risk of individuals inadvertently being burned. But their higher flame speed means that hydrogen fires burn out rapidly. A vehicle fire involving petrol may take 20–30 minutes to burn out; with hydrogen, the equivalent time would be only 1–2 minutes.

The risks entailed in hydrogen fuel use, in short, are different to those associated with other fuels, but most experts consider hydrogen be no more dangerous overall (IEA, 1999c).

14.6 Can fossil fuel use be made more sustainable?

In the preceding sections, we explored various ways in which the carbon emissions from fossil fuel combustion could be sequestered in order to minimize their contribution to climate change; and we looked at methods for converting fossil fuels to hydrogen (with capture and sequestration of carbon produced in the process) which can then react with oxygen in a fuel cell to produce heat and electricity in a clean, combustion-less process. To what extent could such technologies render fossil fuel use more sustainable?

In Chapter 1, we described the main criteria by which the sustainability of energy sources should be judged. These were:

- that they should not be substantially depleted by continued use;
- that their use does not entail pollutant emissions or other hazards to the environment on a substantial scale; and
- that their use does not involve the perpetuation of substantial human health hazards or social injustices.

We also stressed that sustainability in this context is a relative rather than absolute concept; and suggested that although sustainability ought in principle to be assessed over an indefinitely long time scale, for practical purposes a century or so might be a reasonable period over which to make a judgement.

How does 'cleaner' fossil fuel use, incorporating the improved technologies described in this chapter, measure up to our sustainability criteria?

On our first criterion, depletion, even 'clean' fossil fuel is clearly unsustainable in the long term. Fossil fuels, even if used more cleanly, will still eventually run out. Indeed, given that significant additional quantities of energy are required to clean up emissions from fossil fuel combustion, to capture and sequester CO_2, or to reform fossil fuels into cleaner hydrogen, then the rate at which cleanly-used fossil fuels would become depleted is somewhat greater than would be the case if clean-up technologies were not employed.

On the second criterion, pollutant emissions and health hazards, the verdict is less clear. Emissions of SO_2, NO_x and particulates from fossil fuel combustion can be cleaned up to a very considerable extent, but perhaps not eliminated completely.

The use of fuel cells avoids the emission problems of the combustion process, but if the hydrogen for fuel cells is derived from fossil fuels, this still entails CO_2 emissions that need to be dealt with by capture and sequestration.

Carbon dioxide emissions from fossil fuel combustion, or from re-forming of fossil fuels to produce hydrogen, are inevitable, but as we have seen the CO_2 can be captured and sequestered, at least from large installations. However, it would probably not be practical to capture and sequester CO_2 emissions from small-scale or mobile combustion sources, such domestic boilers or car engines. Further research is also required to establish what proportion of the CO_2 emitted from human fossil fuel combustion can safely be sequestered, without subsequent leakage, for the many centuries or millennia that are necessary, in the various 'sinks' that may be available.

On our third sustainability criterion, hazards to human health and perpetuation of social injustices, 'clean' fossil fuels are in many respects obviously an improvement over 'dirty' fossil fuels on health grounds; but in terms of social justice they do not offer any significant improvement. Fossil fuels, whether 'dirty' or 'clean', because of their concentration in a few geologically-favoured locations, are likely to continue to engender the major conflicts and injustices that seem inevitably to arise when humans compete for scarce resources.

Finally, how do 'clean' fossil fuels compare, in terms of sustainability, with renewable energy sources such as solar, hydro, wind, wave and tidal power?

This interesting question is addressed in our companion volume *Renewable Energy*.

References

Beggs, C. (2002) *Energy: Management, Supply and Conservation*, Butterworth-Heinemann, 284pp.

Boyle, G. (ed) (1996, 2003) *Renewable Energy: Power for a Sustainable Future*, Oxford University press and Open University.

Department of Trade and Industry (various issues, 1990–2001) *Energy Trends,* monthly, HMSO.

Department of Trade and Industry (1999) *Cleaner Coal Technologies* UK Department of Trade and Industry, Report DTI/Pub URN 99/959, 28pp.

Department of Trade and Industry (2001) *Digest of UK Energy Statistics 2000*, Annex B; Energy and the Environment, Table B1, p. 251.

Eastop, T. D. and Croft, D. R. (1990) *Energy Efficiency for Engineers and Technologists*, Longman Scientific and Technical.

Eyre, N. J. (1990) *Gaseous Emissions due to Electricity Fuel Cycles in the United Kingdom*, Harwell, Oxfordshire, Energy Technology Support Unit.

Goldemberg, J. (ed.) (2000) *World Energy Assessment: Energy and the Challenge of Sustainability*, New York, United Nations Development Programme, United Nations Department of Economic & Social Affairs and World Energy Council, 508pp.

Hoffman, P. (2001) *Tomorrow's Energy: Hydrogen, Fuel Cells and the Prospects for a Cleaner Planet*, MIT Press, p. 289.

International Energy Agency (1999a) *Cleaner Coal Technologies: Options*, International Energy Agency, Paris, IEA, 29pp.

International Energy Agency (1999b) *Ocean Storage of CO_2*, IEA Greenhouse Gas R&D Programme Report, Paris, IEA, 25pp.

International Energy Agency (1999c) *Hydrogen – Today and Tomorrow*, IEA Greenhouse Gas R&D Programme Report, Paris, IEA, 34pp.

International Energy Agency (2001) *Putting Carbon Back in the Ground*, IEA Greenhouse Gas R&D Programme Report, Paris, IEA, 26pp.

Intergovernmental Panel on Climate Change (2001) *Third Assessment Report*, Cambridge University press.

Laughton, M. (2002) 'Fuel Cells', *IEE Science & Education Journal*, Institution of Electrical Engineers, London.

Royal Commission on Environmental Pollution (2000) *Energy – the Changing Climate*, CMND 4749, The Stationery Office, p. 292.

Appendix

This appendix is designed as a quick reference source for 'energy arithmetic'. Section A1 offers a brief summary of the arithmetic itself; A2 gives conversion factors between the main units used for energy and power throughout this book; and A3 relates some older or non-SI units to their SI equivalents. (There is a brief introduction to the SI system in *'Energy Arithmetic'*, in Chapter 2. Formal definitions and more detailed accounts of the basis of the SI units can be found in *Quantities, units and symbols* published by The Royal Society, 1975.)

A1 Orders of magnitude

Discussions of energy consumption and production frequently involve very *large* numbers, and accounts of processes at the atomic level often need very *small* numbers. Two solutions to the problem of manipulating such numbers are described here: the use of a shorthand form of arithmetic and the use of prefixes.

Powers of ten

Two million is two times a million, and a million is ten times ten times ten times ten times ten times ten – six of them in all. This remark can be written mathematically:

$$2\ 000\ 000 = 2 \times 10 \times 10 \times 10 \times 10 \times 10 \times 10 = 2 \times 10^6$$

The quantity 10^6 is called *ten to the power six* (or *ten to the six* for short), and the 6 is known as the **exponent**. The advantage of using this power-of-ten form is particularly obvious for *very* large numbers. World primary energy consumption in the year 2000, for instance, was 424 000 000 000 000 000 000 joules, and it is certainly easier to write (or say) 424×10^{18} joules than to spell out all the zeros.

The method can also be used for very small numbers, with the convention that one tenth (0.1) becomes 10^{-1}; one hundredth (0.01) becomes 10^{-2}, etc. The separation of two atoms in a typical metal, for instance, might be about 0.13 of a billionth of a metre, which is 0.000 000 000 13 m. In more compact form, it becomes 0.13×10^{-9} m.

Scientific notation is a way of writing numbers that is based on this concept, but with a more specific rule. Any number, whatever its magnitude, is written in scientific notation as *a number between 1 and 10* multiplied by the appropriate power of ten. So in this form, the numbers above would be written slightly differently:

424×10^{18} becomes 4.24×10^{20}

0.13×10^{-9} becomes 1.3×10^{-10}

In practice, styles like those on the left in each of these are commonly used, and a number in this form will be accepted by computers and scientific calculators – but it will be reproduced by them in either 'long decimal' or strict scientific notation, depending on the mode selected.

Prefixes

The powers of ten provide the basis for the prefixes used to indicate multiples (including sub-multiples) of units. The following table, parts of which appeared in the main text as Table 2.1, shows these in decreasing order.

Table A1

Symbol	Prefix	Multiply by	... which is
E	exa-	10^{18}	one quintillion
P	peta-	10^{15}	one quadrillion
T	tera-	10^{12}	one trillion
G	giga-	10^{9}	one billion
M	mega-	10^{6}	one million
k	kilo-	10^{3}	one thousand
h	hecto-	10^{2}	one hundred
da	deca-	10	ten
d	deci-	10^{-1}	one tenth
c	centi-	10^{-2}	one hundredth
m	milli-	10^{-3}	one thousandth
μ	micro-	10^{-6}	one millionth
n	nano-	10^{-9}	one billionth
p	pico-	10^{-12}	one trillionth
f	femto-	10^{-15}	one quadrillionth
a	atto-	10^{-18}	one quintillionth

A2 Units and conversions

Chapter 2 discussed some of the units used in specifying quantities of energy or power in the 'real world'. The three tables in this section summarize the relationships between some of these units.

Energy

The main energy units used in this book are the joule, the kilowatt-hour and the tonne of oil or coal equivalent. Tables A2 and A3 give the conversion factors between the most frequently used multiples of these units, on the 'household' scale (A2) and on the larger scale of national or world data (A3). For the reasons discussed in Chapter 2, the oil and coal equivalents are expressed to two significant figures only, and readers are advised always to take note of the conversion factors adopted by any data source they use.

Table A2

	MJ	GJ	kWh	toe	tce
1 MJ =	1	0.001	0.2778	2.4×10^{-5}	3.6×10^{-5}
1 GJ =	1000	1	277.8	0.024	0.036
1 kWh =	3.60	0.0036	1	8.6×10^{-5}	1.3×10^{-4}
1 toe =	42 000	42	12 000	1	1.5
1 tce =	28 000	28	7800	0.67	1

Table A3

	PJ	EJ	TWh	Mtoe	Mtce
1 PJ =	1	0.001	0.2778	0.024	0.036
1 EJ =	1000	1	277.8	24	36
1 TWh =	3.60	0.0036	1	0.086	0.13
1 Mtoe =	42	0.042	12	1	1.5
1 Mtce =	28	0.028	7.8	0.67	1

Note that the factors for kWh and TWh are direct energy conversions, not the equivalent power station outputs discussed in Chapter 2.

Power

Power is the rate at which energy is consumed, transferred or transformed, and its unit is the watt. It follows that any rate of production or consumption of energy can be expressed as an average rate in watts. Table A4 shows the quantities of energy per hour and per year for different constant rates in watts. Note that it can also be used to show that, for instance, 1 kWh is 3.6 MJ and 1 TWy (terawatt-year) is 31.54 EJ or 750 Mtoe.

Table A4

Rate	Joules ... per hour	per year	Kilowatt-hours per year	Oil equivalent per year	Coal equivalent per year
1 W	3600 J	31.54 MJ	8.76	0.75×10^{-3} toe*	1.1×10^{-3} tce*
1 kW	3.6 MJ	31.54 GJ	8760	0.75 toe	1.1 tce
1 MW	3.6 GJ	31.54 TJ	8.76×10^{6}	750 toe	1100 tce
1 GW	3.6 PJ	31.54 PJ	8.76×10^{9}	0.75 Mtoe	1.1 Mtce
1 TW	3.6 TJ	31.54 EJ	8.76×10^{12}	750 Mtoe	1100 Mtce

* I.e. the energy equivalent of 0.75 kg of oil or 1.1 kg of coal

Other quantities

Table A5 gives the SI equivalents of a few other metric units and some older units that remain in common use. The final column shows the inverse relationships. For brevity, scientific notation is used for numbers greater than 10 000 or less than 0.1.

Table A5

Quantity	Unit	SI equivalent	Inverse
mass	1 oz (ounce)	$= 2.834 \times 10^{-2}$ kg	1 kg = 35.27 oz
	1 lb (pound)	$= 0.4536$ kg	1 kg = 2.205 lb
	1 *ton*	$= 1016$ kg	1 kg $= 0.9842 \times 10^{-3}$ ton
	1 *short ton*	$= 972$ kg	1 kg $= 1.1021 \times 10^{-3}$ short ton
	1 t (*tonne*)	$= 1000$ kg	1 kg $= 10^{-3}$ t
	1 u (*unified mass unit*)	$= 1.660 \times 10^{-27}$ kg	1 kg $= 6.024 \times 10^{26}$ u
length	1 in (inch)	$= 2.540 \times 10^{-2}$ m	1 m = 39.37 in
	1 ft (foot)	$= 0.3048$ m	1 m = 3.281 ft
	1 yd (yard)	$= 0.9144$ m	1 m = 1.094 yd
	1 mi (mile)	$= 1609$ m	1 m $= 6.214 \times 10^{-4}$ mi
speed	1 km hr^{-1} (kph)	$= 0.2778$ m s^{-1}	1 m s^{-1} = 3.600 kph
	1 mi hr^{-1} (mph)	$= 0.4470$ m s^{-1}	1 m s^{-1} = 2.237 mph
area	1 in^2	$= 6.452 \times 10^{-4}$ m^2	1 m^2 = 1550 in^2
	1 ft^2	$= 9.290 \times 10^{-2}$ m^2	1 m^2 = 10.76 ft^2
	1 yd^2	$= 0.8361$ m^2	1 m^2 = 1.196 yd^2
	1 acre	$= 4047$ m^2	1 m^2 $= 2.471 \times 10^{-4}$ acre
	1 ha (hectare)	$= 10^4$ m^2	1 m^2 $= 10^{-4}$ ha
	1 mi^2	$= 2.590 \times 10^6$ m^2	1 m^2 $= 3.861 \times 10^{-7}$ mi^2
volume	1 in^3	$= 1.639 \times 10^{-5}$ m^3	1 m^3 $= 6.102 \times 10^4$ in^3
	1 ft^3	$= 2.832 \times 10^{-2}$ m^3	1 m^3 = 35.31 ft^2
	1 yd^3	$= 0.7646$ m^3	1 m^3 = 1.308 yd^2
	1 litre	$= 10^{-3}$ m^3	1 m^3 = 1000 litre
	1 gal (UK)	$= 4.546 \times 10^{-3}$ m^3	1 m^3 = 220.0 gal
	1 gal (US)	$= 3.785 \times 10^{-3}$ m^3	1 m^3 = 264.2 gal (US)
	1 *bushel*	$= 3.637 \times 10^{-2}$ m^3	1 m^3 = 27.50 bushel
force	1 lbf (weight of 1 lb mass)	$= 4.448$ N	1 N = 0.2248 lbf
pressure	1 lbf in^{-2} (or *psi*)	$= 6895$ Pa	1 Pa $= 1.450 \times 10^{-4}$ psi
	1 *bar*	$= 10^5$ Pa	1 Pa $= 10^{-5}$ bar
energy	1 ft lb (*foot-pound*)	$= 1.356$ J	1 J = 0.7376 ft lb
	1 eV (*electron volt*)	$= 1.602 \times 10^{-19}$ J	1 J $= 6.242 \times 10^{18}$ eV
	1 MeV	$= 1.602 \times 10^{-13}$ J	1 J $= 6.242 \times 10^{12}$ MeV
power	1 HP (*horse power*)	$= 745.7$ W	1.341×10^{-3} HP

Note: Units shown in italics are defined or discussed elsewhere in this book. See the Index for their locations.

Reference

The Royal Society (1975) *Quantities, Units and Symbols*, London, The Royal Society.

Acknowledgements

Grateful acknowledgement is made to the following sources for permission to reproduce material within this product.

Chapter 1

Tables

Table in Box 1.4: Four energy scenarios for the UK in 2050, Royal Commission on Environmental Pollution, 22nd Report, Energy – The Changing Climate, June 2000. Crown copyright material is reproduced under class licence number C01W0000065 with the permission of the Controller of Her Majesty's Stationery Office and the Queen's Printer for Scotland.

Figures

Figure 1.1 © John Isaac/Still Pictures; Figure 1.2 © Ian Hodgson/Reuters/Popperfoto; Figure 1.3 © PA Photos; Figure 1.4 © Accent Alaska; Figure 1.5 © Peter Bennets/Lonely Planet Images; Figure 1.6 © Bryan & Cherry Alexander Photography; Figure 1.7 © Mary Evans Picture Library; Figure 1.10 © Greenpeace/Carr; Figure 1.11, BP Statistical Review of World Energy 2001; Figure 1.15 & 1.16 BP Statistical Review of World Energy 2002; Figure 1.12 © Mary Evans Picture Library; Figure 1.13 © Gerry Gibb Photography; Figure 1.14 © Peter Mueller/Reuters/Popperfoto; Figure 1.18 Used by permission from the Intergovernmental Panel on Climate Change; Figure 1.19 Courtesy of UKAEA; Figure 1.20 Courtesy of Framatome-ANP, C. Pauquet; Figure 1.21 © PA Photos/EPA; Figure 1.22 Courtesy of EFDA-JET; Figure 1.23 Courtesy of ARBRE Energy Ltd, www.arbre.co.uk; Figure 1.24 © Ecoscene/Joel Creed; Figure 1.25 © Ecoscene/Mike Maidment; Figure 1.28 Courtesy of Susan Roaf; Figure 1.29 © Hank Morgan/Science Photo Library; Figure 1.30 © Derek Taylor; Figure 1.31 Courtesy of National Wind Power Ltd; Figure 1.32 © Adam Schmedes/Lokefilm, www.lokefilm.dk; Figure 1.33 Courtesy of AMEC Wind; Figure 1.34 The Crown Estate/British Wind Energy Association; Figure 1.35 Courtesy of Wavegen Co.Uk, www.wavegen.com; Figure 1.36 © Martin Bond/Science Photo Library; Figure 1.37 Courtesy of Marine Current Turbines Limited; Figure 1.38 © Photo: Alinari; Figure 1.41 With permission, from the Annual Review of Energy and the Environment, Volume 25 © 2000 by Annual Reviews www.annualreviews.org; Figure 1.43 © North Wales Newspapers; Figure 1.44 © Ecoscene/Nick Hawkes; Figure 1.45 Digest of UK Energy Statistics Web site: www.dti.gov.uk, Department of Trade & Industry, Crown copyright is reproduced with the permission of the Controller of Her Majesty's Stationery Office; Figure 1.46 © Greenpeace/Visser; © Greenpeace/Hindle; © Greenpeace/Cobbing; Figure 1.47 Courtesy of Chetwood Associates (Architects); Figure 1.48 The Elizabeth Fry Building: Courtesy of University of East Anglia; Figure 1.49 Blunden, J. and Reddish, A. (1991), 'Energy efficiency of different transport modes', figure 5.3, p. 40, *Energy Resources and Environment,* Hodder & Stoughton, by permission of the authors; Figure 1.50 Depart of Transport, Local Government and Regions (DTLR) 2001, *Transport Statistics, Great Britain* (27th edn), Crown copyright material is reproduced by permission of the Controller of Her Majesty's Stationery

Office; Figure 1.51 Courtesy of Toyota (GB) PLC; Figure 1.52 Reproduced with permission from Hypercar, Inc; Figure 1.53 Courtesy of Daimler-Chrysler; Figure 1.54 Courtesy of IEA Greenhouse Gas R & D Programme; Figure 1.55 *World Energy Assessment: Energy and The Challenge of Sustainability*, © 2000 UNDP.

Chapter 2

Figures

Figures 2.1 and 2.2: © Hulton Archive.

Sources

Figure 2.4: BP (2001); Figure 2.5: Times (1980) and BP (various issues); Figure 2.6: Ramage (1997); Figure 2.7: BP (2001) and DTI (2001); Figures 2.8 and 2.9: BP (1991), BP (2001), DTI (various issues), Eurostat (1990) and Ministry of Power (1962); Figure 2.10: BP (2001) and DEA (2001); Figures 2.11 and Figure 2.12: Eurostat (1990) and DEA (2001); Figure 2.13: BP (2001) and EIA (2001); Figures 2.14, 2.15 and 2.16: EIA (2001); Figure 2.17: BP (2001) and MINEFI (2001); Figures 2.18 and 2.19: Eurostat (1990) and MINEFI (2001); Figure 2.20: BP (2001) and EIA (2002); Figures 2.21 and 2.22: BP (2001), EIA (2002) and IEA (1998).

Chapter 3

The authors would like to thank Horace Herring for his assistance with statistical research.

Figures

Figures 3.2 & 3.17: Figures 6.6 and 5.24 from *Energy in World History* by Vaclav Smil. Copyright © 1992 by Westview Press. Reprinted by permission of Westview Press, a member of Perseus Books, L.L.C; Figures 3.6: © Mary Evans Picture Library; Figures 3.11: *Digest of UK Energy Statistics (DUKES),* 2001, DTI. Crown copyright material is reproduced under Class Licence Number C01W0000065 with the permission of the Controller of HMSO and the Queen's Printer for Scotland; Figures 3.14: © Martin Bond/Science Photo Library; Figures 3.15: *UK Energy in Brief,* July 2001, DTI. Crown copyright material is reproduced under Class Licence Number C01W0000065 with the permission of the Controller of HMSO and the Queen's Printer for Scotland; Figures 3.16: *UK Energy Sector Indicators,* 2001, DTI. Crown copyright material is reproduced under Class Licence Number C01W0000065 with the permission of the Controller of HMSO and the Queen's Printer for Scotland.

Sources

Figure 3.18: MINEFI (2001), Eurostat (1990), EIA (2000), TERI (2002), DTI (2001a) and DEA (2001); Figure 3.19: DTI (2001a) and (2001).

Chapter 4

Figures

Figure 4.1 © Science Museum/Science & Society Picture Library; Figure 4.2 © Science Museum/Science & Society Picture Library; Figure 4.3 University of Glasgow; Figure 4.6 Courtesy of Scottish Power; Figure 4.7 © Mary Evans Picture Library; Figure 4.8 © Science Museum/Science & Society Picture Library; Figure 4.9 © Mary Evans Picture Library; Figure 4.12 © Mary Evans Picture Library; Figure 4.13 © Martin Bond/Science Photo Library; Figure 4.14 © Mary Evans Picture Library; Figure 4.15 © The National Portrait Gallery; Figure 4.16 (a) © Hugh Turvey/Science Photo Library; Figure 4.16 (d) © P.M. Northwood; Figure 4.17 © Science Museum/Science & Society Picture Library; Figure 4.18 © Martin Bond/Science Photo Library; Figure 4.21 © National Portrait Gallery; Figure 4.24 © The Mansell Collection/Timepix/Rex Features.

Chapter 5

Figures

Figure 5.2 Antiquarian Images; Figure 5.3 Courtesy of Iron Bridge Gorge Museum Trust; Figure 5.5 © Hulton Archive; Figure 5.6 Courtesy of Professor Paul F. Starrs, University of Nevada; Figure 5.8 © 2002 The Science and Mathematics Teaching Center, University of Wyoming; Figure 5.9 Courtesy of the Spatial Sector, Environmental Unit of Sinclair Knight Merz Pty Limited; Figure 5.11 Courtesy of Corus; Figure 5.14 © Michael Gunther/Still Pictures; Figure 5.16 Courtesy of Hitachi Ltd. (Japan).

Sources

Figures 5.1 and 5.4: Ramage (1997); Figure 5.10: DTI (2001).

Chapter 6

Figures

Figure 6.5: © Science Museum/Science & Society Picture Library; Figures 6.8, 6.10, 6.12, 6.13, 6.14, 6.19: © Science Photo Library; Figure 6.15: © Hulton Archive; Figure 6.20: 'Development of Tandem-Compound 1000-MW Steam Turbine Generator', *Hitachi Review,* vol. 47, 1998, no. 5, courtesy of Hitachi, Japan.

Sources

Figures 5.1 and 5.4: Ramage (1997); Figure 5.10: DTI (2001).

Chapter 7

Text

Box 7.1: Kerr, D. (1994) 'Beginnings: "Paraffin" Young', *Shale Oil Scotland,* published and reproduced with permission of the author, David Kerr

Figures

Figures 7.2, 7.5, 7.8, 7.21, 7.24 & 7.30: © Hulton Archive; Figures 7.3, 7.11: © BP plc; Figure 7.4: © Bettman/Corbis; Figure 7.6: Whitehead, H. (1986) *An A-Z of Offshore Oil and Gas,* Kogan Page; Figures 7.9, 7.13 & 7.15: *Digest of UK Energy Statistics, DUKES,* 2001. Crown © material is reproduced under Class Licence no. C01W0000065 with the permission of the Controller of HMSO and the Queen's Printer for Scotland; Figure 7.12: Courtesy of ExxonMobil; Figure 7.14: © E. O. Hoppé/Corbis; Figure 7.17(a): © Hulton Deutsch Collection/Corbis; (b): © Marcus Enoch; Figure 7.18: Courtesy of Methanex NX Ltd (Motunui); Figure 7.20: © Mary Evans Picture Library; Figure 7.22: Crawford, H. B., Eisler, W. and Strong, L. (1977) *Energy Technology Handbook,* McGraw Hill, Inc.; Figure 7.26: Courtesy of Sasol Ltd; Figure 7.28: Courtesy of Southern Pacific Petroleum, N. L.; Figure 7.29: Courtesy of Suncor Energy; Figure 7.32: Laherrère, Jean, *Forecasting Future Production for Past Discovery,* OPEC Seminar, 28 September, 2001. Courtesy of the author.

Chapter 8

Figures

Figures 8.1, 8.4, 8.5, 8.9, & 8.11: © Science Photo Library; Figure 8.2: Courtesy of Janet Ramage; Figure 8.3: © Rogers and Mayhew (1980) Engineering Thermodynamics, Longman. Reproduced in Stone, R. (1992) Introduction to internal combustion engines, Macmillan.; Figure 8.8, 8.22 & 8.24: Courtesy of Bob Everett; Figure 8.12: Courtesy of MAN B&W Diesel Ltd.; Figures 8.18, 8.20 & 8.21: Reproduced with the kind permission of Rolls-Royce plc, www.rolls-royce.com/education; Figure 8.14: Courtesy of Marcus Enoch; Figure 8.19: © Hulton Archive; Figure 8.25: Courtesy of NREL/PIX.

Chapter 9

Figures

Figures 9.1, 9.2, 9.4, 9.5, 9.9, 9.15, 9.16 left, 9.17 & 9.23: © Science Photo Library; Figure 9.7: ASHRAE Centennial Collection; Figures 9.8 & 9.16 right: Electricity Council; Figures 9.12, 9.18, 9.20, 9.21, 9.22, 9.25, 9.27, 9.28, 9.34, 9.35, 9.36 & 9.41: Courtesy of Bob Everett; Figures 9.24, 9.32: © Hulton Archive; Figure 9.26: Illustrated London News; Figure 9.29: Electrical Association for Women; Figure 9.39: © Electricity Association.

Chapter 10

Figures

Figure 10.5 Archiv zur Geschichte der Max-Planck-Gesselschaft, Berlin-Dahlem; Figure 10.6 © Science Photo Library; Figure 10.9 © US National Archives/Science Photo Library; Figure 10.11 © American Institute of Physics/Science Photo Library; Figure 10.12 Hahn, O. (1950) *New Atoms Progress and Some Memories,* Elsevier Publishing Company Inc., Reproduced by permission from UN Publications Board; Figure 10.21 © Jeri Laber/Physics Today Collection/American Institute of Physics/Science Photo Library.

Chapter 11

Figures

Figure 11.1 World Nuclear Association; Figure 11.5 Courtesy of British Energy Group; Figure 11.6 Used with permission from Swedish Nuclear Fuel and Waste Management Company; Figure 11.7 Courtesy of EFDA-JET; Figure 11.10: © John Harris/Report Digital.

Chapter 12

The author would like to thank Susan Walker for her contribution to an earlier version of this text and Horace Herring for his statistical analysis contributions.

Figures

Figures 12.6 & 12.14 'The death of inflation', *The Guardian,* Saturday 5 May, 2001 © Guardian; Figure 12.7 © Hulton Archive; Figure 12.8 *UK Energy in Brief July 2001*, Energy Policy of the DTI. Crown copyright material is reproduced under Class Licence Number C01W0000065, with the permission of the Controller of Her Majesty's Stationery Office and the Queen's Printer for Scotland; Figure 12.10 Courtesy of Bob Everett; Figure 12.20 © John Harris, www.reportdigital.co.uk

Chapter 13

Text

Boxes 13.9-13.11: Gagnon, L *et al* (2002) 'Life-cycle assessment of electricity generation: The status of research in the year 2001', *Energy Policy,* Vol. 30 No. 10.

Figures

Figure 13.1: *World Energy Assessment; Energy and the Challenge of Sustainability,* United Nations Development Programme, 2000; Figure 13.2: Swanson, J. and Renew, D. C. 'Power-frequency Fields and People', *IEE Engineering Science and Education Journal,* April 1994, Institute of Electrical Engineers; Figure 13.3: Courtesy of Bob Everett; Figures 13.4-

Chapter 14

Figures

Index

A

Aberfan disaster 165
absolute zero of temperature 200, 201
absorption 578
acceleration due to gravity 133
accidents 523
    catastrophic 511, 523, 545–52
    coal mining 165, 527–9
    nuclear 18, 414–16, 438, 439–42,
        548–52
    workplace 527–9
acid deposition (acid precipitation or
    acid rain) 537, 554, 555, 556
actinides (transuranic elements) 402,
    408–9
Active (later Locomotion) 206
activity 401
adsorption 578
advanced gas-cooled reactors (AGRs)
    368, 418–20, 422
aeolipyle 188, 189
aerodynamic drag 301
aerosols 531, 540, 541
affordability 487–9
afterburner 321
agriculture
    consequences of climate change for
        547
    expanding uses of energy 95–6,
        96–100
    fertilizers 97–8
    India 124
    mechanization 98–100
air/fuel ratio 311, 312
air pollution 531–6, 564
air preheaters 217
air resistance 301
air transport 235
    aircraft petrol engines 303
    demand for air transport fuel 253
    diesel engines 307
    energy consumption 117–18
    hydrogen as fuel 592
Airbus Industrie 592
Alaska 240
Algeria 243
alcohol (ethanol) 21
Alkali Inspectorate 107
alkaline fuel cell (AFC) 586
alpha particles (α-particles) 397
alternating current (AC) 339–43
alternators 337
    three-phase 342, 343
aluminium smelting 106
ammonia 268
amp (A) 139
Ampère, André Marie 140, 335
anaerobic decay 22–3
animal dung 20, 123, 124

animals 102
annuitization 496–8
anthracite 164, 173, 174
anthropogenic greenhouse effect 16
anticlines 228, 229
appraisal wells 238
aquifers 33
    saline 48, 580
'Arbre' power station, Eggborough 20
Ardnacrusha hydroelectric scheme 369
Area Boards 377, 380
Argentina 240
armoured face conveyor 165
ash 172, 174
Askew, Norman 468
associated gas 235
Athabasca tar sands, Canada 280–1
Atmospheric Engine 191–2
atmospheric pollution 531–6, 564
atmospheric pressure 193
atomic bombs 407, 410
atomic nucleus 147, 402
atomic number 146, 149
atomic structure 147
atomic theory of matter 136
Australia 167, 169
autoproducers 369
availability 258
aviation spirit 249
aviation turbine fuel (jet fuel) 246, 249,
    253
axial compressor 315, 317

B

background radiation 537–9
Bacon, Francis 136
Bacon, Roger 584
bag filters 182
bar 193
barrages, tidal 31, 32
barrel of oil equivalent (boe) 62
barrels 61, 62
Bathgate Chemical Works 232
batteries 140, 334–5, 480
    rechargeable 335, 356, 357
battery electric vehicles 355–7
Battersea power station 367–8, 371
Becquerel, Henri 349, 350, 396
becquerels (Bq) 401
beach price 478
Bell, Alexander Graham 358
benefit-cost ratio (B/C) 501
Benz, Karl 298, 299
benzene 305
Bergius process 274
beta particles (β-particles) 397, 397–8
Binney, Edward W. 232
bioenergy 8, 9, 19–21, 28, 67
    see also biofuels; biomass

biofuels 19–21, 67
    carbon sequestration in forests
        577–8
    Denmark 77, 78
    India 123, 124
    UK 74, 75
    see also bioenergy; biomass
biological pump 582–3
biomass
    harvesting 525
    impacts of 520, 523–5, 553, 554,
        558, 559, 560
    world energy source 8, 9, 66, 67
    see also bioenergy; biofuels
bitumen 247, 249
    tar sands 280–1
bituminous coal 164, 173, 174
black coal (anthracite) 164, 173, 174
black start 372
blackouts 372, 386
Blake, William 5
blast effect 205
blending 247–8
Blix, Hans 447
Blücher 206, 207
Blyth harbour offshore wind farm 30
Boeing 707 319
Bohr, Niels 403, 406
boilers 176–82
    power-station boilers 177–81,
        217–18, 220
boiling water reactors (BWRs) 418, 422
bombs, atomic 407, 410
Boulton, Matthew 194, 195
BP 290
Brazil 167, 168
brick chimneys 160
brick-making 107, 159
Brighton Electric Light Company 486
Brighton Power Station 339
Britannia 215
British Admiralty 214
British Broadcasting Company (BBC) 359
British Coal 510
British Electricity Trading and
    Transmission Arrangements (BETTA)
    381
British Energy (BE) 380, 385, 444, 464
British Nuclear Fuels (BNFL) 380, 444,
    464, 503
British Railways 307–8
British thermal unit (BTU) 63
bronze-age culture 95, 96
brown coal (lignite) 164, 174
Brown's Ferry reactor accident,
    Alabama 549
Brunel, Isambard Kingdom 208, 209
bubble fusion 432
bubbling fluidized bed combustion
    (BFBC) 180

buildings 39–41
burners (fast reactors as) 429
burning oil 124, 246, 249
buses 108, 299, 592
butane 249
bypass fan 319–21

C

cadmium 521
Calder Hall power station 17, 410, 418
California 27, 386
Calley, John 191
caloric 198
calories 63–4
calorific value 254
can-in-canister plan 429
Canada 119, 240
cancer 530, 538–9
CANDU reactor 420–1, 422, 460
capacity factor 353
capital costs 492, 498
    and coal conversion 264
    nuclear power 442, 499–501
caprock 228, 229, 580
car batteries 140
carbon arc lamps 334, 336, 337
carbon capture 578
carbon dioxide 3, 314, 447
    capturing and sequestering
        emissions 577–83
    and climate change 539–40
    combustion product 174, 175, 181,
        259, 309
    comparing impacts of electricity
        generating systems 553–4
    global concentrations 15–16
    impacts of energy use 521, 522
    reducing emissions from combustion
        574–6
    sustainable forest management 20
carbon monoxide 260, 309, 531–2,
    532–4
carbon sequestration 47, 48, 577–83,
    589–90
    beneath earth's surface 578–80
    forests 48, 577–8
    oceans 580–3
carbonization of coal 267–9
    low-temperature 273–4
carburettor 298
Carno wind farm, Wales 29
Carnot, Sadi 198
Carnot efficiency 199, 200, 382
Carnot engine 198–200
Carnot's formula 199, 200, 382
cars 108, 121, 298–303
    birth of car engine 298–300
    gas turbines for 321–2
    hybrid electric drives 44, 45, 314,
        325, 357
    motorization of US 302–3
    power and speed 300–1

quest for Stirling engine 325–7
cash flow
    balancing investment against 490–3
    DCF analysis 493–508
catalysts 271, 275, 585
catalytic converters 312–13
catalytic cracking 248
catastrophes 511, 523, 545–52
    see also accidents
cause and effect 523
cement 107
Central Electricity Board (CEB) 366, 372
Central Electricity Generating Board
    (CEGB) 367, 377, 380, 443, 510
central heating 101, 255, 362
centrifugal enrichment methods 423
cerium oxide 348
Chadwick, James 402
chain reactions 404, 405
    controlled 408
    runaway 415
Chapelcross power plant 418
charcoal 105, 159, 161, 267
Charlotte Dundas 208
chemical absorption 578
chemical adsorption 578
chemical energy 145–8
chemical equilibrium 271, 310, 311
chemical reactions 145
    reversible 271, 310
Cheney, Dick 448, 469
Chernobyl accident 18, 415, 439–41,
    446, 549, 551–2
chimneys 100, 160
China 167, 536, 537
    Han Dynasty 95, 96
    Three Gorges Dam project 369, 536
chlorofluorocarbons (CFCs) 15, 363, 541
circulating fluidized bed combustion
    (CFBC) 180
cities 102–3
    urban air pollution 532–6
City of London Electric Lighting
    Company 361
Civaux PWR, France 17
Clarke, Arthur C. 325, 359, 360
Clarke, Colin 507
Claude, George 350
Clean Air Act 1956 49, 533
Clean Development Mechanism (CDM)
    466
Clerk, Dugeld 298
Clermont 208
climate change 4, 14–16, 469, 564
    consequences of 544–5, 546–7
    impacts of energy use 539–45
    Kyoto protocol 466, 576
    nuclear power as answer to 447–56
    see also greenhouse gases
Climate Change Levy 466, 480, 511
closed fuel cycle 425, 426
coal 7, 10, 16, 49, 155–84
    combustion 171–5

composition of 171–2
Denmark 76–7
domestic energy 100–1
energy efficiency improvements 37
fires, furnaces and boilers 176–82
formation 10
France 83, 84
history of use 158–63
impacts of use 520, 525–7
India 85, 86, 87, 167, 168
miners' strike in UK 510
mining 164–5
prices 481, 485–6, 487
reserves and production 11, 13,
    166–9
R/P ratio 11, 13
security of supply 509–10
types of 163–5
UK 71, 72, 73, 162, 163, 167, 168–9
UK energy balance 111, 112, 114
units based on 61, 63
USA 79, 81, 163, 167, 168
uses 170
world primary energy 8, 9, 66–9,
    157, 158
coal-bed methane, enhanced recovery
    of 579
coal conversion
    gas from coal 267–73
    obstacles to conversion 262–5
    oil from coal 273–8
coal-fired power stations 37, 38, 368
    boilers 177–81
    external costs 561
    flue gases 181–2
    IGCC 578, 590, 591
    impacts of 555, 558, 558–9, 560
    location 373–4
    thermal efficiency 94
coal gas 101, 103, 171, 231, 236, 267–73
coal holes 63
coal liquefaction 262, 273–8
coal seams 163
coal tar 267, 268, 269
Coalbrookdale 161, 162
Coalene 274
Coalite 273–4
coefficient of performance (COP) 364
co-generation see combined heat and
    power
coke 161–2, 170, 171, 267, 268, 269
    iron and steel manufacture 105, 106
coking coal 164, 170, 171
cold fusion 432
colour rendering 349
combine harvesters 98, 99
combined cycle gas turbines (CCGTs)
    36–7, 38, 222, 257, 321, 573
    costing energy 492, 498, 502–3, 504
    dash for gas 37, 74, 114, 381–4, 445,
        576
    operating system 382
    reducing carbon emissions 575, 576

combined heat and power (CHP) 37, 115, 122, 123, 369–71, 573
  large scale CHP with community heating 370–1
combustion 157
  biofuels 20
  capturing and sequestering emissions from 577–83
  cleanness in 259–60
  coal 171–5
  incomplete 20
  products 174–5
  reducing pollutant emissions from 574–6
combustion chamber 317
Comet 318–19
commercial companies 506
commercial and institutional sector *see* services sector
common mode failure 549
community (district) heating 37, 115, 123, 370–1
community-scale impacts 522, 531–6
compact fluorescent lamps (CFLs) 351, 490–1, 512
compound engine 203–4, 208
compound turbine systems 216–17
compounds 145
compression ignition engine *see* diesel engines
compression ratio 297, 304
compressor 363, 364
concentric piston engine 323–4
condensate 218
  condensers
    power station turbines 218, 220–1
    refrigerators 362, 363, 364
    separate 194, 195
condensing boiler 255
consequences, probabilities and 548–50
conservation of energy, law of 57–8, 131
consumption
  as conversion 57–8
  demand for oil products 250–3
  demand surges for electricity 379–80
  Denmark 75–7
  France 83–5
  India 85–6
  international comparisons 69–71
  natural gas 243–4, 245, 254–7
  per capita 9, 60
  UK 71–4
  USA 79–81
  world primary energy consumption 8–9, 57–60, 62, 68
containment 414
control devices 109
control rods 413
convenience 258–9
conventions 65–6
conversion of appliances for North Sea gas 236
conversion between units 64–5, 598–600

conversion efficiency *see* efficiency
conversion losses 111, 112, 113–15
conversion processes 262–78
  gas from coal 267–73
  gas from oil 265–6
  obstacles to coal conversion 262–5
  oil from coal 273–8
  oil from gas 266–7
Cooke, William 358
cooking 100–2, 361–2
coolant 413, 414
  loss-of-coolant accidents 415
cooling towers 218
Cooper Hewitt, Peter 349
Corliss steam engine 204
correlations 523
cost
  capital costs *see* capital costs
  obstacle to coal conversion 263–4
  and price 489–90
  waste disposal and decommissioning 444, 499–501
cost breakdown 499
costing energy 475–516
  comparing external costs of electricity generating systems 561–4
  externality costs of pollution and disaster 511
  inflation, real prices and affordability 482–9
  investing in energy 489–508
  new technologies, economies of scale and market washing 511–12
  nuclear power 438, 442–5, 464–6
  present-day energy prices 478–82
  real world complications 509–12
  Sasol plant 276–7, 277–8
  security and diversity of supply 509–11
cracking 248
critical mass 406, 416
critical temperature 193
cross-compound systems 204, 216
Cruachan pumped storage system 135
crude oil 227, 245, 247
  see also oil
Curaçao 208
current, electric 137, 139
cutting machines/power loaders 165
cyclone filters 182

D

D-D fusion 431
D-T fusion 430, 431
Daimler, Gottlieb 234–5, 298, 299
Daimler-Chrysler 45, 46
Daimler motor company 299
Dalton, John 145
dams 22, 525, 536
Darby, Abraham 161–2
dash for gas 37, 74, 114, 381–4, 445, 576

data
  interpreting 64–6, 166
  types and impacts of energy use 523
Davy, Humphry 141, 334
de Dion Bouton 300
de Havilland Comet 318–19
debt 503–4
de-carbonization 588
decommissioning and disposal costs 444, 499–501
deep mines 164, 165
deep repository 448, 449, 450
Deeside CCGT power station 38
delivered energy 93–5
  international comparisons 121–4
  UK 110–13
Deltic railway locomotives 307
demand *see* consumption; supply and demand
demand-side efficiency improvements 39–46
  domestic sector 39
  industrial sector 41–2
  services sector 39–41
  transport sector 42–6
dematerialization 42, 117
Denmark 29, 37
  delivered energy 122–3
  electricity 120, 121, 384, 387–8
  energy and GDP 118, 120
  energy prices 479
  primary energy 75–9
  renewables 76, 77, 78–9
density 134
depleted oil/gas wells 578–9, 580
depleted uranium 413, 423
depletion of energy sources 6, 594
Deptford Power Station 345, 346
DERV (DiEsel for Road Vehicles) 247, 249, 309
deuterium 407, 430
deuteron 407
developing countries 466–7, 523–5, 536
development, sustainable 6
deviation drilling 239
Didcot power station 38
Diesel, Rudolf 305
diesel 555
  price 478–9, 485
  see also DERV
diesel engines 295, 305–9, 328
  emissions from 313
  reducing pollution 309–14
  road, rail and air 307–9
  ships 215, 306–7
dinitrogen (nitrous) oxide 521, 539, 540, 541
dioxins 21, 532, 535
direct current (DC) 339, 340
direct disposal 426
direct liquefaction 274, 277
direct methanol fuel cell (DMFC) 586, 587

disasters *see* catastrophes
discharges from nuclear plants 538
discounted cash flow (DCF) analysis 493–508
    acceptable discount rates and investment lifetimes 505–8
    basic discounting formulae 495–8
    calculation 499–501
    dealing with normal financial risk 503–5
    interest rates and discount rates 493–5
    and perceived costs of different forms of energy 502–3
    project lifetime 495
distillation
    of coal 268
    of oil 245–9
distribution
    electricity 37, 371–5
    gas 242–3, 244, 260
    oil 260
district (community) heating 37, 115, 123, 370–1
diversity of supply 509–11
domestic hot water 101
domestic sector 116
    demand for gas 255, 257
    efficiency improvements 39–41
    expanding uses of energy 95, 100–4
    household-scale impacts 522, 523–5
    India 124, 524
    prices 479–80, 485–6
    UK energy balance 110–12, 113
dose of radiation 401
    collective dose 462–3
    health effects 538
double-acting piston 194, 195
double-flow turbine systems 216–17
Doxford International Business Park 28
drag coefficient 301
Drake, Edwin Laurentine 233, 234
drilling for oil 233, 236–7, 238–9
Dundonald, Lord 231, 267
dung 20, 123, 124
duty of an engine 196
dynamic equilibrium 310
dynamos 335, 336–7

E

economic development 466–7, 523–5, 536
economies of scale 511–12
economizers 217
ecosystems 546
Edison, Thomas 34, 338, 339, 358
    Pearl Street electric lighting station 34, 35, 338
Edison Electric Illuminating Company 338
Edison and Swan United Electric Light

Company 338
efficacy (of lighting sources) 104
efficiency 94
    best efficiency for petrol and diesel engines 314
    Carnot 199, 200, 382
    coal conversion 263
    of an engine 196
    gas turbines 321
    improving for steam engines 203–4
    power-station turbines 221–2
    solar cells 144
efficiency improvements 5, 36–46, 47, 573
    demand-side measures 39–46
    rebound effect 46
    supply-side measures 36–8
Einstein, Albert 142, 151
electric charge 137
electric fields 141, 528–30
electric lighting 34, 35, 336, 337–52
    comparison of lighting efficacy 352
    continuing development 348–52
    fluorescent lamp 349–52
    incandescent light bulb 103–4, 337–8, 348–9
    LED 352
    rise of 337–47
Electric Lighting Act 1882 339
electric motors 105, 340, 364
electric traction 353–7
electric transmission 308, 357
electrical energy 136–40
electrical potential energy 137
electrical wiring 343–4
Electricité de France (EdF) 387
electricity 331–92
    AC vs DC 339–42
    comparing external costs of electricity generating systems 561–4
    comparing impacts of electricity generating systems 553–60
    conversion losses 113–15
    electric lighting *see* electric lighting
    electric traction 353–7
    expanding uses 358–65
    gas generation 37, 74, 114, 257, 381–4, 445, 576
    growth in world demand 365, 366
    high voltage vs low voltage 343–7
    international comparisons 120–1, 384–90
    large scale generation 365–71
    metering and tariffs 347
    19th century 334–47
    occupational hazards in producing 526–7
    organization of the electrical supply system 375–81
    percentage generated by nuclear power 437

present-day usage 365
    prices 36, 479–80, 481–2, 486, 487–8
    security of supply 510–11
    transmission *see* transmission of electricity
    UK energy balance 110, 111, 112
Electricity Act 1957 373
electrolysis 334–5, 585, 589
electromagnetic radiation 140–4
electromagnetic spectrum 141, 142–3
electromagnetic waves 141
electromagnetism 335–7
    health effects of electromagnetic fields 528–30
electrons 138, 147
electronvolts (eV) 396
electrostatic precipitator 182
elements 145, 146
    production of new elements 408–9
Elling, Egidius 314–15
embedded generation 376
embodied energy 106
emissions 6
    capturing and sequestering carbon emissions 577–83
    from diesel engines 313, 314
    making fossil fuels more sustainable 594–5
    from petrol engines 311–13, 314
    reducing from fossil fuel combustion 574–6
    Stirling engine 325
    see also under individual types of emission
endothermic reactions 271, 310
energy
    defined as capacity to do work 6
    forms of *see* forms of energy
    law of conservation of energy 57–8, 131
    quantities of 61–4
    units 58–60, 396, 598–9
energy amplifier 459
'energy arithmetic' 58–9, 597–600
energy balance, UK 109–18
energy 'chains' 35–6
energy consumption *see* consumption
energy content of fuels 148
energy density 260
energy-efficient appliances 40
energy/GDP ratio 118–20
energy intensity 119
energy payback ratios 558–60
energy prices *see* prices
energy production *see* production
energy scenarios 49–53
energy services 34–6, 93
    linking supply and demand 35–6
energy sources
    classification of impacts by 519–20
    present sources and sustainability 7–23

renewable sources 23–34
world energy sources 66–71
energy stores
    fuels as 148
    uranium as 152
energy uses 91–128
    changing patterns 47–9
    comparisons of delivered energy 121–4
    domestic sector 95, 100–4
    electricity use 365
    energy balance for UK 109–18
    expanding uses of energy 95–109
    food see food
    industrial sector see industrial sector
    international comparisons 118–24
    past societies 95–6
    present-day 109–24
    primary, delivered and useful energy 93–5
    services sector 95, 108–9
    transport sector 95, 107–8
English Channel electricity link 374
English Electric company 307
enhanced oil recovery (EOR) 278, 578–9, 580
enriched uranium 413
enrichment 406, 423, 452–3
Enron Corporation 386, 389
entropy 201
environment
    acid deposition 537, 554, 555, 556
    climate change see climate change
    emissions see emissions
    impacts on see impacts of energy use
    obstacles to coal conversion 262–3
    pollution see pollution
equilibrium, chemical 271, 310, 311
equity 503
Ericsson, John 208–9
ethane 249
European Pressurized Reactor (EPR) 457
Eurostar 355
eutrophication 98
evaporator 363, 364
exciter 343
exhaust temperature 198
exothermic reactions 271
explosions, steam 204
exponent 597
exposures, discharges and 538
external combustion engines 295
    Stirling engine 295, 322–7
    see also steam engines
external costs 511
    comparing for electricity generating systems 561–4
extraction
    coal 164–5
    nuclear fuel 422
    oil 11, 240–2
Exxon Valdez oil spillage 2

F

Faraday, Michael 140, 141, 335
fast breeder reactors (FBRs) 18, 409, 428–9, 450, 454–5
fast neutrons 406
fast reactors 427–9
fault traps 228, 229
Fawley oil refinery 252
Fermi, Enrico 402, 403–5, 406, 407, 408, 409
Ferranti, Sebastian Ziani de 344–6
fertile isotopes 409
fertilizers 97–8
fields of force 141
financial risk 503–5
Finland 448, 464
fireplaces 100, 101
fires 176–82
First World War 97–8, 101–2
Fischer-Tropsch synthesis 274–7
fissile isotopes 150, 409
fission, nuclear 16, 150, 402–29
    energy from 404
    energy and mass 151–2
    fast reactors 427–9
    reactors and bombs in Second World War 405–11
    thermal fission reactors 411–22
fixed-bed catalysts 275
fixed carbon (FC) 172, 174
flue gas desulphurization (FGD) 182
flue gases 181–2
fluidized bed boilers 178, 179–81, 181–2
fluidized bed combustion (FBC) 178, 179–81
fluids under pressure 193
fluorescent lamps 349–52
    CFLs 351, 490–1, 512
fly ash 179, 181, 182
Flying Scotsman 207
flying shuttle 160
Foden lorry company 307
food 95, 96–100
    cooking 100–2, 361–2
    preserving and processing 102–3
Ford, Henry 302
Ford 45, 46
forests/forestry 525
    carbon sequestration 48, 577–8
former Soviet Union (FSU) 240, 243, 421, 467
    see also Russia
forms of energy 129–53
    chemical 145–8
    electrical 136–40
    electromagnetic radiation 140–4
    energy and mass 151–2
    heat 136, 200–1, 300
    kinetic 131–4, 300, 319
    nuclear 148–50
    potential see potential energy

fossil fuels 3–4, 5, 157, 571–96
    capturing and sequestering carbon emissions 577–83
    'cleaning-up' technologies 47, 48
    comparing external costs of electricity generating systems 561–4
    fuel cell 584–7
    hydrogen economy based on 587–93
    making their use more sustainable 594–5
    present sources and sustainability 7–16
    reducing emissions from combustion 574–6
    world primary energy 8–9, 66–9
    see also under individual fuels
four-stroke engine 296–7
Fowey Consols mine engine 203
fractionation 245–7
fractions, oil 246–7, 249
France 17, 118, 119, 121, 303, 479, 510
    delivered energy 123
    electricity 120–1, 384, 386–7
    nuclear power 83, 84, 85, 446, 455, 467
    primary energy 83–5
free electrons 138
free neutrons 402, 403–5
free protons 402
frequency 340
friction 316
    rolling 301
Frisch, Otto 403, 406
Frisch-Peierls Memorandum 406, 407
fuel cell electric vehicles 45–6, 587
fuel cells 584–7, 589, 591–2
    basic principles 585
    types 586–7
    see also hydrogen economy
fuel economy 44–5
fuel fabrication 413, 423
fuel injection pump 307
fuel oil 247, 249, 250–2
fuel poverty 488–9
fuel switching 47, 574–6
Fulton, Robert 208
Fulton 208
furans 532
furnaces 176–82
fusion, nuclear 19, 429–32, 455–6
    energy from 430
    fusion reactor 431–2
    safety 459–60
future value 493
Futtsu CCGT plant 383

G

Galloway hydroelectric scheme 369
Galvani, Luigi 138
gamma particles (γ-particles) 397, 399
gamma-radiation 399

gamma-rays 399
Gardner company 307
gas
  natural *see* natural gas
  prices 258, 259, 479–80, 481, 485–6,
    487–8
  secondary fuel 262–90
  town gas *see* town gas
gas-cooled reactors 418–20
Gas Light and Coke Company 269
gas lighting 268–9, 337, 348
gas mantle 104, 348
gas oil 247, 249
gas separation membranes 578
gas turbines 295, 314–22, 381
  British jet engine 316–17
  for cars 321–2
  CCGTs *see* combined cycle gas
    turbines
  German jet engine 315
  hydrogen powered 592
  improving power and efficiency 321
  industrial 321
  modern jet engines 319–21
  post-war developments 318–19
  power to weight ratio 327, 328
gaseous diffusion 423
gaseous pollutants 531
gasification 270–3
Gasmotorenfabrik Deutz AG 297, 298
Gasohol 21
gasometers 256
GDP, energy consumption and 118–20
gearbox 314
General Electric Company 349, 350,
  361, 362
General Motors 325
generators 335–7
  see also alternators
geological traps 228, 229, 230
geomagnetic survey 237
geostationary satellites 359–60
geothermal energy 24, 33, 79, 83, 520
Germany 27, 163
  hyperinflation 484
  jet engine 315
  nuclear power 408, 455
  security of supply 509–10
glass making 107
Glen Canyon Dam, Lake Powell,
  Arizona 22
global-scale impacts 522, 539–45
Gloster Meteor 317, 318
Grain oil-fired power station 368, 504–5
Grand Union Canal 373
Grangemouth refinery complex 233,
  279
graphite 407, 412
grate boilers 177–9
gravitational potential energy 133,
  300–1
gravity, acceleration due to 133
gravity (or gravimetric) survey 237

Great Britain 209
Great Fire of London 1666 100
Great Ocean Conveyor Belt 582, 583
Great Western 208
Green River shale formation, US 280
greenhouse effect 14–16
greenhouse gases 3–4, 15–16
  emissions from domestic sector in
    India 524–5
  emissions from electricity generation
    553–4
  global climate change 539–44
  hydropower generation and 22–3
  incomplete combustion 20
  see also climate change; *and under*
    individual greenhouse gases
Grosvenor Art Gallery 344–5
Grove, William 335, 584
Grove fuel cell 335, 584
Guericke, Otto von 193
guide vanes 211, 212
Gulf War 511

H

Haber-Bosch process 97–8
Hahn, Otto 402, 403, 408
half-life 399–400, 401
halocarbons 540, 541
Han Dynasty China 95, 96
hard coal (anthracite) 164, 173, 174
health hazards 3, 6, 594, 595
  see also impacts of energy use
heat energy 136, 300
  atoms in motion 200–1
heat engines 185–223
  age of steam 202–9
  principles of 197–202
  steam engines *see* steam engines
  steam turbines *see* steam turbines
  see also gas turbines; internal
    combustion engines
heat flow 202
heat index 544
heat pumps 79, 202
  and refrigerators 363–4
heat value 164, 166
heating 49, 100–2, 137, 255, 256, 361–2
heating oil 247, 249, 486, 487
heavy distillate 246, 247, 249
heavy oil 281–2
heavy water 407
heavy water reactors 412
Heinkel, Ernst 315
Heisenberg, Werner 408
Hero 188, 189
Hertz, Heinrich 141, 359
high-grade ore 413
high intensity discharge lamp (HID) 349
high-level waste 426, 427, 461
high pressure steam 203, 210–11
high temperature heat process 107
high temperature reactor (HTR) 457

hills, climbing 300–1
Hindenburg airship disaster 592–3
Hoover dam hydroelectric project 369
Hornblower, Jonathan 203–4
horse buses/trams 103, 354
horsepower 196
household-scale impacts 522, 523–5
Hubbert, M. King 284–5
Hubbert's peak 285, 286–90
Human Disruption Index 520–2
hunter-gatherer societies 95, 96
hybrid electric drives 44, 45, 314, 325,
  326, 357
hydrocarbons 10, 521
  see also fossil fuels
hydrocracking 248
hydroelectric power 21–3, 28, 369
  comparing impacts of electricity
    generating systems 555, 558, 559,
    560
  conventions and hydroelectric
    power plants 65–6
  external costs of electricity
    generating systems 561
  France 83
  impacts of 520, 525, 536
  India 85, 87
  UK 74, 75
  US 79, 81, 82
  world primary energy 8, 9, 66–9
hydrogen 407, 430
hydrogen bomb 432
hydrogen economy 587–93
  hydrogen storage and use in
    transport 590–2
  safety 592–3
hydrogenation 274
hypercar 45
hyperinflation 484
hythane 590–1

I

impact pathway approach 561–4
impacts of energy use 5, 517–69
  accidents and risk 545–52
  classification 519–23
  community-scale impacts 522,
    531–6
  comparing external costs of
    electricity generating systems
    561–4
  comparing impacts of electricity
    generating systems 553–60
  global-scale impacts 522, 539–45
  household-scale impacts 522, 523–5
  regional-scale impacts 522, 537–9
  workplace-scale impacts 522,
    525–30
impulse turbines 188, 189, 209–11
incandescent light bulbs 103–4, 337–8,
  348–9
India 118

coal 85, 86, 87, 167, 168
delivered energy 123–4
electricity 120, 121, 384, 388–90
energy prices 479, 480, 481–2
fuel poverty 489
impacts from household fuels 524
primary energy 85–7
renewables 85, 86–7
indigenous production 258
indirect liquefaction 274–7
induced radioactivity 427
induction motor 342
Industrial Revolution 160–2
industrial sector
efficiency improvements 39, 41–2
energy prices 480–2
energy uses today 110–12, 113, 116, 117
expanding uses of energy 95, 104–7
gas turbines 321
India 124
UK demand for gas 255–6, 257
industrial spirit 249
inert gases 348
inertial confinement 432
inflation 482–9, 493, 495
information technology 360–1
infra-red radiation 14–15
injuries 526
see also accidents
input power 134
input temperature 198
Institute for Applied Systems Analysis (IIASA) 50
insults, environmental 520–2
integrated gasification combined cycle (IGCC) plants 578, 590, 591
Intercity 125 units 308
interest rates 493–5
intergenerational equity 506–8
Intergovernmental Panel on Climate Change (IPCC) 50, 542–4
intermediate-level waste 427, 461
internal combustion engines 234–5, 295–322
diesel engine 215, 295, 305–9, 328
gas turbine see gas turbines
hydrogen powered 591–2
petrol engine 295–305, 328
power to weight ratio 327–8
reducing pollution from petrol and diesel engines 309–14
internal rate of return (IRR) 501
international comparisons 118–24
delivered energy 121–4
electricity 120–1, 384–90
energy and GDP 118–20
energy prices 479–82
primary energy 69–71
travel 121
international thermonuclear experimental reactor 455
Internet 361

investment in energy 489–508
balancing investment against cash flow 490–3
DCF analysis 493–508
price and cost 489–90
investment lifetimes 495, 505–8
iodine-131 538, 551
ions 138
Iran 240, 241
Iraq 240, 241
iron manufacture 105–6, 161–2
isotopes 149

J

Japan 119, 455
energy prices 479, 480, 481
natural gas 243–4, 245
Jaray, Paul 301
Jellinek, Emile 299
Jenatzy, Camille 355, 356
jet engines see gas turbines
JET 1 gas turbine car 321
Joint European Torus (JET) 19, 432, 455
Joliot-Curie, Frederic 402
Joliot-Curie, Irene 402
Joule, James 136, 137
joule (J) 58
Joule heating 138
Junkers Jumo 205 aircraft engine 307

K

Kay, John 160
Kelvin temperature scale 200
kerogen 278–80
kerosene (paraffin) 124, 246, 249
kerosene lamps 103
Kidd, Steven 468–9
kilocalorie (kcal or Cal) 63–4
kilowatt-hour (kWh) 59–60, 254
kinetic energy 131–4, 300, 319
kinetic theory of heat 136
King Edward 214
Kirk, Alexander C. 232
knocking (pinking) 304
Kolle, Harold 346
Kutir Jyoti programme 489
Kyoto protocol on climate change 466, 576

L

Laherrère, Jean 289
land use 554, 557–8
landfill gas (LFG) 74, 75
Langen, Eugen 296–7
Langlois, Lucille 469
Larderello geothermal power plant, Italy 33
Laval, Gustav de 211
Lawrence Livermore National

Laboratory, California 432
lead-acid batteries 356, 357
lead emissions 521, 522, 534
leaded petrol 304–5
lean burn engine 312
Leblanc process 107
Leclanché cell 335
Lenoir, Etienne 295
life-cycle assessment (LCA) 560
light, nature of 141
light bulbs, incandescent 103–4, 337–8, 348–9
light distillate 246–7, 249
light emitting diode (LED) 352
light pollution 352
light water reactors (LWRs) 412, 416–18
lighting 103–4, 230–4, 336
electric see electric lighting
gas 268–9, 337, 348
oil for 231–4
lignite (brown coal) 164, 174
lime mortar 107
Linde, Carl von 362
Lindsay, Coutts 344–5
linear systems 545
liquefied natural gas (LNG) 243
liquefied petroleum gas (LPG) 124
liquid hydrocarbons 110, 111
from coal 268, 273–8
see also diesel; oil; petrol
liquid-metal fast breeder reactors (LMFBRs) 422, 428–9
lithium ion batteries 357
Livingston, Robert 208
load factor 353
Locomotion 206
London smogs 49, 533, 535
long-term energy scenarios 49–53
longwall method 165
Los Angeles 533
low-grade uranium ore 452–3
Low Level Radiation Campaign 441
low-level waste 427, 461
lubricating oils 231, 232, 233, 247, 249
lumens 103
Lurgi gasifier 271, 272–3
Luz project, California 27

M

magnetic containment 431
magnetic fields 141, 528–30
magnetism 335–7
Magnox reactors 368, 418, 419, 422
maize 21
MAN 307
Manhattan Project 407
Marconi, Guglielmo 359
marginal barrel 264
marine diesel oil 247, 249
market washing 511–12
mass 151, 300
critical mass 406, 416

unified mass unit 396
mass-energy 151–2, 396
mass entertainment 109
mass flow rate 316
mass number 149
Mauretania 214
Maxwell, James Clerk 140–1, 359
Maybach, Wilhelm 298, 299
meat 102
mechanization of farms 98–100
mediaeval society 95, 96, 100, 107–8
medium of exchange, money as 482
Meitner, Lise 402, 403, 408
Meldrum, Edward 231, 232
Mercedes cars 46, 299, 300
mercury
     emissions to atmosphere 521
     fluorescent lighting 350, 351
mercury vapour lamps 349
merit order 378
Messerschmitt Me 262 315, 318
metering 347
methanation 270, 271
methane 15, 16, 249
     anaerobic decay 22–3
     combustion 145
     emissions 539, 540, 541
     impacts of 521
     incomplete combustion of wood 20
     production from coal 270–3
methanol-to-gasoline (MTG) plant 266, 267
Meucci, Antonio 358
Mexico 240
microwave ovens 361
Middelgrunden wind farm, Denmark 29
middle distillate 246, 247, 249
Middle East 240, 286
Midgely, Thomas 304, 363
million barrels daily (Mbd) 62
millions of tonnes of oil equivalent (Mtoe) 62, 254
mine tailings 422, 539
'Miner's Friend' 190–1
miners' strike 510
mining
     coal 164–5
     nuclear fuel 422
mixed oxide fuel (MOX) 426, 427, 453–4, 462
Mobil MTG plant 266, 267
mobile steam power 205–9
mobile telephones 360, 361
modal shift 44
Model T Ford 302
moderators 407, 412
modern bioenergy power plants 8, 9, 20, 21, 67
moisture, in coal 172, 173, 174
molecules 145
molten carbonate fuel cell (MCFC) 586, 587
momentum 319

money
     store of value over time 482, 489–508
     time preference for 493, 505–8
     time value of 493–508
     uses of 482
     value of 482–6
Morse, Samuel 358
motive power 187
motor buses 299
motors, electric 105, 340, 364
moving-bed catalysts 275
municipal solid wastes (MSW)
     incineration 21, 74, 75, 78, 535
Murdoch, William 231

N

naphtha 247, 249, 262
National Grid 342, 366, 371–5, 380
     map 375
     schematic 374
National Grid Company 377, 379, 380
National Power 380
natural gas 7, 16, 49, 157, 225–91
     CCGT see combined cycle gas turbines
     comparing impacts of electricity generating systems 554, 555, 559, 560
     consumption 243–4, 245
     dash for gas 37, 74, 114, 381–4, 445, 576
     Denmark 76, 77, 78
     distribution 242–3, 244, 260
     electricity generation 37, 74, 114, 257, 381–4, 445, 576
     formation 10, 228, 229
     France 83, 84
     impacts 520, 525–7, 561
     India 85, 87
     key characteristics 258–60
     oil from 266–7
     origins of gas industry 235–6
     as primary fuel 227–61
     reserves 11, 12
     R/P ratio 11, 13
     supply-side efficiency improvements 36–7
     UK 71, 73
     UK demand for 254–7
     UK energy balance 111, 112, 114
     UK national transmission system 243, 244
     USA 79, 81
     world primary energy 8, 9, 66–9
natural nuclear reactors 409
natural uranium 413
neptunium 408–9
Nernst, Walter 201
net present value (NPV) 494, 496, 501
Netherlands 243
neutrons 148–9, 397

decay 398
     experiments with 402
     free 402, 403–5
'new' bioenergy 8, 9, 20, 21, 67
New Electricity Trading Arrangements (NETA) 381
New Plymouth MTG plant, New Zealand 266, 267
Newcomen, Thomas 191–4
newspapers 109
Newton, Isaac 131–2
newton (N) 133
Newton's First Law of Motion 131
Newton's Second Law of Motion 132, 316
Newton's Third Law of Motion 316
Niagara Falls 209, 347
nickel-cadmium (NiCad) batteries 356, 357
nickel-metal hydride (NiMH) batteries 356, 357
nickel-iron (NiFe) batteries 357
Nirex 448
nitrogen 181, 521
nitrogen dioxide 309, 310
nitrogen oxide (nitric oxide) 309–11
nitrogenous fertilizers 97–8
NO$_x$ see oxides of nitrogen
noble fuels 157
     see also natural gas; oil
non-associated gas 235
non-conventional petroleum 278–82
Non Fossil Fuel Obligation (NFFO) 510–11
non-linear systems 544–5
non-renewable energy sources 23
     see also fossil fuels; nuclear power
Nord Pool 377, 388
North Pole 4
North Sea oil and gas 236, 242, 252, 381, 511
Norway 242
notional primary energy 66
nuclear atom 147
Nuclear Electric 499–500, 502
nuclear energy 148–50
     mass and energy 151–2
nuclear fission see fission, nuclear
nuclear fuel 412, 413, 422–7
     enrichment and fuel fabrication 423
     mining and extraction 422
     spent fuel 423–7, 539
nuclear fusion see fusion, nuclear
Nuclear Information and Resource Center (NIRS) 468
Nuclear Installations Act 1965 465
nuclear isomers 399
nuclear power 5, 114, 368, 393–473
     accidents 18, 414–16, 438, 439–42, 548–52
     cheaper 464–6
     'cleaning-up' technologies 47
     comparing impacts of electricity

generating systems 555, 557–8, 558, 560
conflicting views on the future 468–70
costing Sizewell C 499–501, 502–3
development issues and proliferation 466–7
economics 438, 442–5, 464–6
external costs 561
France 83, 84, 85, 446, 455, 467
impacts of 520, 526, 527, 527–8, 537–9
India 85, 87
location of power stations 436
long-term answer to climate change 447–56
new developments 456–67
nuclear fuel cycles 422–7
present sources and sustainability 16–19
radioactivity 396–401
reactors see reactors
reasons for decline 438–46
rise and fall of 437–8
safety 414–16, 438, 439–42, 457–60
subsidies 510
UK 49, 71, 73, 74, 448, 455, 469
USA 79–80, 81, 446, 448, 455, 467, 469
waste management 444, 448, 449–50, 460–4, 499–501
world primary energy 8, 9, 66–9
worldwide decline 446
nuclear reactors see reactors
nuclear weapons proliferation 466–7
nucleons 149
nucleus, atomic 147, 402
nuclides 149

O

ocean sequestration of carbons 580–3
enhancing natural uptake 582–3
injection of carbon dioxide 581–2
octane number 304
Oersted, Hans Christian 140, 335
Office of Gas and Electricity Management (OFGEM) 377
offshore wind farms 29–30
Ohain, Hans von 315, 318
Ohm, Georg Simon 137, 138
ohm (Ω) 139
Ohm's Law 138, 139
oil 7, 16, 37, 157, 225–91, 555
from coal 273–8
Denmark 76–7, 122
external costs 561
extracting 11, 240–2
France 83, 84, 85
from gas 266–7
future for 282–90
gas from 265–6
impacts of 520, 521, 525–7

India 85, 86, 87
key characteristics 258–60
locating 233, 236–9
non-conventional sources 278–82
oil wells on fire in Gulf War 2
origins and geology 10, 228–30
origins of oil industry 230–5
peak production 14, 286–90
prices 3, 68, 258, 259, 263–4, 283, 478–9, 479–80, 481, 490
as primary fuel 227–61
refining 245–9
reserves 3, 11, 12, 286–7, 288
R/P ratio 11, 13
as secondary fuel 262–90
UK 49, 71, 72–4
UK demand for oil products 250–3
UK energy balance 111, 112
units based on 61–2
US 79, 80–1, 240, 241, 284–5, 286, 511
world primary energy 8, 9, 66–9
see also diesel; petrol
oil crises 283
oil-fired power stations 368
oil-fired steam turbines 215
oil lamps 103
oil shales 233, 278–80
OPEC cartel 283
open cast mines 164–5
open fuel cycle 424, 426
opportunity cost 493
optimization (in electricity supply industry) 376
orders of magnitude 597–8
Orimulsion 281–2
Orinoco Basin, Venezuela 281
Osaki Power Station, Japan 180–1
oscillating water columns (OWCs) 31
osmium 348
Otto, Nikolaus August 296–7
Otto cycle engine 296–7
ownership of electricity supply 377, 380–1
Oxford solar house 26
oxidation catalytic converters 313
oxides of nitrogen (NO$_x$) 175, 531
acid deposition 537, 554
comparing impacts of electricity generating systems 554, 555, 556
flue gases 181–2
reducing pollution from petrol and diesel engines 309–11
urban air pollution 532–4
ozone 532, 540, 541
ozone hole 541

P

Pacific Gas and Electric 386
Papin, Denis 191
paraffin (kerosene) 124, 246, 249
paraffin wax 231, 232, 247, 249

Parc Colliery, Cwm Park, Rhondda Valley 10
parcel of coal 165
Parsons, Charles Algernon 211–14, 215
particulates 179, 521, 531, 540, 541
diesel engines 313
flue gases 181, 182
urban air pollution 534, 535, 536
pascal (Pa) 193
passive reactor technology 457–9
payback time 491
peak demands 379–80
peak production of oil
US 284–5
world 14, 286–90
pebble bed modular reactor (PBMR) 457–8, 464
Peierls, Rudolf 406
Pennsylvania oil wells 61, 62
Periodic Table 145
personal computers 360–1
'Petite Voiture' 300
petrol 62, 246, 249
price 478–9, 485
protest in 2000 2
petrol-electric vehicles 44, 45, 314, 325, 326, 357
petrol engines 295–305, 328
aircraft petrol engines 303
best efficiency 314
birth of car engine 298–301
compression ratio and octane number 304
emissions from 311–13
lead additives 304–5
motorization of US 302–3
reducing pollution 309–14
petroleum see natural gas; oil
petroleum coke 247, 249
Petroleum Springs, Dalaki 237
Philips 351
Stirling engine 324–5
phosphoric acid fuel cell (PAFC) 586, 587
phosphorus 97
photons 142
photosynthesis 19
photovoltaic panels 26, 27–8, 144, 512
physical absorption 578
physical adsorption 578
physical labour 101–2, 104–5
phytoplankton 582–3
Pickard, James 195
Pimlico District Heating Undertaking 371
pipelines, gas 243
Piper Alpha gas rig fire 2
pitch 268
Planck, Max 142, 143
plant decommissioning costs 444, 499–501
plasma 431
plutonium 408–9, 429, 453–4, 454, 461–2

PM10 particles 534, 536
PM2.5 particles 534
pneumoconiosis 165
polar ice melting 4, 544
pollutant emissions *see* emissions
pollutants
    classifying impacts of energy use by
        520–2
    forms of 531
pollution
    atmospheric 531–6, 564
    externality costs 511
    reducing from petrol and diesel
        engines 309–14
    see also acid deposition; climate
        change; emissions; impacts of
        energy use
Polo, Marco 230
population, global 51–3
potassium 97
potential difference 137
potential energy 131–5
    electrical potential energy 137
    gravitational potential energy 133,
        300–1
Powell, Lake 22
power
    cars 300–1
    defined as rate of doing work 6
    electric 139
    improving for gas turbines 321
    output from a thermal fission reactor
        416
    units 59, 60, 599
    wind turbine 134
Power Jets Ltd 316–17
power loaders 165
power pool 377, 378–9
power-station boilers 177–81, 217–18,
    220
    flue gases 181–2
power-station turbines 215–22
    660-MW turbine 218–22
power stations 49, 257, 365–71
    CHP 37, 115, 122, 123, 369–71, 573
    coal-fired *see* coal-fired power
        stations
    conventions 65–6
    dash for gas 37, 74, 114, 381–4, 445,
        576
    electricity conversion losses 113–15
    energy efficiency improvements
        36–8
    hydroelectric 369
    primary energy and delivered energy
        93–4
    thermal 65–6, 365–8
    see also electricity
power supply 139
power to weight ratio 327–8
Powergen 380
powers of ten 597
precautionary principle 530

prefixes 59, 598
present value 493, 496
pressure 193
    high pressure steam 203, 210–11
pressurized fluidized bed combustion
    (PFBC) 180–1, 182
pressurized water reactors (PWRs) 368,
    416–18, 419, 422, 457
prices 478–82
    converting energy prices 478
    diesel 478–9, 485
    domestic energy 479–80, 485–6
    electricity 36, 479–80, 481–2, 486,
        487–8
    gas 258, 259, 479–80, 481, 485–6,
        487–8
    industrial energy 480–2
    inflation, real prices and
        affordability 482–9
    oil 3, 68, 258, 259, 263–4, 283,
        478–9, 479–80, 481, 490
    petrol 478–9, 485
    price and cost 489–90
primary cells 335
primary energy 55–89
    data interpretation 64–6
    definition 57
    delivered energy, useful energy and
        93–5
    in Denmark 75–9
    in France 83–5
    in India 85–7
    international comparisons 69–71
    quantities of energy 61–4
    in the UK 71–5
    UK energy balance 110, 111, 112
    in the US 79–83
    world consumption 8–9, 57–60, 62,
        68
    world energy sources 66–71
primary fuels 333
    see also coal; natural gas; oil; wood
primary recovery 278
prime movers 187
private individuals 505–6
privatization 377, 380–1
probabilities, and consequences 548–50
Process Inherent Ultimate Safety (PIUS)
    system 457
production
    coal 166–9
    Denmark 75–8
    France 83–5
    Hubbert's peak for oil 284–90
    India 85–6, 87
    indigenous 258
    international comparisons 69–71
    oil 14, 235, 240–1, 284–90
    UK 71–4
    USA 79–81
production wells 238–9
profit 489–90
project lifetime 495

proliferation of nuclear weapons 466–7
propane 249
propellers, screw 208–9
propulsion 316
proton exchange membrane fuel cell
    (PEMFC) 586, 587
protons 148–9
    free protons 402
proximate analysis 173–4
public opinion 441–2
public-private partnerships (PPP) 506
public sector organizations 506
Puffing Billy 206
pulverized fuel (PF) boilers 178, 179
pumped storage 135, 380
pumps 194, 220
    steam pumps 190–1

Q

quad (quadrillion BTUs) 63

R

radio 359
radioactive wastes 19, 426, 427, 429
    waste management 444, 448,
        449–50, 460–4, 499–501
radioactivity 150, 396–401
    health effects 538
    measuring 401
    radioactive decay and half-life
        399–400
    types of radiation 397–9
railways 105, 108
    diesel locomotives 307–8
    electric traction 354–5
    steam engines 205–7
Rainhill trials 206
Rance Estuary tidal barrage, France 31,
    32
Rankine, William 132
ranks of coal 173–4
Rapid L 458
Rasmussen Report 550
rate of decay 399, 400
rate of rotation 217
RBMK reactor 421, 422
reaction turbines 188, 189
reactors, nuclear
    and bombs in Second World War
        405–11
    fast reactors 427–9
    first reactor 406–8
    thermal fission reactors 411–22
real interest rates 495
rebound effect 46
rechargeable batteries 335, 356, 357
reciprocating motion 194–5
refining, oil 245–9
reforming 265–6
    hydrogen production 589, 590